Thermal-Hydraulics in Nuclear Fusion Technology: R&D and Applications

Thermal-Hydraulics in Nuclear Fusion Technology: R&D and Applications

Editors

Alessandro Del Nevo
Marica Eboli

MDPI • Basel • Beijing • Wuhan • Barcelona • Belgrade • Manchester • Tokyo • Cluj • Tianjin

Editors
Alessandro Del Nevo
Italian National Agency for New
Technologies, Energy and
Sustainable Economic
Development (ENEA)
Italy

Marica Eboli
Italian National Agency for New
Technologies, Energy and
Sustainable Economic
Development (ENEA)
Italy

Editorial Office
MDPI
St. Alban-Anlage 66
4052 Basel, Switzerland

This is a reprint of articles from the Special Issue published online in the open access journal *Energies* (ISSN 1996-1073) (available at: https://www.mdpi.com/journal/energies/special_issues/Thermal_hydraulics_in_Nuclear_Fusion_Technology_RD_and_Applications).

For citation purposes, cite each article independently as indicated on the article page online and as indicated below:

LastName, A.A.; LastName, B.B.; LastName, C.C. Article Title. *Journal Name* **Year**, *Volume Number*, Page Range.

ISBN 978-3-0365-3032-1 (Hbk)
ISBN 978-3-0365-3033-8 (PDF)

Contents

About the Editors

Alessandro Del Nevo Nuclear Engineer (2002) and PhD in Nuclear Safety (2007) at University of Pisa. He worked at University of Pisa (2003–2010) and, then, at the Experimental Engineering Division of ENEA (Brasimone Research Center). Since 2015, he has been the Head of Systems and Components Design Laboratory. He exploits his technical skills in the fields of thermal-hydraulics; code model development and validation; design of systems and components; design of experimental facilities and experiments; and conduction experimental activities. He has been actively involved in NEA/CSNI and NEA/NSC projects and working groups and IAEA projects. Since 2013, he has contributed to Nuclear Fusion Technology design and R&D activities through the participation in the EUROfusion Project: ENEA responsible of the working package Balance of Plant (2014–2020); leader of the WCLL Breeding Blanket design (2014-2020); responsible of the heavy liquid metals/water interaction activities (2013–2020); and since 2021 deputy project leader and ENEA responsible of the working package Breeding Blanket. He also contributes to Nuclear Training Activities for MSc and PhD university students, NRA, TSO, utilities and industries. He is co-author of 111 publications in international journals; 104 papers in peer-reviewed international conferences; and 8 contributions in published content books and International Technical Reports (e.g., IAEA, NEA).

Marica Eboli Master Degree in Nuclear Engineering (2012) and PhD in Nuclear Safety (2017) at University of Pisa. She is a thermal-hydraulic and safety expert involved in nuclear fission and fusion research activities, presently working in ENEA. Since 2014, she has focused her research activity on the design and execution of experimental activities in LIFUS5/Mod3 under the framework of EUROfusion Horizon 2020 and Horizon Europe Projects for the investigation of the in-box LOCA accidental scenario in the WCLL BB, aimed at validating the numerical model of the lithium-lead/water chemical reaction in the SIMMER code for fusion applications. She was awarded the EUROfusion Researcher Grant (2019–2021) on the project "Development of a multi-fluid and multi-field numerical tool for the modelling of WCLL BB water cooling and PbLi systems". Currently, she is the responsible of the new LIFUS5/Mod4 facility of W-HYDRA platform, which will be constructed at ENEA Brasimone Research Center. Moreover, she is the ENEA scientific responsible of the working package Balance of Plant (WPBoP) and the WCLL BOP R&D Lead Engineer. She also has experience in mentoring for MSc and PhD university students. She is co-author of 24 publications in International Journals; and more than 50 international technical reports.

Preface to "Thermal-Hydraulics in Nuclear Fusion Technology: R&D and Applications"

The perspective of having an almost inexhaustible source of energy has driven worldwide research initiatives on nuclear fusion technology, which have the main representation with the construction of the international nuclear fusion research and engineering megaproject (ITER) in Europe. In this framework, thermal hydraulics is a key discipline which is essential for design and safety demands. Thermal hydraulics is employed in the design phase of the systems and components to demonstrate performance, and to ensure their reliability and their efficient and economical operation. ITER is in charge of investigating the transients of the engineering systems; this includes safety analysis aimed at demonstrating that the operation is safe and consistent with the regulatory authority requirements.

In nuclear fusion technology, thermal hydraulics is required for the design and analysis of cooling and ancillary systems, such as the blanket (i.e., breeding, test or shielding), the divertor, the cryogenic, and the balance of plant systems, as well as the tritium carrier, extraction and recovery systems. Thus, the analyses involve different fluids (water, non-condensable gases, liquid metals) in a broad spectrum of operative conditions (from cryogenic to high temperatures, from vacuum to high pressures), which make their compatibility with the materials challenging.

Although the knowledge of thermal hydraulics benefits from the tremendous efforts conducted for the development of nuclear fission technology, it still has remarkable gaps regarding fusion technology needs. Numerical tools such as well-established system codes suffer, because fluids, parameter ranges, and a field of applications are outside their development and validation boundaries. More sophisticated and complex CFD codes are challenged by the geometrical dimension and complexity of the domains, and by the multi-physics requirements of the analyses. A considerable amount of resources are devoted at the international level for constructing experimental infrastructures, and for establishing and conducting experimental programs, in full-scale and scaled-down facilities, which are aimed at demonstrating the technical feasibility of system and component designs, as well as at generating reference databases to support code development and validation.

In view of the above, this Special Issue documents the present scientific and technical status and recent advances in relation to the "Thermal-hydraulics in Nuclear Fusion", which include, but are not limited to: thermal-hydraulic analyses of systems and components, including magneto-hydrodynamics; safety investigations of systems and components; numerical models and code development and application; codes coupling methodology; code assessment and validation, including benchmarks; experimental infrastructures design and operation; experimental campaigns and investigations; scaling issue in experiments.

The Special Issue collects 20 papers, which are divided into different groups, in order of appearance: multi-fluid multi-phase thermo-hydraulic phenomena relevant to safety of lithium lead breeding blankets (four papers); thermo-hydraulic qualification of helium breeding blanket first wall mock-ups with focus on the surface temperature and its measurement (one paper); thermal hydraulic analyses in support of the design activities relevant for the water cooled breeding blanket CFETR of and its water experimental test loop (two papers); system code models development and CFD applications related to magneto-hydrodynamics for the breeding blankets design (four papers); thermal hydraulic system code transient analyses of helium and water breeding blankets

Balance of Plant of EU-DEMO (three papers); thermo-hydraulics of Li and related phenomena for the target of DONES facility (three papers); experimental and theoretical assessment of the hydraulic performances of the divertor outer vertical target mock-up, with focus on the coolant distribution and pressure drops (one paper), calibration and validation of a multi-physics model, based on the coupling of a CFD model with electro-dynamics and thermo-mechanic models for the simulation of the gyrotrons employed for electron cyclotron heating and current drive of magnetic confinement fusion machines (one paper); and numerical simulations on pebble bed sizing and porosity with a focus on heat transfer behavior and thermal-mechanical response of the tritium breeder pebble bed (one paper).

In conclusion, the scientific and technical contributions from the authors provide the readers with useful information related to the aforementioned topics and cover past and current R&D activities dedicated to this Special Issue.

Acknowledgments

The guest editors acknowledge all authors who have submitted papers to this Special Issue. Special thanks are due to colleagues for the kind collaboration in reviewing these papers.

Alessandro Del Nevo, Marica Eboli
Editors

Article

Experimental and Numerical Results of LIFUS5/Mod3 Series E Test on In-Box LOCA Transient for WCLL-BB

Marica Eboli [1], Francesco Galleni [2], Nicola Forgione [2], Nicolò Badodi [3], Antonio Cammi [3] and Alessandro Del Nevo [1,*]

1 Italian National Agency for New Technologies, Energy and Sustainable Economic Development (ENEA), Department of Fusion and Technology for Nuclear Safety and Security, 40032 Camugnano, Italy; marica.eboli@enea.it
2 Department of Civil and Industrial Engineering, University of Pisa, 56122 Pisa, Italy; francescog.galleni@dici.unipi.it (F.G.); nicola.forgione@unipi.it (N.F.)
3 Department of Energy, Politecnico di Milano, 20156 Milano, Italy; nicolo.badodi@polimi.it (N.B.); antonio.cammi@polimi.it (A.C.)
* Correspondence: alessandro.delnevo@enea.it

Abstract: The in-box LOCA (Loss of Coolant Accident) represents a major safety concern to be addressed in the design of the WCLL-BB (water-cooled lead-lithium breeding blanket). Research activities are ongoing to master the phenomena and processes that occur during the postulated accident, to enhance the predictive capability and reliability of numerical tools, and to validate computer models, codes, and procedures for their applications. Following these objectives, ENEA designed and built the new separate effects test facility LIFUS5/Mod3. Two experimental campaigns (Series D and Series E) were executed by injecting water at high pressure into a pool of PbLi in WCLL-BB-relevant parameter ranges. The obtained experimental data were used to check the capabilities of the RELAP5 system code to reproduce the pressure transient of a water system, to validate the chemical model of PbLi/water reactions implemented in the modified version of SIMMER codes for fusion application, to investigate the dynamic effects of energy release on the structures, and to provide relevant feedback for the follow-up experimental campaigns. This work presents the experimental data and the numerical simulations of Test E4.1. The results of the test are presented and critically discussed. The code simulations highlight that SIMMER code is able to reproduce the phenomena connected to PbLi/water interaction, and the relevant test parameters are in agreement with the acquired experimental signals. Moreover, the results obtained by the first approach to SIMMER-RELAP5 code-coupling demonstrate its capability of and strength for predicting the transient scenario in complex geometries, considering multiple physical phenomena and minimizing the computational cost.

Keywords: SIMMER code; RELAP5 code; in-box LOCA; WCLL breeding blanket; LIFUS5/Mod3

Citation: Eboli, M.; Galleni, F.; Forgione, N.; Badodi, N.; Cammi, A.; Del Nevo, A. Experimental and Numerical Results of LIFUS5/Mod3 Series E Test on In-Box LOCA Transient for WCLL-BB. *Energies* **2021**, *14*, 8527. https://doi.org/10.3390/en14248527

Academic Editors: Christian Veje and Dan Gabriel Cacuci

Received: 27 October 2021
Accepted: 26 November 2021
Published: 17 December 2021

Publisher's Note: MDPI stays neutral with regard to jurisdictional claims in published maps and institutional affiliations.

1. Introduction

In the framework of the development of the European DEMO nuclear fusion reactor, Water Cooled Lithium Lead (WCLL) BB is considered a candidate option for the leading breeding blanket technology [1–4], and has been recently considered in the ITER Test Blanket Module (TBM) program [5]. The major safety issue for the design of this component is the interaction between PbLi and water caused by a tube rupture in the breeding zone, the so-called in-box LOCA (Loss of Coolant Accident) scenario. This phenomenon has been investigated in order to obtain robust data for the validation of system code used in deterministic safety analyses. Indeed, a qualified code is of primary importance for the evaluation of the accidental consequences and for choosing possible mitigating countermeasures, besides proposing design solutions to prevent damages to the blanket box structures.

The R&D related to the PbLi/water interaction took several aspects into account. The first was the implementation of the PbLi/water chemical reaction model in SIMMER code [6], so the verification and validation activity required the application of a standard code methodology [7–9] to experimental data with reproducible and defined initial and boundary conditions [10,11] provided by the new LIFUS5/Mod3 campaigns [12]. The facility was commissioned and two separate effects test (SET) campaigns were executed: the first one (Series D) had the main objective of the generation of an experimental database for the validation of the chemical model in the modified version SIMMER codes; the second one (Series E) focused on the investigations of interaction phenomena between PbLi and water, which strictly depends on choked flow instauration and pressure difference as well as the thermo-hydraulic conditions of the injected water [13]. Meanwhile, numerical simulation activities were performed [14–16]. The numerical results were compared with the experimental data, pointing out differences and similarities to analyze capabilities and limits of the SIMMER codes. The comparison was made by qualitative and quantitative accuracy evaluations; the former was based on the identification of phenomenological windows and of the relevant thermo-hydraulic aspects, and the latter was based on a systematic analysis of the deviation of the predicted target variables with respect to the corresponding measured values.

Experimental results also constituted a useful database for the support of a new STH/2D coupling calculation tool (STH—System Thermal Hydraulics codes) [17,18]. The strength of this tool is the possibility of obtaining high-fidelity calculations in complex geometries, considering multiple physical phenomena and minimizing the computational cost. Its development permitted the performance of a preliminary analysis on WCLL TBM and its ancillary systems, investigating the behavior under an in-box LOCA postulated event [19,20]. The developed coupling technique can be defined as a "two-way", "non-overlapping", "online" methodology [21], with the SIMMER and RELAP5 computational domains separated by interfaces and linked by an external script. Through these interfaces, data are exchanged at each time-step between the two codes in both directions in order to provide proper boundary conditions for the advancement of the calculations. The synchronized advancement in the time domain is controlled by means of an implicit coupling methodology [21].

The present work gives a complete and comprehensive analysis of the experimental results of Test E4.1 and the numerical simulations performed both with SIMMER-III standalone, and with the SIMMER-III and RELAP5/Mod3.3 coupling tools. The results are critically discussed, highlighting the shortcomings and potential of the experimental procedures as well as the deficiencies and capabilities of the numerical tools.

2. Materials and Methods: LIFUS5/Mod3 Facility

LIFUS5/Mod3 (Figure 1) is a separate effects test facility, designed and constructed at ENEA CR Brasimone [12]. Its core is composed of two reaction vessels, S1A and S1B, designed to perform experiments over a wide variety of liquid metals, such as eutectic lead-lithium, eutectic lead-bismuth, and pure lead. Each vessel of the facility has a different functionality: S1A has a geometrical capacity of 100 L and was recently used to characterize the leak detection systems in LBE pool [22], whereas the vessel S1B is smaller, with a geometrical capacity of 30 L. This latter vessel was employed to study the PbLi–water interaction, in the framework of the EUROfusion program, with a main objective of investigating the phenomena occurring during the interaction between the two fluids. The data collected are used mainly to validate the chemical model implemented in the SIMMER code, together with the following expected outcomes:

- the generation of detailed and reliable experimental data;
- the investigation of the dynamic effects of energy release, chemical reaction, and hydrogen production on the structures;
- the broadening of the current knowledge of physicochemical behavior of PbLi eutectic alloys, and the understanding of relevant phenomena associated with its use;

- the expansion of the database used for code verification.

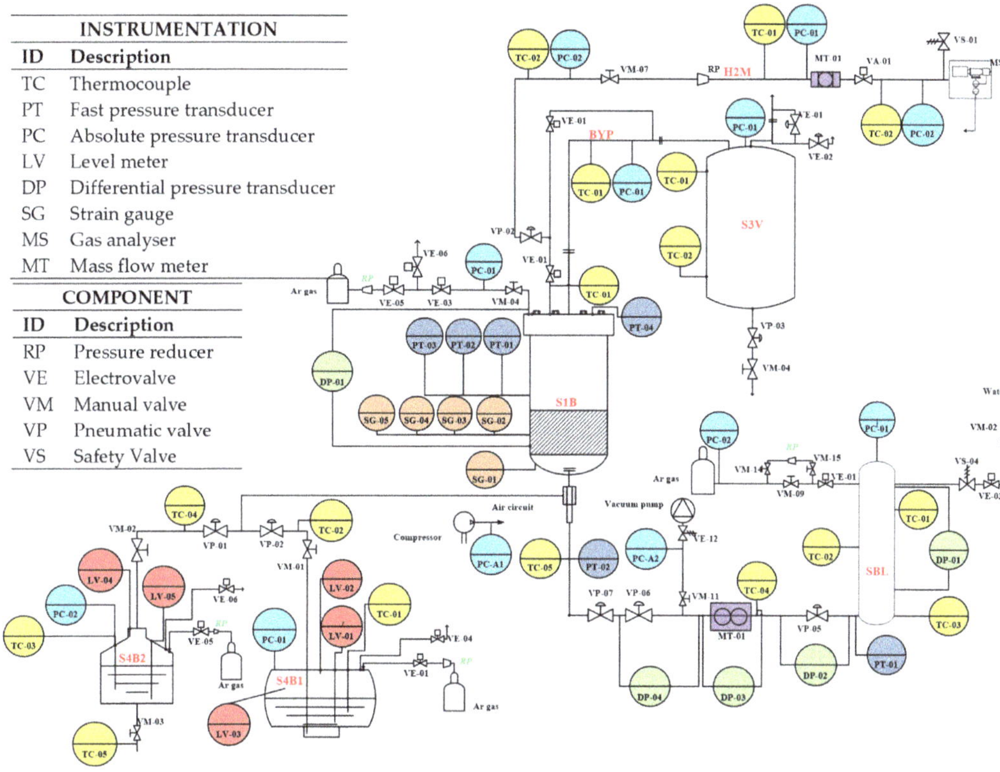

INSTRUMENTATION	
ID	**Description**
TC	Thermocouple
PT	Fast pressure transducer
PC	Absolute pressure transducer
LV	Level meter
DP	Differential pressure transducer
SG	Strain gauge
MS	Gas analyser
MT	Mass flow meter

COMPONENT	
ID	**Description**
RP	Pressure reducer
VE	Electrovalve
VM	Manual valve
VP	Pneumatic valve
VS	Safety Valve

Figure 1. LIFUS5/Mod3 Process and Instrumentation diagram.

2.1. Facility Description

LIFUS5/Mod3 was built by upgrading the previously existing LIFUS5/Mod2 facility [23]. The new experimental plant maintains the expansion vessel (S3V) and the old reaction vessel (S1A), while installing a new vessel, namely, S1B. This new vessel holds a smaller volume than S1A but can withstand higher pressures of up to 200 bar at a temperature of 500 °C, being representative of the WCLL operative conditions, and in accordance with PED directives [24]. The expansion vessel S3V is shared between the old and new branches and is used to collect gasses or other substances released by the test sections in case of rupture disk activation. The facility is composed of five main components (see Table 1):

- The main reaction vessel, S1B, where the water–PbLi interaction takes place;
- The water tank SBL and the injection line, which is used to bring water to test conditions. This is built out of an enlarged section of pipe, connected by a line to the bottom of S1B; pressure is maintained inside this tank by the action of an argon cylinder connected to its top;
- The safety expansion vessel, S3V, connected to S1B by means of two in-series rupture disks, to avoid damaging the rest of the facility in case of overpressure;
- The PbLi storage tanks, S4B1 and S4B2, holding, respectively, fresh and exploited alloys;
- The hydrogen extraction and analysis system.

Table 1. LIFUS5/Mod3 component design features.

Component	Parameter	Value
S1B reaction vessel	Volume (L)	30
	Inner diameter (m)	0.28
	Height (m)	0.56
	Design pressure (bar)	200
	Design temperature (°C)	500
SBL water pipe	Volume (L)	4.05
	Inner diameter (m)	0.04
	Design pressure (bar)	200
	Design temperature (°C)	350
S3V dump vessel	Volume (L)	2000
	Inner diameter (m)	1
	Design pressure (bar)	10
	Design temperature (°C)	400
S4B1 fresh PbLi	Volume (L)	400
	Inner diameter (m)	0.54
	Length (m)	1.56
	Design temperature (°C)	450
S4B2 depleted PbLi	Volume (L)	400
	Inner diameter (m)	0.54
	Length (m)	1.56
	Design temperature (°C)	450

Each of these components is equipped with instrumentation to allow for the acquisition of temperature and pressure data and is heated by means of a heat tracing system surrounded by mineral wool.

2.1.1. Reaction Vessel S1B

The main vessel, S1B, has an internal volume of about 30 L, and during each test, it is filled with 25 L of PbLi. It is completely isolated from the external environment by a top flange, and the remaining internal volume is filled with argon cover gas to avoid a PbLi reaction with moisture and oxygen in the air. The internal geometry of the vessel is composed of a cylinder with a radius of 0.1285 m, and a lower spherical portion with the same radius; the overall height of this vessel is 0.555 m. S1B is connected to the rest of the facility components by a series of penetrations in its top flange, that are shown in the P and ID of Figure 1, and which are:

- A gooseneck seal to allow thermocouples cables passage;
- The housing for the fast pressure transducer;
- The hydrogen extraction line connection;
- The expansion line connecting the vessel to S3V via the rupture disks;
- The housing of a differential pressure meter for level-monitoring and a pressure transducer for absolute pressure-measurement.

Other penetrations are needed on the sides and on the bottom of the vessel to allow extensive data collection; in particular, three more penetrations, at a 120° angle from each other, are made on the walls of the vessel (Figure 2a). These house the fast pressure transducers (PT) for the acquisition of the pressure-wave propagation.

A 2″ penetration is placed at the bottom of the vessel to allow for the loading and unloading of the PbLi alloy, and it also acts as the terminal part of the water injection line. A complete list of the penetrations and instrumentations mounted on the vessel S1B is shown in Table 2.

| (a) | (b) | (c) |

Figure 2. LIFUS5/Mod3 test section. (**a**) External surface of vessel S1B; (**b**) assembly of the test section; (**c**) test section and thermocouples detail.

Table 2. Summary of the penetrations and instrumentation installed on the vessel S1B.

Position	N°	Utilization	Component ID
A	1	Water injection and PbLi charging/discharging system	Steel pipe
B	1	Connection to S3V-expansion/dump vessel	Steel pipe
C	1	Gooseneck sealing system to TC passage	Test section TCs
D	1	Hydrogen measurement system	MS-H2M-01
E	1	DP meter and PC pressure transducer	DP-S1B-01 PC-S1B-01
F	1	PT fast pressure transducer	PT-S1B-04
G-H-I	3	PT on cylindrical shell	PT-S1B-01/02/03
L	1	DP meter	DP-S1B-01
On shell	3	Strain gages (circumferential)	SG-S1B-02/03/04
On shell	1	Strain gages (axial)	SG-S1B-05
On bottom	1	Strain gages (radial)	SG-S1B-01

2.1.2. Test Section

Figures 2 and 3 show the test section inserted into the vessel S1B. This component is welded directly on the top flange of S1B and is designed to be axial-symmetric. The upper holed plate, visible on the left part of Figure 3a, delimits the interaction zone where the chemical reaction can take place, by breaking down the impinging jet of subcooled water from the injector. Simultaneously, the holes allow for the passage of water vapor and hydrogen produced during the tests.

The lateral sides of the test section are, instead, open, to allow for the propagation of the pressure wave generated during the interaction and its measurement through the sensors positioned on the sides of the vessel. These sensors include both strain gages and dynamic pressure transducers to record the vessel deformation and pressure-wave intensity and shape. A total of 74 0.5 mm K-type thermocouples are installed on the test section over six different levels ranging in elevation from the injector to the holed plate, and uniformly distributed in radial directions (Figure 2c). By means of this configuration, the map of the temperature can be identified as well as the layout of the water jet (symmetry, width, and height). The exact positioning of the thermocouples is shown in Figure 3b.

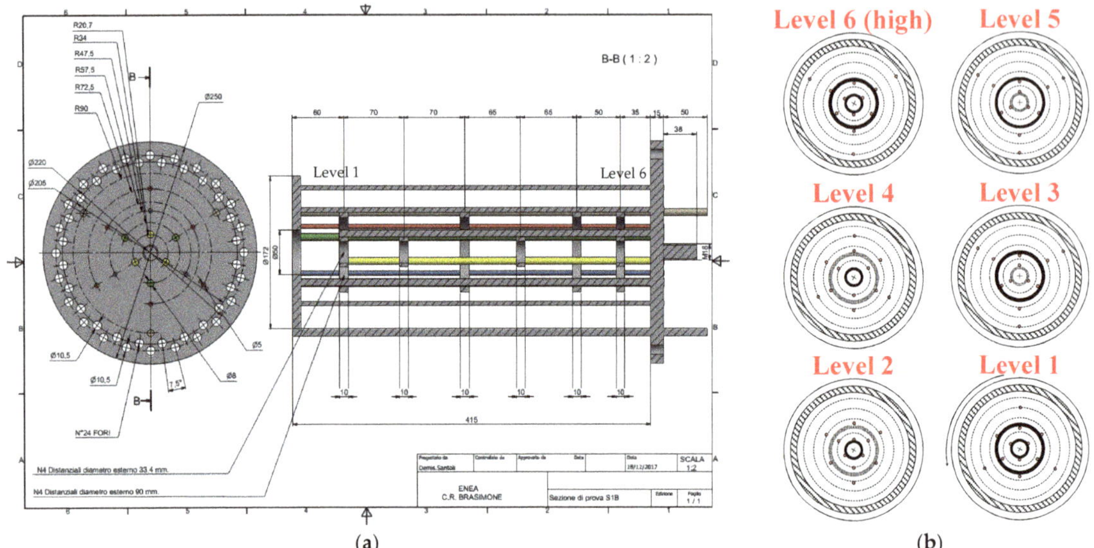

Figure 3. LIFUS5/Mod3 test section. (**a**) Schematic drawing of the test section; (**b**) Layout implementation of the thermocouples.

2.1.3. Injection Line

Before each test, water is loaded and brought to the necessary pressure and temperature conditions inside the SBL tank (Figure 1). The water tank is composed of a 1¹/2″ sch.160 vertical pipe, connected at the top to the water charge line and to the argon pressurization cylinder (Figures 4 and 5a), used to maintain the desired water pressure throughout the test. This tank is instrumented with two thermocouples and a DP meter used to monitor the water level during the charging phase and the heating phase.

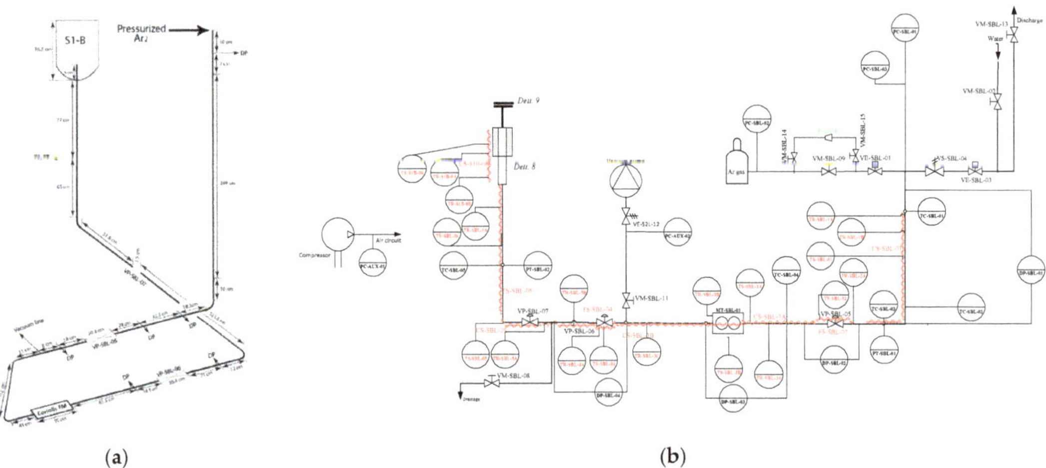

Figure 4. Schematic drawings of the injection line. (**a**) Injection line layout and components positioning; (**b**) injection line P&ID.

Figure 5. Detail of the injection line and device. (**a**) Upper part of the water injection line connected to argon gas cylinder; (**b**) injection line with valves and Coriolis; (**c**) injection piping and connection to the S1B vessel; (**d**) injector cover brass cap with specified notch dimensions.

The water injection line is connected to the bottom of the SBL tank, and it is composed of a 1/2″ Swagelok steel pipe which leads to the S1B vessel, as shown in Figure 5b. This line is instrumented with a series of thermocouples and pressure transducers, and it is heated by means of heating wires surrounded by a thick insulation. Such a configuration allows for the maintenance of the water at the right conditions throughout the test and the measurement of its relevant parameters during the injection phase.

The differential pressure meters allow for the measurement of the pressure-drop across the valves. They are mounted in the following positions (Figure 4b):

- In the SBL vertical section to measure the water level before each test;
- Across valve VP-SBL-05, named: DP-SBL-02;
- Across the Coriolis mass flow meter, named: DP-SBL-03;
- Across valve number VP-SBL-06, named: DP-SBL-03.

The valves VP-SBL-05 and VP-SBL-06 are fast-actuated valves that can reach a fully open position in less than 0.3 s. To achieve this fast actuation time, an independent compressed air line is implemented in the facility to supply the power needed.

Before each test is performed, the whole injection line is evacuated by means of a vacuum pump. This ensures that only water is injected into the reaction vessel, and that no air or argon is present in the system.

The injector device, shown in Figure 5c, is composed of two coaxial pipes, flanged to the bottom of vessel S1B. Water enters the reaction vessel via the innermost line, which is made of $1/2''$ Swagelok pipe, and is forced directly inside the interaction zone. A brass plug with a small hole at its center is fixed to the top of this pipe, with the aim of limiting the amount of water injected during each test. The size of the hole performed on the cap can be varied from test to test, to simulate different types of breakage in the real system.

To avoid the premature injection of the water during the initial test phase, a brass cap is placed on top of the plug and soldered to it. This cap, shown in Figure 5d, is built with a notch, which is engineered to break at the desired test pressure. With this system, the experimenters can regulate, with precision, both the mass flow rate of the injected water (by means of the orifice diameter) and the pressure at which the injection occurs, i.e., the pressure at which the cap ruptures (by means of the notch depth).

The two tanks act, respectively, as a storage tank for unused PbLi and as a dump tank for depleted PbLi. Prior to each test, the alloy is loaded into the reaction vessel from tank S4B1, and after each test, the depleted alloy is discharged into tank S4B2 (Figure 6). This procedure ensures the avoidance of the contamination of fresh PbLi with oxides and other reaction products. The tanks are instrumented to allow for temperature and level control of their content.

2.1.4. Expansion Line

The expansion vessel, S3V, acts as a safety volume to collect reaction products such as water vapor, hydrogen, or PbLi in case of a rupture disk breakage. It has a volume of 2000 L and a design pressure of 10 bar. To protect the facility from overpressure during the tests, the line connecting S1B and the relief tank, S3V, is equipped with two rupture disks mounted in series. The first one has a diameter of $2''$ and is rated for 190 bar at a temperature of 400 °C. The second one has a diameter of $3''$ and is rated for 154 bar at 450 °C.

2.1.5. Hydrogen Extraction Line

Following each test, the gas composition inside vessel S1B will change from the initially pure argon to a mixture of argon, hydrogen, and unreacted water vapor. This mixture will be extracted from two spillage points and quantitively analyzed by means of a gas analyzer (labelled MS-H2M-01 on the P and ID of Figure 1).

The instrument working principle is based on the measurement of the thermal conductivity variation of the gas mixture. This procedure is carried on at a pressure of 2 bar and at a specific mass flow rate, and these conditions are met by using a pressure reducer and a mass flow meter and controller.

Data from the gas analyzer and from the mass flow meter are acquired using a dedicated system, which is, in turn, connected to the main control system.

Figure 6. Storage (S4B1) and drain (S4B2) tanks, loading and discharge lines.

2.1.6. Instrumentation and Control Units

The Data Acquisition and Control Subsystem (DACS) architecture of this facility is subdivided into two separate sections: the real-time control and data acquisition, and the control, interlock, and safety system. The facility is equipped with a wide variety of sensors to acquire all the relevant thermo-dynamical and mechanical data, such as pressure, temperature, and strain. These sensors are constantly acquired at 1 Hz during facility operation, while during each test, thermocouples are acquired at 50 Hz and fast pressure transducers at 10 kHz.

To control and operate the facility, a series of pressure reducers and manual, pneumatic, safety, and electro valves are installed in various points. All these components are operated by the experimenter via the control software.

2.1.7. Test Matrix

The Series-E test matrix is reported in Table 3. This test matrix was designed with the aim of performing tests at conditions relevant to the WCLL BB design [1–4]. In particular, during a postulated WCLL-BB in-box LOCA, the phenomena and processes occurring in the PbLi/water interaction are governed, firstly, by thermodynamic parameters, and then by chemical reaction. All phenomena strictly depend on the amount of water injected into the BB box, and are sensitive, besides the design of the water and PbLi loops, to:

- The pressure difference between the water circuit and the PbLi, which is the driving parameter of the transient: once pressures in the two systems come into equilibrium, the injection stops;
- The size of the break;

- Water flow condition: the instauration of choked flow limits the velocity of water itself, while the void fraction of the jet will affect the actual injected mass.

Table 3. Series-E Test Matrix.

Test Series E	D Orifice (mm)	Water T (°C)	PbLi T (°C)	Injection Time (s)	Injection Pressure (bar)
#1	4	295	330	1	155
#2	4	295	430	1	155
#3	1	295	330	0.5	155
#4	2	295	330	1	155
#5	1	295	330	1.5	155
#6	4	295	380	1	155
#7	2	295	380	1.3	155
#8	1	295	380	2	155

In order to investigate these phenomena, the test matrix was based on executed tests with varying diameters of the injection orifice, injection times, and PbLi temperatures. For all the tests, the injection pressure in the water line was fixed at 155 bar, and the temperature set to 295 °C.

2.2. Code Nodalizations

The post-test analyses have been performed with the beta version of SIMMER-III, modified for fusion application by the University of Pisa, and identified as "SIMMER-III Ver. 3F Mod. 0.1". This version derives from SIMMER code Ver.3F [25,26], and the main difference is the implementation of the chemical reaction between PbLi and Water [6]. The code SIMMER-III was used both in standalone mode and in a coupled mode with RELAP5/Mod3.3; the coupling technique was created, and is still under development, at the University of Pisa [17,18].

SIMMER-III is a two-dimensional, multi-velocity field, multi-phase, multi-component, Eulerian fluid-dynamics code which can be coupled with a structure model (fuel pin) and a spacetime and energy-dependent neutron kinetics model. The fluid-dynamics portion, which constitutes about two-thirds of the code, is interfaced with the structure model through heat and mass transfer at the structure interfaces, while the neutronics portion can provide nuclear heat sources based on the mass and energy distributions calculated by the other code elements.

The basic geometric structure of SIMMER-III is a two-dimensional cylindrical (R-Z) system. Therefore, all the nodalizations developed through SIMMER-III to model the LIIFUS/Mod3 main components assume an axially symmetric cylindrical geometry. It is important to notice that this means losing information on the evolution on the azimuthal direction.

The facility set-up for the nodalization is shown in Figures 7 and 8. It is constituted by 5 main parts:

- The injection line (blue dashed line in Figures 7 and 8):
  - Coupled calculation: fully nodalized with RELAP5;
  - Standalone calculation: reduced to a short vertical section below the S1B;

- The reaction vessel S1B (including Test Section);
- The expansion line (including the first rupture disk);
- The hydrogen extraction line (up to the collecting valve);
- The thermocouple supporting passage (gooseneck).

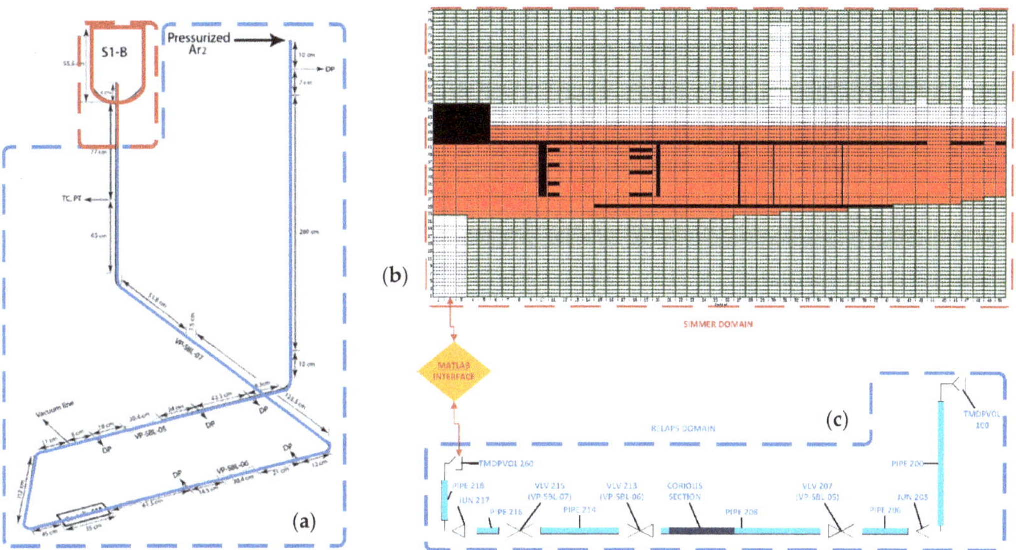

Figure 7. Scheme of coupled codes nodalization. (**a**) LIFUS5/Mod3 facility and coupling tool interface; (**b**) S1B modeled by SIMMER III; and (**c**) injection line modeled by RELAP5/Mod3.3.

The SIMMER-III geometrical domain is obtained by 50 radial and 77 axial mesh cells (Figure 8a). In Figures 7 and 8, the colors distinguish the different fluids and structure materials, as set at the beginning of the transient (t = 0 s). Therefore, the PbLi is represented in red, the water in blue, the argon cover gas (and the hydrogen produced by the reaction) in white, the non-calculation zones are highlighted by a green mesh fence, and SS316 in black as the structural material. The correspondence of the main dimensions of LIFUS5/Mod3 and the SIMMER-III reference model is reported in Table 4.

Table 4. SIMMER-III reference model and LIFUS5/Mod3 facility: main dimensions.

Region	Dimension	SIII	L5/M3 Facility	SIII Cells
S1B	H (m)	0.575	0.555	22–52
	D (m)	0.2614	0.257	1–50
	V (L)	26.575	26.590	-
Free gas	H (m)	-	−0.04	NN-52
(in S1B)	V (L)	3.4	3.554	-
Inj-device	H (m)	0.02	0.02	22
	D (m)	0.0094	0.0094	1–3
	$D_{orifice}$ (m)	0.002	0.002	1
Inj-line	H (m)	(variable)	~6.8	1–22
	D (m)	0.0094	0.0094	1–3
	D (m)	0.0094		1–3

The reference mesh cells for the temperature analysis inside the S1B and for the pressure measuring, representing the position of installed thermocouples and pressure sensors in S1B, are listed in Table 5.

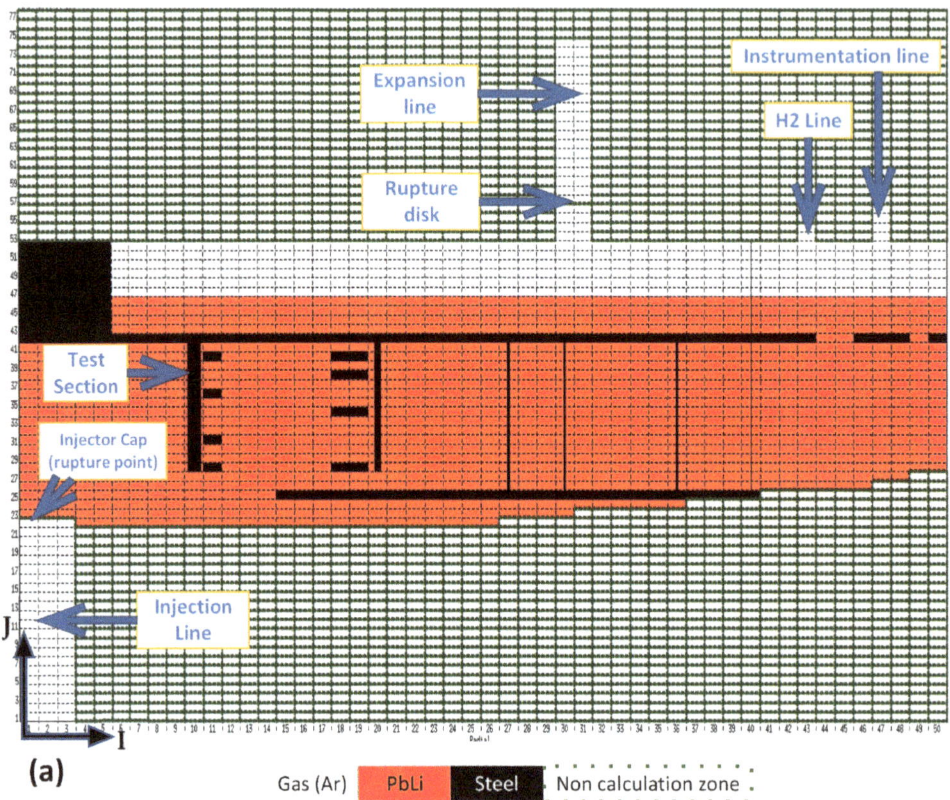

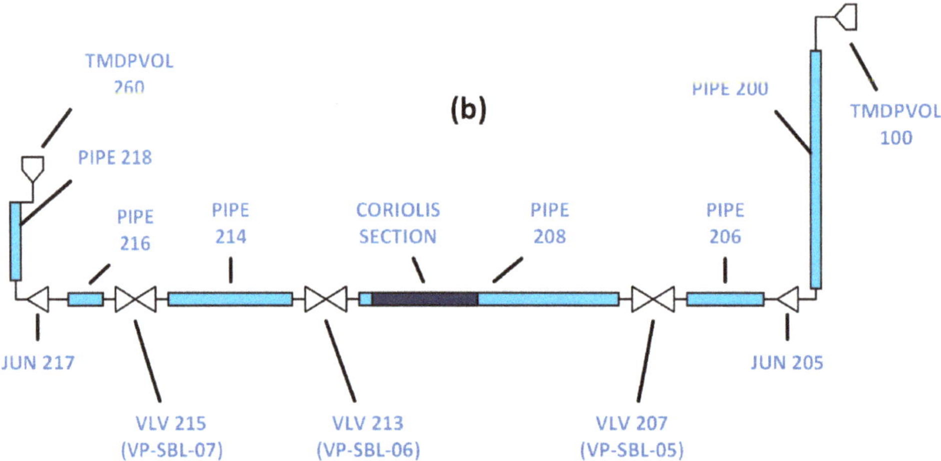

Figure 8. Detailed code nodalization. (**a**) SIMMER-III nodalization (not on scale) for LIFUS5/Mod3 facility, used in Standalone and coupled analysis; (**b**) RELAP5/Mod3.3 nodalization of the injection line (only pipe lengths on scale).

Table 5. Location of the TCs and PTs on the SIMMER-III reference model.

	Ring 1	Ring 2	Ring 3	Ring 3	Ring 4	Ring 5	Ring 6
	Installed (TCs), Inside S1B						
Level 1	(12,28)		(26,28)	(28,28)		(35,28)	
	TC-R11-L1 TC-R13-L1	-	TC-R31-L1 TC-R32-L1 TC-R33-L1	TC-R34-L1 TC-R35-L1 TC-R36-L1	-	TC-R53-L1	-
Level 2	(12,31)	(19,31)	(28,31)		(31,31)		
	TC-R11-L2 TC-R12-L2 TC-R13-L2	TC-R21-L2 TC-R22-L2	TC-R32-L2 TC-R33-L2		TC-R41-L2 TC-R42-L2 TC-R43-L2	-	-
Level 3	(12,34)		(26,34)		(31,34)		(41,34)
	TC-R12-L3 TC-R13-L3	-	TC-R31-L3 TC-R33-L3		TC-R41-L3 TC-R42-L3 TC-R43-L3	-	TC-R61-L3 TC-R62-L3 TC-R63-L3
Level 4		(19,36)	(28,36)		(31,36)	(37,36)	
	-	TC-R21-L4 TC-R22-L4 TC-R23-L4	TC-R31-L4 TC-R32-L4 TC-R33-L4		TC-R41-L4 TC-R42-L4 TC-R43-L4	TC-R52-L4 TC-R53-L4	-
Level 5	(12,38)		(26,38)		(31,38)		(41,38)
	TC-R11-L5 TC-R13-L5	-	TC-R31-L5 TC-R32-L5 TC-R33-L5		TC-R41-L5 TC-R43-L5	-	TC-R61-L5 TC-R62-L5 TC-R63-L5
Level 6	(12,40)	-	(26,40)	(28,40)	-	-	(41,40)
	TC-R11-L6 TC-R12-L6		TC-R31-L6 TC-R33-L6	TC-R34-L6 TC-R35-L6 TC-R36-L6			TC-R61-L6 TC-R62-L6 TC-R63-L6

Installed (TC, PTs)	
Location	Cell
Injection line—vacuum part, TC-SBL-05	(1–3,17)
Injection line—pressurized part, TC-SBL-01, @ temperature BC	(1–3,1)
Injection line—pressurized part, PC-SBL-01, @ pressure BC	(1–3,1)
Injection line—vacuum part, PT-SBL-02	(3,17)
Expansion line, rupture disk, PT-S1B-04	(30,52)
Reaction vessel (S1B), PT-S1B-01/03	(50,33)
Reaction vessel (S1B), PC-S1B-01	(47,53)

The standalone and coupled version of the SIMMER-III nodalization are essentially the same, the only difference being the vertical length of the cells reproducing the injection line; in the coupled version, this region is reduced to only 30 cm in order to minimize the influence of SIMMER in the simulation of the pressurization, and therefore, fully exploit the capability of RELAP5 in simulating 1-D pipelines.

2.2.1. Reaction Vessel S1B

The reaction vessel, S1B, is modeled by two separate regions; at initial conditions, the primary region includes only liquid PbLi ($I = 1$–50, $J = 45$–NN). The second region is related to argon as the cover gas at the top and dominated by the cells ($I = 6$–50, $J =$ NN); see Figure 8a. NN depends upon the initial conditions of PbLi level. The hemispherical part of the vessel is modeled by adding 6 non-calculation regions as well.

2.2.2. Test Section

The test section is installed to support the instrumentation inside the S1B vessel, specifically for the thermocouples. The test section is divided into two different parts, the upper holed disk-shape part, which is considered as 3 different regions with steel as the cladding material at J = 65 (S-III), and the second part, which contains 6 sets of rods with different lengths for each, plus the spacer rings. The whole test section is represented in 6 different levels by 10 vertical rods and 6 horizontal rings. In Figure 8a, the black cells without green fences show the test parts for the SIMMER reference model.

2.2.3. Injection Line

The injection line configuration consists of 4 horizontal and 2 vertical parts, and it is too complex to be modeled with a full domain as a SIMMER-III nodalization; therefore, it was decided to cut it and consider it as an integrated vertical pipe which is connected from the top to the reaction vessel and from the bottom to the boundary cell. With this assumption, all of the bends and relevant losses are translated into orifice coefficients and transferred to the new positions in the vertical direction. In the SIMMER-III nodalization, the injection line started from cell I = 1–3, radially, and J = 1–13, axially. In the standalone version, it contained 3 different regions, starting from valves VP-SBL-06, and including VP-SBL-07 and 2 pipes; see Figure 8a. The upstream part to the water region was cut and excluded from the reference nodalization, and the acquired data from PC-SBL-01 was used as the boundary condition for post-test calculations.

In the RELAP5 nodalization, the injection line is fully nodalized, reproducing all the main features of the line, including the Coriolis section (Figure 8b). The nodalization consists of volume elements, with a specified area and length. The different volume elements and their properties are listed in Table 6 and shown in Figure 8b. The volume elements are connected to each other using junctions, to which a pressure-drop coefficient can be assigned. The junctions are used here to reproduce the change of orientation of the pipes. The properties of these junctions can be found in Table 6. The injection line starts with SBL, which is modeled as a pipe with 28 volume elements. The water tank is pressurized by a time dependent volume, representing the argon tank. They are connected together by a single junction. The bottom of SBL is connected via a single junction to a smaller pipe containing 10 volume elements. This pipe is connected to a motor valve, representing VP-SBL-05. The valve can be opened or closed, and it is also possible to assign the timing of action. A valve is only a junction, so the volume elements of this valve are incorporated in the next and previous pipes. After other four pipes, two motor valves, representing VP-SBL-06 and VP-SBL-07, and a junction, the line is connected to a trip valve.

2.2.4. Expansion Line

The pipe connecting S1B to S3V (expansion tube) is modeled by an annular tube with an equivalent diameter of 0.0428 m, approximately, at a similar distance from the central Z axis to preserve the flow area. This toroidal tube is located at I = 30–31, J = 76–97. The rupture disk was specified by adding a virtual walls. However, during post-test calculations, the virtual wall remained closed since no signal was recorded in the expansion line during Series D Tests, which means the rupture disk kept closed during the transient. Since the S3V volume does not take an action during the tests, the S3V expansion vessel was not considered in the post-test analysis.

2.2.5. Hydrogen Extraction Line

The hydrogen line is supposed to collect the produced H2 gas from S1B and conduct it through the hydrogen-analyzer system. In the present models, this part is not completely modeled, but its first part, up to the collecting valve, is considered. The cells (43,76–77) represent the volume as an equivalent region for the reference model.

Table 6. Characteristics of RELAP5/Mod3.3 nodalization.

Component Number	Hydrodynamic Component	Description	Length (m)	Area (m^2)
100	Time-dependent volume	Argon tank	(-)	(-)
200	Pipe	SBL	2.8	9.069×10^{-4}
205	Single junction		(-)	
206	Pipe		1.096	6.936×10^{-5}
207	Motor valve	VP-SBL-05	(-)	6.93×10^{-5}
208	Pipe	Normal section	2.094	6.936×10^{-5}
		Coriolis section	1.48	6.12×10^{-4}
213	Motor valve	VP-SBL-06	(-)	6.93×10^{-5}
214	Pipe		1.742	6.936×10^{-5}
215	Motor valve	VP-SBL-07	(-)	6.93×10^{-5}
216	Pipe		0.518	6.936×10^{-5}
217	Single junction		(-)	
216	Pipe		1.4	6.936×10^{-5}
260	Time-dependent volume	Coupling interface	(-)	(-)

2.2.6. Boundary and Initial Conditions (BIC)

The reference calculation starts at t = 0 s, which represents the valve VP-SBL-06 opening. The injector cap rupture is simulated by disappearance of the virtual wall at cell I = 1, J = 22, which recreates the 2-mm orifice. The time at which the injector breaks up is obtained from the specific test experimental data. Since the closest PT to the injector cap is PT-SBL-02, the injection pressure trend is assumed to be the pressure recorded by PT-SBL-02, applied in cells (1–3,1); on the other hand, since, in the coupled calculation, the whole line is simulated, the BC set-up pressure and temperature, obtained from PC-SBL-01 and TC-SBL-01, is imposed at TMPVOL 100. Moreover, the boundary condition of the continuous inflow of injected water is applied in the same mesh cells. The initial conditions of pressure, temperature, filling level of lithium-lead in S1B, and the amount of water are set coherently with the experimental data reported in Table 7.

After setting all BICs, the compiled executive file of SIMMER code was used to, firstly, run the reference model, and later, the other cases, for sensitivity analysis. The calculations lasted for 3 s.

3. Results

In the following, the experimental data of Test E4.1 are reported, together with a critical analysis of the results, as well as the numerical post-test simulations obtained in SIMMER-III standalone and RELAP5/SIMMER-III coupled configurations.

3.1. Test E4.1 Experimental Results

The test E4.1 was executed on 4 November 2020 at 11:23:42. It corresponded to Test #4 of the test matrix proposed in the framework of EUROfusion project (Table 3). The injection procedure (see P&ID in Figure 1) is foreseen to open the valves connecting the argon cylinder to the SBL water line (VM-SBL-09 and VE-SBL-01); thus, the system is pressurized and maintained in this condition to reach the thermal equilibrium. Then, valve VP-SBL-05 is opened to permit the filling of the line up to the injection valve VP-SBL-06. The acquisition system is in stand-by, waiting for the injection signal. This signal is activated by the operator when all parameters are correctly checked. Then, the fast acquisition system starts to record the data, and the valve VP-SBL-06 is opened up to the closing signal. The closure of valve VP-SBL-06 is automatically activated after 1 s. The water injection in S1B lasts for 0.93 s (from cap rupture to VP-SBL-06 fully closed). Once the pressure inside injector reaches rupture limit, the injector breaks and the water starts to flow into reaction vessel S1B and the transient starts. The pressure from the argon cylinder is kept for the whole transient. A valve called VP-SBL-07 is installed to protect VP-SBL-06 from the PbLi drop. The acquisition system continues to record data (pressures, temperatures, levels, and so on) up to the end of the test. The acquisition system worked properly. The strain gages and pressure transducers signals were acquired at 10 kHz, and the thermocouple signals at 50 Hz.

From the phenomenological point of view, the test can be divided into four different phases:

1. Pressurization of water injection line (0 ms to 229.4 ms), from the pressure rise in the injection line (valve VP-SBL-06 starts to open, start of test) to the rupture cap occurrence (start of injection).
2. Water–PbLi interaction (229.4 ms to 1156.8 ms), from the cap rupture occurrence to the full closure of valve VP-SBL-06 (end of injection). This phase can be divided into three different sub-phases:

    a. Flashing of injected water (229.4 ms to 245 ms), from cap rupture occurrence to the ending of the first pressure peaks.
    b. Pressurization dominated by the two-phase thermodynamic interaction (245 ms to 420 ms), from the ending of the first pressure peaks to the pressure slope change.
    c. Pressurization dominated by the single-phase thermodynamic interaction (420 ms to 1156.8 ms), from the pressure slope change to valve VP-SBL-06 fully closed.

3. Pressurization dominated by the chemical reaction (1156.8 ms to ~20,000 ms), from the full closure of valve VP-SBL-06 to the S1B pressure stabilization.
4. Ending phase (~20,000 ms to the End of Test), from the S1B pressure stabilization to the End of Test.

In each of these phenomenological phases, the main parameter trends, such as the pressure in the water line (Figure 9), the pressure (Figure 10) and the strain in the reaction vessel (Figure 11), and the temperatures in the test section (Figure 12), showed different behaviors.

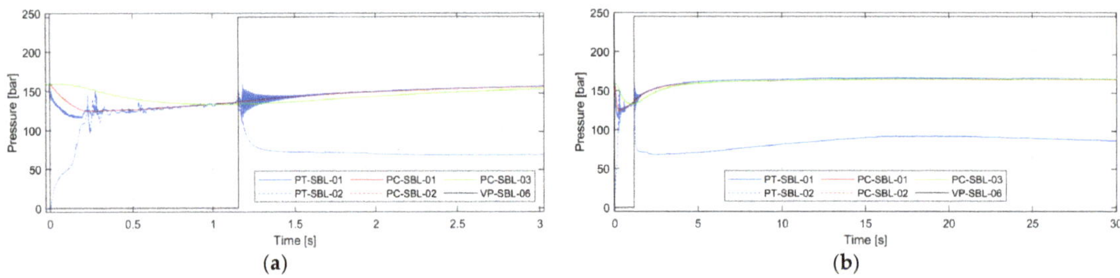

Figure 9. Pressure time trends in SBL injection line and valve position. (**a**) zoom on [0–3] s, (**b**) zoom on [0–30] s.

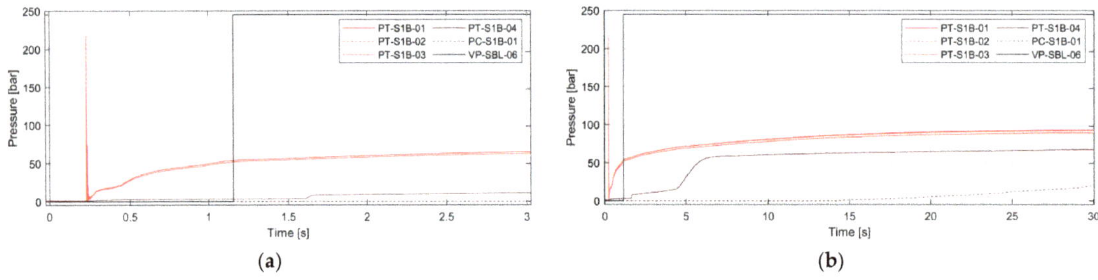

Figure 10. Pressure time trends in S1B reaction vessel and valve position. (**a**) zoom on [0–3] s, (**b**) zoom on [0–30] s.

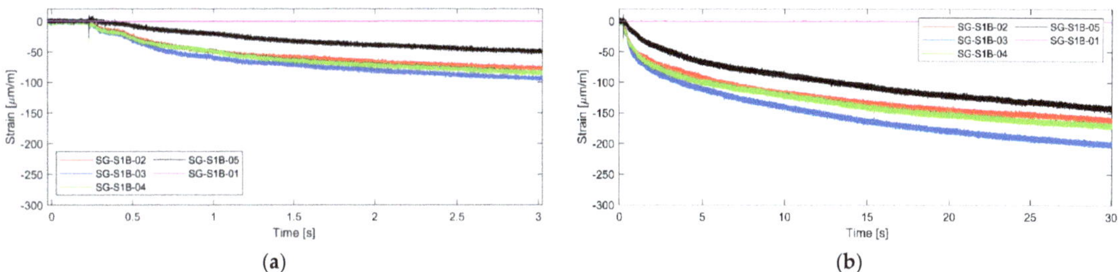

Figure 11. Strain time trends on external surface of S1B reaction vessel. (**a**) zoom on [0–3] s, (**b**) zoom on [0–30] s.

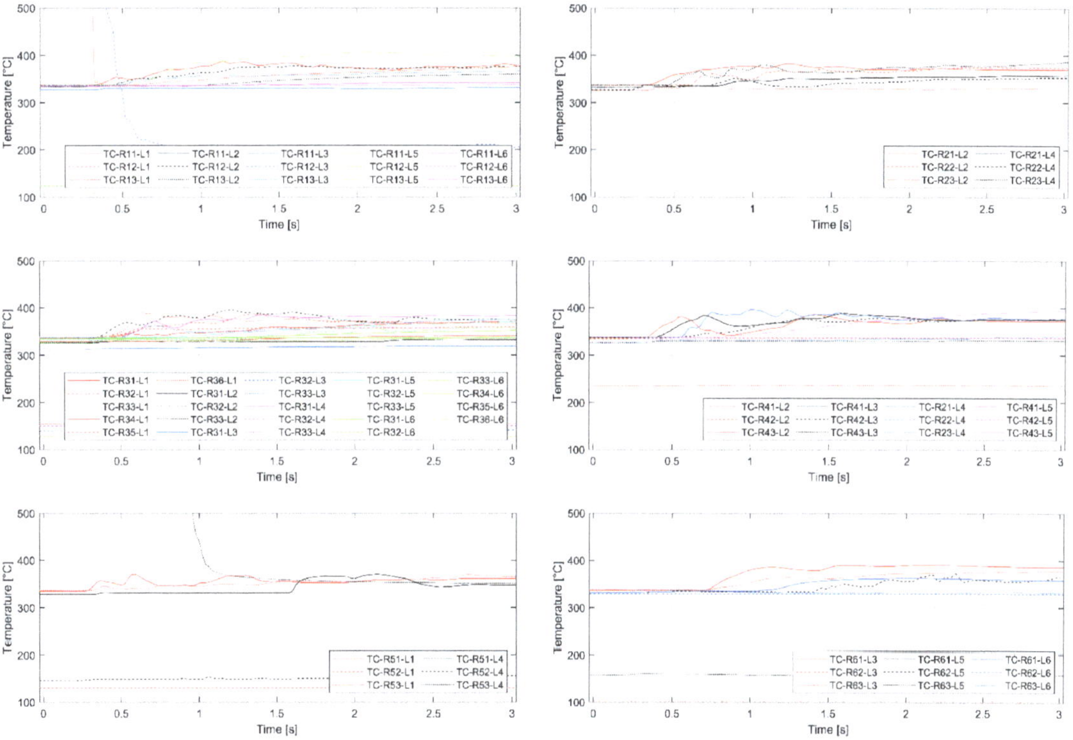

Figure 12. Temperature time trends in S1B at different rings (from Ring 1 to Ring 6, see Figure 3b).

3.1.1. Pressure Time Trends

Two fast pressure transducers acquired the pressure trend in the injection line PT-SBL-01, positioned in the water tank, and PT-SBL-02, positioned just below the injector. Moreover, two different absolute-pressure transducers, PC-SBL-01 and PC-SBL-03, are installed (Figure 9). The former is able to follow the fast pressure change during the first phase of the test; meanwhile, the latter is not, even if it can be installed at a high temperature. The sensor PC-SBL-02 recorded the pressure in the argon cylinder. In the first phase of the transient, the PT-SBL-02 sensor recorded a pressurization rate of 633.3 bar/s. Once valve VP-SBL-06 was fully opened, the water filled the injection line between valve VP-SBL-06 and the injector cap inside S1Buntil cap rupture, which occurred at a pressure of 146.3 bar. As soon as the cap broke, the second phase started, and water was injected in the reaction vessel and the pressure in the injection line decreased rapidly. PT-SBL-02 recorded a decrease from 146.3 bar to about 100 bar, due to the instantaneous increase of volume; then, its decrease was stopped, remaining constant for few tenths of a millisecond, probably due to the two-phase water injection. The pressure in the injection line rose again, up to the constant value of about 125 bar, equal to the pressure recorded by PT-SBL-01 in the SBL tank, until the valve closure instant. After that, the third phase started. The water tank, SBL, and the reaction vessel, S1B, were isolated; therefore. The pressure in SBL increased due to the connection with the upstream argon cylinder, and PT-SBL-02 measured a sudden pressure decrease as a consequence of a disconnected pressurization system. The pressure equalization between S1B and the injection line downstream VP-SBL-06 occurred after the valve closure at about $t = 3$ s.

Concerning the pressure inside S1B reaction vessel (Figure 10), during the first phase, the initial relative pressure after PbLi filling was about 0.7 bar. Once the injector cap broke (the second phenomenological phase), the pressurization of S1B was characterized by several peaks, lasting for about 15 ms, due to the water-flashing inside the tank, followed by steam and gas expansion through the perforated top plate of the test section and towards the gas plenum (decrease of pressure), and a new water-flashing. For each peak, different signals were recorded by the fast pressure transducers, reflecting the pressure-wave propagation inside the reaction vessel. The PT set in S1B cover gas (PT-S1B-04) showed a delayed pressure increase, due to the gas expansion towards the upper part of the S1B reaction vessel. Additionally, the sensor was partially plugged by PbLi splashes; indeed, a difference of about 20 bar was recorded in respect of the other PTs. Up to about 420 ms, the pressure recorded a monotonous increase to almost 20 bar, dominated by the two-phase water interaction, the water evaporation, and partially by the hydrogen generation. Then, the pressure started to increase, again, up to 53 bar at the closure of valve VP-SBL-06. This phase was dominated by the pressurization due to single-phase water interaction, and by hydrogen production and temperature increases due to the exothermic chemical reaction between PbLi and water. Then, during phase 3, the S1B reaction vessel and the injector were in connection, the pressure in the system continued to increase due to the chemical reaction between PbLi and the injected water, and the hydrogen generation increased, from 53 bar to almost 90 bar.

3.1.2. Strain Time Trends

The strain time trends, measured on the outer surface of S1B, are depicted in Figure 11. It is possible to highlight several peaks due to pressure waves propagated in the liquid alloy as a result of cap-breaking and water-flashing. SG-S1B-01, positioned in radial direction on the bottom part of the reaction tank, was defective. On the contrary, the other strain gages positioned on the cylindrical shell of S1B recorded different peaks. The higher values of the first peaks was recorded by SC, positioned in circumferential direction, whilst the axially positioned SG-S1B-05 measured lower values. The SGs showed deformation trends that perfectly overlapped with the pressure trends behavior recorded by PT. At valve closure, the SGs measured values in the range of -26.8 to -68.1 μm/m. Then, during phase 4, the SG recorded a monotonous increase of the deformation rate, until the strains reached an almost-stationary condition. The sensors showed a residual strain at the End of Test (EoT) that could be ascribed to the different working temperature, in respect to the initial one, and to the pressurization of S1B. The SGs set on a circumferential position measured their maximum strain value (respectively, -173.1, -217.6, and -191.4 μm/m for SG-S1B-02/03/04). The strain gage positioned in an axial position on the cylindrical shell of the vessel measured a lower strain of about -160.2 μm/m.

3.1.3. Temperature Time Trends

The initial temperature of the PbLi in S1B was about 333.9 °C. During the experimental campaigns, an unexpected phenomenon was observed. Indeed, after each test, a variable number of thermocouples modified their readout behavior, providing unreliable data. It is noted that the involved thermocouples tended to give readouts of temperatures that were lower than the expected and differed substantially from the average readout of the thermocouple bundle mounted in the test section. This readout degradation follows a unique behavior, and considering the harsh environment in which these sensors must operate, its cause has been hypothesized to lie in a phenomenon commonly known as "green rot" [27], the common name for referring to the oxidation of the chromium in K-type thermocouples, which typically occurs in high-temperature and reducing environments. K-type thermocouples are partially made of a Ni-Cr alloy, which is normally protected from oxidation by a thin layer of oxide that forms over its surface. However, the presence of a reducing agent (such as hydrogen, generated by the PbLi–water interaction) and the high temperature environment can substantially accelerate the corrosion of the sensing tip.

Due to these considerations, during the tests performed on the LIFUS5/Mod3 facility, an extensive phenomenon of green rot occurred on the thermocouples of the test section, which were exposed to a high-temperature and hydrogen-rich environment. Moreover, the small size of the thermocouples themselves (0.5 mm in diameter), and the mechanical stress due to the pressure waves generated by the tests, might have accelerated the degradation. Nevertheless, in the subsequent data analysis, anomalous thermocouples have been individuated, and their data discarded, to avoid the degradation of the results. The classification of anomalous thermocouples was performed by carefully analyzing the signal provided during the PbLi charging phase, when the test section was in thermal equilibrium with the PbLi. In this condition, in fact, the high thermal conductivity of the alloy caused all of the thermocouples at the same depth to read the same temperature values and allowed for the individuation of the broken ones.

The temperature time trends measured by all the TCs set on the levels and rings of the test section are shown in Figure 12. The nomenclature TC-RXX-LY defined the position of the thermocouples in the ring and in the level, i.e., the code TC-R31-L2 specifies the first TC (starting from the established position shown in Figure 2b, in counterclockwise direction) positioned in Ring 3 and Level 2. Each figure reports the temperature time trends of TCs installed in the same ring. Ring 1 and Level 1 are the closest to the injector, Ring 6 and Level 6 are the farthest from it. In test E4.1, the effect of the thermodynamic interaction between PbLi and water, characterized by the cooling of the melt due to the water jet expansion, did not appear, which is clear evidence that the jet was almost spread in S1B. On the contrary, at the interface between two fluids, the chemical reaction occurs, generating hydrogen and heat. Indeed, the temperatures increased, as recorded by all the TCs installed in the test section. The temperature trends showed that the chemical reaction prevailed in the middle levels (Level 3, Level 4, and Level 5) and was almost spread in a radial direction; indeed, not-significant peaks are recorded (the highest value of 439.67 °C was reached by TC-R61-L3 at 16.54 s), but the heat generated by the chemical reaction led to a general increase of temperature in the whole system with an average difference of about 20 °C (up to valve closure). Additionally, after the valve closure, the temperatures in the reaction zone increased again as a result of the continuous chemical reaction between the water already injected and the PbLi.

3.1.4. Hydrogen Production

Finally, concerning the measurement of the hydrogen produced by the reaction, it was necessary to obtain the amount of injected water during the test in order to verify the reliability of the hydrogen gas analyzer. This amount is evaluated a posteriori, considering the integration of the mass flow rate evaluated by the Coriolis mass flow meter (Figure 13a). The mass flow meter recorded a delayed measure, which was not fully able to follow the fast dynamic of the injection. Indeed, the instrument continued to measure once the injecting valve was already closed. However, the amount of injecting water was evaluated considering the following rationale:

- At SoT, the facility condition was: valve VP-SBL-05 opened, and valve VP-SBL-06 closed; DP-SBL-01 (Figure 13b) measured 90.8 mbar, which corresponded to an amount of water in the SBL line equal to 944 g;

- During the injection, DP-SBL-01 measured 0.0 bar; therefore, all the water was injected through the line;
- At the end of the test, a part of the water that remained trapped in all the injection lines was collected and weighted, with a result of 750 g;
- This result, considering the differential pressure sensor measurement, permitted the obtaining of the injected water, with a result of 944 − 750 g = 194 g;
- The Coriolis mass flow meter measured an integral value of 395 g;
- Some of the water remained trapped between valve VP-SBL-06 and valve VP-SBL-07. The volume of this section of the line is equal to 0.14 L, which corresponds to a mass of 111 g (at 250 °C, temperature effectively recorded);
- The amount of injected water, considering the Coriolis mass flow meter measure, permitted the obtaining of the injected water, with a result of 395 − 111 = 284 g;
- Considering the double-check of the injected water, the estimation gave a range between 194 and 284 g.

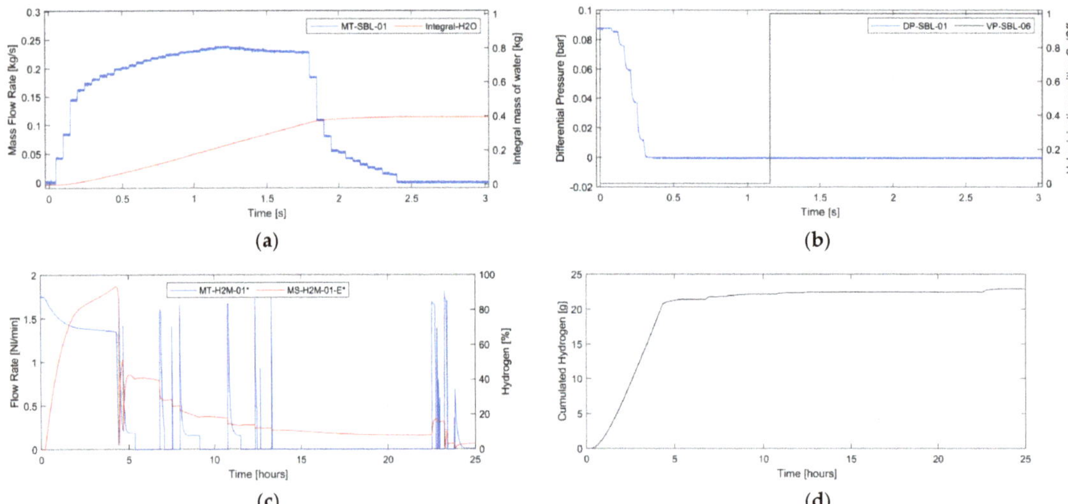

Figure 13. (**a**) Injected water mass flow rate and integral value; (**b**) differential pressure measured in the SBL tank; (**c**) gas mass flow rate (MT-H2M-01) and hydrogen concentration percentage (MS-H2M-01) time trends; (**d**) cumulated hydrogen time trend.

Since, in test E4.1, the rupture disk did not fail, the hydrogen measurement proceeded from the extraction line, opening the valve VE-S1B-01, and then moved towards the hydrogen-measurement system by opening VP-S1B-02 and VP-S1B-07 (see P and ID in Figure 1). The analysis of the gas proceeded for 24 h, injecting argon gas in reaction vessel S1B once the pressure decreased, to try of remove all the hydrogen produced by the chemical reaction. Since the gas-flow controller was calibrated to a gas mixture of 50% Ar and 50% H_2, the mass flow rate recorded by MT-H2M-01 was corrected a posteriori by a conversion factor based on FLUIDAT® Bronkhorst database, and on the basis of the amount of volume concentration calculated by the analyzer (the symbol * in Figure 13c highlights the correction). The maximum value of recorded concentration was 93.4%, and the maximum mass flow rate was 1.8 nl/min. Considering that, a total of 510.6 nl of gas passed through the flow meter, containing 22.9 g of hydrogen (Figure 13d). This result is in the foreseen range and was calculated by the stoichiometry. Indeed, for a range of 194–284 g of injected water, the hydrogen produced by the reaction shall be in the range between 10.8 and 31.6 g, according to the predominant reaction between PbLi and water.

3.2. Numerical Results in Standalone and Coupled Configurations

This section summarizes and discusses the major results obtained with the numerical simulations of test E4.1, comparing the results of the two configurations of the codes (standalone and coupled) with the experimental data and trends. Table 7 provides an overview of the main conditions of the experiment and simulations.

Figures 14–18 show the evolution of the pressures and temperatures in the injection line and in the S1B vessel, as predicted by the codes in the standalone and coupled configurations (SIMMER-III and SIMMER-III/RELAP5, respectively). The simulated time spans from the starting of the transient (time 0, opening of valve VP-SBL-06) to 3 s. It is important to notice that the whole experiment lasts much longer (60 s). However, this part of the numerical work was focused on the simulation of the injection transient; therefore, a time duration up to 3 s was considered sufficient for the evaluation of the performances of the codes.

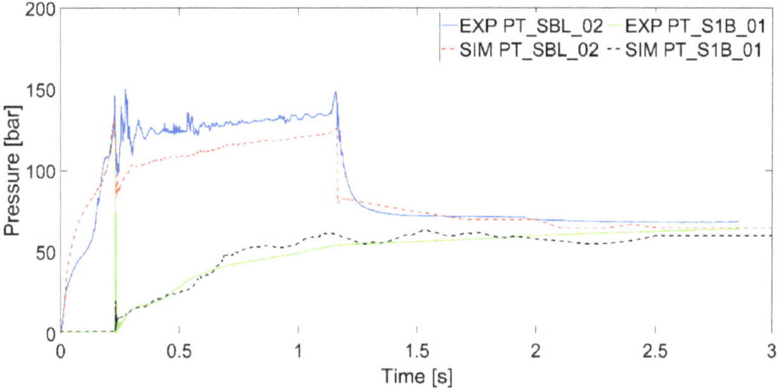

Figure 14. Standalone simulation—pressure evolution in the injection line and in S1B, compared with experimental data (PT).

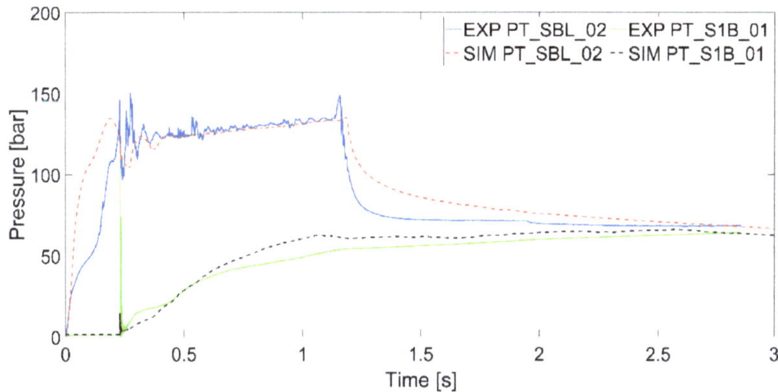

Figure 15. Coupled simulation—pressure evolution in the injection line and in S1B, compared with experimental data (PT).

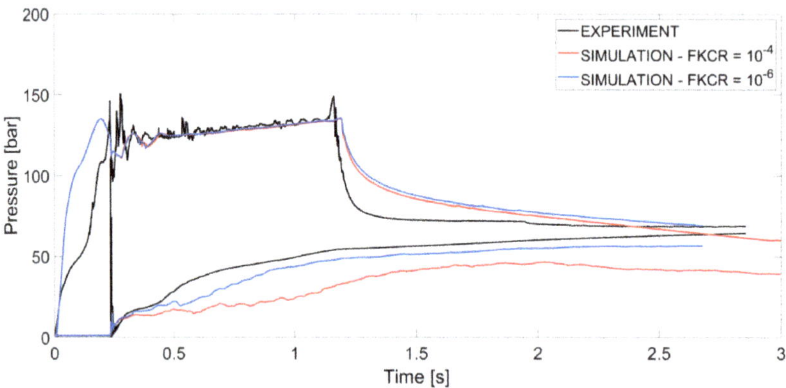

Figure 16. Coupled simulation—pressure evolution in the injection line and in S1B, compared with experimental data at different reaction rates.

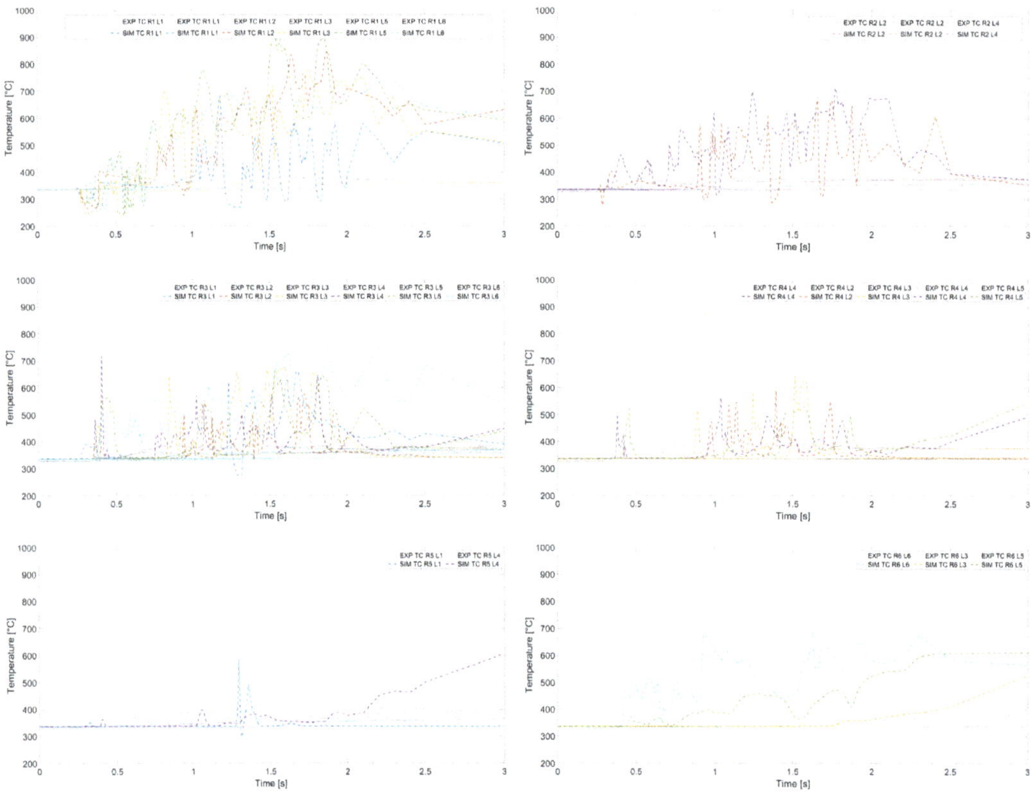

Figure 17. Temperature evolution in S1B for standalone simulations at different radial and axial positions (refer to Table 5).

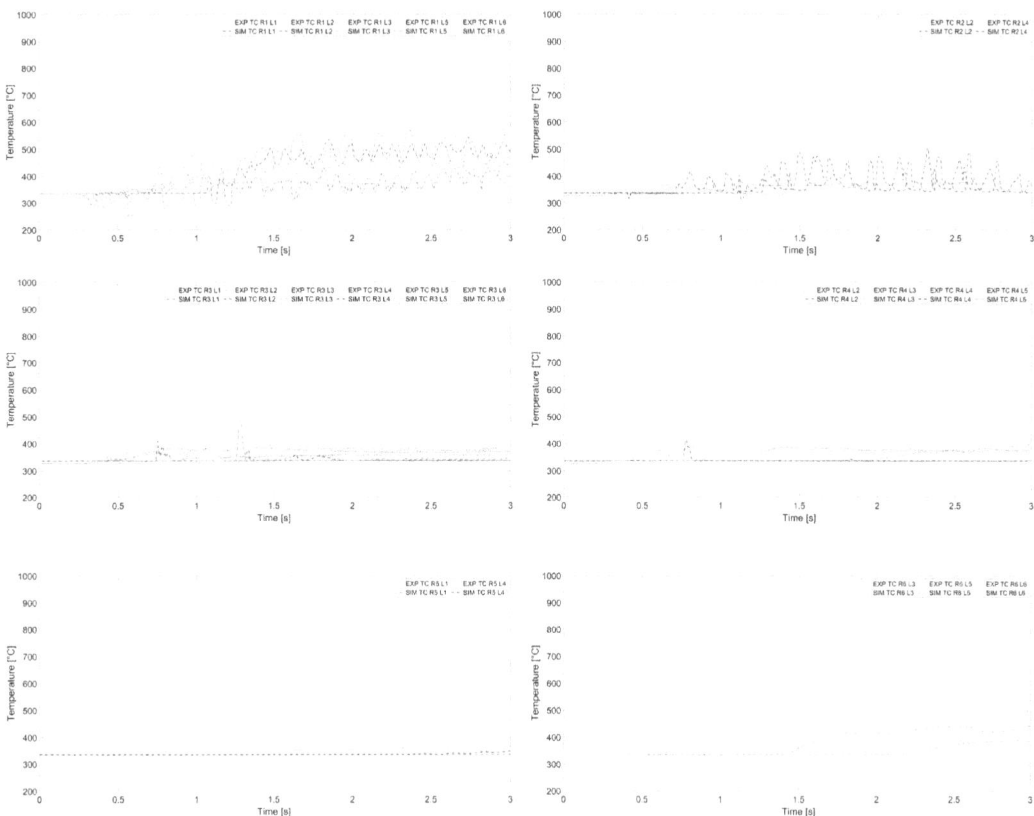

Figure 18. Temperature evolution in S1B for coupled simulations at different radial and axial positions (refer to Table 5).

3.2.1. Pressure Evolution

The evolution of the pressure is presented in Figures 14 and 15, compared with experimental data. Regarding the first transient period (i.e., the pressurization of the line, from 0 to 0.2 s), the pressure increase rate was well-captured only for the very first moments; afterwards, for up to 0.2 s, both the configurations show a significant overprediction of the pressure inside the line.

For the standalone configuration, the possibility of analysis remains very limited, with the only "tunable" parameter being the so-called orifice coefficients, which affect the pressure-drops through the numerical cells; furthermore, the significant discrepancies between the standalone nodalization and the real geometry of the line make it quite difficult to understand the causes of this overprediction. For instance, since SIMMER standalone simulates the whole injection line as a vertical pipe, it is difficult to perform analyses on the effect of water position and the water quality in the line. However, the critical point identified is the absence of any kind of delay regulation of the opening time of the injection valve in SIMMER: the valve VP-SBL-06 is only represented by a virtual wall, which instantaneously disappears at the valve-rupture time, without simulating opening-time intervals; this might lead to a not-sufficiently precise simulation of the pressurization.

On the other hand, in the coupled configuration, exploiting RELAP5 features, several attempts were carried out to improve the quality of the pressurization, through sensitivity analyses (not shown here) of the pressure-drop coefficients of the junctions and valves, the position and quality of the water inside the line, and the opening rate of the VP-SBL-06 valve. However, it was not possible to correct the overprediction of the pressure increase rate and, as a conclusion of the sensitivity analysis, it is presumed that the most probable cause is the opening law of the valve, which is modeled linearly by default in RELAP5. Actually, it is possible to impose a customized opening law on RELAP5, but this study of the real opening law was deemed beyond the scope of this work for the time being. It will certainly be considered for further investigation.

For all these reasons, it was chosen to simulate the rupture of the cap at the time in which the pressure in the SIMMER injection line reached a value close to both the experimental rupture pressure and time.

At a time around 0.2 s, the virtual wall (SIMMER) and the valve (RELAP5) representing the cap in the simulations are set to open and the injection starts; the water continues to flow for about 1 s, and then the injection is stopped by closing valve VP-SBL-06 (t = 1.2 s).

As can be seen from Figures 14 and 15, the pressurization rate of the injection line (PT-SBL-02) is well-captured for both the configurations for the whole interval; however, the two simulations differ in the prediction of the absolute values of the pressure: the standalone one shows a remarkable underprediction of about 20 bars all along the transient, whilst the coupled configuration perfectly matches the value of the experimental pressure. Compared against the experimental data, the coupled RELAP5 was also able to capture the oscillations of the pressure in the line caused by the abrupt rupture of the cap, and it also matched the damping time (~0.4 s). As stated above, the error in the standalone case is due to the limited capability of SIMMER in reproducing the geometry of the injection line and, consequently, the behavior of the injected water. At the end of the injection (~1.2 s) the pressure drops sharply in the standalone configuration, while it decreases slowly—more similar to the experimental data—in the coupled configuration.

The continuous green line shown in Figures 14 and 15 represents the pressurization of the vessel S1B, as captured, during the experiment, by the pressure transducer PT-S1B-01; the dashed black line, instead, is the pressure calculated by SIMMER during the simulation. It is clear, comparing the two lines, that both the configurations correctly predict this slow increase in the pressure, with a slightly better match for the coupled codes towards the end of the transient. Indeed, the larger and most critical difference with the experimental data is seen at the very beginning of the pressurization, immediately after the rupture. In the experiment, an extremely fast and high-pressure peak was detected, but this peak was not predicted in either of the two simulations; instead, a much smaller peak is seen—almost 10-times smaller. Since both the signals (experimental and numerical) are recorded at the same frequency (10^4 Hz), this error cannot be attributed to a difference in the sampling time. This might indicate that the fast pressure peaks in the experiment could derive from the interaction of the solid material (steel components) with the liquid metal, which SIMMER cannot simulate.

Comparing the pressure, one last important observation can be drawn from the numerical work, given the final scope of the whole analysis, since the main objective is the study of the chemical interaction between PbLi and water. The kinetic of this reaction is still scarcely understood, and the experimental and numerical campaign of Lifus5/Mod3 I s also aimed at a better understanding the real impact of this parameter. The chemical module in SIMMER does not provide a direct control for the kinetic of the reaction, but it is possible to use a parameter which limits the amount of hydrogen produced by the reaction in each time-step, thereby effectively adjusting the velocity of the reaction. During this work, it was found that relatively high velocities of the reaction would bring about a slower pressurization and significant underpredictions of the pressure inside the S1Bs; this might sound counterintuitive, since a faster reaction should cause a faster increase of the pressure. However, the effect of the chemical reaction is to cause high pressure peaks only locally, where the reaction takes place, and especially close to the injection point, consequently slowing down the flow of the water and the pressurization of the vessel. Figure 16 provides an example of this behavior: the blue and red lines are the results of two simulations with different reaction velocities (FKCR coefficient, as explained above). The velocity of the reaction does not have any significant impact on the pressure in the injection line, but it strongly affects the pressure inside the S1B vessel, with the lower rate (blue line, FKCR 10^{-6}) leading to a higher pressure during the whole transient. Overall, the simulation seems to suggest that the chemical reaction can be assumed relatively slow and as having a small impact on the first seconds of the transient.

3.2.2. Temperature Evolution

Figures 17 and 18 collect the temperature trends simulated by the standalone and coupled simulations, respectively, plotted against the experimental data. The trends are collected by ring location and level (i.e., radial and axial position); the reader can refer to Table 5 for a better understanding of the positions. It is important to notice that, as already stated above, this part of the numerical work was focused on the initial seconds of the experiment, chiefly on the transients related to the injection of the water and the pressurization of the S1B vessel, whilst it is clear from the conclusion of this work that a much longer time interval is needed to have a comprehensive evaluation of the codes for what concerns the temperatures. However, some important observations can be drawn from these results.

First of all, it is evident from the comparison with the experimental data that both the configurations of the codes predict a significantly more chaotic behavior for the transients in all the analyzed locations. Indeed, in the experiment, the temperatures at all the positions showed a relatively slow and smooth increase, which never presented significant peaks and never reached beyond 100 °C above the start temperatures; the simulation in both the configurations revealed many fast peaks, especially in the radial positions close to the injection cap, with the coupled simulations showing an interesting behavior with a regular oscillation frequency. Furthermore, it is interesting to notice that SIMMER simulations show a significant drop in the temperatures close to the injection point immediately after the cap rupture: this trend is not observed in the experimental data.

These behaviors are still under investigation, but they might be related to a balance between the evaporation (through flashing) of the water injected in the vessel and the energy released by the chemical reaction.

A certain degree of uncertainty remains regarding the choice of the temperature to be compared with experimental data, since SIMMER is able to provide the temperatures of the different liquids and the temperature of the gas, but it is not fully clear what would be the proper comparison with experimental thermocouples measuring in a multiphase environment.

However, as a general conclusion, the coupled configuration seems to provide a substantially better estimation of the evolution of the temperatures, both in terms of the quality of the transients and in terms of the absolute values. Given the difference between the two configurations, this is most likely due to an improved control of the behavior of the water in the injection line.

3.2.3. Mass Flow Rate and Hydrogen Production

Figures 19–21 show, respectively, the liquid water flow rate, the total amount of water injected, and the total mass of hydrogen generated inside the S1B vessel; the last two values are also compared against the estimation obtained through the experimental data.

The mass flow rate of injected water is significantly higher in the standalone simulation, with two abrupt changes shown at the instants in which the cap rupture and the closing of valve VP-SBL-06 occur. This behavior is quite unphysical, and it corroborates once again the shortcomings of SIMMER in properly simulating long pipelines and valve actions. However, the total mass of water injected during the whole transient remains inside the estimated experimental value, as shown in Figure 20.

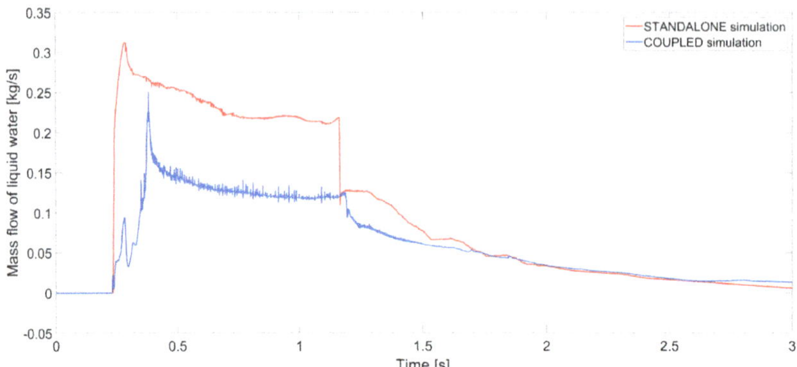

Figure 19. Standalone and Coupled simulations—liquid water mass flow rate in the injection line.

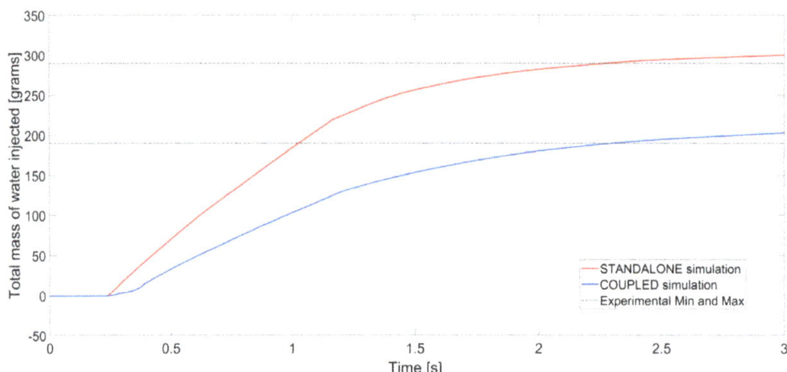

Figure 20. Standalone and Coupled simulations—total mass of liquid water injected in S1B.

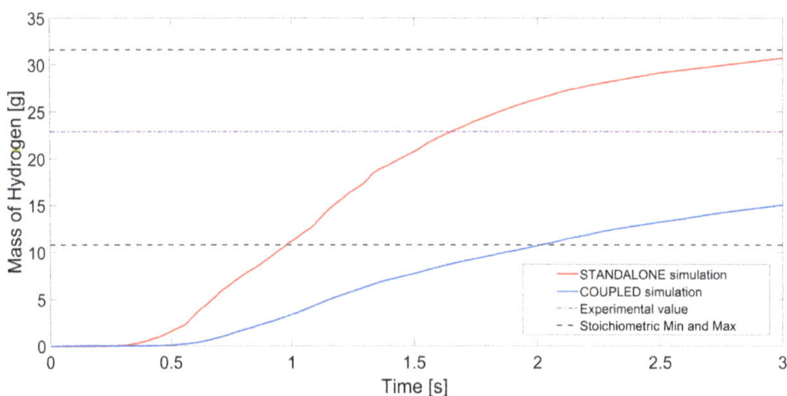

Figure 21. Standalone and Coupled simulations—total mass of generated hydrogen.

Regarding the amount of hydrogen produced (Figure 21), the two simulations showed different results, with higher values predicted by the standalone configuration: this was, nonetheless, expected, since the final hydrogen production depends on the total mass of injected water. Furthermore, both the simulations showed that the hydrogen generation did not reach an entirely steady state value, with the chemical reaction still active after

3 s of simulation; however, the hydrogen production seemed to nearly reach a plateau, with the final level close to the experimental one, especially for the coupled configuration. Even though this predicted time evolution is useful for supporting the necessity of further analysis with longer transients, it is important to stress that these results cannot be used to validate a specific hydrogen production rate and, consequently, provide an estimation of the kinetic of the reaction, since the experimental value was calculated a posteriori, and no experimental time trends are available concerning the hydrogen generation.

Table 7. Test E4.1, main parameters.

#	LIFUS5/Mod3			Test E4.1		
	Parameter	ID	Unit	Experiment	Standalone	Coupled
	SYSTEM S1B					
S1-1	P @ SoT	PC-S1B-01	bar	0.7	1.00	1.28
S1-2	T_{PbLi} @ SoT	average	°C	333.9	337.00	337.00
S1-3	PbLi LVL (from S1B top flange)	TC-S1B-L1/L2	mm	−50	−50	−50
S1-4	Vol. gas	−	l	4.073	4.6	4.1
S1-5	First P peak at the injector cap rupture	PT-S1B-03	bar	217.7	19.84	14.76
S1-6	Second delayed P peak	PT-S1B-01	bar	74.2	16.04	9.40
S1-7	Ratio between $P_{peak}1$ and $P_{injection\_line}$ at the injector cap rupture	−	−	1.49	0.15	0.11
S1-8	Ratio between $P_{peak}2$ and $P_{injection\_line}$ at the injector cap rupture	−	−	0.51	0.12	0.07
S1-9	P at EoT	PT-S1B-01/03	bar	94.0	44.33	62.43
S1-10	Rupture disk open time	PC-BYP-01	s	−	−	0.00
S1-11	Min T_{PbLi}	−	°C	−	508.00	235.00
S1-12	Max T_{PbLi}	TC-R61-L3	°C	439.67	1210.42	1709.87
S1-13	T_{PbLi} at EoT (Level 6)	TC-RXY-L6	°C	345–384	750.61	527.29
	SYSTEM SBL					
S2-1	P in gas line at SoT	PC-SBL-02	bar	158.8	158.70	158.70
S2-2	P at SoT	PC-SBL-01	bar	158.7	158.70	158.70
S2-3	T at SoT	TC-SBL-04	°C	274.7	304.78	302.51
S2-4	Min. P during injection	PT-SBL-01	bar	115.73	117.01	110.00
S2-5	P at EoT	PC-SBL-01	bar	163.5	155.00	155.00
S2-6	T at EoT	TC-SBL-01/05	°C	263.0	319.72	277.02
S2-7	Mass of water injected	MT-SBL-01	g	194–285	300	202.08
	INJECTION SYSTEM					
I-1	Start of injection (valve opening instant)	−	s	0	0	0
I-2	Injection time (from cap rupture)	−	s	0.9274	0.92	0.93
I-3	Injector cap rupture instant	−	s	0.2294	0.23	0.22
I-4	Pressure of cap rupture	PT-SBL-02	bar	146.3	132.03	134.98
I-5	Pressurization rate	PT-SBL-02	bar/s	633.3	571.18	698.51
I-6	Injection valve fully closed instant	−	s	1.1568	1.148	1.150
	SYSTEM H2					
H-1	H2 generated	−	g	22.91	30.73	15.05

4. Discussion and Conclusions

The objective of the Test E4.1 was successfully achieved and the acquired measurements contributed to the enlargement of existing databases and the increasing of the comprehension of the phenomena occurring during PbLi/water interactions, as well as provided experimental data with defined initial and boundary conditions. The transient can be divided into four phases, according to the pressure trend.

The test confirmed the dynamic of the PbLi/water interaction, where the thermodynamic interaction is the predominant process occurring in the first hundreds of milliseconds after the cap rupture, followed by the secondary process, involving the chemical reaction, which generates hydrogen and an increase in temperature. During the injection of water into the reaction vessel S1B, the pressure sensors recorded several initial narrow peaks, lasting for a few milliseconds, due to the instantaneous flashing of the injected water, followed by the expansion of steam and gas through the perforated top plate of the test section towards the gas plenum. The maximum pressure peak, due to the flashing of water, was 217.7 bar. Moreover, the pressure data overlapped with the strain data permitted the recognition of the pressure-wave propagation and the investigation of the dynamic effects of energy release on the structures.

Concerning the temperature behavior, in test E4.1, the cooling effect due to the flashing of water and its expansion disappeared. This is due to the shape of the water jet, which was almost spread when it was injected into the melt. Moreover, the spread of the jet led to a general increase of temperature in the whole system, due to the heat generated by the chemical reaction occurring at the interface between fluids, without significant temperature peaks.

Regarding the quantification of the hydrogen, which is one of the crucial data to be used in the code validation activity, the experimental result (22.91 g) was in the foreseen range and was calculated by the stoichiometry, according to the predominant reaction between PbLi and water, and considering the evaluation of the injecting water (194–284 g).

Finally, the experimental results of test E4.1 permitted the obtaining of reliable data to be used for the validation of the modified-version SIMMER-III code for fusion applications, and of the coupled RELAP5/SIMMER approach.

As for the numerical viewpoint, the experimental test E4.1 was simulated using two different methodologies. In the first case, SIMMER-III was used in a standalone configuration to simulate the S1B vessel and a rough geometrical approximation of the whole injection line of the LIFUS5/Mod3 facility; in the second case, SIMMER-III was kept for the simulation of the S1B, but it was also coupled with the code RELAP5 (version 3.3), which allowed for the creation of a nodalization which was significantly closer to the real geometry of the injection line. The numerical results were then compared with the experimental data in order to provide a better understanding of the performance of the two methodologies, and a first evaluation of the capability of SIMMER-III code in simulating the transients characterizing the Series E.

The numerical results presented here show that the code SIMMER-III is capable of correctly capturing the main phenomena involved in the experiments, providing good estimations, especially for the long pressure transients in the S1B vessel. Furthermore, the coupled technique worked smoothly, and outperformed the standalone simulation, with a significantly superior prediction of the pressurization in the injection line, allowing for more flexibility on the handling of the behavior of the valves and the imposition of the boundary conditions.

Regarding the chemical reaction, and considering the final hydrogen generation, the code results seem satisfactory and coherent with the experimental results and the stoichiometric calculations; nevertheless, more work is needed to improve the quantitative prediction of the temperatures, even though the coupled technique already shows a substantial improvement. Another important conclusion for the chemical reaction is an early indication that the reaction appears to have a much stronger impact on long transients and a lesser effect at the early stages of the experiment; while further investigation is certainly necessary to confirm this trend, this might be an important advancement in a comprehensive definition of the kinetic of the reaction.

The achievements reached during the experimental activities performed under FP8 EUROfusion Horizon 2020 will be the starting point of the new R&D plan in the framework of FP9 EUROfusion Horizon Europe. A new LIFUS5/Mod4 Integral Test Facility (ITF) has been designed and will be installed at ENEA CR Brasimone. The facility is a full-scale representative of the WCLL TBM PbLi loop and it will be coupled with the Water Loop facility, a full-scale representative of the WCLL TBM Water Cooling System. The objective of the experimental campaign is to investigate the phenomenology, the behavior, and the response of the WCLL Test Blanket Systems under in-box LOCA at integral levels and in relevant operative conditions. Moreover, the facility will be able to reproduce and assess the effectiveness of the safety functions and procedures implemented in such scenarios. This involves continuing the validation activity on the code-coupling and the procedure for its application, together with the validation of SIMMER against ITF experiments.

Author Contributions: Conceptualization, M.E. and A.D.N., methodology, M.E. and F.G.; software, M.E. and F.G.; validation, F.G. and M.E.; investigation, M.E., F.G. and N.B.; data curation, M.E. and F.G.; formal analysis, M.E. and F.G.; writing—original draft preparation, M.E., F.G. and N.B.; writing—review and editing, M.E.; supervision, A.D.N., N.F. and A.C. All authors have read and agreed to the published version of the manuscript.

Funding: This work has been carried out within the framework of the EUROfusion Consortium and has received funding from the Euratom research and training programmes 2014–2018 and 2019–2020, under grant agreement No 633053. The views and opinions expressed herein do not necessarily reflect those of the European Commission.

Institutional Review Board Statement: Not applicable.

Informed Consent Statement: Not applicable.

Data Availability Statement: Data sharing is not applicable to this article.

Conflicts of Interest: The authors declare no conflict of interest. The funders had no role in the design of the study; in the collection, analyses, or interpretation of data; in the writing of the manuscript, or in the decision to publish the results.

References

1. Del Nevo, A.; Arena, P.; Caruso, G.; Chiovaro, P.; Di Maio, P.; Eboli, M.; Edemetti, F.; Forgione, N.; Forte, R.; Froio, A.; et al. Recent progress in developing a feasible and integrated conceptual design of the WCLL BB in EUROfusion project. *Fusion Eng. Des.* **2019**, *146*, 1805–1809. [CrossRef]
2. Martelli, E.; Del Nevo, A.; Arena, P.; Bongiovi, G.; Caruso, G.; Di Maio, P.A.; Eboli, M.; Mariano, G.; Marinari, R.; Moro, F.; et al. Advancements in DEMO WCLL breeding blanket design and integration. *Int. J. Energy Res.* **2017**, *42*, 27–52. [CrossRef]
3. Del Nevo, A.; Martelli, E.; Agostini, P.; Arena, P.; Bongiovì, G.; Caruso, G.; Di Gironimo, G.; Di Maio, P.; Eboli, M.; Giammusso, R.; et al. WCLL breeding blanket design and integration for DEMO 2015: Status and perspectives. *Fusion Eng. Des.* **2017**, *124*, 682–686. [CrossRef]
4. Tassone, A.; Del Nevo, A.; Arena, P.; Bongiovi, G.; Caruso, G.; Di Maio, P.A.; Di Gironimo, G.; Eboli, M.; Forgione, N.; Forte, R.; et al. Recent Progress in the WCLL Breeding Blanket Design for the DEMO Fusion Reactor. *IEEE Trans. Plasma Sci.* **2018**, *46*, 1446–1457. [CrossRef]
5. Cismondi, F.; Spagnuolo, G.; Boccaccini, L.; Chiovaro, P.; Ciattaglia, S.; Cristescu, I.; Day, C.; Del Nevo, A.; Di Maio, P.; Federici, G.; et al. Progress of the conceptual design of the European DEMO breeding blanket, tritium extraction and coolant purification systems. *Fusion Eng. Des.* **2020**, *157*, 111640. [CrossRef]
6. Eboli, M.; Forgione, N.; Del Nevo, A. Implementation of the chemical PbLi/water reaction in the SIMMER code. *Fusion Eng. Des.* **2016**, *109-111*, 468–473. [CrossRef]
7. Eboli, M.; Forgione, N.; Del Nevo, A. Assessment of SIMMER-III code in predicting Water Cooled Lithium Lead Breeding Blanket "in-box-Loss of Coolant Accident". *Fusion Eng. Des.* **2020**, *163*, 112127. [CrossRef]
8. D'Auria, F.; Galassi, G.M. *Code Assessment Methodology and Results*; IAEA Technical Workshop/Committee on Computer Aided Safety Analyses: Moscow, Russia, 1990.
9. Bonuccelli, M.; D'Auria, F.; Debrecin, N.; Galassi, G.M. A Methodology for the Qualification of Thermalhydraulic Code Nodalizations. In Proceedings of the NURETH-6 Conference, Grenoble, France, 5–8 October 1993.

10. Eboli, M.; Del Nevo, A.; Pesetti, A.; Forgione, N.; Sardain, P. Simulation study of pressure trends in the case of loss of coolant accident in Water Cooled Lithium Lead blanket module. *Fusion Eng. Des.* **2015**, *98–99*, 1763–1766. [CrossRef]
11. Eboli, M.; Del Nevo, A.; Forgione, N.; Porfiri, M.T. Post-test analyses of LIFUS5 Test#3 experiment. *Fusion Eng. Des.* **2017**, *124*, 856–860. [CrossRef]
12. Eboli, M.; Moghanaki, S.K.; Martelli, D.; Forgione, N.; Porfiri, M.T.; Del Nevo, A. Experimental activities for in-box LOCA of WCLL BB in LIFUS5/Mod3 facility. *Fusion Eng. Des.* **2019**, *146*, 914–919. [CrossRef]
13. Eboli, M.; Crugnola, R.M.; Cammi, A.; Khani, S.; Forgione, N.; Del Nevo, A. Test Series D experimental results for SIMMER code validation of WCLL BB in-box LOCA in LIFUS5/Mod3 facility. *Fusion Eng. Des.* **2020**, *156*, 111582. [CrossRef]
14. Moghanaki, S.K.; Eboli, M.; Forgione, N.; Martelli, D.; Del Nevo, A. Validation of SIMMER-III code for in-box LOCA of WCLL BB: Pre-test numerical analysis of Test D1.1 in LIFUS5/Mod3 facility. *Fusion Eng. Des.* **2019**, *146*, 978–982. [CrossRef]
15. Moghanaki, S.K.; Galleni, F.; Eboli, M.; Del Nevo, A.; Paci, S.; Forgione, N. Analysis of Test D1.1 of the LIFUS5/Mod3 facility for In-box LOCA in WCLL-BB. *Fusion Eng. Des.* **2020**, *160*, 111832. [CrossRef]
16. Moghanaki, S.K.; Galleni, F.; Eboli, M.; Del Nevo, A.; Paci, S.; Forgione, N. Post-test analysis of Series D experiments in LIFUS5/Mod3 facility for SIMMER code validation of WCLL-BB In-box LOCA. *Fusion Eng. Des.* **2021**, *165*, 112268. [CrossRef]
17. Gonfiotti, B.; Moghanaki, S.K.; Eboli, M.; Barone, G.; Del Nevo, A.; Martelli, D. Development of a SIMMER-III/RELAP5 coupling tool. *Fusion Eng. Des.* **2019**, *146 Pt B*, 1993–1997. [CrossRef]
18. Galleni, F.; Moghanaki, S.; Eboli, M.; Del Nevo, A.; Paci, S.; Ciolini, R.; Frano, R.L.; Forgione, N. RELAP5/SIMMER-III code coupling development for PbLi-water interaction. *Fusion Eng. Des.* **2020**, *153*, 111504. [CrossRef]
19. Galleni, F.; Moscardini, M.; Eboli, M.; Del Nevo, A.; Martelli, D.; Forgione, N. Preliminary analysis of an in-box LOCA in the breeding unit of the WCLL TBM for the ITER reactor with SIMMER-IV code. *Fusion Eng. Des.* **2021**, *169*, 112472. [CrossRef]
20. Moscardini, M.; Galleni, F.; Pucciarelli, A.; Eboli, M.; Del Nevo, A.; Paci, S.; Forgione, N. Thermo-hydraulic analysis of PbLi ancillary system of WCLL TBM undergoing in-box LOCA. *Fusion Eng. Des.* **2021**, *168*, 112614. [CrossRef]
21. Pucciarelli, A.; Toti, A.; Castelliti, D.; Belloni, F.; Van Tichelen, K.; Moscardini, M.; Galleni, F.; Forgione, N. Coupled system thermal Hydraulics/CFD models: General guidelines and applications to heavy liquid metals. *Ann. Nucl. Energy* **2020**, *153*, 107990. [CrossRef]
22. Eboli, M.; Del Nevo, A.; Forgione, N.; Giannetti, F.; Mazzi, D.; Ramacciotti, M. Experimental Characterization of Leak Detection Systems in HLM Pool Using LIFUS5/Mod3 Facility. *Nucl. Technol.* **2020**, *206*, 1409–1420. [CrossRef]
23. Pesetti, A.; Del Nevo, A.; Forgione, N. Experimental investigation and SIMMER-III code modelling of LBE–water interaction in LIFUS5/Mod2 facility. *Nucl. Eng. Des.* **2015**, *290*, 119–126. [CrossRef]
24. PED Directive. Available online: https://ec.europa.eu/growth/sectors/pressure-equipment-and-gas-appliances/pressure-equipment-sector/pressure-equipment_en (accessed on 27 October 2021).
25. AA.VV. *SIMMER-III (Version3.F) Input Manual*; O-arai Engineering Center, Japan Nuclear Cycle Development Institute: Ibaraki, Japan, May 2012.
26. AA.VV. *SIMMER-IV (Version 3.F) Input Manual*; O-arai Engineering Center, Japan Nuclear Cycle Development Institute: Ibaraki, Japan, May 2012.
27. Green Rot Effects. Available online: https://blog.wika.us/products/temperature-products/green-rot-affects-type-k-thermocouples/ (accessed on 27 October 2021).

Article

Numerical Simulations with RELAP5-3D and RELAP5/mod3.3 of the Second Experimental Campaign on In-Box LOCA Transients for HCLL TBS

Alessandro Venturini [1,*], Marco Utili [1] and Nicola Forgione [2]

[1]   ENEA Brasimone, 40032 Camugnano, BO, Italy; marco.utili@enea.it
[2]   Dipartimento di Ingegneria Civile e Industriale, University of Pisa, Largo Lucio Lazzarino 2, 56122 Pisa, PI, Italy; nicola.forgione@unipi.it
*   Correspondence: alessandro.venturini@enea.it

Abstract: In-box LOCA was identified as one of the worst accidental scenarios for the HCLL TBS (Helium Cooled Lithium-Lead Test Blanket System). Aiming to experimentally analyze the consequences of this transient, ENEA designed and built THALLIUM (Test HAmmer in Lead LIthiUM), a facility that reproduces the *LiPb* loop of the HCLL TBS. Two experimental campaigns were carried out by simulating the rupture of a stiffening plate and the related helium injection in the *LiPb* loop. The obtained experimental data were used to check the capabilities of RELAP5 system code to reproduce the pressure wave propagation that follows this accident. The first simulations were made with RELAP5-3D using *LBE* (Lead–Bismuth Eutectic) as a system fluid, as the thermophysical properties of *LiPb* are tabulated only up to a maximum value of 40 bar in this version of the code. Then, *LiPb* properties were implemented in RELAP5/mod3.3, after selecting the proper correlations from a literature review. This work summarizes the numerical simulations of the second experimental campaign, which was simulated with both versions of the code. The simulations highlight that the code is able to accurately reproduce the experimental results and that RELAP5-3D is slightly more precise than RELAP5/mod3.3 in predicting the pressure trends.

Keywords: lead-lithium eutectic; In-box LOCA; RELAP5; HCLL TBS

Citation: Venturini, A.; Utili, M.; Forgione, N. Numerical Simulations with RELAP5-3D and RELAP5/mod3.3 of the Second Experimental Campaign on In-Box LOCA Transients for HCLL TBS. *Energies* **2021**, *14*, 4544. https://doi.org/10.3390/en14154544

Academic Editor: Dan G. Cacuci

Received: 27 April 2021
Accepted: 25 July 2021
Published: 27 July 2021

1. Introduction

This work reports the numerical simulations with RELAP5 system code (Idaho National Laboratoriy, Idaho, USA), in the 3D [1] and mod3.3 [2] versions, of the second experimental campaign performed in THALLIUM (Test HAmmer in Lead LIthiUM). THALLIUM is a facility operated at ENEA R.C. Brasimone that accurately reproduces the *LiPb* (Lead-Lithium Eutectic) loop of the HCLL TBS (Helium Cooled Lithium-Lead Test Blanket System) [3,4]. The *LiPb* loop allows the circulation of the liquid breeder through the Test Blanket Module (TBM) by means of a mechanical pump. It also hosts the system that extracts the generated tritium from the *LiPb* and the system to purify the alloy from corrosion products and other impurities (cold trap). THALLIUM was designed and built in 2015–2016 within the framework of the agreement F4E-FPA-372, which dealt with experimental tests in support of the preliminary design of the European TBS.

The main purpose of THALLIUM is to study the pressure wave propagation that follows the injection of high pressure helium in the *LiPb* loop during the transient named In-box LOCA (Loss Of Coolant Accident). This accidental transient is considered one of the worst conceivable accidents for the HCLL TBS by the ITER (International Thermonuclear Experimental Reactor) classification system [3]. The preliminary safety report of the HCLL TBS [5] lists the In-box LOCA among the six accidents which were considered as the reference ones "because they have the highest expected consequences—i.e., those that establish the system safety design requirements, or because they present some peculiarities from the design point of view". In particular, the In-box LOCA is considered in this group of six reference accidents because it can lead to consequential leaks into the Vacuum Vessel.

THALLIUM, described in detail in [6], was used to perform two experimental campaigns, whose results are reported in [7] and in [8]. The rupture of a stiffening plate of the HCLL TBS and the resulting injection of helium at high pressure were simulated. Two injection valves with different flow area and opening times were used in the experimental campaigns. Helium for the injection was supplied by the facility HeFUS3 (Helium for FUSion 3, [9,10]), at 400 °C and up to 80 bar, while *LiPb* was loaded in THALLIUM by means of the IELLLO (Integrated European Lead Lithium LOop, [11,12]) facility.

The two experimental campaigns in THALLIUM produced the first results on In-box LOCA for HCLL TBS, while few numerical analyses were found from a literature review on HCLL TBS and on other helium-cooled liquid Breeding Blanket concepts ([4,13]). Instead, a lot of efforts have been recently devoted to experimentally and numerically investigating the In-box LOCA transient for the WCLL (Water Cooled Lithium-Lead) Breeding Blanket e.g., [14–16]. The simulations with RELAP5-3D of the THALLIUM first experimental campaign are described in [17].

RELAP5 is a thermo-hydraulic system code based on a non-homogenous non-equilibrium six-equation model for two-phase systems, solved by a semi-implicit numerical scheme for transient analysis. The idea behind the development of RELAP5, carried out by the Idaho National Laboratory, is to have a code that is able to predict at a system level how operational and accidental transients evolve. Parametric and sensitivity analyses are also possible with this code. The code has been extensively validated for light water nuclear reactors, and it can also work with several system fluids (e.g., Heavy Liquid Metals, helium, heavy water, etc.). The code includes specific models, often coming from experimental observations that allow for simulating complex phenomena and also models of components that are common in power plants (e.g., pumps, valves, etc.).

The work presented in this paper aims to assess the capability of RELAP5 to be used for LiPb/helium systems, as almost no previous activities with this code have been found from a literature review. In particular, the possibility to accurately simulate the consequences of an In-box LOCA transient is demonstrated by showing the good agreement between numerical and experimental pressure trends in different parts of the THALLIUM facility on the six injections whose results are reported in [8]. This outcome might be useful for the design or the safety of Breeding Blankets that use a combination of helium and *LiPb* (besides the HCLL, there are the American Dual-Coolant Lithium Lead and the Chinese Dual-functional Lithium Lead). Generally, the availability of a system code that is accurate in reproducing the behavior of *LiPb* systems might prove to be useful for many Breeding Blankets and/or Test Blanket Systems.

2. Brief Description of the Facility and of the Tests

This section gives a general overview of the facility layout, aiming to give enough details to make the following analyses understandable without the need to read the previous activities on THALLIUM. However, a full description of the facility and its systems is shown in the paper [6] or in the ENEA internal report [18].

Figure 1 shows the main components, the valves, and the instruments of THALLIUM (the pressure transducers are indicated in red). THALLIUM aims to mimic the *LiPb* loop of the HCLL TBS, in the 2014 configuration [3].

The main components of THALLIUM are:

- the TBM (Test Blanket Module) mock-up: a steel box which reproduces 1:1 a breeding unit of the HCLL TBM. It is internally divided by a horizontal plate with a hole that connects the lower chamber to the upper one. The plate simulates the stiffening plate that breaks giving rise to the accident.
- an expansion tank: a reduced mock-up of the storage/recirculation tank of the HCLL TBS, without the pumping system. A relief valve (VS205) is installed on a branch above the expansion tank. The valve is controlled by a pre-loaded spring.

- upper and lower leg: these pipes reproduce the pipe forest of the HCLL TBS, with the same dimensions and the same geometry, with the aim to simulate as precisely as possible the pressure wave propagation. The two legs are almost symmetrical, apart from the final part, in which the upper leg bifurcates.
- the bypass line: one of the ends of the upper leg. The bypass line is composed of two segments, divided by a rupture disc. The segment upstream of the rupture disc is initially filled with LiPb, while helium initially fills the segment connected to the gas dome of the expansion tank. When the pressure wave breaks the disc, part of the energy of the pressure wave is dissipated. *LiPb* starts to flow from the bypass to the gas dome of the expansion tank.
- two isolation valves (IV): valves with closing time lower than 0.2 s are installed on both legs.
- the injection line: this line supplies helium to the facility. A Venturi flow meter allows for monitoring the injected flow rate, while the injection pressure is measured by a pressure transducer. A fast opening valve has the task to start the injection. This valve has been kept as close as possible to the TBM mock-up (about 26 cm), so that the possibility of *LiPb* plugs in the line is minimum, while the injection also does not lose too much energy before reaching the mock-up.

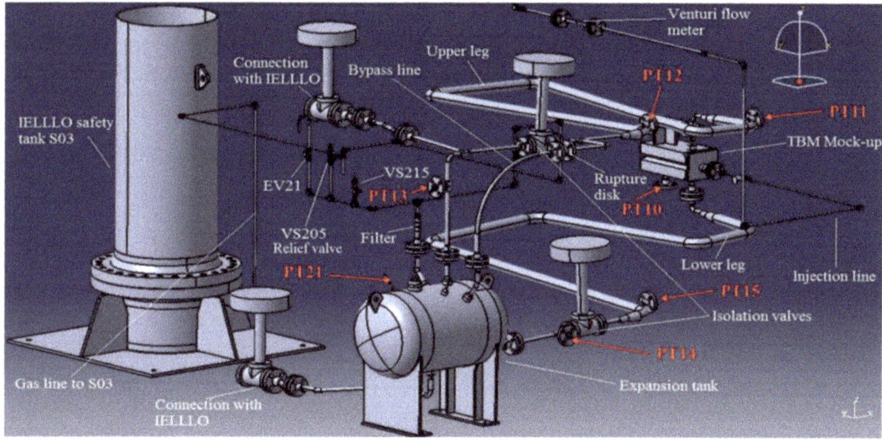

Figure 1. Sketch of the facility.

With respect to the first experimental campaign, THALLIUM was equipped with a bigger injection valve with the idea of analyzing the consequences of a transient with larger injected flow rates (in the range 60–300 g/s). The orifice of this valve has a flow area of $9.503 \cdot 10^{-5}$ m^2.

Six tests were performed in the second experimental campaign (Table 1). Each test begins with the opening of the injection valve: high pressure helium is put into contact with low pressure LiPb. In the first part of the transient, *LiPb* is still practically stagnant, while a single-phase pressure wave propagates at about 1.1 km/s [6] in the facility. Then, helium forces *LiPb* to move towards the expansion tank by passing through the lower and upper legs, as well as through the bypass line after having broken the rupture disc. The relief valve opens at about 9 bar, discharging helium into a dedicated tank. The transient ends when the injection valve closes. The isolation valves were kept open during the six tests of the second experimental campaign.

The results of the first five experiments showed that the transient ends in less than 15 s. For this reason, this was chosen to be the duration of the experiments of the second campaign. However, tests #9 and #10 have different durations. An issue with the injection valve caused test #9 to be halted after 9 s. Instead, the idea behind the different duration of test #10 was to investigate the different transient behavior with shorter timescales.

Table 1. Main parameters used in the second experimental campaign.

Test Number	Injection Time [s]	He Pressure [bar]
6	15	50
7	15	60
8	15	70
9	9	50
10	1.5	60
11	15	60

As the five tests of the first experimental campaign were performed with a helium pressure of 80 bar, the injection pressure was varied in this second campaign with the aim to have some data on the impact of helium pressure on the transient evolution. This was further motivated by the fact that the helium pressure could be lower than 80 bar in some operational transients foreseen for the HCLL TBS.

A total of seven absolute pressure transducers measure the propagation of the pressure wave throughout the facility. The total measuring error of these instruments has been evaluated to be about 1.1 bar, with the exception of the transducer in the gas dome of the expansion tank. This transducer is different from the other six as it works in gas. Its measuring error has been evaluated to be about 0.65 bar. Including the error bars in the following figures would completely jeopardize their readability and thus error bars were not shown in the plots. The six transducers and their use in *LiPb* are better presented in [11]. Details on the experimental results, including an evaluation of the experimental uncertainties, are described in [8].

3. RELAP5 Nodalization

This section briefly describes the nodalization used in the simulations, as its full description has been already reported in [17]. The nodalization (Figure 2) is composed of 269 nodes, each measuring between 0.1 and 0.2 m. This length is a compromise between a detailed representation of the system and a reasonable calculation time. Heat structures were not included in the inputs, as the transient is fast enough to make neglecting heat transfers acceptable.

Time steps of 10^{-5} s were necessary in the first part of the transient (first 5 s of the injection); however, even smaller time steps (down to 10^{-6} s) were needed for some critical parts of the transient. The plot frequencies were set to a value that, multiplied by the time step of the corresponding time period, would have given a time resolution of 1 ms, the same given by the pressure transducers used in the experimental campaign.

The key difference between the input files used for RELAP5-3D and for RELAP5/mod3.3 is the working fluid. Indeed, the fact that the thermodynamic tables are limited to 40 bar for *LiPb* in RELAP5-3D forced to adopt *LBE* (Lead–Bismuth Eutectic) as working fluid. Instead, *LiPb* could be used as system fluid in RELAP5/mod3.3 as its properties were implemented in the code [19] to make possible a comparison of the simulations with the two versions. A brief comparison of the properties of *LBE* and *LiPb* that have the biggest impact on the In-box LOCA transient (density and speed of sound) is reported in the following paragraph.

The non-condensable gas model was used to simulate the injected He. The pressure at the beginning of the transient was set to 3 bar for the liquid metal and to a pressure corresponding to the experimental one for helium. The temperature of the entire system was 400 °C at the beginning of the transient.

The valves and the rupture disc were simulated through motor valves. The opening time of the injection valve was checked with the experimental one test by test (thanks to micro-switches installed on the valve). The relief valve is opened when the pressure in the pipe 154 exceeds 9 bar. A difference of 9 bar in the pressures of the two volumes across the rupture disc was the input for its opening.

The handbook by Idelchik [20] was used to derive the coefficients for the pressure losses of common geometries. The values of the coefficients can be found in [17].

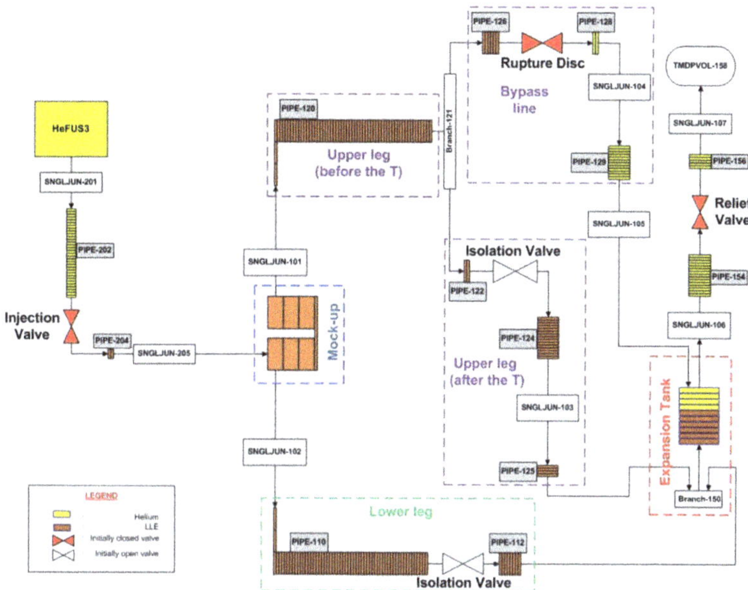

Figure 2. RELAP5 nodalization adopted for the simulations.

4. Modification of RELAP5/Mod3.3 and Discussion over the Expected Differences with RELAP5-3D

The property library for *LiPb* in RELAP5-3D has a limited range of pressures, with an upper threshold of 40 bar. Moreover, discrepancies were found between the properties implemented in the code and correlations available in literature, including some properties of particular importance for the In-box LOCA transient (density and speed of sound). For this reason, a literature review of the *LiPb* thermophysical properties was carried out in order to select a correlation for each property among the many proposed by different authors. RELAP5/mod3.3 was then modified at the University of Pisa by implementing the chosen *LiPb* properties, as described in [19]. The literature review is described in detail in [21].

As density and speed of sound are the two properties that primarily affect the results of these calculations, it is interesting to assess how much they differ for *LiPb* and *LBE* in the thermodynamic conditions typical of the injections (400 °C, ~room pressure). Among the many correlations for density as a function of temperature available in literature, the ones chosen for this comparison are:

$$\rho_{LiPb}\left(\mathrm{kg/m^3}\right) = 10,520.35 - 1.19051 \cdot T \; (508 \leq T[\mathrm{K}] \leq 880) \tag{1}$$

$$\rho_{LBE}\left(\mathrm{kg/m^3}\right) = 10,981.7 - 1.1369 \cdot T \; (410 \leq T[\mathrm{K}] \leq 726). \tag{2}$$

These correlations, plotted in Figure 3, were found in the work by Stankus et al. [22] for *LiPb* and in the OECD-NEA Handbook [23] for *LBE*. From these, it is possible to evaluate the relative difference:

$$diff_{density} = \left(\frac{\rho_{LiPb} - \rho_{LBE}}{\rho_{LiPb}} \right)_{T=400\ °C} \cong 5\ \% \tag{3}$$

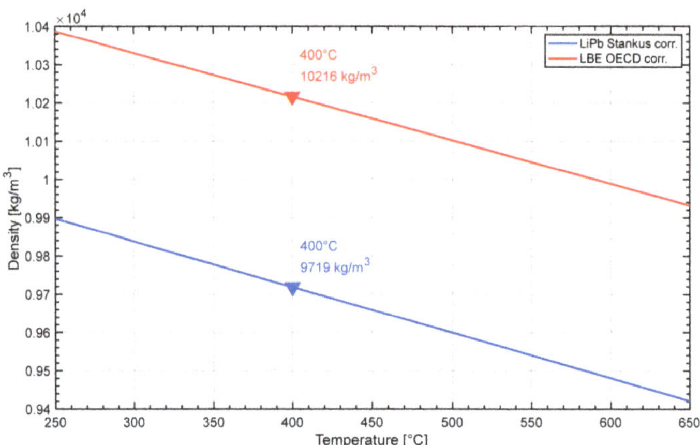

Figure 3. Density of *LiPb* and *LBE* as a function of temperature.

As far as the speed of sound is concerned, only one reference was found for *LiPb* (in the work by Ueki et al. [24]), while the correlation suggested in the work by Sobolev [25] was chosen for LBE:

$$c_{LLE}(\text{m/s}) = 1876 - 0.306{\cdot}T\ (513 \leq T[\text{K}] \leq 783) \tag{4}$$

$$c_{LBE}(\text{m/s}) = 1855 - 0.212{\cdot}T\ (433 \leq T[\text{K}] \leq 1073) \tag{5}$$

The two correlations are plotted in Figure 4 in the range 250–650 °C. The relative difference for speed of sound is:

$$diff_{sound} = \left(\frac{c_{LLE} - c_{LBE}}{c_{LLE}} \right)_{T=400\ °C} \cong 2.5\ \% \tag{6}$$

As a consequence of the results of Equations (3) and (6), a large discrepancy in the results of the simulations is not expected because of the use of a different system fluid (*LiPb* or *LBE*).

However, to further assess the differences in the results caused by the use of different lead alloys, two simulations were performed with initial He pressure of 35 bar, in order to be able to use both *LiPb* and *LBE* with the same version of the code. Figure 5 presents some of the results of these simulations, showing the pressure trend in the TBM mock-up and in the expansion tank. The difference due to the use of *LBE* or *LiPb* is practically negligible and in accordance with the results of Equations (3) and (6).

Moreover, as shown in the next paragraphs, the average difference of the numerical simulations performed using *LBE* in RELAP5-3D and using *LiPb* in RELAP5/mod3.3 is lower than 1 bar, much smaller than the difference between the numerical and experimental results. Indeed, as shown in the next section, the pressures predicted by RELAP5-3D had an average discrepancy of about 3 bar with respect to the experimental values, while RELAP5/mod3.3 was slightly less accurate, with an average discrepancy of about 4 bar.

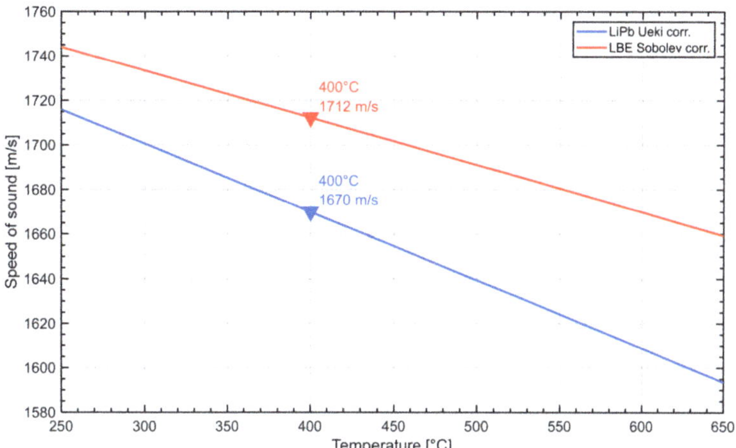

Figure 4. Speed of sound of *LiPb* and *LBE* as a function of temperature.

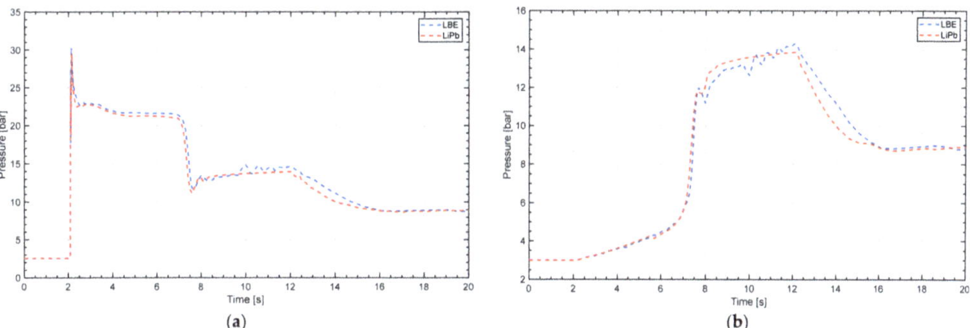

(a)   (b)

Figure 5. Pressure trends in the TBM mock-up (**a**) and in the expansion tank (**b**) with *LBE* and *LiPb* as system fluids.

Hence, the use of *LBE* to simulate this transient proved to introduce a discrepancy which is, at worst, similar to the experimental error of measurement (i.e., about 1 bar) and, thus, it can be considered acceptable.

5. Numerical Simulation of the Second Experimental Campaign of THALLIUM

The same nodalization, time steps, and boundary conditions were used to perform the simulations with RELAP5-3D and RELAP5/mod3.3. The only difference was the system fluid used in the two versions of the code.

The plots shown from Figures 6–30 highlight that RELAP5 is able to simulate the In-box LOCA transient, presenting a general good agreement with the experimental data. In particular, the average discrepancy between RELAP5-3D and the experimental data turned out to be about 3.2 bar, considering the six injections and all the transducers, except PT10 in the injection #10 because of its malfunctioning in that particular test. Instead, the same evaluation for RELAP5/mod3.3 resulted in an average discrepancy of about 4.0 bar with the experimental data. The percentage difference between these two values is about 21.0%, taking RELAP5/mod3.3 as a reference. Considering the maximum pressure measured in these injections, which is about 65 bar by pressure transducer PT12 in test#8, the percentage discrepancy is 4.9% for RELAP5-3D and 6.2% for RELAP5/mod3.3. The data reported from Tables 2–7 show the rounded-up average discrepancies between RELAP5-3D and the experimental data and between the two versions of the code for each point of measurement.

However, RELAP5 fails to simulate the pressure peaks which appear during the final pressure increase. These peaks, positive in the trends measured by the transducers upstream of the isolation valves and negative downstream of the isolation valves, are less evident in the trends of the second experimental campaign, but still present in several injections.

The two discrepancies highlighted in simulating the first experimental campaign between the qualitative trends of RELAP5-3D and the experimental data are still relevant in these simulations with both versions of the code:

(1)　the numerical simulations show a delay in the phenomena that follow the opening of the relief valve. The effect of the relief valve is observed at about 5 s in the experimental trends and consists of a depressurization for the transducers located before the isolation valves or in a pressurization downstream of the isolation valves. These effects are also present in the simulations, but they happen with almost 1 s delay. This delay can be motivated by two events: the pressure which causes the valve to open is reached later in the simulations than in the experiments or the numerical opening time of the relief valve is too long. Unfortunately, both parameters are uncertain as there are no pressure meters close to the relief valve, and the opening time is also not monitored with micro-switches. However, a sensitivity has been performed on the opening time, showing that this parameter has little impact on the delay. For this reason, it is deemed that the delay in the simulation of the relief valve effect is caused by a slow pressure increase in the piping that connects the expansion tank to the valve itself.

(2)　the second experimental peak is not present in the numerical results. This peak has an unclear origin, but it was linked with the propagation of a pressure wave produced by the elasticity of the piping or of the expansion tank. Of course, RELAP5 cannot simulate such a phenomenon.

Moreover, a third difference between experimental and numerical results is new to the simulations of the second experimental campaign: RELAP5-3D, and in part also RELAP5/mod3.3, tends to underestimate the minimum following the opening of the relief valve (i.e., the minimum before the final pressure increase).

The particular behaviors in the simulation of each injection are showed and described in the following sections. Moreover, particular discrepancies in some pressure peaks, usually the first peak in the locations of PT11 and PT14, are highlighted.

5.1. Test VI

Figures 6, 9 and 10 show the comparison between the experimental pressure trends and the numerical trends evaluated by RELAP5-3D and RELAP5/mod3.3 for test #6. The simulations qualitatively agree with the experimental trends, even though the codes tend to generally overestimate the pressure in the whole transient, with an average discrepancy of less than 2.5 bar. Table 2 shows the rounded-up average discrepancies between RELAP5-3D and the experimental data and between the two versions of the code for each point of measurement. The value of the first peak in the TBM mock-up, the most important value for the design team, is about the same in the comparison between the experimental value and the value calculated by RELAP5-3D (about 29 bar). Instead, the difference is about 3.5 bar in the comparison between the experimental value (29 bar) and the value calculated by RELAP5/mod3.3 (25.5 bar). This difference corresponds to about the 12% of the experimental value. It is impossible to distinguish the first peak from Figure 6; thus, Figure 7 shows a zoom of the very first phase of the injection.

Table 2. Average discrepancies between RELAP5-3D, RELAP5/mod3.3, and the experimental data (test #6).

Transducer	RELAP5-3D vs. Exp. [bar]	RELAP5-3D vs. RELAP5/mod3.3 [bar]
PT10	2.0	1.3
PT21	2.6	1.3
PT11	1.6	1.1
PT12	1.8	1.2
PT15	1.6	0.8
PT13	1.5	1.2
PT14	1.8	1.0

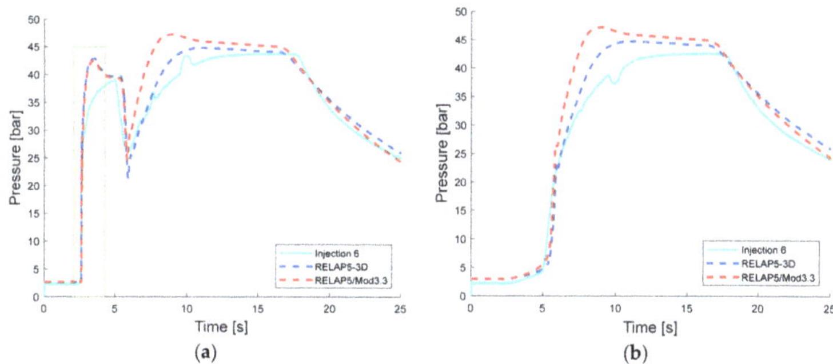

Figure 6. Experimental pressure trends in the TBM mock-up (**a**) and in the expansion tank (**b**), compared with results from RELAP5-3D and RELAP5/mod3.3 (test #6). The green ellipse and rectangle highlight the zoomed portions of the plot that are shown in Figure 7a,b, respectively.

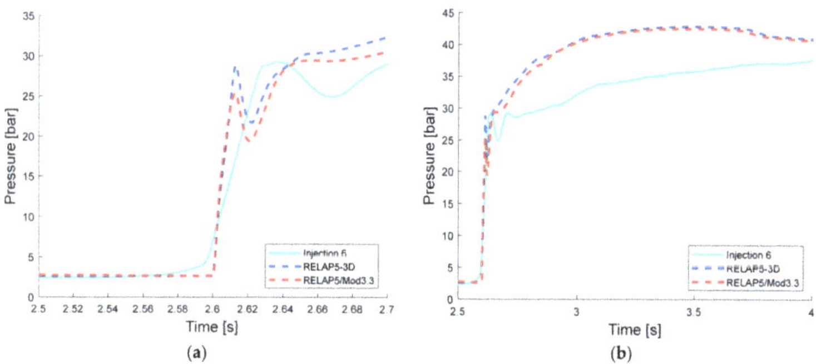

Figure 7. Zooms of the first peak (**a**) and of the first seconds following the peak (**b**) in the TBM mock-up (test #6).

Figure 8 shows the time evolution of the difference between experimental and numerical values for four transducers. The behavior in the locations of the other three transducers is very similar and they are not plotted here for the sake of brevity. In the simulations of the first experimental campaign, the two general differences mentioned in Section 5 (the one-second delay on the effect of the relief valve and the absence of the second experimental peak) were the biggest contributors to the value of the average discrepancy [17]. Here, these two contributions are flanked by a third one related to the pressurization in the very first part of the transient: indeed, the codes move slightly up the first pressurization, causing the behavior shown in Figure 8.

The largest average discrepancy shown by RELAP5/mod3.3 with respect to RELAP5-3D is motivated by the large hump exhibited between 5 and 9 s.

Similar considerations also apply to tests from #7 to #11. The discrepancy vs. time plots for those tests were omitted to not burden the readability of the paper.

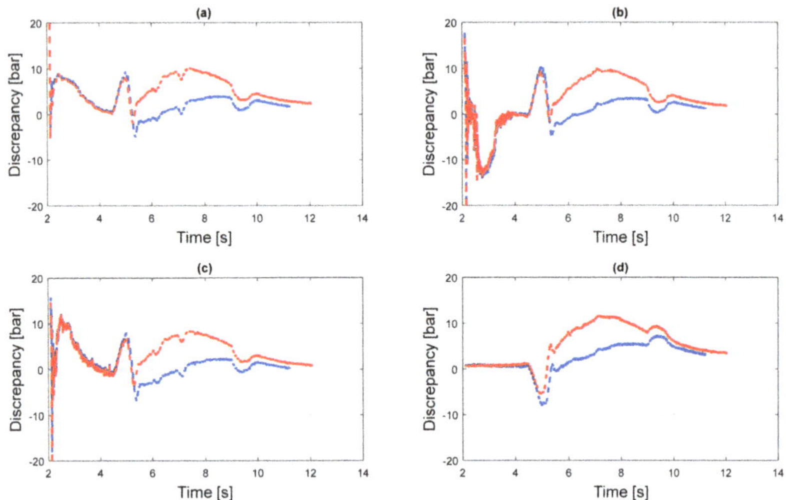

Figure 8. Time evolution of the discrepancy between the experimental pressure trends in four locations of the facility and the corresponding numerical trends with RELAP5-3D (blue curve) and with RELAP5/mod3.3 (red curve). (**a**) TBM mock-up—PT10; (**b**) upper leg—PT12; (**c**) lower leg—PT15; (**d**) expansion tank—PT21.

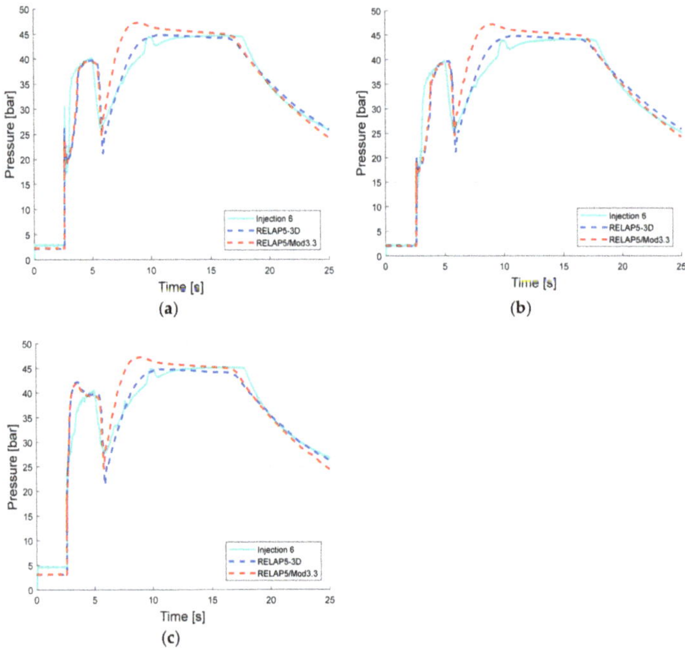

Figure 9. Experimental pressure trends in the upper and lower legs upstream of the isolation valves measured by PT11 (**a**), PT12 (**b**), and PT15 (**c**), compared with results from RELAP5-3D and RELAP5/mod3.3 (test #6).

The first peaks in two locations highlight high discrepancies. The first peak in the location of PT11 is about 30.5 bar in the experimental measurement, about 26 bar simulations by RELAP5-3D (−15% of the experimental value) and about 24 bar in the simulation by RELAP5/mod3.3 (−21%). Instead, in the location of PT14, the experimental value is about 13 bar, while the value calculated by RELAP5-3D is about 10 bar (−23% of the experimental value) and the one calculated by RELAP5/mod3.3 is about 9.5 bar (−27%).

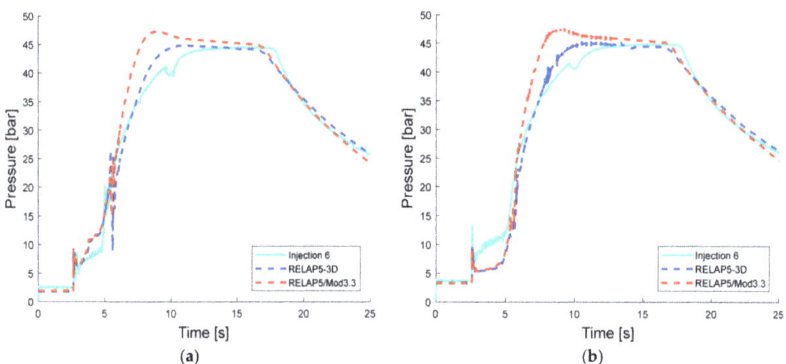

Figure 10. Experimental pressure trends in the upper and lower legs downstream of the isolation valves measured by PT13 (**a**) and PT14 (**b**), compared with results from RELAP5-3D and RELAP5/mod3.3 (test #6).

5.2. Test VII

Figures 11, 13 and 14 show the comparison between the experimental pressure trends and the numerical trends evaluated by RELAP5-3D and RELAP5/mod3.3 for test #7. Even in this test, the codes tend to generally overestimate the pressure trends, with an average discrepancy of slightly more than 4.5 bar, the largest among the tests of the second experimental campaign.

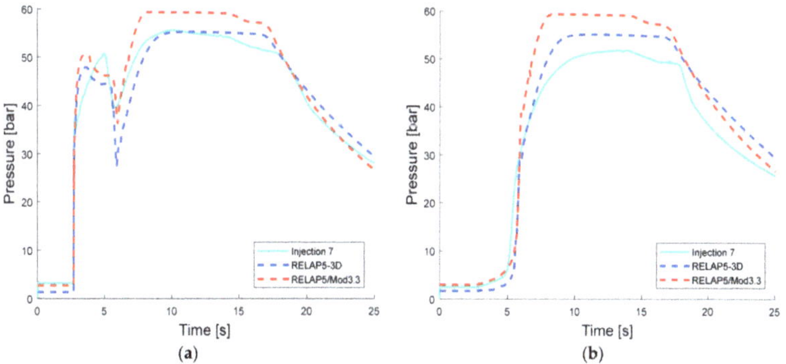

Figure 11. Experimental pressure trends in the TBM mock-up (**a**) and in the expansion tank (**b**), compared with results from RELAP5-3D and RELAP5/mod3.3 (test #5).

The experimental value of the first peak in the TBM mock-up is about 37 bar. The values calculated by the two versions of RELAP5 are about 31 bar for RELAP5-3D (−16% of the experimental value) and about 29 bar for RELAP5/mod3.3 (−21.5%). As it is impossible to distinguish, the first peak from Figures 11 and 12 shows a zoom of the very first phase of the injection.

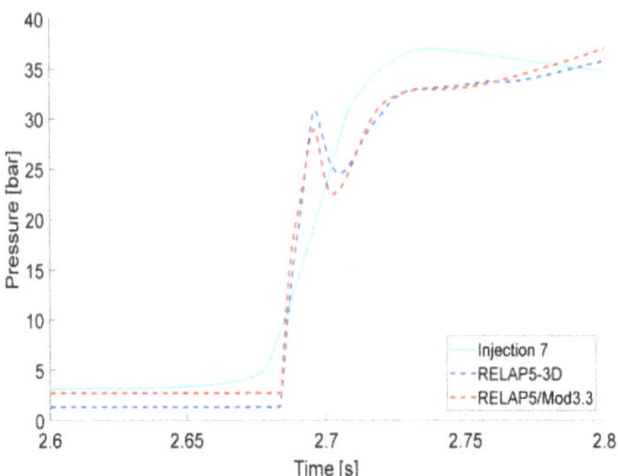

Figure 12. Zoom of the first peak in the TBM mock-up (test #7).

Table 3 shows the rounded-up average discrepancies between RELAP5-3D and the experimental data and between the two versions of the code for each point of measurement.

The discrepancy between the experimental and numerical values for the first peaks in PT11 and PT14 proved to be relatively high also in test #7. The first peak measured by PT11 is about 37 bar, RELAP5-3D predicted about 27 bar (-27% of the experimental value), and RELAP5/mod3.3 about 26 bar (-30%). Instead, in the location of PT14, the experimental value is about 16 bar, while the value calculated by RELAP5-3D is about 11 bar (-31% of the experimental value), and the one calculated by RELAP5/mod3.3 is about 10 bar (-37.5%).

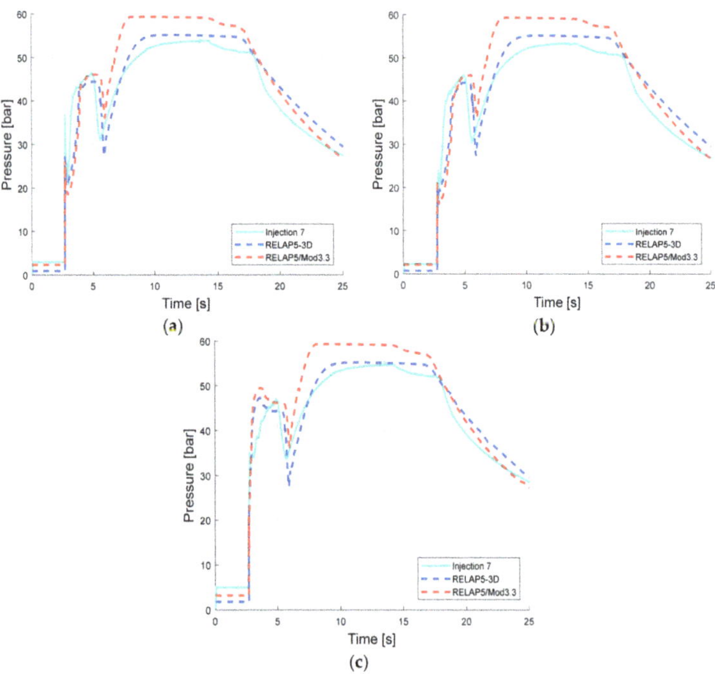

Figure 13. Experimental pressure trends in the upper and lower legs upstream of the isolation valves measured by PT11 (**a**), PT12 (**b**), and PT15 (**c**), compared with results from RELAP5-3D and RELAP5/mod3.3 (test #7).

Table 3. Average discrepancies between RELAP5-3D, RELAP5/mod3.3, and the experimental data (test #7).

Transducer	RELAP5-3D vs. Exp. [bar]	RELAP5-3D vs. RELAP5/mod3.3 [bar]
PT10	2.9	0.8
PT21	5.3	1.3
PT11	3.9	1.7
PT12	4.2	1.8
PT15	3.4	1.5
PT13	4.1	1.5
PT14	3.6	1.4

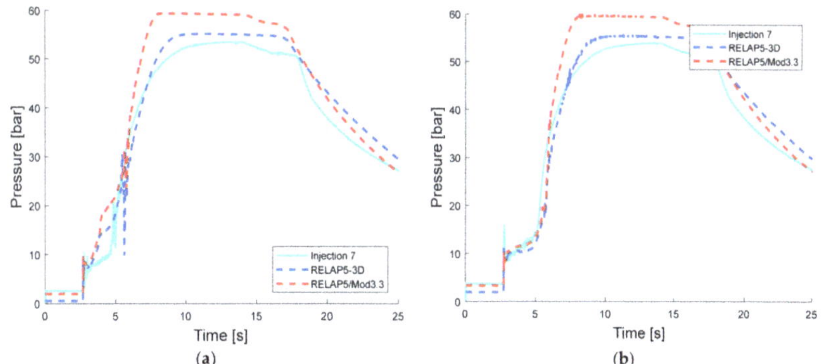

Figure 14. Experimental pressure trends in the upper and lower legs downstream of the isolation valves measured by PT13 (**a**) and PT14 (**b**), compared with results from RELAP5-3D and RELAP5/mod3.3 (test #7).

5.3. Test VIII

Figures 15, 17 and 18 show the comparison between the experimental pressure trends and the numerical trends evaluated by RELAP5-3D and RELAP5/mod3.3 for test #8. Even in this test, the codes tend to generally overestimate the pressure trends, with an average discrepancy of about 4 bar.

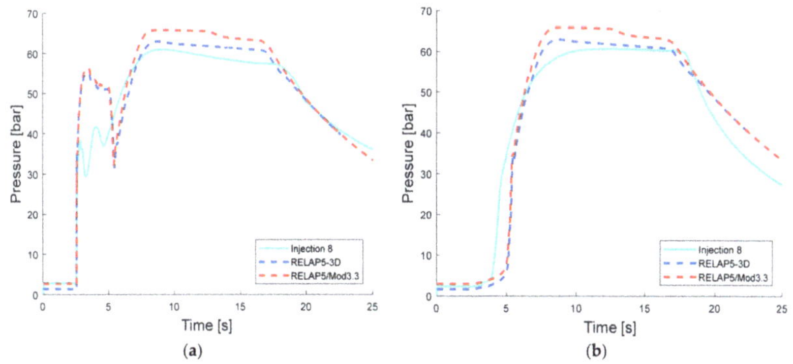

Figure 15. Experimental pressure trends in the TBM mock-up (**a**) and in the expansion tank (**b**), compared with results from RELAP5-3D and RELAP5/mod3.3 (test #8).

The experimental value of the first peak in the TBM mock-up is about 38 bar. The values calculated by the two versions of RELAP5 are about 34 bar for RELAP5-3D (-10% of the experimental value) and about 30.5 bar for RELAP5/mod3.3 (-19.5%). Moreover, the two codes predict the first peak at about 2.6 s, approximately the same time of tests #6 and #7. Instead, the experimental peak occurs at about 2.9 s, likely because of a slightly slower opening of the injection valve. As it is impossible to distinguish the first peak from Figures 15 and 16 shows a zoom of the very first phase of the injection.

Table 4 shows the rounded-up average discrepancies between RELAP5-3D and the experimental data and between the two versions of the code for each point of measurement.

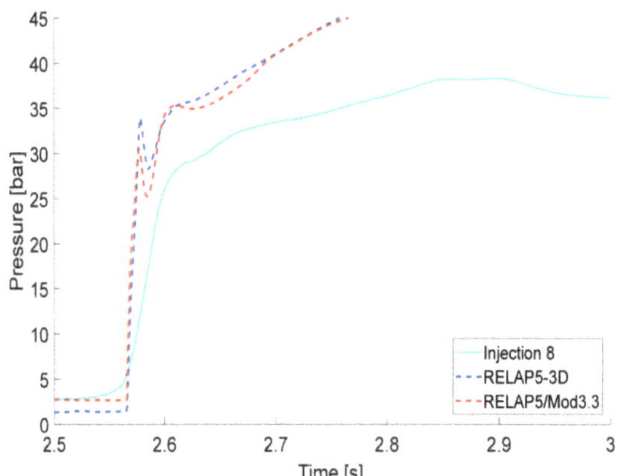

Figure 16. Zoom of the first peak in the TBM mock-up (test #8).

Table 4. Average discrepancies between RELAP5-3D, RELAP5/mod3.3, and the experimental data (test #8).

Transducer	RELAP5-3D vs. Exp. [bar]	RELAP5-3D vs. RELAP5/mod3.3 [bar]
PT10	4.6	0.7
PT21	4.5	0.4
PT11	3.6	0.5
PT12	3.7	−0.3
PT15	4.0	0.1
PT13	3.3	0.6
PT14	3.5	0.5

As far as the first peaks in the other parts of the loop are concerned, in test #8, the codes predict lower first peaks in PT14, while higher values in PT11. The first peak measured by PT11 is about 11 bar, while RELAP5-3D predicted about 30 bar (+172% of the experimental value) and RELAP5/mod3.3 about 28 bar (+154%). The low experimental value and the discrepancies being so much higher than the average ones raises doubts about the correct measurement of this particular peak. The experimental value measured by PT14 is about 16 bar, while the value calculated by RELAP5-3D is about 10 bar (-37.5% of the experimental value) and the one calculated by RELAP5/mod3.3 is about 11 bar (-31%). Moreover, in the case of PT12, the pressurization happens earlier in the codes with respect to the experiment. The experimental trend seems to lack of the first peak, reaching directly the pressure typical of the second peak, almost 50 bar. This anomalous behavior makes useless a comparison with the codes on the value of the first peak.

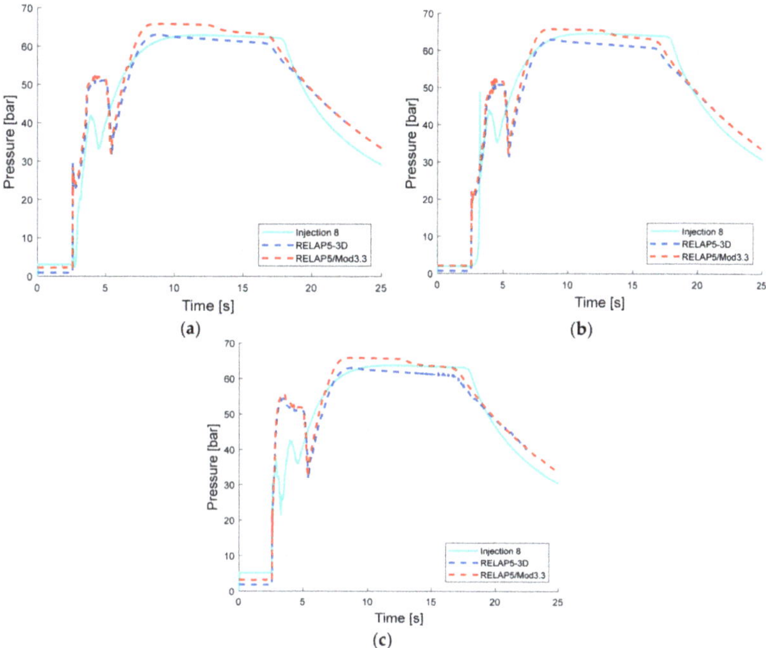

Figure 17. Experimental pressure trends in the upper and lower legs upstream of the isolation valves measured by PT11 (**a**), PT12 (**b**), and PT15 (**c**), compared with results from RELAP5-3D and RELAP5/mod3.3 (test #8).

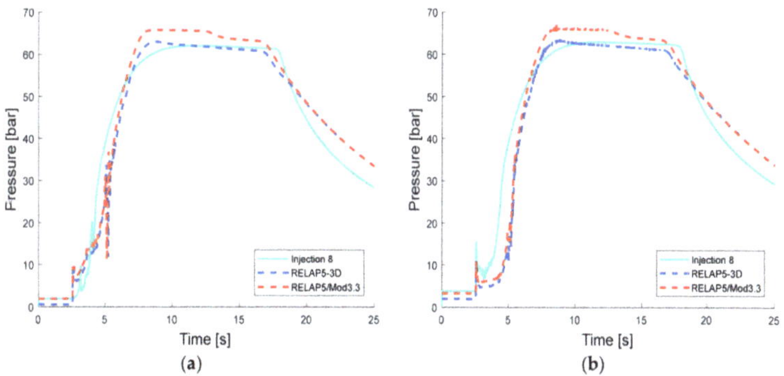

Figure 18. Experimental pressure trends in the upper and lower legs downstream of the isolation valves measured by PT13 (**a**) and PT14 (**b**), compared with results from RELAP5-3D and RELAP5/mod3.3 (test #8).

5.4. Test IX

Figures 19, 21 and 22 show the comparison between the experimental pressure trends and the numerical trends evaluated by RELAP5-3D and RELAP5/mod3.3 for test #9. Even in this test, the codes tend to generally overestimate the pressure trends, with an average discrepancy of less than 4 bar.

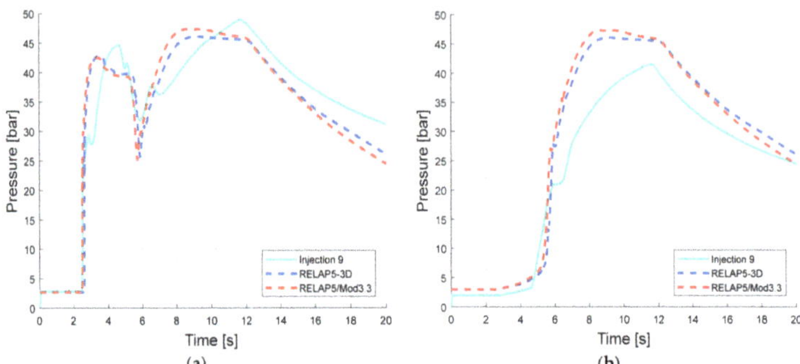

Figure 19. Experimental pressure trends in the TBM mock-up (**a**) and in the expansion tank (**b**), compared with results from RELAP5-3D and RELAP5/mod3.3 (test #9).

The experimental value of the first peak in the TBM mock-up is about 30 bar. The values calculated by the two versions of RELAP5 are about 29 bar for RELAP5-3D (-3% of the experimental value) and about 25.5 bar for RELAP5/mod3.3 (-15%). Even in this injection, the two codes predict the first peak earlier than in the experimental trend. The reason is likely the same: the slightly slower opening of the injection valve. As it is impossible to distinguish the first peak from Figures 19 and 20 shows a zoom of the very first phase of the injection.

Table 5 shows the rounded-up average discrepancies between RELAP5-3D and the experimental data and between the two versions of the code for each point of measurement.

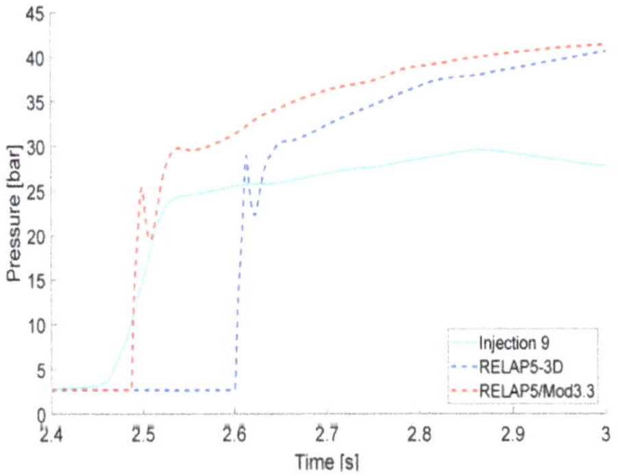

Figure 20. Zoom of the first peak in the TBM mock-up (test #9).

Table 5. Average discrepancies between RELAP5-3D, RELAP5/mod3.3, and the experimental data (test #9).

Transducer	RELAP5-3D vs. Exp. [bar]	RELAP5-3D vs. RELAP5/mod3.3 [bar]
PT10	4.0	0.6
PT21	5.0	−0.1
PT11	3.5	0.4
PT12	3.3	0.6
PT15	3.5	0.4
PT13	3.5	0.4
PT14	3.4	0.5

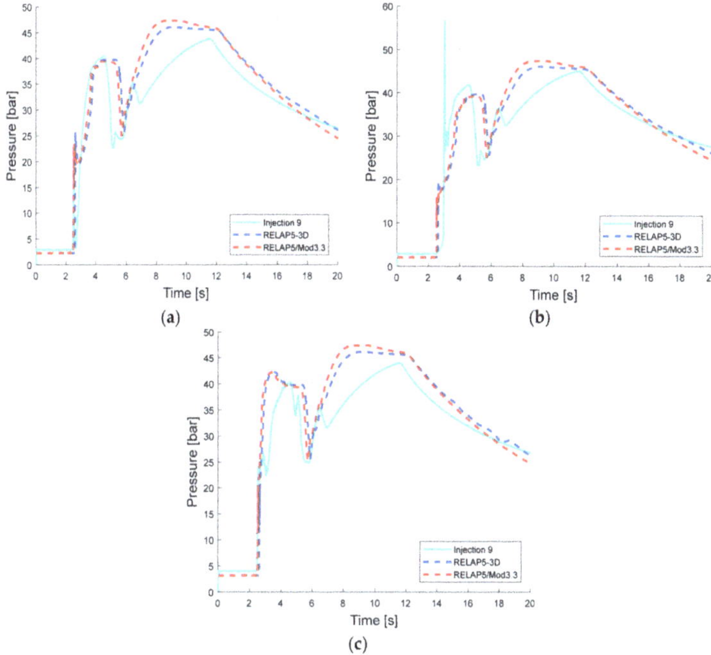

Figure 21. Experimental pressure trends in the upper and lower legs upstream of the isolation valves measured by PT11 (**a**), PT12 (**b**) and PT15 (**c**), compared with results from RELAP5-3D and RELAP5/mod3.3 (test #9).

As far as the first peaks in the other parts of the loop are concerned, in test #9, the codes predict lower first peaks in PT14, while higher values in PT11. The first peak measured by PT11 is about 19 bar, while RELAP5-3D predicted about 25.5 bar (+34% of the experimental value) and RELAP5/mod3.3 about 23 bar (+154%). The low experimental value and the discrepancies being so much higher than the average ones raises doubts to the correct measurement of this particular peak. The experimental value measured by PT14 is about 13 bar, while the value calculated by RELAP5-3D is about 10 bar (−23% of the experimental value) and the one calculated by RELAP5/mod3.3 is about 8 bar (−38%).

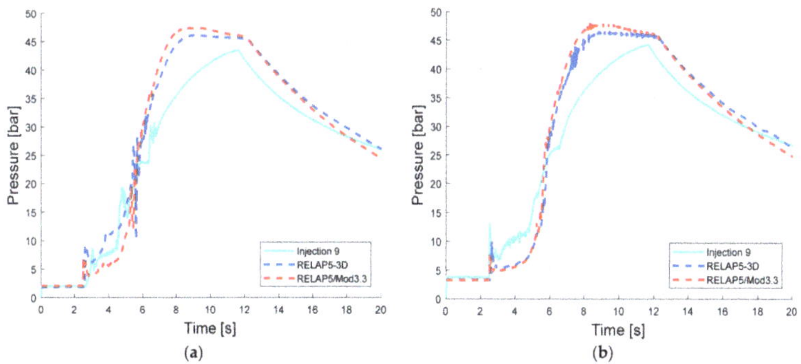

Figure 22. Experimental pressure trends in the upper and lower legs downstream of the isolation valves measured by PT13 (**a**) and PT14 (**b**), compared with results from RELAP5-3D and RELAP5/mod3.3 (test #9).

5.5. Test X

Figures 23, 25 and 26 show the comparison between the experimental pressure trends and the numerical trends evaluated by RELAP5-3D and RELAP5/mod3.3 for test #10. Even in this test, the codes tend to generally overestimate the pressure trends, with an average discrepancy of about 2.5 bar. An error in the Data Acquisition and Control System prevented measuring the pressure trend in the mock-up; for this reason, Figure 23a shows only the pressure trends computed by the two versions of RELAP5.

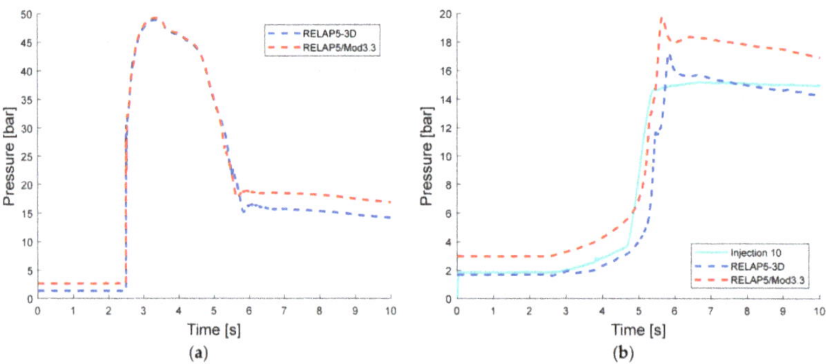

Figure 23. Numerical pressure trend in the TBM mock-up (**a**) and experimental pressure trend in the expansion tank (**b**), compared with results from RELAP5-3D and RELAP5/mod3.3 (test #10).

The values of the first peak calculated by the two versions of RELAP5 are about 30.5 bar for RELAP5-3D and about 28 bar for RELAP5/mod3.3, with a discrepancy of about 8% using RELAP5-3D as a reference. Figure 24 shows a zoom of the very first phase of the injection.

As far as the first peaks in the other parts of the loop are concerned, also in test #10, the codes predict lower first peaks in PT14, while higher values in PT11. The first peak measured by PT11 is about 22 bar, while RELAP5-3D predicted about 27 bar (+23% of the experimental value) and RELAP5/mod3.3 about 26 bar (+18%). The experimental value measured by PT14 is about 12 bar, while the value calculated by RELAP5-3D is about 9 bar (−13% of the experimental value), and the one calculated by RELAP5/mod3.3 is about 10 bar (−9%). Table 6 shows the rounded-up average discrepancies between RELAP5-3D and the experimental data and between the two versions of the code for each point of measurement.

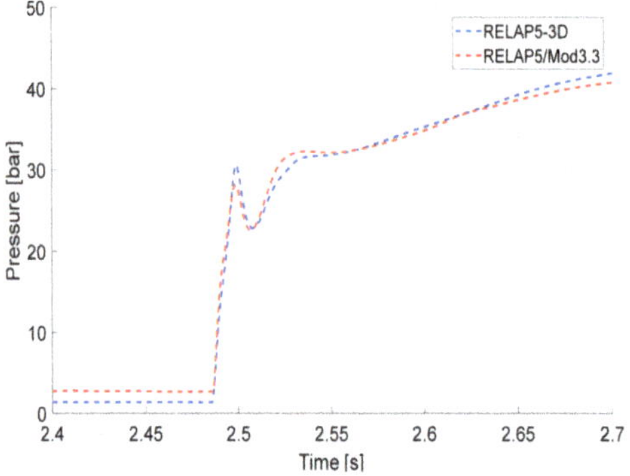

Figure 24. Zoom of the first peak in the TBM mock-up (test #10).

Table 6. Average discrepancies between RELAP5-3D, RELAP5/mod3.3, and the experimental data (test #10).

Transducer	RELAP5-3D vs. Exp. [bar]	RELAP5-3D vs. RELAP5/mod3.3 [bar]
PT10	-	−0.4
PT21	1.6	0.6
PT11	3.3	−0.1
PT12	3.8	−0.7
PT15	3.6	−0.4
PT13	2.4	−0.2
PT14	1.1	0.8

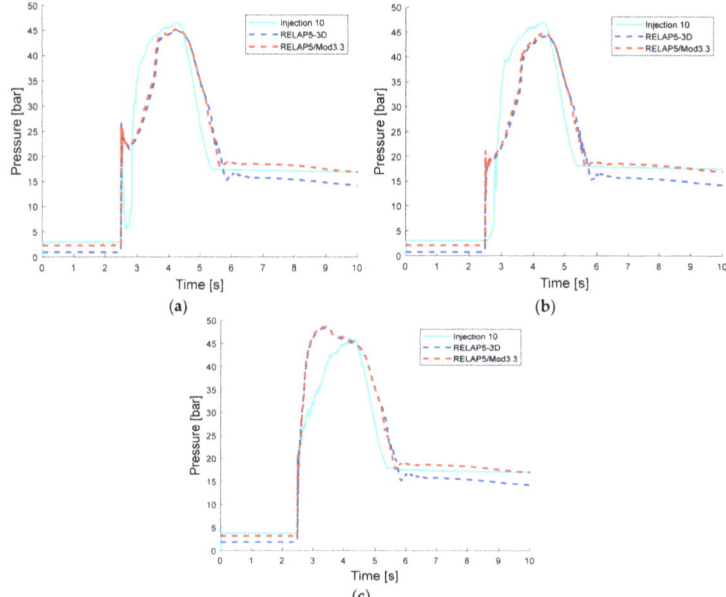

Figure 25. Experimental pressure trends in the upper and lower legs upstream of the isolation valves measured by PT11 (**a**), PT12 (**b**), and PT15 (**c**), compared with results from RELAP5-3D and RELAP5/mod3.3 (test #10).

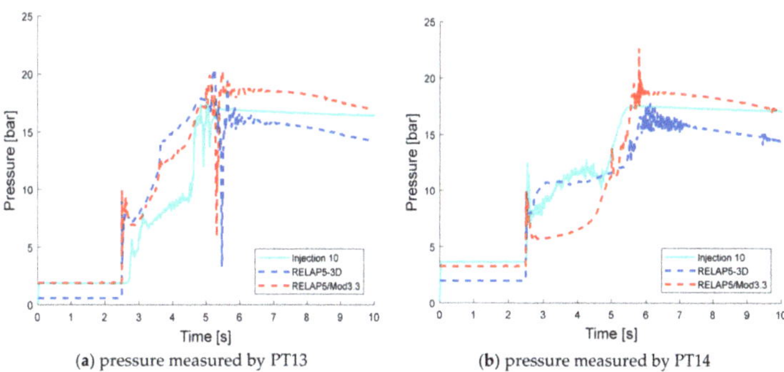

Figure 26. Experimental pressure trends in the upper and lower legs downstream of the isolation valves measured by PT13 (**a**) and PT14 (**b**), compared with results from RELAP5-3D and RELAP5/mod3.3 (test #10).

5.6. Test XI

Figures 27, 29 and 30 show the comparison between the experimental pressure trends and the numerical trends evaluated by RELAP5-3D and RELAP5/mod3.3 for test #11. Even in this test, the codes tend to generally overestimate the pressure trends, with an average discrepancy of about 4 bar.

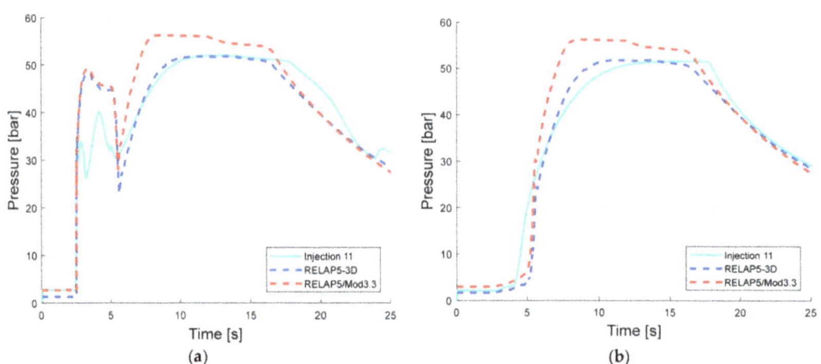

Figure 27. Experimental pressure trends in the TBM mock-up (**a**) and in the expansion tank (**b**), compared with results from RELAP5-3D and RELAP5/mod3.3 (test #11).

The experimental value of the first peak in the TBM mock-up is about 35 bar. The values calculated by the two versions of RELAP5 are about 30 bar for RELAP5-3D (-14% of the experimental value) and about 28.5 bar for RELAP5/mod3.3 (-18.5%). Figure 28 shows a zoom of the very first phase of the injection.

Table 7 shows the rounded-up average discrepancies between RELAP5-3D and the experimental data and between the two versions of the code for each point of measurement. This test highlighted the biggest difference between RELAP5-3D and RELAP5/mod3.3, with an average of about 1.7 bar.

As far as the first peaks in the other parts of the loop are concerned, in test #8, the codes predict lower first peaks in PT14, while higher values in PT11. The first peak measured by PT11 is about 14 bar, while RELAP5-3D predicted about 26.5 bar ($+89\%$ of the experimental value) and RELAP5/mod3.3 about 26 bar ($+93\%$). The experimental value measured by PT14 is about 15 bar, while the value calculated by RELAP5-3D is about 9.5 bar (-36.5% of the experimental value) and the one calculated by RELAP5/mod3.3 is about 10 bar (-33%).

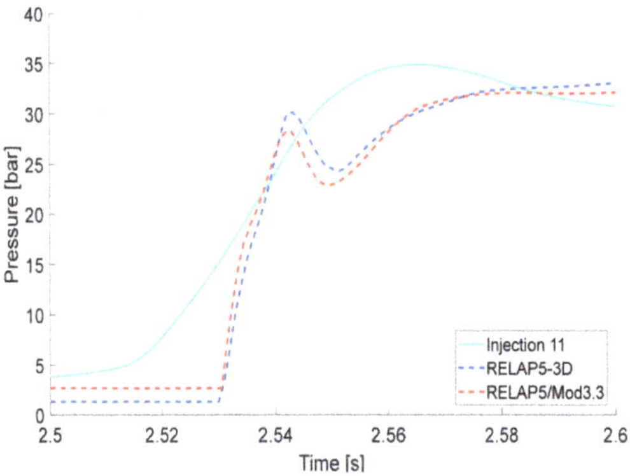

Figure 28. Zoom of the first peak in the TBM mock-up (test #11).

Table 7. Average discrepancies between RELAP5-3D, RELAP5/mod3.3, and the experimental data (test #11).

Transducer	RELAP5-3D vs. Exp. [bar]	RELAP5-3D vs. RELAP5/mod3.3 [bar]
PT10	3.6	2.2
PT21	2.6	2.2
PT11	3.3	1.6
PT12	3.8	1.0
PT15	3.7	1.5
PT13	2.1	2.0
PT14	2.8	1.4

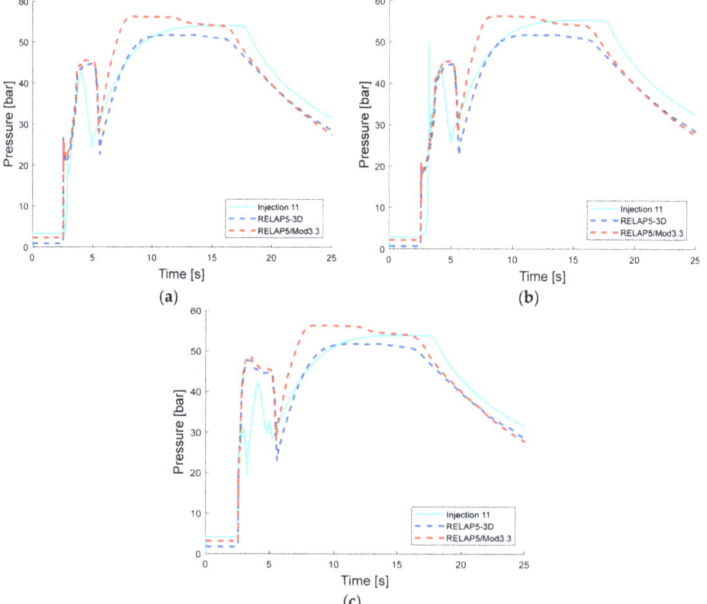

Figure 29. Experimental pressure trends in the upper and lower legs upstream of the isolation valves measured by PT11 (**a**); PT12 (**b**); and PT15 (**c**), compared with results from RELAP5-3D and RELAP5/mod3.3 (test #11).

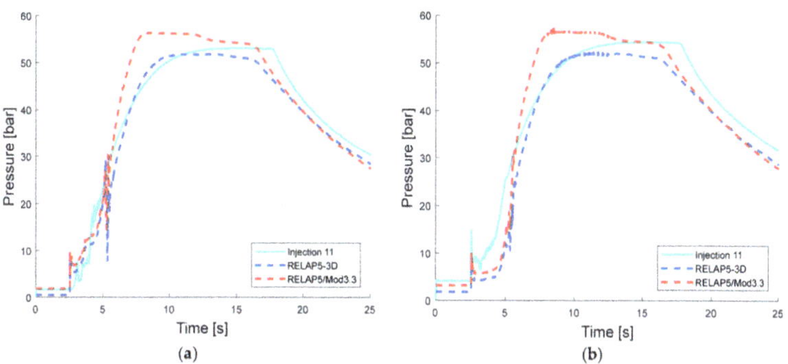

Figure 30. Experimental pressure trends in the upper and lower legs downstream of the isolation valves measured by PT13 (**a**) and PT14 (**b**), compared with results from RELAP5-3D and RELAP5/mod3.3 (test #11).

6. Conclusions

Simulations with RELAP5-3D were carried out using *LBE* instead of LiPb, as the property library of *LiPb* only reaches 40 bar in this version of the code. As a consequence, and also because errors were found in some properties (including density and speed of sound, particularly important for this transient), *LiPb* was added in RELAP5/mod3.3 at the University of Pisa.

The experimental campaign, the second one with the THALLIUM facility, allowed for strengthening the knowledge on the In-box LOCA evolution for the HCLL TBS. The results confirmed that the transient can be roughly divided into two parts: the first one, which goes from the injection up to the opening of the relief valve, is dominated by the fast pressure increase linked with the propagation of the pressure wave. This part of the transient is heavily influenced by the presence of the isolation valves: the small orifice of these valves protects the downstream section of the facility from the pressurization, by dampening the pressure wave. Instead, the second part of the transient sees a new pressurization linked with the arrival of high pressure helium from the injection line. This part of the transient is ruled by the ratio between the relief valve and the injection valve flow areas. In THALLIUM, the former is smaller than the latter, leading to a pressurization of the system. For this reason, a careful choice of the relief valve might help to protect the Test Blanket System, or similar systems, against these kinds of scenarios.

This experimental campaign has also underlined the impact of the opening time of the injection valve on the first pressure peak. Indeed, if the pressure peak in the first campaign, whose injection valve opened in about 40 ms, was about 70% of the helium pressure, and the slower opening time of the new injection valve (about 500 ms) caused this percentage to decrease to 60% of the injection pressure.

The numerical simulations of the second experimental campaign showed an excellent agreement in the comparison of the two versions of the code (average discrepancy of about 0.8 bar) and a general good agreement with the experimental data (average discrepancy of about 3.2 bar and about 4 bar. The two general discrepancies highlighted during the simulations of the first experimental campaign (i.e., delayed effect of the relief valve and absence of the second pressure peak) are still present here, even though they are less evident.

The comparison between RELAP5-3D and RELAP5/mod3.3 showed that the first one is slightly more accurate in reproducing the experimental results of the second campaign (average discrepancy of about 3.2 bar vs. about 4 bar), even though *LiPb* was implemented and used as system fluid in RELAP5/mod3.3. This accessorily confirms that the differences in the properties of *LiPb* and *LBE* have little impact on this type of transient.

These results demonstrate that RELAP5 accurately predicts the pressure trends. This confirms that its use in the design of the *LiPb* loops can be taken into consideration.

Author Contributions: Conceptualization, M.U. and A.V.; methodology, A.V.; software, A.V.; validation, A.V.; formal analysis, A.V.; investigation, A.V.; resources, M.U. and N.F.; data curation, A.V.; writing—original draft preparation, A.V.; writing—review and editing, A.V.; visualization, A.V.; supervision, M.U.; project administration, M.U.; funding acquisition, M.U. All authors have read and agreed to the published version of the manuscript.

Funding: This research was partially funded by Fusion for Energy, Grant No. FPA-372-SG02.

Institutional Review Board Statement: Not applicable.

Informed Consent Statement: Not applicable.

Data Availability Statement: Not applicable.

Acknowledgments: The first author wants to express his gratitude to EURO fusion for funding his Engineering Grant.

Conflicts of Interest: The authors declare no conflict of interest.

References

1. INL. *RELAP5-3D© Code Manual Volume I: Code Structure, System Models and Solution Methods*; INL/MIS-15-36723; Idaho National Laboratories: Idaho Falls, ID, USA, 2015.
2. INL. *RELAP5/MOD3 Code Manual Volume I: Code Structure, System Models, and Solution Methods*; Idaho National Laboratories: Idaho Falls, ID, USA, 2003.
3. ITER. *ITER D2.3: Final STR_56.A1.LP. (v.2.1 2014/08/07)*; ITER PbLi Loop—Final Technical Report; ITER: St. Paul-lez-Durance, France, 2014.
4. Aiello, G.; De Dinechin, G.; Forest, L.; Gabriel, F.; Puma, A.L.; Rampal, G.; Rigal, E.; Salavy, J.F.; Simon, H. HCLL TBM design status and development. *Fus. Eng. Des.* **2011**, *86*, 2129–2134. [CrossRef]
5. Panayotov, D. *HCLL-TBS Preliminary Safety Report, Internal F4E Report Number 239WLJ (v 2.4)*; F4E: Barcelona, Spain, 2013.
6. Utili, M.; Venturini, A.; Lanfranchi, M.; Calderoni, P.; Malavasi, A.; Zucchetti, M. THALLIUM: An experimental facility for simulation of HCLL In-box LOCA and validation of RELAP5-3D system code. *Fus. Eng. Des.* **2017**, *123*, 102–106. [CrossRef]
7. Venturini, A.; Utili, M.; Martelli, D.; Ricapito, I.; Malavasi, A. Experimental campaign on pressure wave propagation in LLE. *Fus. Eng. Des.* **2018**, *136*, 809–814. [CrossRef]
8. Venturini, A.; Utili, M.; Martelli, D.; Malavasi, A.; Ricapito, I.; Tarantino, M. Experimental investigation on HCLL-TBS In-box LOCA. *Fus. Eng. Des.* **2019**, *146*, 173–177. [CrossRef]
9. Barone, G.; Martelli, D.; Forgione, N.; Utili, M.; Ricapito, I. Experimental campaign on the upgraded He-FUS3 facility. *Fus. Eng. Des.* **2017**, *123*, 181–185. [CrossRef]
10. Barone, G.; Coscarelli, E.; Forgione, N.; Martelli, D.; Del Nevo, A.; Tarantino, M.; Utili, M.; Ricapito, I.; Calderoni, P. Development of a model for the thermal-hydraulic characterization of the He-FUS3 loop. *Fus. Eng. Des.* **2015**, *96*, 212–216. [CrossRef]
11. Venturini, A.; Utili, M.; Gabriele, A.; Ricapito, I.; Malavasi, A.; Forgione, N. Experimental and RELAP5-3D results on IELLLO (Integrated European Lead Lithium LOop) operation. *Fus. Eng. Des.* **2017**, *123*, 143–147. [CrossRef]
12. Venturini, A.; Papa, F.; Utili, M.; Forgione, N. Experimental Qualification of New Instrumentation for Lead-Lithium Eutectic in IELLLO Facility. *Fus. Eng. Des.* **2020**, *156*, 111683. [CrossRef]
13. Lee, D.W.; Jin, H.G.; Shin, K.I.; Lee, E.H.; Kim, S.K.; Yoon, J.S.; Ahn, M.Y.; Cho, S. Investigation into the In-box LOCA consequence and structural integrity of the KO HCCR TBM in ITER. *Fus. Eng. Des.* **2014**, *89*, 1177–1180. [CrossRef]
14. Moghanaki, S.K.; Galleni, F.; Eboli, M.; Del Nevo, A.; Paci, S.; Forgione, N. Analysis of Test D1.1 of the LIFUS5/Mod3 facility for In-box LOCA in WCLL-BB. *Fus. Eng. Des.* **2020**, *160*, 111832. [CrossRef]
15. Eboli, M.; Forgione, N.; Del Nevo, A. Assessment of SIMMER-III code in predicting Water Cooled Lithium Lead Breeding Blanket "in-box-Loss Of Coolant Accident". *Fus. Eng. Des* **2021**, *163*, 112127. [CrossRef]
16. Eboli, M.; Crugnola, R.M.; Cammi, A.; Khani, S.; Forgione, N.; Del Nevo, A. Test Series D experimental results for SIMMER code validation of WCLL BB in-box LOCA in LIFUS5/Mod3 facility. *Fus. Eng. Des.* **2020**, *156*, 111582. [CrossRef]
17. Venturini, A.; Utili, M.; Forgione, N. Numerical simulations with RELAP5-3D of the first experimental campaign on In-box LOCA transient for HCLL TBS. *Fus. Eng. Des.* **2021**, *163*, 112160. [CrossRef]
18. Utili, M.; Venturini, A.; Martelli, D.; Malavasi, A.; Laffi, L. *Tests Simulating LLE-Loop Pressurisation due to LOCA Accident*; ENEA Internal Report IT-T-R 228; ENEA: Rome, Italy, 2016.
19. Barone, G.; Martelli, D.; Forgione, N. Implementation of the lead lithium eutectic properties in RELAP5/Mod.3.3 for nuclear fusion system applications. *Fus. Eng. Des.* **2019**, *146*, 1308–1312. [CrossRef]
20. Idelchik, I.E. *Handbook of Hydraulic Resistance*, 3rd ed.; Jaico Publishing House: Mumbai, India, 2008.
21. Martelli, D.; Venturini, A.; Utili, M. Literature review of lead-lithium thermophysical properties. *Fus. Eng. Des.* **2019**, *138*, 183–195. [CrossRef]
22. Stankus, S.V.; Khairulin, R.A.; Mozgovoi, A.G. An Experimental Investigation of the Density and Thermal Expansion of Advanced Materials and Heat–Transfer Agents of Liquid–Metal Systems of Fusion Reactor: Lead–Lithium Eutectic. *High Temperature* **2006**, *44*, 829–837. [CrossRef]
23. *Handbook on Lead-Bismuth Eutectic Alloy and Lead Properties, Materials Compatibility, Thermal-Hydraulics and Technologies*, 2015 ed.; OECD: Paris, France; NEA: Paris, France, 2015; Available online: https://www.oecd-nea.org/jcms/pl_14972/handbook-on-lead-bismuth-eutectic-alloy-and-lead-properties-materials-compatibility-thermal-hydraulics-and-technologies-2015-edition?details=true (accessed on 27 April 2021).
24. Ueki, Y.; Hirabayashi, M.; Kunugi, T.; Yokomine, T.; Ara, K. Acoustic Properties of Pb-17li Alloy for Ultrasonic Doppler Velocimetry. *Fus. Sc. Tech.* **2009**, *56*, 846–850. [CrossRef]
25. Sobolev, V. *Database of Thermophysical Properties of Liquid Metal Coolants for GEN-IV*; Scientific Report SCK●CEN-BLG-1069; Belgian Nuclear Research Centre: Mol, Belgium, 2010; Available online: https://inis.iaea.org/search/search.aspx?orig_q=RN:43095088 (accessed on 27 April 2021).

Article

Numerical Simulation of an Out-Vessel Loss of Coolant from the Breeder Primary Loop Due to Large Rupture of Tubes in a Primary Heat Exchanger in the DEMO WCLL Concept

Francesco Galleni [1,*], Marigrazia Moscardini [1], Andrea Pucciarelli [1], Maria Teresa Porfiri [2] and Nicola Forgione [1]

[1]   Laboratory of Numerical Simulations for Nuclear Thermal-Hydraulics, Department of Civil and Industrial Engineering, University of Pisa, 56122 Pisa, Italy; marigrazia.moscardini@dici.unipi.it (M.M.); andrea.pucciarelli@dici.unipi.it (A.P.); nicola.forgione@unipi.it (N.F.)

[2]   ENEA-CR Frascati, Dipartimento Fusione e Tecnologie per la Sicurezza Nucleare, 00044 Roma, Italy; mariateresa.porfiri@enea.it

*   Correspondence: francescog.galleni@dici.unipi.it

Abstract: This work presents a thermohydraulic analysis of a postulated accident involving the rupture of the breeder primary cooling loop inside a heat exchanger (once through steam generator). After the detection of the loss of pressure inside the primary loop, a plasma shutdown is actuated with a consequent plasma disruption, isolation of the secondary loop, and shutoff of the pumps in the primary; no other safety counteractions are postulated. The objective of the work is to analyze the pressurization of the primary and secondary sides to show that the accidental overpressure in the two sides of the steam generators is safely accommodated. Furthermore, the effect of the plasma disruption on the FW, in terms of temperatures, should be analyzed. Lastly, the time transients of the pressures and temperatures in the HX and BB for a time span of up to 36 h should be obtained to assess the effect of the decay heat over a long period. A full nodalization of the OTSG was realized together with a simplified nodalization of the whole PHTS BB loop. The code utilized was MELCOR for fusion version 1.8.6. The accident was simulated by activating a flow path which directly connected one section of the primary with the parallel section of the secondary side. It is shown here that the pressures and the temperatures inside the whole PHTS system remain below the safety thresholds for the whole transient.

Keywords: WCLL-BB; MELCOR; PHTS; safety analysis; DEMO

Citation: Galleni, F.; Moscardini, M.; Pucciarelli, A.; Porfiri, M.T.; Forgione, N. Numerical Simulation of an Out-Vessel Loss of Coolant from the Breeder Primary Loop Due to Large Rupture of Tubes in a Primary Heat Exchanger in the DEMO WCLL Concept. *Energies* **2021**, *14*, 6916. https://doi.org/10.3390/en14216916

Academic Editor: Hyungdae Kim

Received: 17 August 2021
Accepted: 18 October 2021
Published: 21 October 2021

Publisher's Note: MDPI stays neutral with regard to jurisdictional claims in published maps and institutional affiliations.

1. Introduction

In the Roadmap to Fusion Electricity Horizon 2020, the European DEMOnstration Fusion Reactor (DEMO) is expected to be a nuclear fusion power plant with the aim of showing the feasibility of the production of electrical power through the conversion of around 2 GW_{th}, generated continuously by the fusion reaction. The operational sequence is a pulsed operation, which consists of 11 pulses per day; each pulse comprises a burn time of 2 h (power pulse period; 100% of fusion power) and a dwell time of 10 min (1% of fusion power generated due to the decay heat) [1,2].

The work presented here aims to investigate the consequences of a loss of coolant accident in the primary heat transport system (PHTS) of the water-cooled lithium lead (WCLL) breeding blanket (BB) DEMO concept [2–4]. In particular, the objective of this work is the simulation of a rupture of pipes of the primary system into the PCS (power conversion system or secondary system) inside the once through steam generator (OTSG) which acts as interface between the two systems [5–7]. In normal operations, the energy transferred from the breeding zone (BZ) and from the first wall (FW) to PCS through steam generators is used to produce the main steam at condition suitable to feed the steam turbine.

The postulated accident foresees the rupture of the primary cooling loop inside the OTSG with a discharge of primary coolant in the secondary side.

In this work, a detailed nodalization of the primary and secondary system of the OTSG is presented together with a simplified nodalization of the whole BZ loop. A connection between the two sides of the OTSG is created in order to simulate the LOCA scenario, which comprises a mitigated plasma disruption, and the results of the consequent transients are analyzed and discussed.

The nodalization of the involved systems was created with MELCOR for fusion 1.8.6, and all the simulations were performed with the same code. The MELCOR code has been under development for fusion applications for many decades [8,9]. Chiefly because of its capabilities of assessing thermal–hydraulic transients of fusion reactor systems and the transport of radionuclides, MELCOR was chosen, together with other system thermal hydraulics codes, to be used to perform safety analyses for the ITER project and, consequently, for the DEMO project [10]. Especially in the last 10 years, MELCOR was widely used for accident analysis related to ITER reactor safety and in the preliminary design phase of the DEMO reactor ([11–17], among many others).

2. DEMO WCLL-BB PHTS System

2.1. General Parameters and Power Data

The thermodynamic cycle used as reference for the design of the WCLL BB PHTS DEMO reactor was mainly based on parameters similar to pressurized water reactors: the coolant was water at 15.5 MPa with inlet and outlet temperatures equal to 295 °C and 328 °C, respectively [3,7–9]. The main working parameters of the BB PHTS are given in Table 1, whilst the power parameters are reported in Table 2. Furthermore, Table 3 collects the mass flow rates of the FW and the BZ cooling systems. All the parameters and characteristics summarized here can be found in several technical and published studies [18].

Table 1. WCLL DCD BOP BB cooling system parameters [3].

Description	Units	Parameters
Typology of coolant		Water
Pressure	MPa	15.5
Temperature range	°C	295–328
Coolant density (average)	kg/m^3	701.3
Design pressure	MPa	17.8
Design temperature	°C	345
Total flow rate	kg/s	9936.0
FW flow rate		
Inboard (IB) blanket	kg/s	17.8
Outboard (OB) blanket		35.5
BZ flow rate		
IB blanket	kg/s	48.2
OB blanket		127.5

Table 2. DEMO and WCLL DCD BOP BB power balance [3].

Description	Unit	WCLL 2018 Design
Total nuclear heating	MW	1650.3
Total FW Heat Flux	MW	272.7
Neutron Wall Load	MW	167.0
Total FW power	MW	439.8
Total BZ power	MW	1483.2
Total power	MW	1923.2

Table 3. WCLL DCD BOP BB power and coolant flow rates (TD 295–328 °C) [3].

Description	Power (MW)			Mass Flow Rate (kg/s)		
----	Total	FW	BZ	Total	FW	BZ
IB segment	12.780	3.449	9.331	66.0	17.8	48.2
IB sector Total	25.560	6.898	18.662	132.0	35.6	96.4
IB sectors (16)	409.960	110.368	**298.592**	2112.0	569.6	1542.4
OB segment	31.545	6.867	24.678	163.0	35.5	127.5
OB sector Total	94.635	20.601	74.034	489.0	106.5	382.5
OB sectors (16)	1514.160	329.616	**1184.54**	7824.	1704.	6120.
Total reactor	**1923.120**	**439.984**	**1483.136**	**9936.0**	**2273.6**	**7662.4**

The *WCLL DCD BOP* BB PHTS constitutes two independent primary systems:

- The BZ primary system (BZ PHTS);
- The FW primary system (FW PHTS).

The main components of the WCLL BB PHTS are indicated in the 3D CAD model (see Figure 1). Table 4 reviews the system-relevant data [2,7,19,20]. Details of the system architecture can be found in [18].

During normal operations, in pulse time, BZ and FW PHTSs transfer power to the PCS, through two once through steam generators (OTSGs) per system (i.e., four SGs in total). The BZ PHTS power is 1483 MW, and the FW PHTS power is 439.8 MW. A total of six main coolant pumps (MCPs) are installed to allow the circulation of the primary coolant (four pumps for BZ PHTS and two for FW PHTS). Each PHTS is equipped with a pressurizer (PRZ).

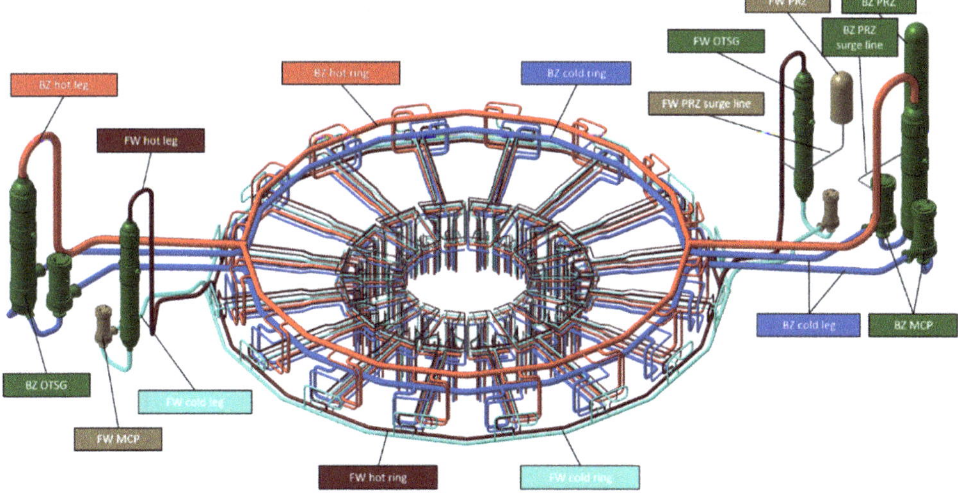

Figure 1. Overview of the DEMO WCLL BB PHTS.

Table 4. System-relevant parameters.

WCLL DCD BOP BB PHTS Design and Operating Parameters	BZ	FW
General		
Thermal power (MW)	1483	439.8
Operating pressure (MPa)	15.5	15.5
Reactor vessel inlet temperature (°C)	295	295
Reactor vessel outlet temperature (°C)	328	328
Overall volume (m^3)	~563	~159
Overall PHTS piping length (km)	~3.2	~3.7
Number of loops	2	2
Loop data		
Piping length (km)	~1.25	~1.52
Hot/cold manifolds per loop and size	8/8 DN-150, DN-200, DN-350	8/8 DN-100, DN-125, DN-200
Hot/cold legs per loop and size	1/2 DN-850, DN-650	1/1 DN-500, DN-500
Hot/cold ring header per loop and size	1/2 DN-650	1/1 DN-350
Pump		
Number of pumps per loop	2	1
Type	Centrifugal, vertical single-stage (RSR)	Centrifugal, vertical single-stage (RSR)
Effective pump power to coolant (MW)	3.03	1.79
Pressurizer		
Number of units	1	1
Total volume (m^3)	101.4	32.5
Liquid volume (m^3)	44.8	16.3
Heat Exchanger (Steam Generator)		
Number of units	2	2
Steam generator power (MW_{th}/unit)	742	219.9
Type	OTSG	OTSG
Heat transfer area (m^2/unit)	4903	1423
Feedwater temperature (°C)	238	238
Exit steam pressure (MPa)	6.4	6.4
Steam flow per SG (kg/s)	404	119.9
Flow rate per SG (kg/s)	3831.2	1136

2.2. BZ Once through Steam Generator

In the BZ Primary System, each OTSG removes 742 MW_{th} of thermal power, with a mass flow rate of coolant equal to 3831.2 kg/s. On the secondary side, the water is assumed to be at a pressure of 6.4 MPa, and the feedwater coolant inlet temperature is expected to be at 238 °C. The feedwater flow rate is imposed at 404 kg/s, in order to produce the same amount of superheated steam at 299 °C. It is important to underline that the OTSG is still in a design phase [21,22]. A simplified scheme of a generic OTSG is shown in Figure 2 and the main characteristics are summarized in Table 5.

The selected OTSG is characterized by 7569 tubes, with a length of 12.987 m.

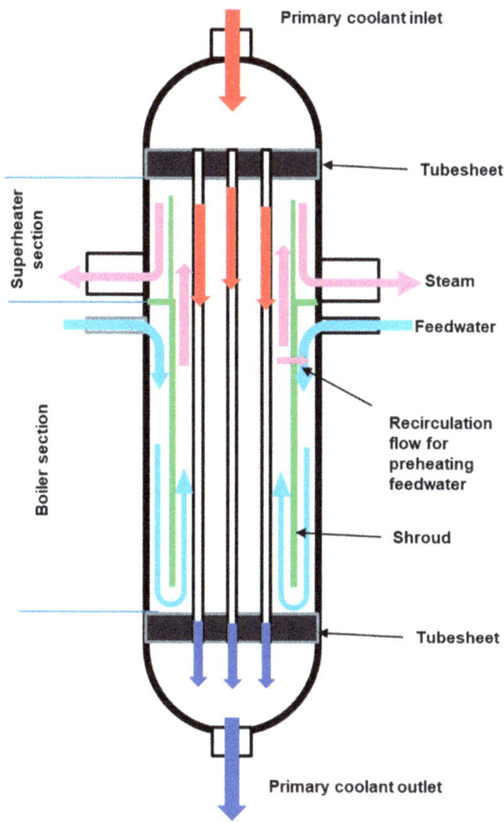

Figure 2. Conceptual scheme for once through steam generator [21].

Table 5. Main features of an OTSG.

SG Power	MW$_{th}$	742
Primary side pressure	MPa	15.5
Primary side water T$_{in}$	°C	328
Primary side water T$_{out}$	°C	295
Secondary side pressure	MPa	6.41
Secondary side water T$_{in}$	°C	238
Secondary side water T$_{out}$	°C	299
No. of tubes	--	7569
Tube OD	mm	15.88
Tube thickness	mm	0.864
Tube length	m	12.987
Tubesheet lattice	--	Square
Tubesheet p/D	--	1.28
Heat transfer area	m^2	4903
V water tubes	m^3	20
D$_{ext}$ vessel	m	2.9

2.3. BB and PHTS Integration

The WCLL DEMO 2018 blanket is composed of 16 sectors (22.5°) in the toroidal direction [4,23]. Each sector consists of three poloidal segments in the OB (outboard) blanket and two poloidal segments in the IB (inboard) blanket. The geometrical features and the inventory of the WCLL-BB are provided in Table 6.

The inner structure of each segment comprises a stack of about 100 toroidal–radial so-called breeding units (BUs). The BU is a sort of elementary cell repeated along the poloidal direction of each segment. It features the integrated FW–SW complex, the BZ, and the corresponding part of manifolds and BSS.

The coolant feeding pipes deliver cold water to FW and BZ systems, whilst the outlet pipes collect hot water and distribute it to the PHTS.

Considering the FW and BZ systems, the volume of water is estimated taking into account the BUs (assuming a rough value of 100 for each sector), the manifolds, and the feeding pipes. Furthermore, the velocities in the inlet feeding pipes and distributors were calculated assuming a density of 737.1 kg/m^3 (15.5 MPa and 295 °C), while a density of 657.5 kg/m^3 (15.5 MPa and 328 °C) was taken for the outlet pipes and collectors. The main geometrical data are provided in Table 6.

Table 6. WCLL BB In-Vessel inventory.

Description	Pipe Size	OD (mm)	Thick. (mm)	MF Total (kg/s)	v (m/s)	L (m)	L Tot (m)	V (m³)	H$_2$O Mass Inventory (kg)
Inlet IB FW distributor	DN-100	114.3	8.8	35.6	6.6	17.6	281.2	2.07	1522.21
Inlet IB FW LIBS/RIBS	DN-100	114.3	8.8	17.8	3.3	5.8	186.5	1.37	1009.43
Outlet IB FW LIBS/RIBS	DN-100	114.3	8.8	17.8	3.7	6.4	203.9	1.50	984.62
Outlet IB FW collector	DN-100	114.3	8.8	35.6	7.4	17.6	281.2	2.07	1357.82
Inlet IB BZ distributor	DN-200	219.1	17.5	96.4	4.9	17.5	280.3	7.46	5500.35
Inlet IB BZ LIBS/RIBS	DN-150	168.3	12.7	48.2	4.1	4.0	128.1	2.05	1513.83
Outlet IB BZ LIBS/RIBS	DN-150	168.3	12.7	48.2	4.6	4.7	149.1	2.39	1572.12
Outlet IB BZ collector	DN-200	219.1	17.5	96.4	5.5	17.5	279.9	7.45	4898.97
Inlet OB FW distributor	DN-200	219.1	17.5	106.5	5.4	11.3	180.8	4.81	3548.13
Inlet OB FW COBS	DN-125	139.7	11.0	35.5	4.4	7.9	126.4	1.38	1013.89
Inlet OB FW LOBS/ROBS	DN-125	139.7	11.0	35.5	4.4	9.2	293.9	3.20	2357.06
Outlet OB FW COBS	DN-125	139.7	11.0	35.5	5.0	8.3	133.1	1.45	952.47
Outlet OB FW LOBS/ROBS	DN-125	139.7	11.0	35.5	5.0	9.4	301.6	3.28	2157.42
Outlet OB FW collector	DN-200	219.1	17.5	106.5	6.1	11.5	183.3	4.88	3208.51
Inlet OB BZ distributor	DN-350	355.6	28.0	382.4	7.4	10.6	170.3	12.00	8847.80
Inlet OB BZ COBS	DN-200	219.1	17.5	127.5	6.5	7.0	111.5	2.97	2187.86
Inlet OB BZ LOBS/ROBS	DN-200	219.1	17.5	127.5	6.5	7.4	236.1	6.29	4633.48
Outlet OB BZ COBS	DN-200	219.1	17.5	127.5	7.3	7.7	122.7	3.27	2147.62
Outlet OB BZ LOBS/ROBS	DN-200	219.1	17.5	127.5	7.3	8.4	267.9	7.13	4688.31
Outlet OB BZ collector	DN-350	355.6	28.0	382.4	8.2	11.0	176.3	12.43	8172.15
Total							*4094.1*	*89.44*	*62,274.06*

3. Postulated Accidental Scenario

In this section, the specifications of the accident—such as the event sequence and the main objectives—are reported and described in Tables 7–10.

Table 7. Accidental scenario general specifications.

Parameter	Specification
Name of event	LBO3 out-vessel loss of coolant from the breeder primary loop due to large rupture of tubes in a primary HX in WCLL concept
Category	Accident
Objectives	Show that accidental overpressure in the secondary loop of PHTS will be safely accommodated. Show that post-accident cooling of the decay heat removal system of the VV is sufficient to remove decay heat during the 32 h of the offsite loss of power. Show that radioactive releases, if any, are adequately confined.
Scope of analysis	Integrated breeder blanket PHTS thermohydraulic analysis. VV, VVPSS(W) pressure transient analysis if melting occurs due to mitigated plasma disruption. Analysis of ACPs and tritium transport in the containment volumes as a consequence of the plasma disruption, if melting occurs.
Acceptance criteria	Maximum Eurofer temperature < 550 °C Confinement integrity: BB module safety assessment pressure < 18 MPa Confinement integrity: VV safety assessment pressure < 200 kPa

Table 8. Accidental scenario main events sequence.

Parameter	Specification
Definition of initiating event	Break in the primary cooling loop side (**break size 0.0028 m^2, corresponding to the rupture of 9 tubes of the primary side**) inside the HX toward the secondary side.
Possible transient sequence	**Fusion power is terminated by loss of pressure (-20% of operating pressure) in the BB cooling loop inside the HX. The initiating event is followed by a fast plasma shutdown (FPS) actuated 3 s from the low signal, which leads to a mitigated plasma disruption for 5 ms.** The disruption could cause failure of the FW cooling pipes in a BB module if temperature melting is reached in FW Eurofer. In such a case, the break flow area to be considered is reported in [SDL19], chapter 2.1. The VV decay heat removal cooling loops will cool down the in-vessel components post accident. Ingress of coolant and radioactive inventories (tritium, dust, and suspended products) will be mobilized. The rupture discs toward the VVPSS(W) open upon reaching the set VV pressure point. Mobilized radioactivity is transported into the VVPSS(W). After the coolant inventory is lost, the FW/breeder blanket modules will be cooled by steam convection and thermal conduction/radiation to the VV. The DV components are accounted for in this analysis as heat structures at the initial temperature and without cooling during the transient.
Aggravating failures	None.
Loss of power	A loss of offsite power occurs at the same time of the plasma disruption.

Table 9. Accidental scenario system assumptions.

Parameter	Specification
Process system assumptions	VV decay heat removal (DHR) will remove the decay heat. The temperature of the VV is maintained at 40 °C by DHR.
Safety systems assumptions	VV pressure limit is 0.2 MPa. VVPSS(W) rupture discs open upon reaching the VV pressure set point.
Source term	Tritium, dust in the VV, ACP products in the PHTS, tritium and activated products in the breeder materials and/or purge gas.

Table 10. Expected results from the analysis of the accidental scenario.

Parameter	Specification
General	The output locations, parameters, and time trace should show the results of safety analyses to support objectives and purposes. The time span should cover until the transients are stabilized.
FW temperature	Transient curve for FW and BB module temperature of the failed loop.
Confinement response	Pressures and atmosphere temperature in BB cooling loop, HX. Pressures and atmosphere temperature in VV and VVPSS(W) only in case of FW Eurofer structure melting.
Cooling system conditions	Water break flow versus time in HX and in VV and VVPSS(W) only in case of FW Eurofer structure melting. Water inventory inside the affected FW/breeder blanket cooling system.
Radioactive transport	Transient curves for tritium, dust, and sputtering concentration (airborne, deposited) in VVPSS(W) if FW structure melts. Bookkeeping of mobilized tritium, dust, and sputtering products if FW structure melts.

4. Nodalization of the WCLL-PHTS

Figure 3 shows the whole nodalization of the BZ loop. Since the loop is symmetrical with respect to the two OTSGs, only half of the loop is represented, without losing consistency. All the geometrical characteristics of the components of the nodalization were retrieved from the references presented above and summarized in the previous sections of this document. A list of all the main characteristics can be found in Tables 7 and 8.

The whole nodalization can be roughly divided into two sections: the main one, more complex, which is the nodalization of the BZ OTSG, and the simpler one, which is the nodalization of the BB volumes; the two regions are highlighted in Figure 3, whilst a focus on the BZ OTSG is shown in Figure 4. The cold and hot rings provide a connection between the two regions. The pressurizer is directly connected to the hot ring.

The OTSG is partitioned in the primary and secondary side. Each side is divided into several control volumes (CVH in MELCOR), which correspond to the hemispherical top and bottom, as well as to the sections created by the support plates. The division into different sections is required to achieve a temperature gradient as close as possible to the actual one. All the control volumes are connected by flow paths (FL). CVH 10 and 11 (see Figure 4) are dummy volumes used only to impose the BCs to the secondary side at normal operations. The CHV and FL geometrical characteristics are summarized in Tables 11 and 12, respectively.

In normal conditions, the primary and secondary side are connected only by heat structures (HSs), which allows heat transfer between the volumes but do not permit any passage of fluids. Between the primary and secondary side, there are a total of 17 equal heat structures. Each HS has a cylindrical geometry divided into four nodes with a total thickness of 0.8 mm and a height of 0.7825 m, which corresponds to the height of the connected volumes. The multiplicity of the HS is equal to 7569, which means that the HS is equivalent to 7569 heat structures of the same type, for a total surface area of 4795 m^2. All HSs are made of stainless steel, and a convective condition is imposed on both sides of the HS.

The thermal power of the OTSG is simulated by means of two heat structures connected with the inboard and outboard sections of the BB regions. These two HSs provide 592 MW of thermal power to the outboard section (CVH 15) and 149 MW to the inboard section (CVH 14).

The accident is simulated by activating dummy flow paths which connect one volume of the primary side with the corresponding volume of the secondary side. The section area of this FL is equal to the rupture area (0.00312 m^2 for nine tubes on the OTSG primary side).

These FLs are located at different heights to allow sensitivity analysis on the location of the rupture.

Table 11. CVH number and characteristics.

Nodalization CVH	Volume Water (m^3)	Height/Length (m)	Section (m^2)	Description
Primary side				
1	3.22	1.15	4.18	Top hemisphere
2	0.72	0.61	1.18	Top tubesheet
31–46	15.36	0.7825	1.18	Shroud tubes primary
4	0.72	0.61	1.18	Bottom tubesheet
5	3.22	1.15	4.18	Bottom hemisphere
Secondary side				
6	12.65	7.63	1.66	Shell boiler
71–86	34.09	12.70	2.68	Shroud secondary
8	1.81	0.30	6.03	Steam chamber
9	8.42	5.08	1.66	Shell SH
Rings and BB				
12	36.04	146.5	0.246	Hot ring
13	36.04	146.5	0.246	Cold ring
16	7.73	17.9	0.027	Cold ring **distributors IB**
18	14.67	13.1	0.07	Cold ring **distributors OB**
17	7.99	18.5	0.027	Hot ring **collectors IB**
19	15.9	14.2	0.07	Hot ring **collectors OB**
14	73.7	14.7	-	BB inboard
15	121.616	12.3	-	BB outboard

Table 12. FL number and characteristics.

Nodalization FL	From CVH	To CVH	From Height (m)	To Height (m)	Section (m^2)	Hydraulic Diameter (m)	Length (m)
Primary side							
FL00100	12	1	4.25	16.5272	0.4117	0.724	45
FL00200	1	2	15.3732	15.3732	1.1818	1.2267	0
FL00300	2	31	14.7636	14.7636	1.1818	1.2267	0
FL03100 TO FL04600	(31)	(46)	13.9811	13.9491	1.1818	0.0141	0.032
FL00400	46	4	1.7636	1.7636	1.1818	1.2267	0
FL00500	4	5	1.154	1.154	1.1818	1.2267	0
Secondary side							
FL00600	10	6	9	9	0.198639	0.0889	0
FL07100 TO FL08600	(71)	(86)	13.9491	13.9811	0.89466	0.012636	0.032
FL00700	6	71	2.2	2.2	4.330349	0.84579	0
FL00800	86	8	14.4636	14.4636	2.6846	0.0279	0

Table 12. Cont.

Nodalization FL	From CVH	To CVH	From Height (m)	To Height (m)	Section (m²)	Hydraulic Diameter (m)	Length (m)
Rings and BB							
FL01200	5	13	0	2	0.246	0.56	23
FL01300	13	18	1.724	1.724	0.62	0.3	10.6
FL01400	18	15	−7.746	−7.746	0.564	0.1841	10.6
FL01500	15	19	−7.746	−7.746	0.564	0.1841	10.6
FL01600	19	12	3.784	3.784	0.62	0.3	10.6
FL01700	13	16	1.724	1.724	0.156	0.184	17.5
FL01800	16	14	−4.446	−4.446	0.255	0.1429	4
FL01900	14	17	−4.446	−4.446	0.255	0.1429	4
FL02000	17	12	3.784	3.784	0.156	0.184	17.5
Accident							
FL10101	33	84	12.7	12.7	0.0028	0.0141	0
FL10102	39	78	7.9	7.9	0.0028	0.0141	0
FL10103	45	72	2.2	2.2	0.0028	0.0141	0

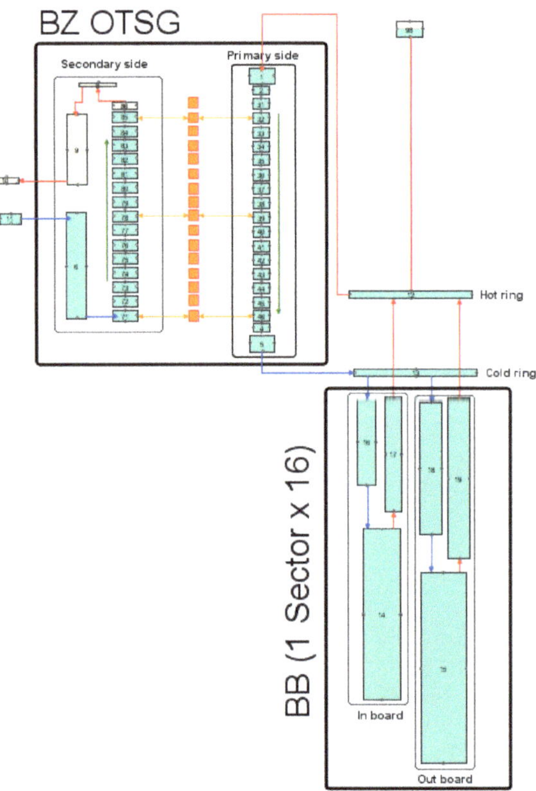

Figure 3. MELCOR nodalization of PHTS loop.

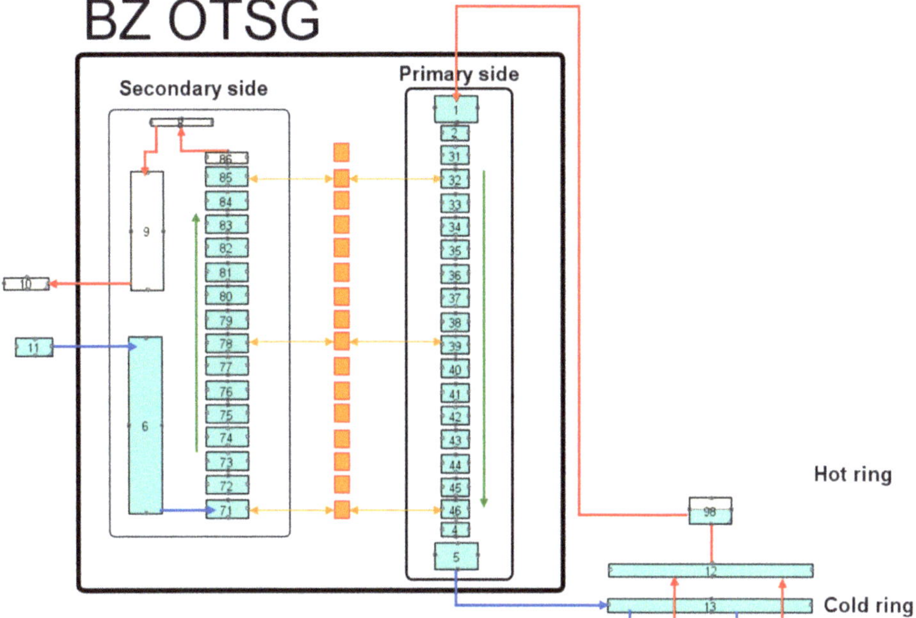

Figure 4. Detail of BZ OTSG nodalization.

4.1. Boundary Conditions

In order to impose the required primary coolant mass flow rate of 3831.2 kg/s, a fluid velocity was imposed at FL12, since MELCOR does not allow to impose directly the mass flowrate. Furthermore, only for the initial steady condition, before the onset of the accident, the temperature on the bottom of the OTSG (CVH 5) was imposed to be 295 °C and the pressure was set at 15.5 MPa.

As mentioned in the previous section, two dummy CVHs (10 and 11) were used to impose the BCs at the secondary side. A pressure of 6.4 MPa and a feedwater coolant inlet temperature of 238 °C were imposed. The feedwater flow rate is kept at 404 kg/s by means of imposing an equivalent constant velocity FL, in order to produce the same amount of superheated steam at 299 °C.

The velocity (mass flow rate) imposed at FL12 was set to 0.0 at the FPS since the pumps were shut down due to the loss of offsite power. Furthermore, at the FPS, the FLs connecting the secondary side of the OTSG to the secondary loop are closed (thus acting as isolation valves), also interrupting the feed water flow in the secondary side. At the current stage, the design of the secondary loop is not completed.

4.2. Decay Heat

The decay heat was imposed using volumetric heat generation from heat structures connected with the outboard and inboard CVH. The total values and time trends, calculated using the most up-to-date volumetric nuclear heating data given in [24], are reported in Table 13.

Table 13. Decay heat densities (MW/m^3) integrated into the WCLL full reactor [24].

Entire Reactor						
Name of Zone	Nuclear Heating	Cooling Time				
	MW/m^3	MW/m^3				
First wall (FW)		0 s	1 s	1 h	1 day	1 week
W	2.18E+01	4.95E−01	4.93E-01	4.63E-01	2.27E-01	5.66E-03
Eurofer	7.14E+00	1.76E-01	1.76E-01	1.01E-01	1.34E-02	6.91E-03
Breeder module (BM)						
BM caps and lateral walls	1.81E+00	1.08E-02	1.08E-02	7.32E-03	8.82E-04	2.57E-04
BM material mixture	1.17E+00	1.23E-02	9.35E-03	3.60E-03	1.07E-03	9.01E-04
BM backwall	1.06E-01	1.50E-03	1.50E-03	1.02E-03	1.20E-04	4.24E-05
BM back support/manifold	5.64E-02	4.31E-04	4.28E-04	2.90E-04	3.45E-05	1.32E-05
Sum (MW/m^3)	1.65E+03	2.21E+01	1.91E+01	9.68E+00	2.26E+00	1.25E+00

However, it is important to note that the values reported in [24] did refer to a different configuration of the WCLL, namely, the one which was composed of modules and, therefore, had a different material ratio; in this work, the total nuclear heating was recalculated scaling the values with an estimation of the total volume obtained from the most recent configuration of the inboard and outboard segments, and this may represent a source of error. Furthermore, it is difficult to estimate the role of the FW in cooling the decay heat of the breeding unit under this particular scenario, since—as shown in the results—the FW loop is not damaged by the plasma disruption, and it can be assumed still functional. As a first approach in this work, only the decay nuclear heating relative to the BM back wall and the manifolds is assumed to have an effect on the damaged BZ cooling loops.

5. Numerical Analyses

In the simulation, the system was initially left to run for 1000 s, to reach a stable steady state in operational conditions, and then the FL connecting the volumes involved in the rupture was opened. Three seconds after the low-pressure signal, i.e., when the pressure in the loop fell below 12.4 MPa, the FPS was activated; plasma disruption occurred, the pumps were stopped, and the secondary loop was isolated.

In total, three different cases were run, at three different rupture locations (top, middle, and bottom). Table 14 summarizes the different cases.

Table 14. Summary of simulation parameters.

	Rupture Location	Number of Pipes	Rupture Size (m^2)	FL Involved
Case 1	Top OTSG	9	0.0028	FL10101
Case 2	Middle OTSG	9	0.0028	FL10102
Case 3	Bottom OTSG	9	0.0028	FL10103

The event sequence of the accident is reported in Table 15.

Table 15. Event sequence.

Event		Time (s)
Steady state		0–1000
	Case 1	1016.5
Low-pressure signal	Case 2	1008.5
	Case 3	1007.5
FPS	Case 1	1019.5
	Case 2	1011.5
	Case 3	1010.5
End of the simulation		150,000

5.1. Pressure Evolution

Figures 5 and 6 show the pressure trend after the rupture, at different time scales. The pressure started decreasing and then showed a sharp peak when the low-pressure signal was reached. This was due to the rapid shutdown of the pump and the isolation of the loop. The same behavior was seen in all the cases, but it is worth noting that, for Case 1, the pressure drop was slower and the peak remained significantly lower than the other two cases. This was due to the different rupture conditions; in Case 1, the primary side discharged in a region of the secondary, which was filled with superheated vapor, whereas, in Cases 2 and 3, it was discharged in regions containing liquid water. Figure 7 shows the time trend of the vapor quality calculated in CVH 85 for Case 1.

Figures 8 and 9 show the pressure trend at the rupture locations of both the primary and the secondary side. In the secondary side of the OTSG, no pressure peaks were predicted, and the pressure increased to an equilibrium value by around 100 s in all the cases. Then, 200 s after the rupture, a steady state was reached, with the pressure lying between 12 and 13 MPa.

All pressures remained well below the threshold of 18 MPa, with the pressure peak for Cases 2 and 3 having a value of about 15 MPa, and that for Case 1 having a value of about 12.5 MPa.

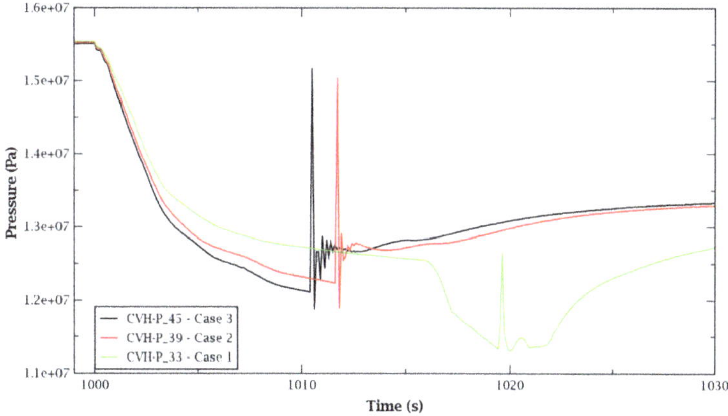

Figure 5. Pressure trend after the rupture: primary side, 1000–1030 s.

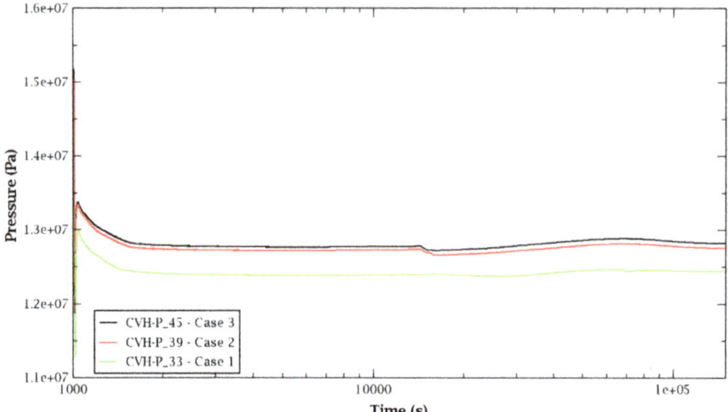

Figure 6. Pressure trend after the rupture: primary side.

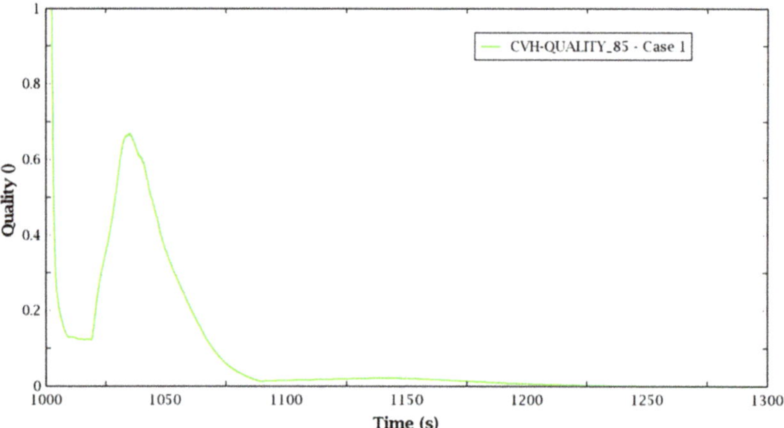

Figure 7. Vapor quality in volume 85 for Case 1.

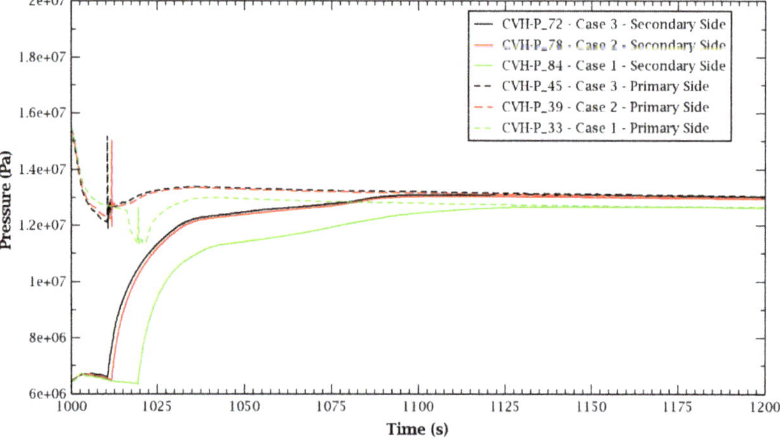

Figure 8. Pressure trend after the rupture: primary and secondary sides, 1000–1200 s.

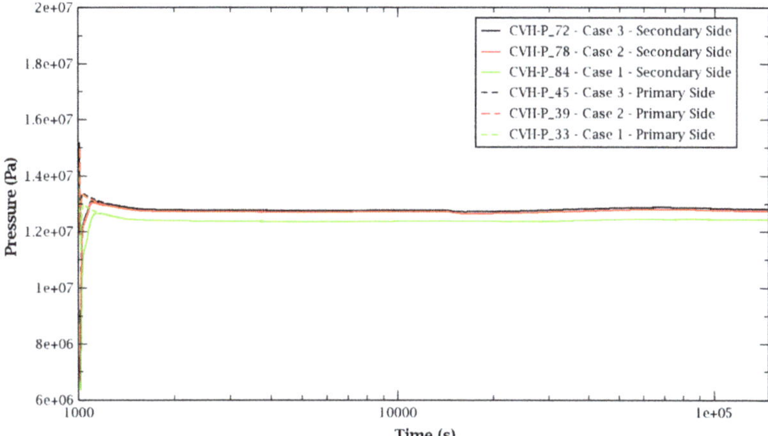

Figure 9. Pressure trend after the rupture: primary and secondary sides.

5.2. Temperature Evolution in the BB

Figures 10–12 show the time evolution of the temperature at the inlets and outlets of the BB, for the whole transient up to 36 h. Because of the effect of the decay heat, the temperatures did not reach a steady state, but tended to increase. The faster temperature increase after 10^4 s was due to the onset of weak natural circulation in the loop, with the water flow even changing direction several times in Cases 2 and 3, as shown in Figure 13. However, the increase could be considered slow throughout the whole transient, and all temperatures remained well below acceptable values.

As mentioned above, the difference between Case 1 and Cases 2 and 3 was due to a different height of the rupture in the primary side of OTSG for the three cases, with a jump of around 5 m between each case. The transient for Case 1 showed a different behavior from the very beginning, because, in this case, the primary discharged, through the rupture, into a region of the secondary which was filled with superheated vapor and not liquid water, as in Cases 2 and 3. This early difference consequently led to a radically different transient of the pressure and general behavior of the loop, compared with the other two cases.

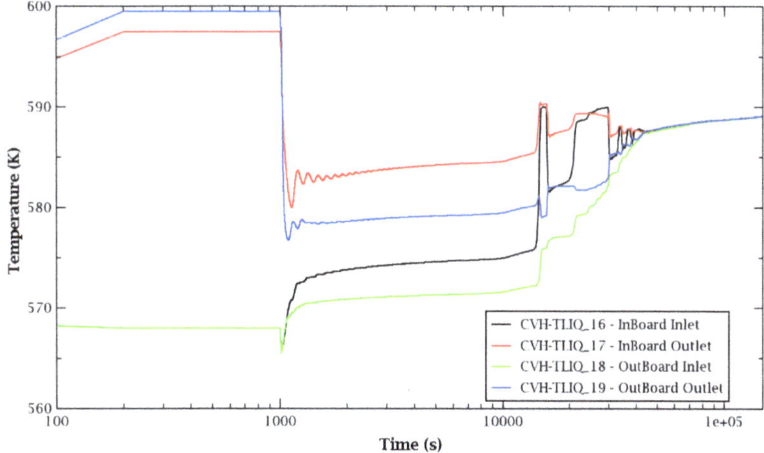

Figure 10. Temperature evolution in the BB: Case 3.

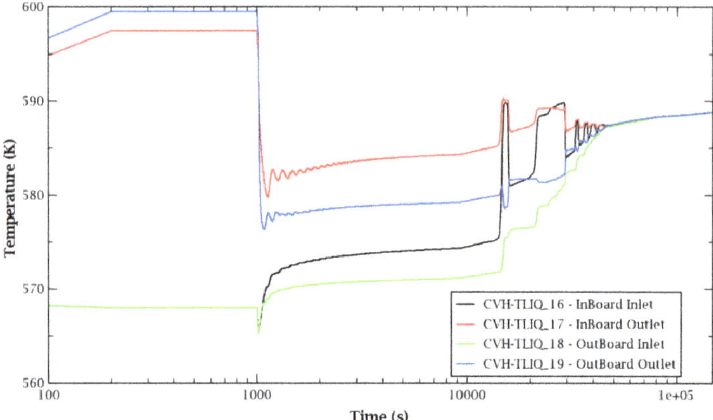

Figure 11. Temperature evolution in the BB: Case 2.

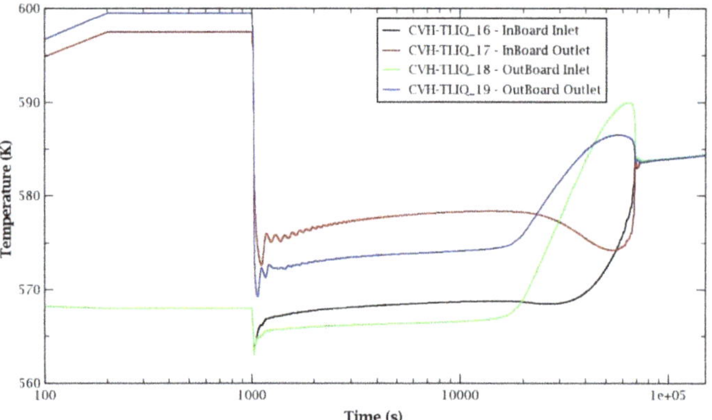

Figure 12. Temperature evolution in the BB: Case 1.

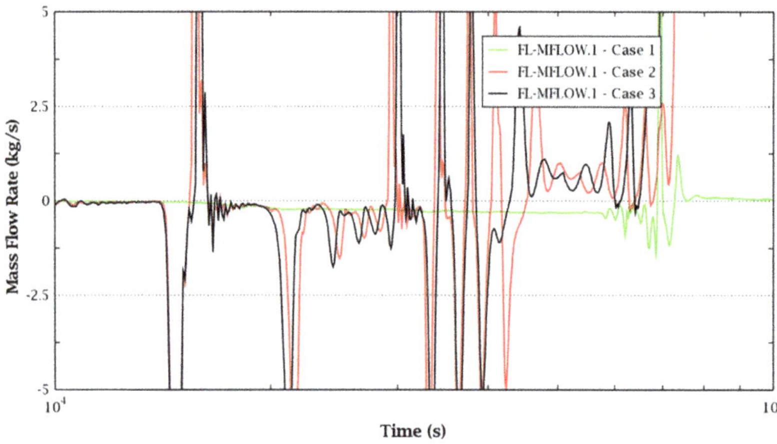

Figure 13. Water mass flow rate in the loop: 10^4–10^5 s.

6. Conclusions

This work presented a thermohydraulic analysis of a postulated accident involving the rupture of a breeder primary loop inside an HX (once through steam generator). After the detection of the loss of pressure inside the primary cooling loop, an FPS is actuated with a consequent plasma disruption, isolation of the secondary loop, and shutoff of the pumps in the primary; no other safety counteractions are postulated.

The objective of the work was to analyze the pressurization of the primary and secondary sides to show that the accidental overpressure in the two sides of the steam generators is safely accommodated. Furthermore, the effect of the plasma disruption on the FW, in terms of temperatures, should be analyzed. Lastly, the time transients of the pressures and temperatures in the HX and BB for a time span of up to 36 h should be obtained to assess the effect of the decay heat over a long period.

A full nodalization of the OTSG was realized together with a simplified nodalization of the whole PHTS BB loop. The code utilized was MELCOR for fusion version 1.8.6. The accident was simulated by activating a flow path which directly connected one section of the primary with the parallel section of the secondary side.

It was shown here that the pressures and the temperatures inside the whole PHTS system remained below the safety thresholds for the whole transient.

The only caveat of this analysis is due to the calculation of the effect of the decay heat, since the tables used for the estimation of this parameter referred to an old version of the WCLL-BB, as no other references are available on this issue.

It is believed that, for a fully comprehensive evaluation of the impact of the decay heat, a coupled calculation should be realized, connecting a nodalization of the FW PHTS loop and the PbLi loop with the nodalization of the BZ PHTS loop presented here.

Author Contributions: Conceptualization, F.G., M.M., A.P. and M.T.P.; methodology, F.G.; validation, F.G.; formal analysis, F.G.; investigation, F.G.; resources, N.F.; data curation, F.G.; writing—original draft preparation, F.G.; writing—review and editing, F.G., M.M., A.P., M.T.P. and N.F.; visualization, F.G.; supervision, M.T.P. and N.F.; project administration, M.T.P. and N.F.; All authors have read and agreed to the published version of the manuscript.

Funding: This research was funded by Euratom research and training program 2014–2018 and 2019–2020, grant number 633053.

Acknowledgments: This work was carried out within the framework of the EUROfusion Consortium and received funding from the Euratom research and training program 2014–2018 and 2019–2020 under grant agreement No 633053. The views and opinions expressed herein do not necessarily reflect those of the European Commission.

Conflicts of Interest: The authors declare no conflict of interest.

Abbreviations

ACP	Activate corrosion product
BB	Breeder blanket
BC	Boundary condition
BOP	Balance of plant
BSS	Back supporting structure
BZ	Breeder zone
CAD	Computer=aided design
CVH	Control volume hydrodynamics
DHR	Decay heat removal
DCD	Direct coupling design
DEMO	DEMOnstration power plant
FL	Flow path
FPS	Fast plasma shutdown

FW	First wall
HT package	Heat structure package
IB	Inboard
LOCA	Loss of coolant accident
OB	Outboard
OTSG	Once through steam generator
PCS	Power conversion system
PHTS	Primary heat transport system
PbLi	Lead–lithium
PRZ	Pressurizer
VV	Vacuum vessel
VVPSS	Vacuum vessel pressure suppression system
WCLL	Water-cooled lithium lead
WPSAE	Work Package Safety and Environment

References

1. Federici, G.; Bachmann, C.; Barucca, L.; Biel, W.; Boccaccini, L.; Brown, R.; Bustreo, C.; Ciattaglia, S.; Cismondi, F.; Coleman, M.; et al. DEMO design activity in Europe: Progress and updates. *Fusion Eng. Des.* **2018**, *136*, 729–741. [CrossRef]
2. Barucca, L.; Bubelis, E.; Ciattaglia, S.; D'Alessandro, A.; Del Nevo, A.; Giannetti, F.; Hering, W.; Lorusso, P.; Martelli, E.; Moscato, I.; et al. Pre-conceptual design of EU DEMO balance of plant systems: Objectives and challenges. *Fusion Eng. Des.* **2021**, *169*, 112504. [CrossRef]
3. Narcisi, V.; Ciurluini, C.; Giannetti, F.; Del Nevo, A. WCLL BB Phts Ddd (Direct Coupling Option with Small ESS). 2020. Report No.: IDM Ref. EFDA_D_2NURWJ. Available online: https://idm.euro-fusion.org/ (accessed on 13 October 2021).
4. Del Nevo, A.; Arena, P.; Caruso, G.; Chiovaro, P.; Di Maio, P.A.; Eboli, M.; Edemetti, F.; Forgione, N.; Forte, R.; Froio, A.; et al. Recent progress in developing a feasible and integrated conceptual design of the WCLL BB in EUROfusion project. *Fusion Eng. Des.* **2019**, *146*, 1805–1809. [CrossRef]
5. Zaupa, M.; Palma, M.D.; Del Nevo, A.; Moscato, I.; Tarallo, A.; Barucca, L. Preliminary Thermo-Mechanical Design of the Once Through Steam Generator and Molten Salt Intermediate Heat Exchanger for EU DEMO. *IEEE Trans. Plasma Sci.* **2020**, *48*, 1726–1732. [CrossRef]
6. Martelli, E.; Giannetti, F.; Ciurluini, C.; Caruso, G.; Del Nevo, A. Thermal-hydraulic modeling and analyses of the water-cooled EU DEMO using RELAP5 system code. *Fusion Eng. Des.* **2019**, *146*, 1121–1125. [CrossRef]
7. Ciurluini, C.; Giannetti, F.; Martelli, E.; Del Nevo, A.; Barucca, L.; Caruso, G. Analysis of the thermal-hydraulic behavior of the EU-DEMO WCLL breeding blanket cooling systems during a loss of flow accident. *Fusion Eng. Des.* **2021**, *164*, 112206. [CrossRef]
8. Merrill, B.; Moore, R.; Polkinghorne, S.; Petti, D. Modifications to the MELCOR code for application in fusion accident analyses. *Fusion Eng. Des.* **2000**, *51–52*, 555–563. [CrossRef]
9. Merrill, B.J. *Recent Updates to the MELCOR 1.8. 2 Code for ITER Applications*; Idaho National Laboratory (INL): Idaho Falls, ID, USA, 2007.
10. Merrill, B.J.; Humrickhouse, P.; Moore, R.L. A recent version of MELCOR for fusion safety applications. *Fusion Eng. Des.* **2010**, *85*, 1479–1483. [CrossRef]
11. D'Ovidio, G.; Martín-Fuertes, F. Accident analysis with MELCOR-fusion code for DONES lithium loop and accelerator. *Fusion Eng. Des.* **2019**, *146*, 473–477. [CrossRef]
12. Grief, A.; Owen, S.; Murgatroyd, J.; Panayotov, D.; Merrill, B.; Humrickhouse, P.; Saunders, C. Qualification of MELCOR and RELAP5 models for EU HCLL TBS accident analyses. *Fusion Eng. Des.* **2017**, *124*, 1165–1170. [CrossRef]
13. Panayotov, D.; Grief, A.; Merrill, B.J.; Humrickhouse, P.; Trow, M.; Dillistone, M.; Murgatroyd, J.T.; Owen, S.; Poitevin, Y.; Peers, K.; et al. Methodology for accident analyses of fusion breeder blankets and its application to helium-cooled pebble bed blanket. *Fusion Eng. Des.* **2015**, *109–111*, 1574–1580. [CrossRef]
14. Nakamura, M.; Tobita, K.; Someya, Y.; Utoh, H.; Sakamoto, Y.; Gulden, W. Safety research on fusion DEMO in Japan: Toward development of safety strategy of a water-cooled DEMO. *Fusion Eng. Des.* **2016**, *109–111*, 1417–1421. [CrossRef]
15. Pescarini, M.; Mascari, F.; Mostacci, D.; De Rosa, F.; Lombardo, C.; Giannetti, F. Analysis of unmitigated large break loss of coolant accidents using MELCOR code. *J. Phys. Conf. Ser.* **2017**, *923*, 012009. [CrossRef]
16. D'Onorio, M.; Giannetti, F.; Porfiri, M.T.; Caruso, G. Preliminary safety analysis of an in-vessel LOCA for the EU-DEMO WCLL blanket concept. *Fusion Eng. Des.* **2020**, *155*, 111560. [CrossRef]
17. D'Onorio, M.; Giannetti, F.; Porfiri, M.T.; Caruso, G. Preliminary sensitivity analysis for an ex-vessel LOCA without plasma shutdown for the EU DEMO WCLL blanket concept. *Fusion Eng. Des.* **2020**, *158*, 111745. [CrossRef]
18. Narcisi, V.; Giannetti, F.; Del Nevo, A. WCLL BB PHTS Architecture Description & BOM (Direct Coupling Option with Small ESS). Report No.: IDM Ref. EFDA_D_2PC2N9. Available online: https://idm.euro-fusion.org/ (accessed on 13 October 2021).
19. Tarallo, A. WCLL BB PHTS CAD Model Description (Direct Coupling Option with Small ESS). Report No.: IDM Ref. EFDA_D_2P9QFM. 2020. Available online: https://idm.euro-fusion.org/ (accessed on 13 October 2021).

20. Tarallo, A. WCLL BB PHTS Drawings (Direct Coupling Option with Small ESS). Report No.: IDM Ref. EFDA_D_2P95DT. 2020. Available online: https://idm.euro-fusion.org/ (accessed on 13 October 2021).
21. Tarallo, A. Preliminary Mechanical Design and Verification of WCLL BB PHTS Once through Steam Generator (OTSG). Report No.: IDM Ref EFDA_D_2MT7PL. 2020. Available online: https://idm.euro-fusion.org/ (accessed on 13 October 2021).
22. Tarallo, A. WCLL OTSG CAD Model. Report No.: IDM Ref. UID 2NJQM8. 2020. Available online: https://idm.euro-fusion.org/ (accessed on 13 October 2021).
23. Catanzaro, I.; Arena, P.; Basile, S.; Bongiovì, G.; Chiovaro, P.; Del Nevo, A.; Di Maio, P.A.; Forte, R.; Maione, I.A.; Vallone, E. Structural assessment of the EU-DEMO WCLL Central Outboard Blanket segment under normal and off-normal operating conditions. *Fusion Eng. Des.* **2021**, *167*, 112350. [CrossRef]
24. Porfiri, M.T.; Mazzini, G. DEMO BB Safety Data List (SDL). Report No.: IDM Ref. EFDA_D_2MF8KU. 2018. Available online: https://idm.euro-fusion.org/ (accessed on 13 October 2021).

Article

Study of the EU-DEMO WCLL Breeding Blanket Primary Cooling Circuits Thermal-Hydraulic Performances during Transients Belonging to LOFA Category

Cristiano Ciurluini [1,*], Fabio Giannetti [1], Alessandro Del Nevo [2] and Gianfranco Caruso [1]

[1] Dipartimento di Ingegneria Astronautica, Elettrica ed Energetica, Sapienza University of Rome, 00186 Rome, Italy; fabio.giannetti@uniroma1.it (F.G.); gianfranco.caruso@uniroma1.it (G.C.)

[2] ENEA FSN-ING-SIS, ENEA CR Brasimone, Località Brasimone, 40032 Camugnano, Italy; alessandro.delnevo@enea.it

* Correspondence: cristiano.ciurluini@uniroma1.it

Abstract: The Breeding Blanket (BB) is one of the key components of the European Demonstration (EU-DEMO) fusion reactor. Its main subsystems, the Breeder Zone (BZ) and the First Wall (FW), are cooled by two independent cooling circuits, called Primary Heat Transfer Systems (PHTS). Evaluating the BB PHTS performances in anticipated transient and accident conditions is a relevant issue for the design of these cooling systems. Within the framework of the EUROfusion Work Package Breeding Blanket, it was performed a thermal-hydraulic analysis of the PHTS during transient conditions belonging to the category of "Decrease in Coolant System Flow Rate", by using Reactor Excursion Leak Analysis Program (RELAP5) Mod3.3. The BB, the PHTS circuits, the BZ Once Through Steam Generators and the FW Heat Exchangers were included in the study. Selected transients consist in partial and complete Loss of Flow Accident (LOFA) involving either the BZ or the FW PHTS Main Coolant Pumps (MCPs). The influence of the loss of off-site power, combined with the accident occurrence, was also investigated. The transient analysis was performed with the aim of design improvement. The current practice of a standard Pressurized Water Reactor (PWR) was adopted to propose and study actuation logics related to each accidental scenario. The appropriateness of the current PHTS design was demonstrated by simulation outcomes.

Keywords: DEMO; primary heat transfer system; balance of plant; RELAP5; loss of flow accident; once through steam generators

Citation: Ciurluini, C.; Giannetti, F.; Del Nevo, A.; Caruso, G. Study of the EU-DEMO WCLL Breeding Blanket Primary Cooling Circuits Thermal-Hydraulic Performances during Transients Belonging to LOFA Category. *Energies* **2021**, *14*, 1541. https://doi.org/10.3390/en14061541

Academic Editor: Guglielmo Lomonaco

Received: 5 February 2021
Accepted: 8 March 2021
Published: 11 March 2021

Publisher's Note: MDPI stays neutral with regard to jurisdictional claims in published maps and institutional affiliations.

1. Introduction

In the European DEMO (EU-DEMO) fusion reactor, the Breeding Blanket (BB) component accomplishes several functions [1,2]. Firstly, it acts as a cooling device. The nuclear interactions between the neutrons produced within the plasma and the lithium contained in the breeder allow to convert the neutron kinetic energy in thermal power to be removed. The same nuclear reactions are supposed to be used to produce the tritium fuel needed to reach the self-sufficiency. Moreover, the breeding blanket serves as shielding, preventing the high-energy neutrons from escaping outside the reactor and protecting from damage the more radiation-susceptible components, like the superconducting magnets.

In the framework of the EUROfusion Programme, two breeding blanket concepts were selected for the EU-DEMO R&D strategy: Water-Cooled Lithium-Lead (WCLL) and Helium-Cooled Pebble Bed (HCPB) [1]. These two technologies will also be tested in the ITER fusion reactor, according to the goals of the ITER Test Blanket Module (TBM) programme [1]. The main outcome of this experimental campaign will be the Return of Experience for the EU-DEMO Breeding Blanket Programme [3]. The computational activity presented in this paper deals with the WCLL option. It foresees the usage of water at typical Pressurized Water Reactor (PWR) thermodynamic conditions (295–328 °C and 15.5 MPa) as coolant [1,2]. The blanket relies on liquid lithium-lead as breeder, neutron multiplier and tritium carrier and on Eurofer as structural material. An armour, consisting of a thin tungsten layer is assumed to cover the First Wall (FW) component (plasma-facing surface).

The cooling systems associated to the principal blanket subsystems, namely the FW and the Breeder Zone (BZ), are called Primary Heat Transfer Systems (PHTS) [2,4]. Their main function is to provide primary coolant at the required thermodynamic conditions. The thermal power they remove is then delivered to the Power Conversion System (PCS) to be converted into electricity [5,6].

With the aim of the design improvement, the evaluation of the BB PHTS thermal-hydraulic (TH) behavior during anticipated transient and accident conditions is a key issue. To achieve this goal, computational activities can be performed by using best estimate system codes. Principally, system codes adopt a one-dimensional approach to solve the balance equations. For this reason, they are more recommended for simulations involving circuits, where the fluid main stream direction can be clearly identified. They allow to simulate the overall primary cooling system, including the pipelines and all the vessel components (pumps, heat exchangers, pressurizer). However, in some of them, also 3D approaches are partially implemented, such as in RELAP5-3D [7], CATHARE-3 [8], SAM [9], and so they can also be used for components characterized by more complex fluid flow paths. Throughout decades, these codes have been validated for Light Water Reactors (LWR), simulating a wide range of transient and accidental scenarios. Hence, their usage can also be envisaged for WCLL blanket, whose primary coolant has similar thermodynamic conditions.

In the last years, a large experience was matured in the simulation of transients involving fusion reactors. Referring to EU-DEMO WCLL PHTS, both the in-vessel [10] and ex-vessel [11] Loss Of Coolant Accidents (LOCA) were investigated with MELCOR code, [12]. The main simulation purpose was assessing the hydrogen production and the radiological source term mobilization in order to demonstrate the consistency of the EU-DEMO design with the safety and environmental criteria. MELCOR code was also used for a parametrical study in support of the reactor Vacuum Vessel Pressure Suppression System design, as described in [13]. A preliminary analysis of the Loss Of Flow Accident (LOFA) is reported in [14]. In this case, RELAP5/Mod3.3 code [15] was used to perform a TH-oriented transient calculation aimed at the sizing of the flywheel to be adopted for the PHTS Main Coolant Pumps (MCPs).

For what concerns the EU-DEMO HCPB PHTS, RELAP5-3D code was properly integrated with a computational fluid-dynamic code in order to investigate the thermal-hydraulic performances of the primary circuits during an Ex-Vessel LOCA scenario, [16]. With the same code, multiple LOFA scenarios were also studied [17]. LOCA transients were also simulated with MELCOR code [18]. The activity goal was to perform a parametric study on the break size and to assess its impact on some reactor relevant parameters, such as containment pressure and FW component maximum temperature.

System codes were largely adopted also in the framework of research activities related to China Fusion Engineering Test Reactor (CFETR) and Korean DEMO (K-DEMO) Reactor. CFETR design foresees a Water-Cooled Ceramic Breeder (WCCB) blanket concept. RELAP5/Mod3.3 code was employed for transient analysis involving LOFA, [19], and Loss of Heat Sink (LOHS), [20], scenarios. The calculations allowed an in-depth evaluation of the WCCB blanket behavior. As initial conditions, different fusion power modes were considered.

One of the blanket concepts proposed for K-DEMO reactor is the water-cooled multiple-layer breeding blanket. It consists of a sandwich of multiple layers of breeder (Li_4SiO_4) and multiplier ($Be_{12}Ti$) mixtures, cooling channels, and structural materials. They are stacked in the radial direction, parallel to the first wall. MELCOR was adopted to investigate the reactor response after a vacuum vessel rupture, mainly focusing on hydrogen production and dust explosions, [21].

System codes also allow to study operational transients and to conceptually design the machine control system. This is a relevant design issue for fusion reactors where a plasma pulsed operating regime is foreseen (including both pulse and dwell phases). Similar studies were conducted for all the aforementioned fusion reactor concepts: EU-DEMO HCPB, [14], with RELAP5-3D; EU-DEMO WCLL, [16], with RELAP5/Mod3.3; CFETR, [22], with RELAP5/Mod3.3; K-DEMO, [23], with Multi-dimensional Analysis of Reactor Safety (MARS-KS) [24].

The calculations presented in this paper were performed within the framework of the EUROfusion Work Package Breeding Blanket, by using a modified version of RE-LAP5/Mod3.3 code, [25]. This new extended version was developed at the Department of Astronautical, Electrical and Energy Engineering (DIAEE) of Sapienza University of Rome, in order to enhance the code capability in simulating fusion reactors. New features implemented include new working fluids (lithium-lead, HITEC® molten salt), new heat transfer correlations, etc. The selected transients to be investigated belong to the category of "Decrease in Coolant System Flow Rate". The considered Postulated Initiating Events (PIE) consist in both the partial and complete LOFA occurring either in BZ or in FW PHTS. In addition, the influence of the loss of off-site power, occurring in combination with PIE, was studied.

In the following, Section 2 offers a brief description of EU-DEMO WCLL reactor configuration. In Section 3, it is described the RELAP5/Mod3.3 model developed to simulate the blanket component and the related primary heat transfer systems. Calculation results are collected in Section 4. Full plasma power state is commented in Section 4.1, while transient simulations are fully analyzed in Section 4.2. A final discussion on the main outcomes of the computational activity is reported in Section 5. The conclusive remarks related to the current work are contained in Section 6. Moreover, at the end of the paper, it is provided a list of the main acronyms used in the text.

2. Short Overview of EU-DEMO WCLL Reactor Configuration

DEMO reactor normal operations are characterized by a pulsed operating regime. It consists in eleven pulses per day, each one made up of a full-power burn time (pulse) of two hours and a dwell time of 10 min [6]. The reference parameters and baseline are those of DEMO 2017 concept [2]. The reactor Computer Aided Design (CAD) model is shown in Figure 1 [2,4,5], including all the PHTS components located inside and outside the Vacuum Vessel.

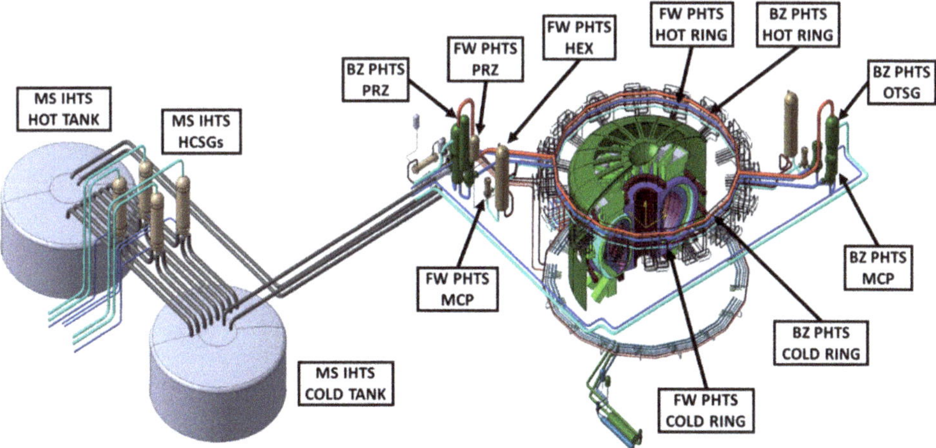

Figure 1. Reactor CAD model, DEMO 2017 baseline [2,4,5]. Overview of the tokamak, the primary cooling systems and the molten salt intermediate circuit and storage tanks.

The reference design adopted for DEMO blanket is the WCLL2018.v0.6, based on Single Module Segment (SMS) approach [2,4]. The overall component is constituted by 16 identical sectors, each one occupying 22.5° in the toroidal direction. Each sector is furtherly divided in In Board (IB) and Out Board (OB) blankets, located radially inwards and outwards with respect to the plasma chamber. At its time, OB is toroidally composed by 3 SMSs named Left OB (LOB), Central OB (COB) and Right OB (ROB), while IB is partitioned only in two SMSs, called Left IB (LIB) and Right IB (RIB). In conclusion, five segments are associated to each DEMO sector.

Each single segment is made up of about 100 breeding cells (BRC), distributed along the poloidal (vertical) direction. The BRC layout is differentiated between segments (especially between OB and IB segments) and, in the same segment, varies according to the poloidal position. The BRC design used as reference for modelling purposes is the one of the COB equatorial cell. Its detailed description can be found in [26,27]. In the BRC, the component facing the plasma chamber is called First Wall (FW). It is protected by a tungsten armor and cooled with water flowing in square channels equally distributed along the poloidal height. The liquid lead-lithium (LiPb) acts as breeder. It enters the BRC from the bottom, flows in the radial direction, from the BRC Back Plate (BP) to the FW, rises poloidally and then turns back radially, from the FW to the BP, exiting through an outlet pipe. The breeder zone refrigeration is assured by a batch of radial-toroidal C-shaped Double Walled Tubes (DWTs). They are displaced in horizontal planes at different poloidal elevations and are split into three arrays along the radial direction. In this way, their cooling capability is uniformly distributed in all the BZ volume. The back part of the breeding cell in the radial direction is devoted to house both LiPb and water manifolds. Finally, the back supporting structure is a continuous steel plate in poloidal direction representing the backbone of the blanket segment. The layout for COB equatorial cell is shown in Figure 2 [2,4].

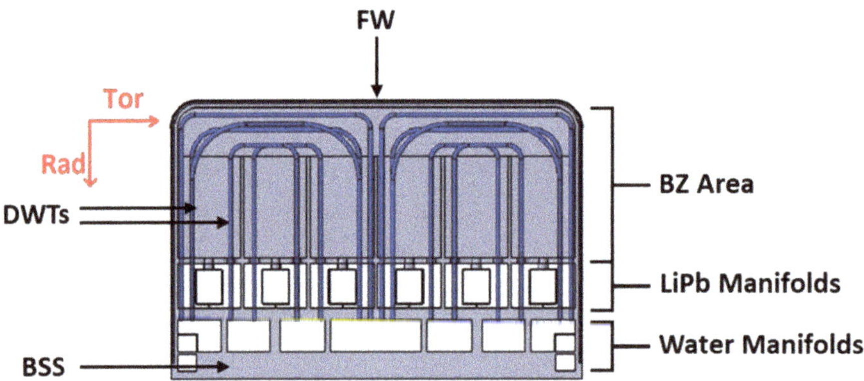

Figure 2. EU-DEMO Water Cooled Lithium-Lead (WCLL) blanket component: detail of the Central Outboard (COB) equatorial breeding cell, WCLL2018.v0.6 design [2,4].

The blanket component is provided with two independent cooling systems: the BZ PHTS and the FW PHTS. The former removes the nuclear heat generated in the breeder zone by the interactions between the lead-lithium and the neutrons coming from the plasma. The latter cools the FW component which is heated up by the incident Heat Flux and by the neutron wall load. The simulation activity presented in this paper is related to the indirect coupling option [4,6]. In this configuration, the BZ PHTS delivers thermal power directly to the PCS, by means of two Once-Through Steam Generators (OTSG). Instead, the FW PHTS, thanks to two water/molten salt Heat EXchangers (HEXs), is connected to an Intermediate Heat Transfer System (IHTS) provided with an Energy Storage System (ESS). The ESS function is to flatten the pulsed source term (plasma power), according to the design requirement of continuous and nearly constant electrical power delivered to the grid. The ESS accumulates a fraction of the FW thermal power during the plasma pulse and delivers it to the PCS during the dwell time. The power fraction to be accumulated during pulse is calculated to obtain a constant turbine load during the overall operating regime (pulse and dwell). The energy storage is constituted by a system of two tanks filled with molten salt at different temperatures. During pulse, there is a net HITEC® flow rate going from the cold tank to the hot one and here accumulated. During dwell, the hot molten salt flows through four Helicoidal Coil Steam Generators (HCSGs) and power is delivered to the PCS. The current PHTS design foresees two loops for each system. They are symmetrically disposed along the tokamak circumference (i.e., toroidal direction). The main PHTS components (for both BZ and FW cooling circuits) are:

- The hot and cold rings, circular collectors (hot) and distributors (cold) of the overall PHTS mass flow from/to the loops and to/from each of the tokamak sectors, respectively.
- The sector manifolds, differentiated in collectors and distributors, respectively connecting the tokamak sectors to the hot ring and the cold ring to the tokamak sectors.
- The loop piping (hot legs, cold legs, loop seals), linking the main vessel components.
- The BZ OTSGs and the FW HEXs.
- The MCPs, providing the primary coolant flow.
- The pressurizer system, one per PHTS, ensuring the pressure control function.

The location of each component in the overall cooling systems is shown in Figure 1. For modelling purposes, the PCS and IHTS system sections considered are only the BZ OTSGs and FW HEXs secondary sides.

3. RELAP5 Thermal-Hydraulic Model

Referring to the reactor configuration outlined in the previous section, a full model of the DEMO WCLL BB PHTS was prepared to perform transient calculations. The main modelling approach considered while developing the input deck was the "slice nodalization" technique. This means that a common vertical mesh was used for all the system components at the same elevation. In addition, the node-to-node ratio, defined as the ratio between the length of two adjacent control volumes (CVs), was kept below 1.25 in the entire model. The respect of this upper limit represents an important criterion to avoid numerical errors due to an inhomogeneous mesh. For all the vessel components and piping, actual design elevations were strictly maintained to avoid inconsistencies mainly in the evaluation of the natural circulation. Fluid and material inventories were rigorously maintained for both BB and PHTS cooling systems.

3.1. Blanket Model

From the hydrodynamic point of view, the BZ and FW cooling circuits were independently simulated. Nevertheless, the two systems are thermally coupled inside the BRC. For this, RELAP5 heat structure components were used to simulate in detail the heat transfer phenomena taking place within the breeding cell. During transient simulations, the BZ and FW thermal coupling has a significant influence on the circuit TH behavior.

As already pointed out, each DEMO sector is constituted by five poloidal segments (three for OB and two for IB). The BZ and FW cooling circuits here contained were collapsed in some equivalent pipe components, three for each PHTS. The OB and IB segments were grouped as following: LOB/ROB, COB, LIB/RIB.

For both BZ and FW PHTS, the equivalent pipes model the overall water flow path inside the vacuum vessel. The components associated to each segment and considered for simulation purposes are: (1) inlet Feeding Pipe (FP); (2) inlet spinal water manifold; (3) DWTs or FW channels; (4) outlet spinal water manifold; (5) outlet FP. The CVs belonging to the equivalent pipes are characterized by different hydraulic properties (flow area, hydraulic diameter, etc.) in order to properly simulate all the aforementioned components. For the equivalent pipes corresponding to LOB/ROB and LIB/RIB, the CVs flow area and hydraulic diameter, as well as the water mass flow, were evaluated considering the reference data belonging to both segments. In this way, the pressure drops through these components were correctly modelled. The PHTS sector collectors and distributors, mentioned in Section 2, are connected to the FPs thanks to inlet and outlet manifolds, closing the overall PHTS circuit. In conclusion, for each PHTS (either BZ or FW) and for each sector, five equivalent pipes and two branches were used. Pipe components correspond to: sector distributor (P1); water circuit inside LOB and ROB (P2); water circuit inside COB (P3); water circuit inside LIB and RIB (P4); sector collector (P5).

Regarding the BRC, the most studied design belongs to the cell located at the equatorial plane of COB [2,26,27]. For this reason, it was adopted as reference and also used for all the other BRCs poloidally distributed along the overall segment. Concerning the BRCs of ROB, LOB, LIB, RIB segments, the reference layout was scaled by using the material inventories derived from the CAD model [2,4,5].

About the DWTs, since these components are in parallel within the BRC, they were collapsed and modelled by using the central batch of CVs of P2, P3, P4 equivalent pipes, the ones related to BZ PHTS. As discussed in Section 2, they are split into three arrays along the radial direction. Moreover, their C-shape in the radial-toroidal plane changes according to the array they belong to [26,27]. The complexity of the geometry requires the choice of a reference DWT layout. For this purpose, the second array was selected, that is the mid-one along the radial direction. It was considered sufficiently representative of the average geometrical features of all the DWTs present in the BRC.

BZ and FW inlet/outlet spinal water manifolds consist in rectangular channels running along the back of the segment, radially inwards with respect to the back supporting structure (see Section 2). They follow the SMS curved profile. In the TH model prepared, the design height difference between heat source and heat sink (the BZ OTSGs and the FW HEXs) thermal centers was maintained. This parameter is of primary importance in all the transients concerning natural circulation, such as LOFA. Manifold-simulating CVs are located before (inlet) and after (outlet) the ones modelling the DWTs/FW channels. In [2], the COB manifold layout is described. In a first approximation, this design was also used for the pipes simulating LOB/ROB and LIB/RIB segments. For any segment, CVs flow area was calculated to maintain the BZ and FW water manifolds inventory. CVs hydraulic diameter was evaluated based on the effective manifold layout.

The RELAP5 heat structure (HS) components were used in the input deck to accomplish several functions: account for the BB solid material inventories (tungsten and EUROFER97); simulate the breeder (simplifying the input); introduce the power source terms (heat flux and nuclear heating); represent the heat transfer phenomena taking place within the BRC; model the pipeline thermal insulation (for sector collectors/distributors and inlet/outlet FPs).

The lithium-lead flow path through the blanket was not modelled in this work from a hydrodynamic point of view. The breeder velocity inside the component is very low [2]. Within the BRC, where the thermal exchange between LiPb and DWTs/FW channels is significant, the breeder convective Heat Transfer Coefficient (HTC) was neglected and only the conductive heat transmission was considered, simulating the lithium-lead as a layer of structural material in the RELAP5 HS components.

A HS was used to simulate the FW front surface. A tungsten layer and a Eurofer thickness were modelled. The Eurofer thickness is the one between the plasma chamber and the FW cooling channels. The heat flux reported in [26] was applied as boundary condition for the plasma-facing surface. An average value was adopted since no poloidal differentiation was considered in the model. The radial segments of the FW component were simulated with a separate HS. In this case, only a Eurofer thickness was considered since the tungsten armor is present only in the front surface. To take into account the heat transfer between FW channels and DWTs inside the BRC, a HS was added. As already discussed, in the radial-toroidal plane, DWTs are divided into three arrays with different layouts. The same DWT reference layout chosen for the hydrodynamic model was used in the thermal problem also. The radial distance between the FW cooling channels and the selected DWT is composed by: a first Eurofer layer, representing the FW thickness between FW cooling channels and FW internal surface; a LiPb layer, corresponding to the radial distance between the FW internal surface and the selected DWT layout; a second Eurofer layer, modelling the DWTs thickness. This HS allows to thermally couple the BZ and FW cooling circuits. Heat transfer between DWTs and LiPb inside BRC was also modelled with a dedicated HS. Two further HSs were used to account for the Eurofer inventory in the water and LiPb manifold region and in the back supporting structure, respectively.

Nuclear heating associated to the aforementioned HSs was computed thanks to the power density radial profiles presented in [27] and by considering the actual materials inventory distribution within the BRC. It was introduced in the input deck as an internal power source term, differentiated for each HS. For each sector, the batch of HSs described so far (six) was replicated for LOB/ROB, COB and LIB/RIB (for a total of 18).

The pipeline heat losses were modelled considering a constant containment temperature (30 °C), and a constant heat transfer coefficient (8 W/m^{2}K). A schematic view of the BB nodalization is provided in Figure 3. The model shown refers to only one of the sixteen identical toroidal sectors. For the correspondent hydrodynamic components, the figure reports also the identification numbers used in the input deck.

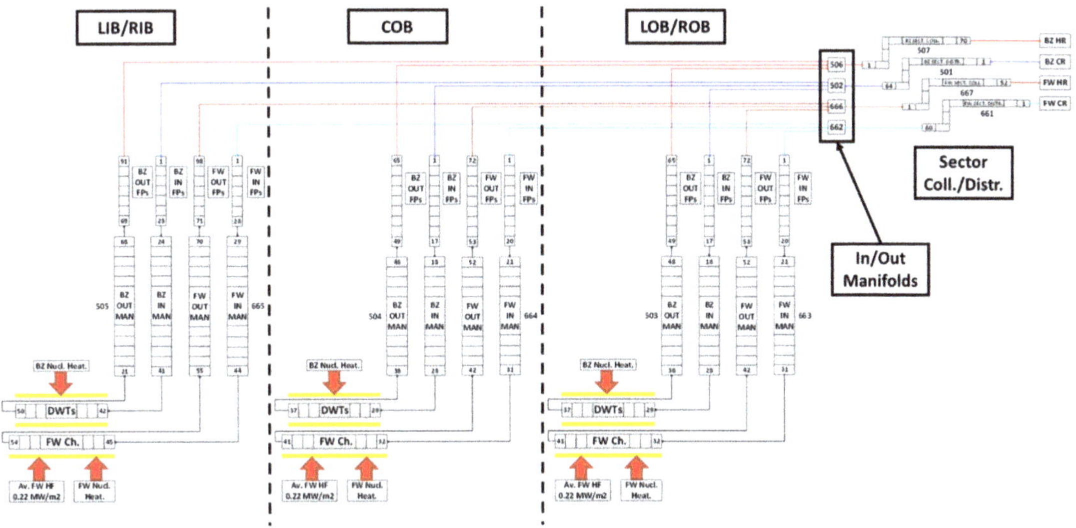

Figure 3. Schematic view of the blanket model, sector one of sixteen. All the Breeder Zone (BZ) and First Wall (FW) Primary Heat Transfer Systems (PHTS) in-vessel components are represented. In addition, red arrows indicate the power source terms.

3.2. PHTS Model

The routing of the BB PHTS pipelines was derived from the current CAD model [2,4,5]. K-loss coefficients for tees, elbows and area changes were calculated by using formulas in [28]. They were associated to pipe component internal junctions to correctly evaluate these minor head losses, when present. To each pipeline corresponds a RELAP5 pipe component, except for the hot and cold rings. For them, four pipes and two multiple junctions were used. Each pipe simulates a quarter of the ring (90°). One multiple junction component manages the connections between pipes (to close the ring) and between the rings and the hot/cold legs. The other multiple junction component links the hot/cold rings with the sector collectors/distributors. These connections are equally distributed along the overall ring length to maintain the toroidal symmetry characterizing the DEMO reactor. Pipeline modelling is contained in Figures 4 and 5, respectively related to the BZ PHTS loop 1 and FW PHTS loop 1. An example of ring nodalization is shown in Figure 6. For the hydrodynamic components reported in each figure, the identification numbers used in the input deck are also indicated. Pipeline thermal insulation was modelled associating a heat structure to each pipe component. The external surface boundary condition for these HSs is the tokamak building atmosphere, modelled with a constant temperature and HTC, as already discussed in Section 3.1.

The BB PHTS pump system consists of six (four for the BZ and two for the FW) centrifugal single stage pumps. They are equally divided in the two loops constituting each PHTS. The MCPs were modelled by using RELAP5 pump components provided with a proportional-integral (PI) controller to set the design mass flow value.

The BZ OTSGs design foresees PHTS water flowing inside the tube bundle and PCS water flowing in shell side. A mesh length of 0.26 m was selected for these components in both (primary and secondary) sides. The details about the nodalization are reported in Figure 4.

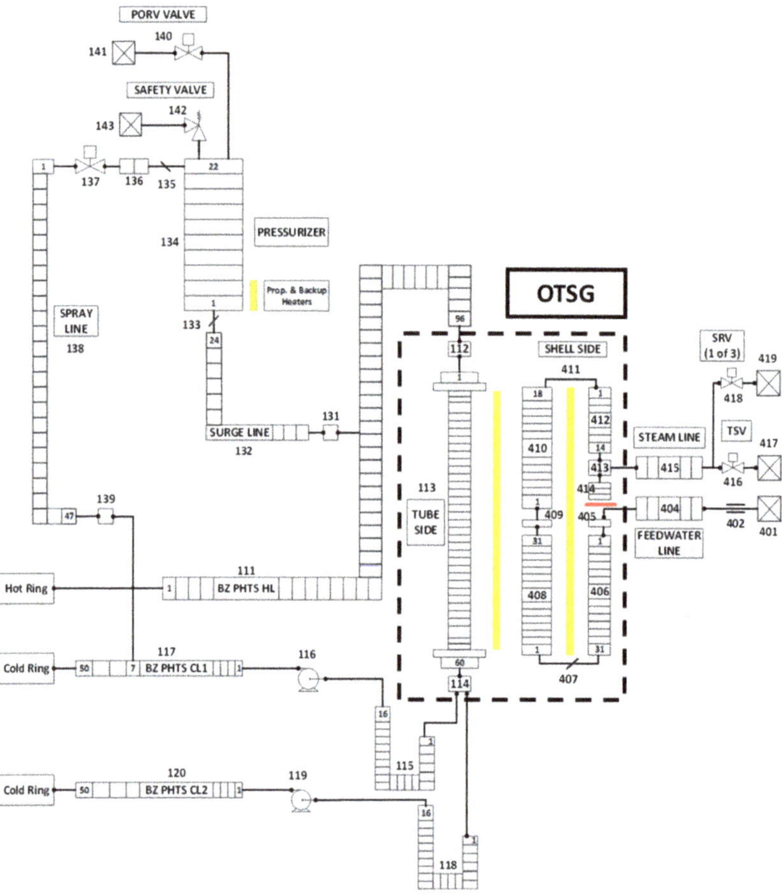

Figure 4. Schematic view of BZ PHTS model, loop one of two. Pressurizer system components are unique and connected only to the represented loop.

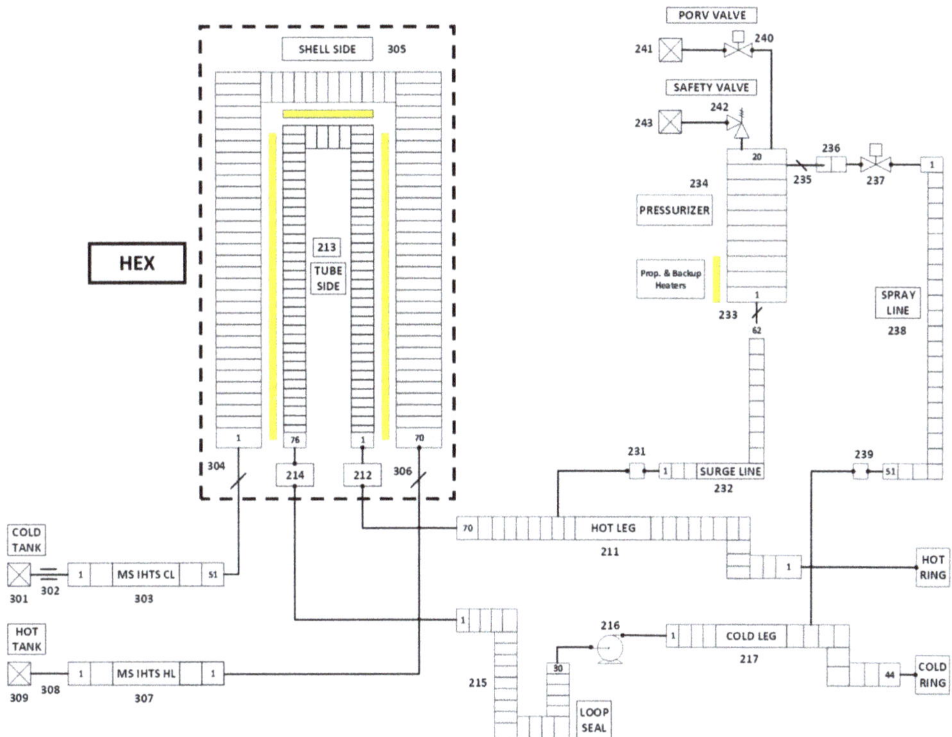

Figure 5. Schematic view of FW PHTS model, loop one of two. Pressurizer system components are unique and connected only to the represented loop.

Each OTSG is provided with two steam lines to avoid excessive pipeline pressure drops due to steam velocity. Feedwater line was simulated with a time-dependent volume and a time-dependent junction to set the PCS water inlet thermodynamic conditions, and with a pipe to simulate the pipeline section before the OTSG entrance. Steam lines were modelled up to the Turbine Stop Valves (TSVs) and equipped with steam line Safety Relief Valves (SRVs). PCS SRVs consists in three steps of relief valves provided with increasing setpoint: 90%, 95% and 100% of the PCS system design pressure (115% of the operating pressure reported in [4–6]). The step 1 relief valves were sized to discharge the 75% of the OTSG steam mass flow, considering chocked flow occurring in the valve throat section, while step 2 and step 3 to discharge the 37.5%. Hence, the full set of SRVs is able to discharge the overall OTSGs steam mass flow with an additional conservative margin of 50%. Main data related to PCS SRVs are collected in Table 1. A schematic view of the BZ OTSGs nodalization is shown in Figure 4. RELAP5 heat structures were used to simulate the thermal transfer taking place within steam generators, as well as the component heat losses. Furthermore, they allow to account for the OTSGs steel inventory (i.e., thermal inertia).

FW HEXs are pure countercurrent heat exchangers with PHTS water flowing inside tube bundle and IHTS molten salt flowing in shell side. The adopted CV length is 0.41 m. For each FW HEX, also the IHTS hot and cold legs were modelled. Cold leg was connected to a boundary condition to set the HITEC® inlet temperature and mass flow rate. The FW HEXs nodalization is shown in Figure 5. Also in this case, heat structures were used to simulate the heat transfer phenomena, the heat losses and the steel inventory related to each heat exchanger. The molten salt HTC was calculated with Sieder-Tate correlation, [29].

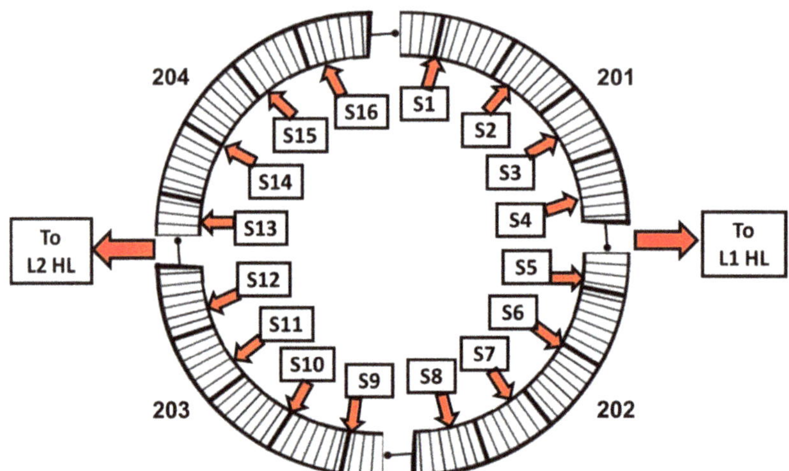

Figure 6. Example of ring nodalization, FW hot ring. The component connections with the sector collectors and loop hot legs are reported.

Table 1. Power Conversion System (PCS) steam line Safety Relief Valves (SRV) features.

Parameter	Unit	Step 1 SRVs	Step 2 SRVs	Step 3 SRVs
Throat section	m^2	1.80×10^{-2}	9.00×10^{-3}	7.50×10^{-3}
Area change rate [1]	s^{-1}	10	10	10
Opening setpoint	bar	66.4	70.0	73.7
Closing setpoint	bar	64.1	64.1	64.1

[1] Valve area change rate is the reciprocal of the valve opening/closing time.

The time-dependent junctions located on the BZ OTSGs feedwater lines and FW HEXs IHTS side cold legs were provided with temperature control systems. They are required to obtain the design PHTS water temperature at BB inlet [2,4–6]. The BZ OTSGs and FW HEXs designs were performed considering that they must exchange their nominal power when operating at End Of Life (EOL) conditions. For this, both tube fouling and tube plugging phenomena were taken into account. At Beginning Of Life (BOL) conditions, when no tube plugging and fouling factors are foreseen, the OTSGs and HEXs exchanged power exceeds the nominal value. This causes a significant alteration of the temperature field in the overall PHTS system and, in particular, at BB inlet. To keep the PHTS parameters at the design values in BOL condition, a control system is required. It was developed to ensure constant water thermodynamic conditions at BB inlet in any operational condition. PHTS temperature is read at OTSG outlet and then compared with a temperature target setpoint [2,4–6], producing an error. The error signal is scaled by using a PI controller. The controller output range goes from zero to 110% of rated PCS feedwater mass flow at EOL condition, [4–6]. The resulting output is the mass flow imposed by the time-dependent junction simulating the BZ OTSG secondary side inlet. The same control logic is applied to loop 2 OTSG and to both FW HEXs.

In each PHTS circuit, the pressurizer system guarantees the pressure control function, maintaining the water pressure at the required value independently on the temperature variations of the coolant induced by the pulsed plasma operation and, in general, by other transient conditions. The main component of this system is the steam bubble pressurizer (PRZ), connected to the loop 1 hot leg by means of a surge line. Since the water thermodynamic conditions are similar, for both BZ and FW PHTS, the pressurizer volume was scaled from PWR design, [30]. The scaling factor adopted was based on the ratios between circuit total inventories and reactor total thermal power. A further safety margin was applied and the resulting component size increased. The tank and the surge line were both simulated with a pipe component. The associated heat losses were modelled with passive heat structures. The pressurizer is equipped with On/Off and proportional electric heaters and a spray line connected to the loop 1 cold leg and controlled by a valve. These systems are installed to face, respectively, under and overpressure transients occurring during both normal operations and abnormal conditions. The proportional heaters are set to operate in a range of pressure around the PHTS loop reference one. These heater banks are supplied by a varying input current that is a function of the pressure deviation signal. Normally, these components are energized at half current when pressure is at nominal value (null error), are cut off when this parameter reaches the higher setpoint and are at full power with pressure at lower setpoint. Instead, pressurizer backup heaters are normally de-energized heater banks turning on if pressure drops below the setpoint adopted for this component (lower than the one of the proportional heaters). They are simply on-off type with no variable control. The heaters electrical power was scaled from PWR design, [30], by using a scaling factor based on reactor thermal power and applying a safety margin. Pressurizer heaters were simulated with active heat structures. The spray valve controller is set to modulate the valve flow starting from a lower setpoint up to a higher one correspondent to the fully open status. Pressurizer sprays operate to prevent lifting of the relief valve. The cold leg water admitted through these components is extremely effective in limiting pressure increases during transient or accident conditions. The correspondent flow capacity was sized by scaling from PWR design [30]. The surge line and spray line routing was derived from CAD model [2,4,5], and rigorously maintained. In case of abnormal transients, if spray nozzles fail in reducing pressure, at the top of pressurizer is also foreseen the presence of a Pilot (Power)-Operated Relief Valve (PORV) and an SRV. A dedicated line connects these components to the pressure relief tank, allowing the discharge of steam. The PORV is provided for plant operational flexibility and for limiting the number of challenges to the pressurizer SRV. For this reason, the former is provided with a lower setpoint than the latter. PORV and SRV were modelled with RELAP5 valve components.

The overall nodalization used for BB PHTS pressurizer system is shown in Figures 4 and 5, for BZ and FW, respectively. The main design data related to both BZ and FW PHTS pressurizer systems are contained in Table 2. The pressure control function setpoints, chosen considering the PWR design [31], are gathered in Table 3.

Table 2. BZ and FW PHTS pressurizer system features.

Parameter	Unit	BZ PHTS	FW PHTS
Pressurizer volume	m^3	101.5	39.3
Proportional heater bank power	kW	1200	800
On/Off heater bank power	kW	2400	1600
Spray line flow capacity	kg/s	36.2	17.9
PORV throat section	m^2	1.84×10^{-3}	1.52×10^{-3}
SRV throat section	m^2	1.84×10^{-3}	1.52×10^{-3}
PORV/SRV area change rate [1]	s^{-1}	10	10

[1] Valve area change rate is the reciprocal of the valve opening/closing time.

Table 3. BZ and FW PHTS pressure control function.

Parameter	Unit	BZ PHTS	FW PHTS
Reference Pressure	bar	155	155
Proportional heater bank lower setpoint	bar	154	154
Proportional heater bank higher setpoint	bar	156	156
Back-Up heater bank on/off setpoint	bar	154	154
Spray Valve start opening setpoint	bar	157	157
Spray valve fully open setpoint	bar	160	160
PORV valve opening setpoint	bar	170	170
PORV valve closing setpoint	bar	165	165
SRV opening setpoint	bar	178	178
SRV closing setpoint	bar	173	173

4. Results

4.1. Full Plasma Power State

The RELAP5 model described so far was used to perform a steady-state simulation of full plasma power state at Beginning Of Life (BOL) condition. During DEMO normal operations, this is the most challenging scenario for the BB PHTS, as confirmed by results presented in [14]. For this reason, such state was chosen as initial condition for the accidental transient calculations discussed in the following sections. The full thermal-hydraulic characterization of BZ and FW primary cooling systems during this scenario is reported in Table 4. The parameters with the indication "BC" were imposed as boundary conditions for the calculation. The mass flow and temperature control systems implemented in the input deck are able to guarantee the required thermodynamic conditions at the BB inlet. Table 4 also indicates the pump head provided by MCPs and the power terms associated with each PHTS. Simulation outcomes are in good accordance with reference data derived from [2,4–6]. Minor discrepancies in the OTSGs/HEXs secondary side parameters are due to the fact that the sizing of these components was performed at EOL, as discussed in Section 3.2. A time step sensitivity was carried out, varying this parameter from 1.0×10^{-3} s to 1.0×10^{-2} s. No sensible differences were observed in the results. Values in Table 4 are for a time step of 5.0×10^{-3} s.

Table 4. Full plasma power state: main thermal-hydraulic parameters related to BZ and FW PHTS, Intermediate Heat Transfer System (IHTS) and PCS. Comparison between simulation results and reference data derived from [2,4–6].

	Parameter	Unit	BZ PHTS		FW PHTS	
			Simulation Result	Reference Data	Simulation Result	Reference Data
PHTS	Mass Flow (per MCP)	kg/s	1915.6	1915.6	1136.8	1136.8
	Hot Leg Temperature	°C	328.2	328	328.1	328
	Cold Leg Temperature	°C	295	295	295	295
	MCP Head	MPa	0.94	0.95	0.83	0.84
	PRZ pressure	MPa	15.6	15.5	15.6	15.5
PCS/IHTS	Feedwater/HITEC® Mass Flow	kg/s	392.8	404	2955.6	3524
	Feedwater/HITEC® Inlet Temp. (BC)	°C	238	238	280	280
	Steam/HITEC® Outlet Temperature	°C	314.6	299	327.7	320
Power Terms	Power removed from blanket	MW	1482.4	1483.2	438.3	439.8
	Power exchanged at OTSG/HEX(per component)	MW	744.5	741.6	219.8	219.9
	MCPs Total Power	MW	11.9	12.1	3.0	3.2
	PRZ Heaters Power	MW	1.08×10^{-2}	-	4.87×10^{-3}	-
	Total System Heat Losses	MW	0.6	-	0.45	-

4.2. Transient Analysis

4.2.1. Selected Cases

The BB PHTS response during accidental conditions was investigated. The calculations are system analyses aimed at understanding the primary cooling circuits TH behavior during such transients. As previously stated, full plasma power state was used as initial condition. The selected PIEs are partial and complete Loss of Flow Accident (LOFA). These accidental scenarios were studied when occurring in both BZ and FW PHTS. Simulations were replicated also considering the influence of loss of off-site power, occurring in combination with PIE. The matrix of all the transient simulations performed in the framework of the current computational activity is represented in Table 5.

Table 5. Matrix of transient simulations performed.

Case ID	PIE	System Involved With PIE	Loss of Off-Site Power [Yes/No]
LF1	Partial LOFA	FW PHTS	no
LF2	Complete LOFA	FW PHTS	no
LF3	Partial LOFA	BZ PHTS	no
LF4	Complete LOFA	BZ PHTS	no
LF5	Partial LOFA	FW PHTS	yes
LF6	Complete LOFA	FW PHTS	yes
LF7	Partial LOFA	BZ PHTS	yes
LF8	Complete LOFA	BZ PHTS	yes

4.2.2. Selected Boundary Conditions and PHTS Actuation Logic

The LOFA PIE is the partial or complete loss of primary coolant flow in BZ or FW PHTS, according to the case considered (see Table 5). Primary pumps coast-down is ruled by the torque-inertia equation reported below.

$$T_{em}\,(\omega) - T_{hyd}\,(\omega) - T_{fr}\,(\omega) = I{\cdot}d\omega/dt \tag{1}$$

In the previous equation, $T_{em}\,(\omega)$ is the motor electromagnetic torque, that during coast-down is zero, $T_{hyd}\,(\omega)$ is the hydraulic torque due to system pressure drops, $T_{fr}\,(\omega)$ is the pump frictional torque due to losses inside the MCP component, ω is the rotational velocity and I is the pump moment of inertia. In the framework of Work Package Balance Of Plant 2020 computational activity, [14], a complete LOFA in both BZ and FW systems (worst possible scenario) was studied. The analysis was aimed at evaluating the required flywheel to be added to BB MCPs in order to obtain the best PHTS and blanket TH performances during the accidental evolution. For this reason, a sensitivity was carried out on this parameter. The selected values for pump moment of inertia were: 3000 kg·m^2 for BZ MCPs and 1573 kg·m^2 for FW MCPs (case 4 in [14]). These parameters were adopted for all the transient simulations involved in the current transient analysis.

An actuation logic, involving some components of the DEMO reactor, was proposed and preliminary investigated. It is inspired by the one used for Generation III + nuclear power plants. The following features were implemented:

- Plasma termination (PT) is actuated by one of the following signals: (i) low flow on BB MCPs (<80% of rated value); (ii) high pressure on BB PRZs (>167 bar); (iii) high temperature at BZ/FW outlet FPs (2 °C below the saturation temperature at the PHTS reference pressure).
- Turbine Trip (TT) is triggered by one of the following signals: (i) PT signal; (ii) low steam flow at OTSGs outlet (<85% of rated value); (iii) low steam temperature at OTSGs outlet (2 °C above the saturation temperature at the PCS reference pressure).
- TT is followed by: (i) PCS feedwater ramp down; (ii) TSVs closure.
- PHTS pressurizer heaters are cut off: (i) on low-level signal in BB PRZs; (ii) following TT signal.

- Spray line flow is interrupted only when all the MCPs belonging to a primary cooling system are off. The hypothesis is that redundant spray lines are connected to both PHTS loops.

The margin adopted for the temperature signals was selected to take into account the typical uncertainty related to a thermocouple reading. For what concerns the BB MCPs trip, different strategies were considered whether or not the loss of off-site power is assumed. If not, for a BZ or FW primary pump, MCP trip can occur following: (i) PIE event; (ii) high-temperature signal at pump inlet (5 °C below the saturation temperature at the PHTS reference pressure). The margin was chosen to avoid cavitation in the component in any transient scenario. If loss of off-site power is assumed, to the previous conditions it is also added the TT signal, since, in this scenario, the turbine is the only element ensuring the Alternating Current (AC) power needed for the MCPs operation. The PI controller associated to BZ and FW primary pumps and used in the full plasma power steady-state simulation is disabled. The rotational velocity is imposed as a constant boundary condition until the MCP trip is not triggered. From this moment, the component coast-down is ruled by the torque-inertia equation reported above.

Also, the management strategy for MS IHTS mass flow was differentiated according to the presence or not of off-site power. If available, HITEC® mass flow is ramp down 10 s after the PIE. Conservatively, it is assumed that the PIE occurs at the end of plasma pulse when the ESS cold tank is nearly empty. Hence, also the HITEC® mass flow must be stopped shortly after the Start Of Transient (SOT). If off-site power is lost, IHTS mass flow is ramp down also following the TT signal (the previous condition is still used). In fact, in this scenario, the turbine is the only element ensuring the AC power needed for the molten salt pumps operation.

The temperature control systems adopted for the full plasma power scenario and related to PCS feedwater and IHTS mass flow are disabled. These parameters are imposed by means of time-dependent junctions and respond to the actuation logics previously described. As a preliminary tentative, their ramp-down is simulated with a linear trend going from nominal value to zero in 10 s. Steam line TSVs are supposed to close in 0.5 s. The plasma ramp-down curve is derived from [32] and reported in Table 6. The relative trend should be applied to both nuclear heating and incident heat flux. It lasts 42 s, after which only decay heat is left (nearly 1% of the reactor rated power).

Table 6. Plasma ramp-down curve: tabulation of relative power values vs time.

Time from Plasma Shutdown [s]	Rel. Power [1] [–]	Time from Plasma Shutdown [s]	Rel. Power [–]
0	1.000	26	0.382
2	0.943	28	0.348
4	0.887	30	0.315
6	0.832	32	0.284
8	0.779	34	0.256
10	0.728	36	0.229
12	0.678	38	0.205
14	0.631	40	0.182
16	0.584	42 [2]	0.162
18	0.540	44	0.019
20	0.498	45	0.017
22	0.457	1 h	0.009
24	0.419	1 day	0.002

[1] Relative values refer to nominal power in full plasma power state. [2] This is the end of the ramp down curve. Next value belongs to decay heat trend.

The initiating event occurs after 100 s of full plasma power state (grey background in the figures of Sections 4 and 5). Timeline was reset in the plots to have PIE at 0 s. Transient calculation was run for 9000 s (2.5 hr), for an overall simulation time of 9100 s. Different time steps were adopted in the calculation. In the first part of the transient, when thermal excursions are expected to be more significant, a lower time step was used (5.0×10^{-3} s). In the final part, this parameter was increased (1.0×10^{-2} s) to speed up the simulation.

4.2.3. LOFA Transients Involving FW Cooling Circuit
FW System Transient Evolution

After PIE, FW PHTS primary flow starts to decrease. In LF1 and LF5 cases, initiating event involves only loop 1 MCP (partial LOFA), instead, in LF2 and LF6 sequences, both loop pumps are stopped (complete LOFA). Low flow is detected shortly after the SOT and plasma termination is triggered. Consequently, also turbine trip is actuated. In LF5 scenario, where loss of off-site power is assumed, this causes the stop of the loop pump not interested from PIE. For this reason, in LF2, LF5 and LF6 transients, the coast-down of both loop pumps is nearly contemporaneous and these cases have a quite similar accidental evolution. Case LF1 differs from the others since loop 2 MCP continues to provide primary flow up to nearly the End of Transient (EOT). A summary of the transient calculations characterized by PIE involving FW pumps is offered by Table 7.

Table 7. Summary table for Loss of Flow Accidents (LOFA) involving FW PHTS Main Coolant Pumps (MCPs).

Event/Parameter	Unit	LF1	LF2	LF5	LF6
PIE (LOFA)	-	Partial (FW)	Complete (FW)	Partial (FW)	Complete (FW)
Loss of off-site power	yes/no	no	no	yes	yes
PT signal occurrence	s	1.5	1.5	1.5	1.5
TT signal occurrence	s	1.5	1.5	1.5	1.5
TSVs start to close	s	1.5	1.5	1.5	1.5
Start of PCS feedwater ramp-down	s	1.5	1.5	1.5	1.5
Start of IHTS mass flow ramp-down	s	10	10	1.5	1.5
Time of FW PHTS water temperature peak	s	15	24	23	23
FW PHTS water temperature peak [1]	°C	329	332	332	332
Time of BZ PHTS water temperature peak	s	-	-	60	60
BZ PHTS water temperature peak [1]	°C	-	-	339	339
FW MCP Trip occurrence (pump not interested by PIE)	s	7696	-	1.5	-
BZ MCPs Trip occurrence	s	7112	7084	1.5	1.5
Time to evacuate BZ OTSGs secondary side inventory	s	500	500	2200	2200
Water mass discharged from BZ OTSGs sec. side	kg	13,281 (per OTSG)	14,718 (per OTSG)	12,795 (per OTSG)	15,465 (per OTSG)
FW PORV first opening time (Long Term)	s	7344	1284	2200	1988
Total FW PHTS water mass discharged at EOT	kg	1301	2094	1398	1352
BZ PORV first opening time (Long Term)	s	6776	6736	4512	4192
Total BZ PHTS water mass discharged at EOT	kg	6724	5831	5482	4417

[1] For all the sixteen sectors, both the BZ and FW PHTS water temperatures were detected at the outlet of COB segment. For each PHTS, peak temperature reported in the table is the maximum among all the temperature readings.

Case LF1

As already stated, loop 2 MCP continues to provide primary flow. The transient results dissymmetrical with respect to the toroidal dimension. The sixteen sectors experience different flows (Figure 7a), with higher values in the ones nearest to the active pump. Consequently, also the PHTS temperatures at BB inlet/outlet are differentiated. Figure 7b reports the values referred to all sixteen sectors. COB segment was chosen as reference to plot simulation results. Forced flow due to loop 2 MCP significantly smooths the temperature peak at BB outlet. The maximum increase (associated to the sectors nearest to the failed pump) is of only one degree (Table 7) with respect to rated value. The temperature excursion is quite negligible.

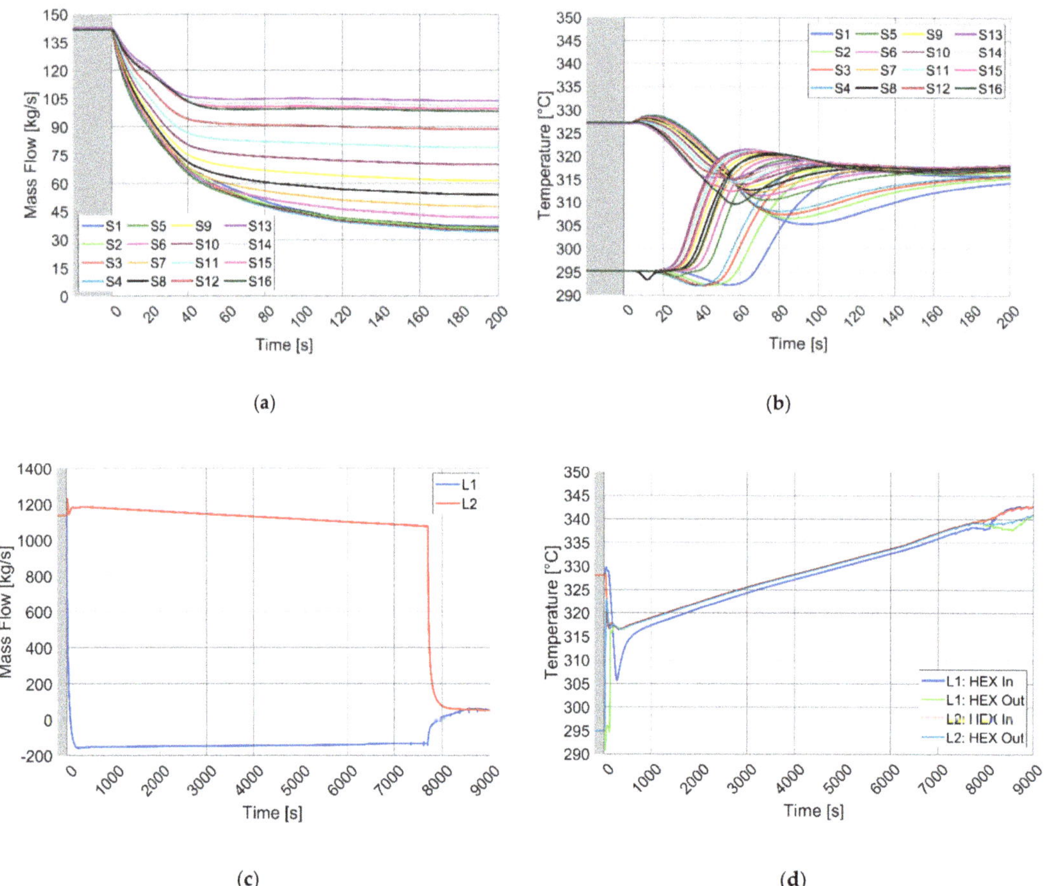

Figure 7. Partial LOFA on FW PHTS without loss of off-site power (LF1 transient): (**a**) Mass flow in FW sectors (early time); (**b**) FW PHTS water temperatures at BB inlet & outlet (all sectors, early time, COB segment); (**c**) Mass flow elaborated by FW PHTS MCPs (full range); (**d**) FW PHTS water temperatures at Heat EXchangers (HEXs) inlet & outlet (full range).

Another interesting effect is the flow inversion in loop 1 (negative mass flow in Figure 7c). The pressure drops related to the blanket component are so high that a part of the flow provided by loop 2 MCP goes through loop 1 in reverse direction instead of flowing in the BB sectors. The reverse flow also causes a temperature inversion in the correspondent loop. After the trip of loop 2 pump (Figure 7d and Table 7), forced circulation is lost and the establishment of natural circulation restores the original temperature field in loop 1. Instead, in the other loop, the forced circulation provokes a quick convergence of the system temperatures. Later, they start to positively drift since BB decay heat overwhelms the system heat losses. The temperature slope is of nearly 12 °C/hr (25 °C in 7500 s). In the case of forced circulation (LF1), the curve slope is higher than the one associated to sequences dominated by natural circulation (LF2, LF5 and LF6, Figure 9c). This can be justified considering that the PHTS coolant is also heated by pumping power. This contribute is of the same order of magnitude of the decay heat. Once loop 2 pump is stopped, when forced circulation is lost and natural circulation establishes, if simulation time were increased, the temperature slope for LF1 scenario would become the same as other transients.

For what concerns the FW PHTS pressure, the presence of the forced circulation (even if reduced with respect to rated value) avoids the challenging of PRZ PORV at SOT (Figure 8a). In the mid-long term, since loop 2 pump is active also pressurizer sprays are still available. The system pressure is kept constant for a long time interval (Figure 8b). During it, with the increase of the system temperature, spray intervention in reducing pressure becomes less and less effective. In fact, from time to time, they introduce in the pressurizer control volume water at higher enthalpy. The level in the component increases almost linearly, as shown in Figure 8c. At a certain point, sprays are unable to perform the pressure control function and the system pressure start to rise triggering the PORV (Figure 8b, for the timing see Table 7). The valve opens when the pressurizer is nearly solid (Figure 8c). From this moment, pressure in the PHTS follows a sawtooth trend due to the PORV periodical openings. This is the way used by FW system to dissipate the decay heat produced in the BB. The total water mass discharged from FW PHTS at EOT is reported in Table 7.

The trend of the maximum Eurofer temperature in the FW component is shown by Figure 8d. After plasma shutdown, the material temperature drops driven by PHTS water temperatures. Instead, in the mid-long term, FW component is heated up by the decay heat and experiences the same temperature slope of PHTS water.

Cases LF2, LF5 and LF6

The FW PHTS mass flows through blanket sectors follow the pump coast-down. It is shown for LF6 sequence in Figure 9b. For all the considered accidental scenarios, as already discussed before, the coast-down of both MCPs is nearly contemporaneous. Hence, these transients result symmetrical with respect to the toroidal dimension. This is clearly visible in Figure 9a reporting the FW PHTS temperatures at BB inlet/outlet (COB segment). Values are plotted for all the sixteen sectors, with a single color for each case considered. Outlet temperatures experience a slight increase due to the short time interval between the occurrence of PIE (i.e., start of pump/pumps coast-down) and the detection of PT signal. After that, since pump coast-down advances more slowly than plasma shutdown (Table 6), outlet temperatures decrease. Peak temperature is the same for all the sectors and for all the cases (Table 7). In LF5 and LF6 scenarios, where loss of off-site power is assumed, IHTS mass flow is ramp down following the turbine trip, while in LF2 sequence it is available for the first 10 s of the transient. As a result, in this latter case, BB inlet temperatures initially decrease (Figure 9a) and restart to increase only after the mass flow ramp down. Instead, in LF5 and LF6 transients, they start immediately to increase, since secondary flow is lost shortly after the SOT. However, apart from this initial difference, the inlet temperatures have a quite similar trend for all the cases.

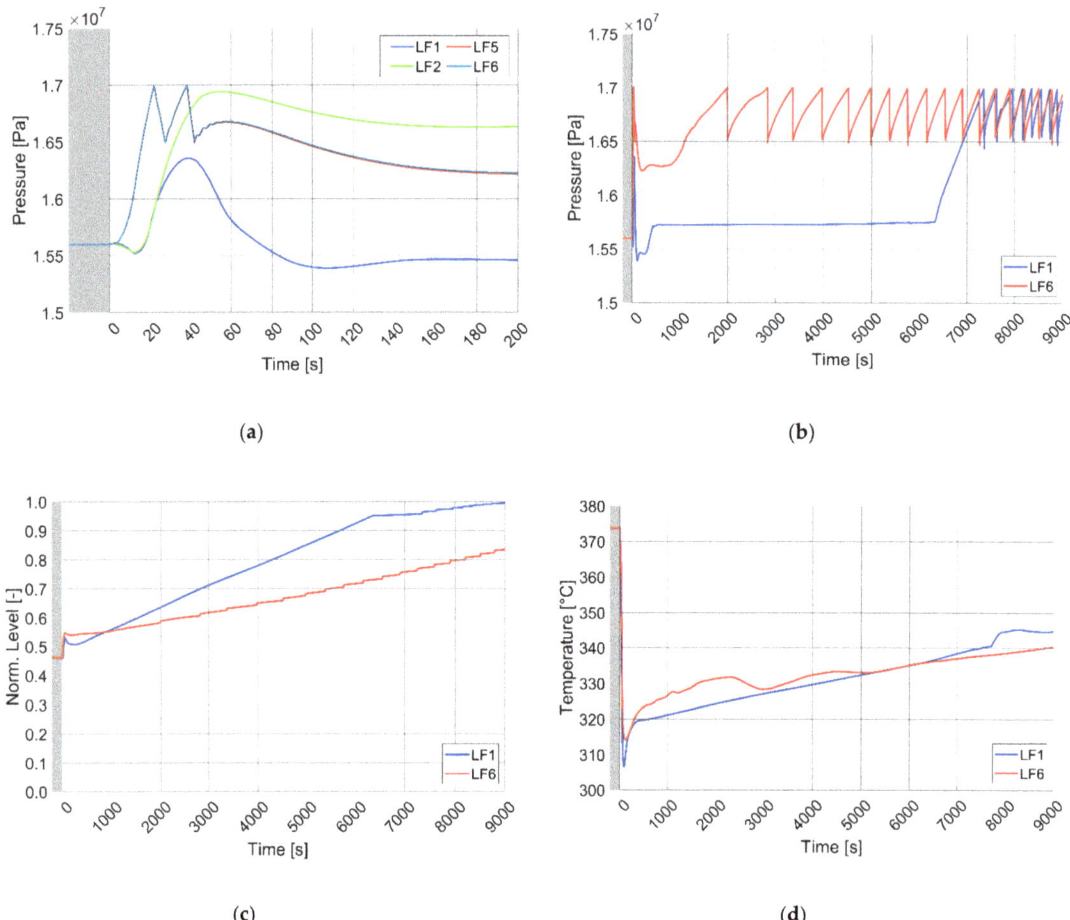

Figure 8. Comparison between LF1 and LF6 transients: (**a**) Pressure in FW PHTS (early time); (**b**) Pressure in FW PHTS (full range); (**c**) Collapsed level in FW pressurizer (normalized, full range); (**d**) Maximum Eurofer temperature in FW component (full range).

The loss of the heat sink also produces a sudden increase in the FW PHTS pressure, as shown by Figure 8a. For LF5 and LF6 sequences, the pressure rise is managed by the PRZ PORV. Instead, in LF1 and LF2 scenarios, the availability of the IHTS mass flow avoids the opening of this component. In the long term, referring to FW PHTS parameter trends, no sensible differences are detected between cases LF2, LF5, LF6. For this reason, only results associated to LF6 sequence were plotted in the figures reported in this section.

During FW pump coast-down system reaches a quite uniform temperature (Figure 9c). It takes a long time interval before the natural circulation establishes in the system. During it, FW temperatures also experience an inversion. Once the natural circulation is completely established, FW temperatures start to positively drift due to the residual decay heat produced in the blanket. The system heat losses are not able to counterbalance this source term. The PHTS temperatures rise of 10 °C in the last 4000 s of simulation with a slope of nearly 9 °C/hr. As discussed before, this parameter is lower than the one observed for case LF1.

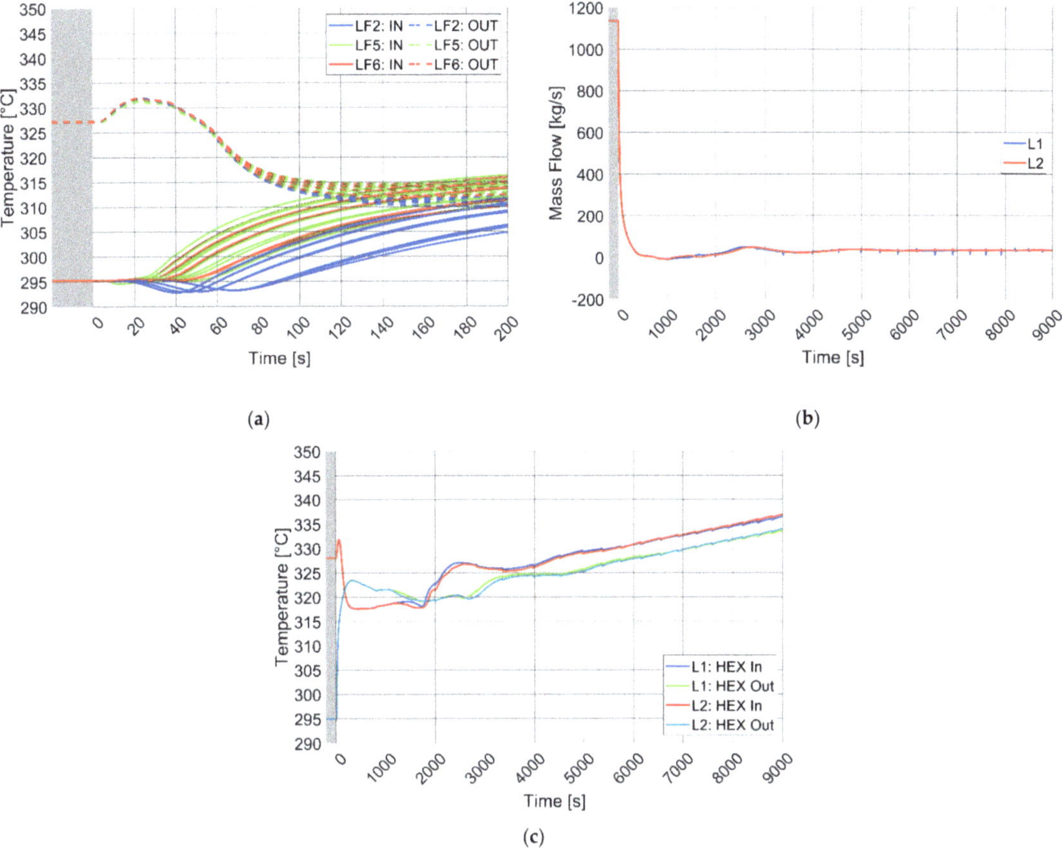

Figure 9. Parameter trends in FW PHTS for LOFA transients characterized by natural circulation (LF2, LF5, LF6): (**a**) FW PHTS water temperatures at BB inlet & outlet (all sectors, early time, COB segment); (**b**) Mass flow elaborated by FW PHTS MCPs (full range, only LF6); (**c**) FW PHTS water temperatures at HEXs inlet & outlet (full range, only LF6).

During accidental evolution, pressure in FW PHTS system increases (Figure 8b). Pressurizer sprays are disabled since all the system pumps are off. Pressure rise continues up to the PORV opening setpoint. With respect to LF1 sequence, the timing of this event is significantly anticipated (Table 7). Later, the system pressure begins to cycle accordingly with the valve component multiple openings. Discharging mass through the PORV is the way adopted by the FW system to dissipate the decay heat produced in the BB. The total amount of water evacuated from FW PHTS at EOT is reported in Table 7. The level in the pressurizer is shown in Figure 8c, normalized with respect to the total height of the component. Pressure rise produces a continuous mass insurge (i.e., level increase) in the component. Furthermore, a step up in the water level is experienced any time PORV opens to discharge mass. At EOT the component is nearly solid.

Finally, Figure 8d reports the trend of the maximum Eurofer temperature in the FW component. The peak present in the PHTS water BB outlet temperatures (Figure 9a) is not visible in the material temperature trend. The FW thermal inertia, even if low, completely smooths this temperature excursion. In the long term, the trend follows that of the PHTS water.

BZ System Transient Evolution

The BZ PHTS performances are strongly influenced by the presence of off-site power. If available, as in LF1 and LF2 sequences, system pumps continue to provide primary flow (Figure 10d is referred to loop 1 MCP 1). Among the interested cases, LF1 was selected to represent the scenarios characterized by the presence of off-site power and only its parameters are plotted in the following figures. Initially, a continuous slight decrease can be detected in the flow trend. It is due to the rise of system average temperature. This causes the decrease of water density in the pump component and also an increase of the loop pressure drops. These two combined effects produce the reduction of the mass flow elaborated by BZ MCPs. When the temperature at pump inlet reaches the setpoint, MCPs trip occurs and forced circulation is lost (for the timing see Table 7). If loss of off-site power is assumed, as in LF5 and LF6 scenarios, BZ MCPs trip occurs following the turbine trip and forced circulation is lost shortly after the SOT (Figure 10d). Natural circulation establishes in the BZ system. LF6 was selected as reference case to plot simulation results related to the absence of off-site power. The presence or not of the forced circulation is the main element affecting the BZ PHTS behavior during such transients.

Forced Circulation (LF1 and LF2 Cases)

When plasma shutdown and turbine trip are triggered, BZ system loses the power source (plasma pulse) and the heat sink (PCS feedwater) at the same time, while maintaining primary flow at nearly nominal value. This combination of factors produces the convergence of the system hot and cold temperatures to a common value (Figure 10a). No temperature peak is detected at BB outlet in any sector. Figure 10a is related to COB segment, but this is still valid for LOB/ROB and LIB/RIB.

The plasma shutdown takes more time (nearly 40 s, Table 6) with respect to PCS feedwater ramp down (10 s) and, above all, TSVs closure (0.5 s). This leads to a power unbalance and a consequent pressure spike in both BZ PHTS and PCS. In BZ PHTS, Figure 10b, the power surplus is dissipated by multiple openings of the pressurizer PORV. In the same way, the PCS pressure transient is managed by the steam line SRVs (Figure 10c). All three steps of this valve system are forced to intervene to limit the pressure increase. The maximum value experienced is slightly above the PCS design pressure. This demonstrates the appropriateness of the current valve design.

In the mid-term, BZ system is cooled down by the OTSGs (Figure 10e, related to BZ loop 1). Their residual cooling capability is due to the flow circulating in the steam generators any time the SRVs open to reduce the PCS pressure. This cooling system is available until a significant water inventory is present in the OTSGs secondary side. As shown by Figure 10f (loop 1 OTSG), the water level in the steam generator riser drops to zero at SOT in correspondence with the power surplus due to plasma shutdown. After that, water level is still present only in the lower downcomer. This is the water inventory available in the mid-term at the OTSGs secondary side. Any time SRVs open to reduce PCS pressure, level decreases. Once the lower downcomer has been completely evacuated, (for the timing and the total amount of mass discharged see Table 7), the dominant effect on the BZ temperatures is the presence of the decay heat. System heat losses are unable to dissipate such thermal power. Temperatures start to positively drift (Figure 10e) with a slope of nearly 12 °C/hr (22 °C in 6500 s, from 500 s to 7000 s). Even for the BZ system, the curve slope related to the forced circulation (LF1 and LF2 sequences) is higher than the one associated to cases dominated by natural circulation (LF5 and LF6). The difference is due to the pumping power, acting as an additional source term of the same order of magnitude of the decay heat. After BZ MCPs trip, whose timing is reported in Table 7, when forced circulation is lost and natural circulation establishes, if simulation time were increased, the same temperature slope would be observed for all the cases.

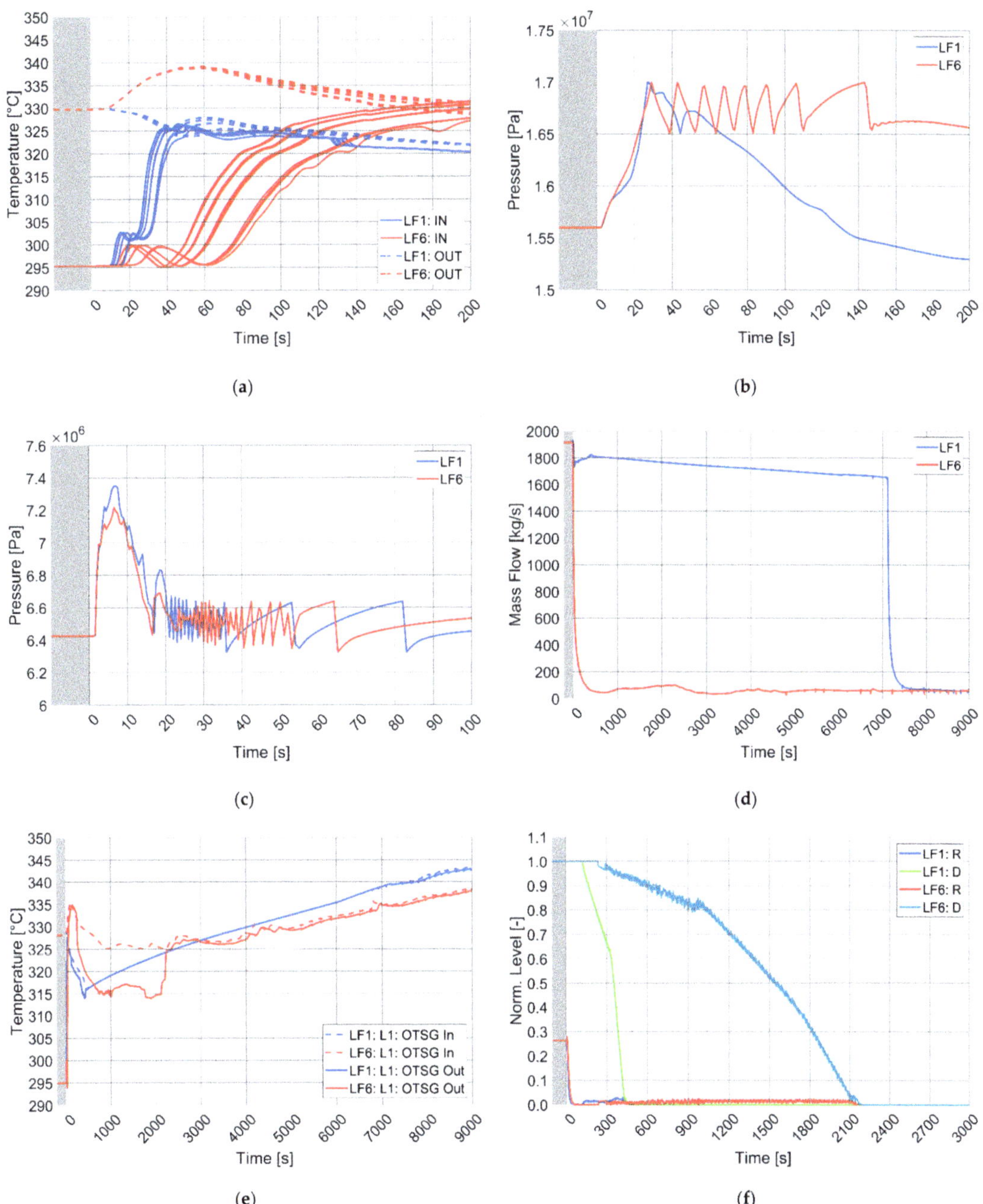

Figure 10. *Cont.*

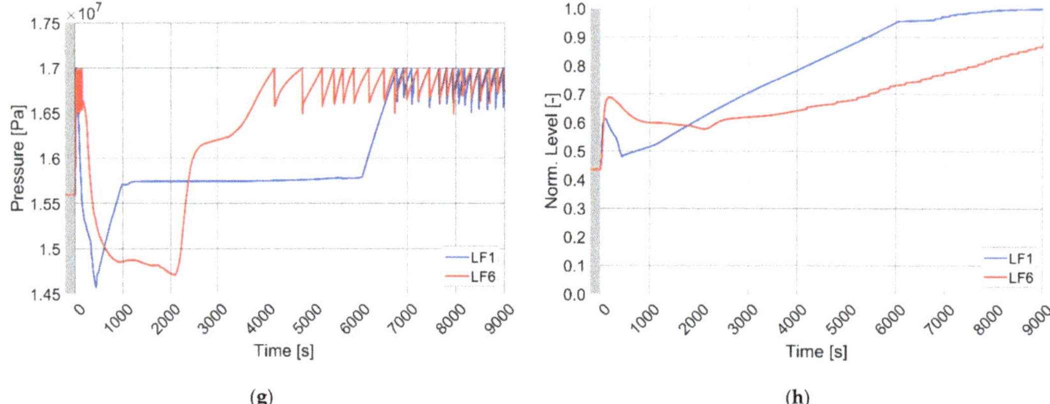

Figure 10. Comparison between LF1 and LF6 transients: (**a**) BZ PHTS water temperatures at BB inlet & outlet (all sectors, early time, COB segment); (**b**) Pressure in BZ PHTS (early time); (**c**) PCS pressure at OTSGs secondary side outlet (early time); (**d**) Mass flow elaborated by BZ loop 1 MCP 1 (full range); (**e**) BZ PHTS water temperatures at OTSG inlet & outlet (loop 1, full range); (**f**) Collapsed level in loop 1 OTSG secondary side riser (R) and lower downcomer (D) (normalized, full range); (**g**) Pressure in BZ PHTS (full range); (**h**) Collapsed level in BZ pressurizer (normalized, full range).

The BZ pressure goes down during the cooling transient provided by the OTSGs in the mid-term (Figure 10g). Its value drops even below the nominal one. This is possible because the pressurizer heaters are offline due to turbine trip. After the complete blowdown of OTSGs secondary side inventory, the system pressure rise following the temperature trend. This increase is limited by the pressurizer sprays that are still active since their operation depends on the BZ pumps. With the increase of the system temperature, they introduce in the pressurizer control volume water at higher enthalpy, reducing the effectiveness of their pressure control action. The pressurizer level also increases almost linearly during this time interval. It is reported in Figure 10h normalized with respect to the component height. When the pressurizer is nearly solid, sprays are unable to perform the pressure control function and the system pressure restart to rise, triggering the PORV. The timing of this event is in Table 7. From this moment, PHTS pressure starts to cycle. In this way, PORV component dissipates the decay heat produced in the blanket. The total PHTS mass discharged at EOT is shown in Table 7.

Natural Circulation (LF5 and LF6 Cases)

In these cases, with PT and TT signals, also the BZ MCPs trip is triggered. The BZ system loses at the same time: the power source (plasma shutdown), the heat sink (turbine trip) and the primary flow (MCPs trip). The PHTS water temperature trends at BB inlet/outlet (Figure 10a, COB segment) result from the relative balance between these decreasing parameters. Initially, the plasma power is dominant and a temperature spike can be detected at the blanket outlet. The peak value is reported in Table 7. Then, the primary pump coast-down, which lasts more than the plasma shutdown curve, becomes prevalent and the system temperatures converge.

The initial power surplus produces a pressure spike in both BZ PHTS and PCS. In the former, Figure 10b, it is managed by the pressurizer PORV, while in the latter, Figure 10c, the pressure transient is limited by the steam line SRVs. All three steps are necessary to limit the pressure rise. The observed maximum value is slightly above the PCS design pressure, proving the effectiveness of the SRVs design even in these scenarios.

In the mid-term, BZ system is cooled down by the OTSGs, as shown in Figure 10d regarding loop 1. As already discussed, their residual cooling capability is available until a significant water inventory is present at OTSGs secondary side. The presence of natural circulation (with respect to forced circulation) increments the time needed to the SRVs to evacuate the OTSGs secondary side inventory (see different timing collected in Table 7 and trends reported in Figure 10f). The lower primary flow in the steam generator (with respect to the one ensured by forced circulation) decreases the overall heat transfer coefficient and, consequently, the thermal power removed by PCS. This slows down the pressure rise in the secondary system and increases the time interval between two subsequent SRVs openings. With natural circulation, the OTSGs cooling capability lasts more than cases dominated by forced circulation.

Terminated the water inventory in the OTSGs secondary side lower downcomer, the dominant effect on the BZ temperatures is the presence of the decay heat. They start to drift positively. The temperature slope is lower than the one due to forced circulation because of the absence of pumping power. Temperatures rise of 10 °C in the last 4000 s of simulation (nearly 9 °C/hr).

During the cooling transient provided by the steam generators, system pressure decreases unlimited by pressurizer heaters (Figure 10g). They are disabled from the occurrence of turbine trip. Later, once evacuated the OTSGs secondary side inventory (the total mass discharged is provided by Table 7), the system pressure starts to rise. Pressurizer sprays are off since no pumps are available in the circuit. The PORV opening setpoint is reached quite faster (compare timing gathered in Table 7). From this moment, PHTS pressure follows the sawtooth trend already discussed. The trend of water level in the pressurizer (Figure 10h) is similar to the one reported in Figure 8c for LF6 sequence. The parameter evolution and the phenomenology occurring in the component are the same. At the end of the transient, the tank is nearly solid. The total BZ PHTS water mass discharged by PORV at EOT is indicated in Table 7.

4.2.4. LOFA Transients Involving BZ Cooling Circuit
BZ System Transient Evolution

Once PIE occurs, the primary flow elaborated by interested pump/pumps starts to decrease. In LF3 and LF7 transients, only loop 1 MCP 1 is stopped (partial LOFA), while, in LF4 and LF8 sequences, all system pumps are involved in the accident (complete LOFA). Low flow takes few seconds to be detected, actuating the plasma shutdown. Consequently, also turbine trip is triggered. In case LF7, where a loss of off-site power is assumed, TT causes the stop of all the system pumps not interested from initiating event. For this reason, in LF4, LF7 and LF8 scenarios, the coast-down of all the BZ pumps is nearly contemporaneous and these cases have a similar accidental evolution. The only different sequence is LF3, where loop 1 MCP 2 and loop 2 MCPs continue to provide primary coolant flow. They are stopped on high-temperature signal at nearly EOT. Summarizing, for what concerns BZ PHTS, the selected cases can be grouped in the same way already seen for FW PHTS in Section 4.2.3. Main events and parameters related to the transient simulations characterized by PIE involving BZ MCPs are collected in Table 8.

Case LF3

In this case, the loop 1 MCP 2 and the loop 2 MCPs are still active after the turbine trip (off-site power is available). The loop 1 MCP 2 increases the mass flow provided (Figure 11c). The loop 1 branch hosting the failed pump becomes an alternative flow path for the mass flow provided by loop 1 MCP 2. The pressure drops related to this path is less than the ones associated to a BB sector (even with the failed pump acting as a minor head loss). Hence, for loop 1 MCP 2 the curve of the hydraulic resistance decreases and, being a constant rotational velocity imposed as a boundary condition for the component, the result is an increase of the mass flow provided and a drop of the pump head. Instead, the operation of loop 2 pumps is only slightly altered with respect to the nominal state. The transient is dissymmetrical with respect to the toroidal dimension. The sixteen sectors experience different flows (Figure 11a) and, consequently, inlet/outlet COB temperatures (Figure 11b). Higher mass flows (i.e., lower outlet temperatures) correspond to the sectors located in diametrically opposite position with respect to the failed pump (four of sixteen). However, the forced flow availability significantly smooths the temperature peaks at COB outlet (only few degrees above the nominal value).

Table 8. Summary table for LOFA transients involving BZ PHTS MCPs.

Event/Parameter	Unit	LF3	LF4	LF7	LF8
PIE (LOFA)	-	Partial (BZ)	Complete (BZ)	Partial (BZ)	Complete (BZ)
Loss of off-site power	yes/no	no	no	yes	yes
PT signal occurrence	s	2.5	2.5	2.5	2.5
TT signal occurrence	s	2.5	2.5	2.5	2.5
TSVs start to close	s	2.5	2.5	2.5	2.5
Start of PCS feedwater ramp-down	s	2.5	2.5	2.5	2.5
Start of IHTS mass flow ramp-down	s	10	10	2.5	2.5
Time of FW PHTS water temperature peak	s	-	-	24	23
FW PHTS water temperature peak [1]	°C	-	-	331	331
Time of BZ PHTS water temperature peak	s	30	58	59	59
BZ PHTS water temperature peak [1]	°C	333	340	340	340
FW MCPs Trip occurrence	s	7496	8440	2.5	2.5
BZ MCPs Trip occurrence (pump not interested by PIE)	s	7260	-	2.5	-
Time to evacuate BZ OTSGs secondary side inventory	s	600 (L1) 460 (L2)	2150	2150	2150
Water mass discharged from BZ OTSGs secondary side	kg	15,017 (L1) 17220 (L2)	15,037 (per OTSG)	14,871 (per OTSG)	12,885 (per OTSG)
FW PORV first opening time (Long Term)	s	7332	8576	2248	2284
Total FW PHTS water mass discharged at EOT	kg	1034	247	1444	1356
BZ PORV first opening time (Long Term)	s	6952	4412	4164	4168
Total BZ PHTS water mass discharged at EOT	kg	6113	4592	4541	5025

[1] For all the sixteen sectors, both the BZ and FW PHTS water temperatures were detected at the outlet of COB segment. For each PHTS, peak temperature reported in the table is the maximum among all the temperature readings.

As observed in FW system for case LF1, a flow inversion can be detected in the BZ system branch where the failed pump is located. The pressure drops related to the blanket component are so high that a part of the flow provided by loop 1 MCP 2 is recirculated through this alternative flow path. Differently from LF1 sequence, the reverse flow does not cause a temperature inversion in loop 1. In fact, each loop pump is hosted in a branch going from the OTSG outlet plenum to the cold ring. Even if there is a reverse flow in one of these branches, the primary flow through the hot leg and the steam generator is ensured in the right direction by the operation of the MCP still active. The effect of the failed pump is visible in Figure 11d. The reduced flow in loop 1 with respect to loop 2, slows down the cooling transient provided by the OTSGs in the mid-term. Loop 2 steam generator runs out its cooling capability one hundred seconds earlier than the correspondent in loop 1 (see Table 8 for timing and water mass discharged). From this moment, no sensible differences are detectable between the TH performances of the two loops.

BZ temperatures positively drift since blanket decay heat overwhelms the system heat losses. The temperature slope is of nearly 11 °C/hr (25 °C in 8000 s). This is the same value obtained for BZ system in LF1 and LF2 scenarios, when LOFA transients involve FW PHTS and off-site power is available to ensure the BZ pumps operation. Forced circulation confirms to produce a higher curve slope than the one associated to natural circulation (see Figure 12 related to case LF8). As already discussed, the PHTS coolant additional heating is caused by pumping power. MCPs trip, whose timing is reported in Table 8, is triggered by a high-temperature signal at the pump inlet. Later, forced circulation is lost and natural circulation establishes (Figure 11c). The temperature slope starts to decrease accordingly.

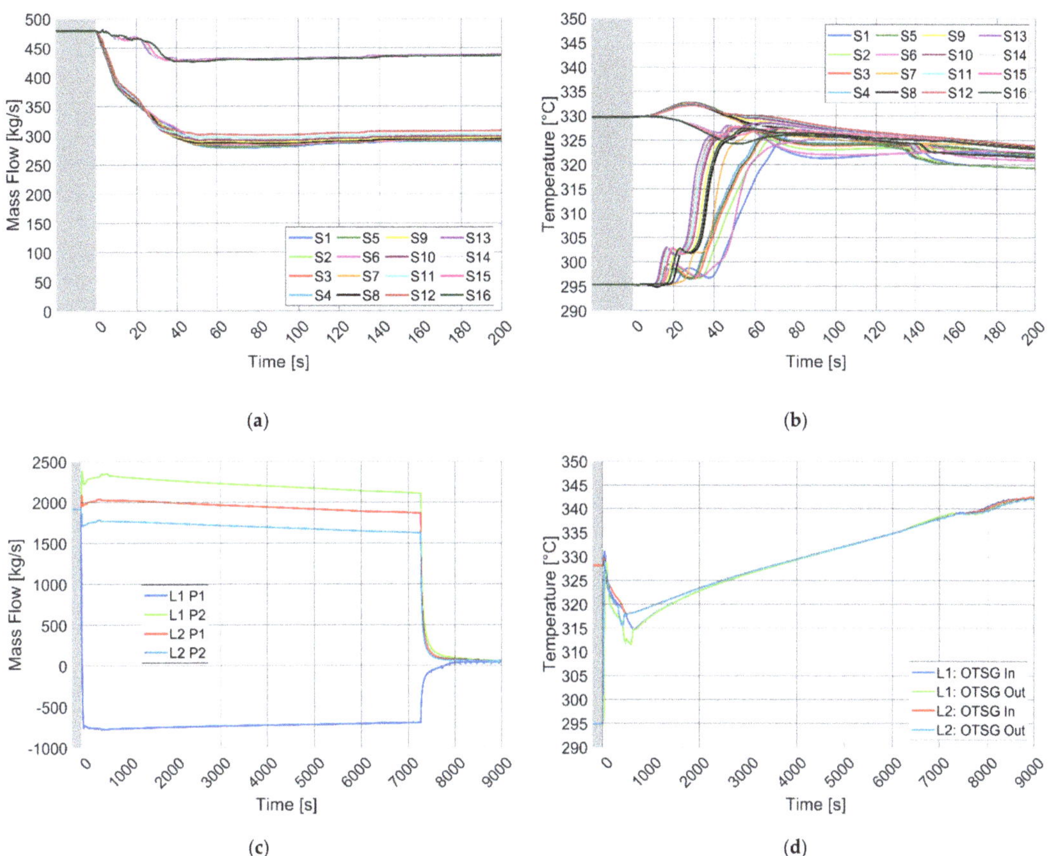

Figure 11. Partial LOFA on BZ PHTS without loss of off-site power (LF3 transient): (**a**) Mass flow in BZ sectors (early time); (**b**) BZ PHTS water temperatures at BB inlet & outlet (all sectors, early time, COB segment); (**c**) Mass flow elaborated by BZ PHTS MCPs (full range); (**d**) BZ PHTS water temperatures at OTSGs inlet & outlet (full range).

The plot of BZ pressure trend is not included in the following since it is the same of LF1 and LF2 transients (see Figure 10g). The presence of pressurizer sprays, ensured by the BZ pumps still active, allows to control the system pressure for nearly two hours. Then, the decay heat is evacuated by discharging PHTS water through the PORV. The relevant parameters are contained in Table 8.

Cases LF4, LF7 and LF8

The considered cases have an accidental evolution very similar to the one described in Section 4.2.3 for LF5 and LF6 sequences. In these scenarios, trip occurs for all the BZ pumps after few seconds from the SOT (see Tables 7 and 8), albeit for different reasons. The resulting transients are quite symmetrical with respect to the toroidal dimension. PHTS temperatures at BB inlet/outlet are the same for all the sectors. They are reported in Figure 12a for LF4, LF7 and LF8 scenarios. Among the different cases, no sensible differences are detectable in the temperature peak at COB outlet. The maximum values, indicated in Table 8, are close to the ones observed for LF5 and LF6 transients (Table 7). Also the BZ system long-term behavior is nearly the same. As an example, the PHTS water temperatures at OTSGs inlet/outlet are plotted for case LF8 in Figure 12b. The trend is very similar to the analogous contained in Figure 10e for LF6 sequence. After pump coast-down, natural circulation establishes in the system, influencing the BZ thermal-hydraulic performances. A detailed description of the transient evolution is provided in Section 4.2.3, in the paragraph referring to BZ PHTS. A quantitative comparison between all the interested cases can be performed looking at the main timing and TH parameters related to the BZ system contained in Tables 7 and 8.

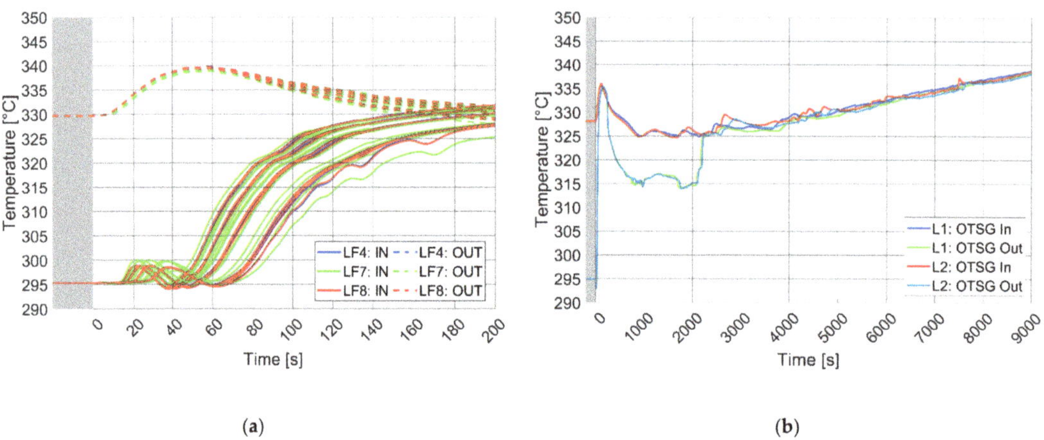

Figure 12. Parameter trends in BZ PHTS for LOFA transients characterized by natural circulation (LF4, LF7, LF8): (**a**) BZ PHTS water temperatures at BB inlet & outlet (all sectors, early time, COB segment); (**b**) BZ PHTS water temperatures at OTSGs inlet & outlet (full range, only LF8).

FW System Transient Evolution

Considerations related to FW system are of the same kind of the ones done in Section 4.2.3 about BZ PHTS. FW pumps are not interested from PIE and the system performances are strongly influenced by the presence of off-site power. If available, as in LF3 and LF4 scenarios, FW pumps continue to provide primary flow. The slight parameter decrease is due to the increase of the system average temperature. (Figure 13c). MCPs trip occurs after more than two hours from PIE (Table 8). It is triggered by a high-temperature signal at the pump inlet. The simulation is characterized by the presence of the forced circulation. Instead, if the loss of off-site power is assumed, as in cases LF7 and LF8, FW MCPs trip occurs following the turbine trip and forced circulation is lost few seconds after SOT (Figure 13c). Natural circulation establishes in the FW system, influencing its TH behavior during the overall simulation.

Forced Circulation (LF3 and LF4 Cases)

Due to the presence of forced circulation, FW temperatures converge very quickly to an average value (Figure 13d). Transient is symmetrical with respect to toroidal dimension and, for all the BB sectors, no temperature peak is present at blanket outlet (Figure 13a). HITEC® secondary flow is available for the first 10 s after PIE. This element, combined with the suitability of forced circulation in the primary system, avoids the opening of the pressurizer PORV in the early time (Figure 13b).

In the long term, FW HEXs are not able to provide any cooling capability and system heat losses do not counterbalance the blanket decay heat. An additional source term is represented by the pumping power. FW temperatures start to drift positively (Figure 13d). The associated temperature slope is of nearly 11 °C/hr (20 °C in 7000 s). The Eurofer maximum temperature in the FW component follows the same time trend of the PHTS water (Figure 13e). Once MCPs trip is triggered, the forced circulation is lost and the natural circulation establishes. The temperature slope decreases to the value related to simulations characterized by natural circulation (LF7 and LF8 scenarios).

Pressure transient for the considered cases (Figure 13f) is similar to the one described for LF1 sequence (see Section 4.2.3 and Figure 8b). After the heat sink loss, FW pressure is limited by pressurizer sprays. When sprays become unable to perform their control function (due to system temperature increase), the management of system pressure switches to PORV component (timing of this event is reported in Table 8). The total mass discharged from the valve at EOT is indicated in Table 8. The plot of pressurizer level related to cases LF3 and LF4 is not included in the following since very similar to the one reported in Figure 8c for LF1 transient.

Natural Circulation (LF7 and LF8 Cases)

For the considered cases, plasma shutdown, turbine trip, FW MCPs trip and IHTS mass flow ramp-down occur at the same time. The PHTS water temperatures at COB inlet/outlet are collected, for all the sectors, in Figure 13a. Their trends result from the relative balance between plasma power, primary flow and secondary flow, all decreasing parameters but with different timing. COB outlet temperatures experience a slight increase since initially the plasma power is prevalent. Then, since the pump coast-down (Figure 13c) takes more time than the plasma shutdown (Table 6) the outlet temperatures start to decrease. Peak value is the same for all the sectors and for all the cases, as reported in Table 8.

Due to the unavailability of forced circulation in both primary and secondary systems, the initial power surplus produces a sudden increase in the FW PHTS pressure, Figure 13b. Pressurizer PORV intervenes to manage this pressure transient.

During the FW pump coast-down, system reaches a quite uniform temperature (Figure 13d). Later, while natural circulation establishes, system temperatures experience an inversion. In the long term, the original temperature field is restored and FW temperatures positively drift. The temperature slope is lower (nearly 9 °C/hr) than the one observed for cases LF3 and LF4, since the additional source term due to pumping power is missing.

After FW MCPs trip, pressurizer sprays are disabled. System Pressure increase can be only limited by the PORV intervention (Figure 13f). The valve opening occurs quite earlier with respect to LF3 and LF4 sequences (compare different timing reported in Table 8). From this moment, the system pressure begins to cycle accordingly with the valve component multiple interventions. The PHTS mass discharged at the EOT is indicated in Table 8.

Figure 13e reports the trend of the maximum Eurofer temperature in the FW component. The peak related to PHTS water present at blanket outlet (Figure 13a) here is not visible. Temperature excursion is smoothed by the FW thermal inertia, even if low. After plasma termination, material temperature drops driven by PHTS water temperature. Instead, in the long term, FW component is heated up by the decay heat. The temperature slope is the same of the PHTS water trend.

Summarizing, the considered cases have accidental evolutions very similar to the one described in Section 4.2.3 for LF2, LF5 and LF6 transients. The common factor to all these scenarios is the occurrence of FW MCPs trip after few seconds from the SOT (see Tables 7 and 8), albeit for different reasons. Hence, the forced circulation is immediately lost and the natural circulation influences the system TH performances during the overall simulation. A qualitative comparison between the interested cases can be performed by looking at the parameter trends collected in Figures 8 and 9 (where LF6 sequence was used as reference) and Figure 13 (using LF8 as selected scenario). For the same purpose, but from a quantitative point of view, parameters and timing contained in Tables 7 and 8 can be used.

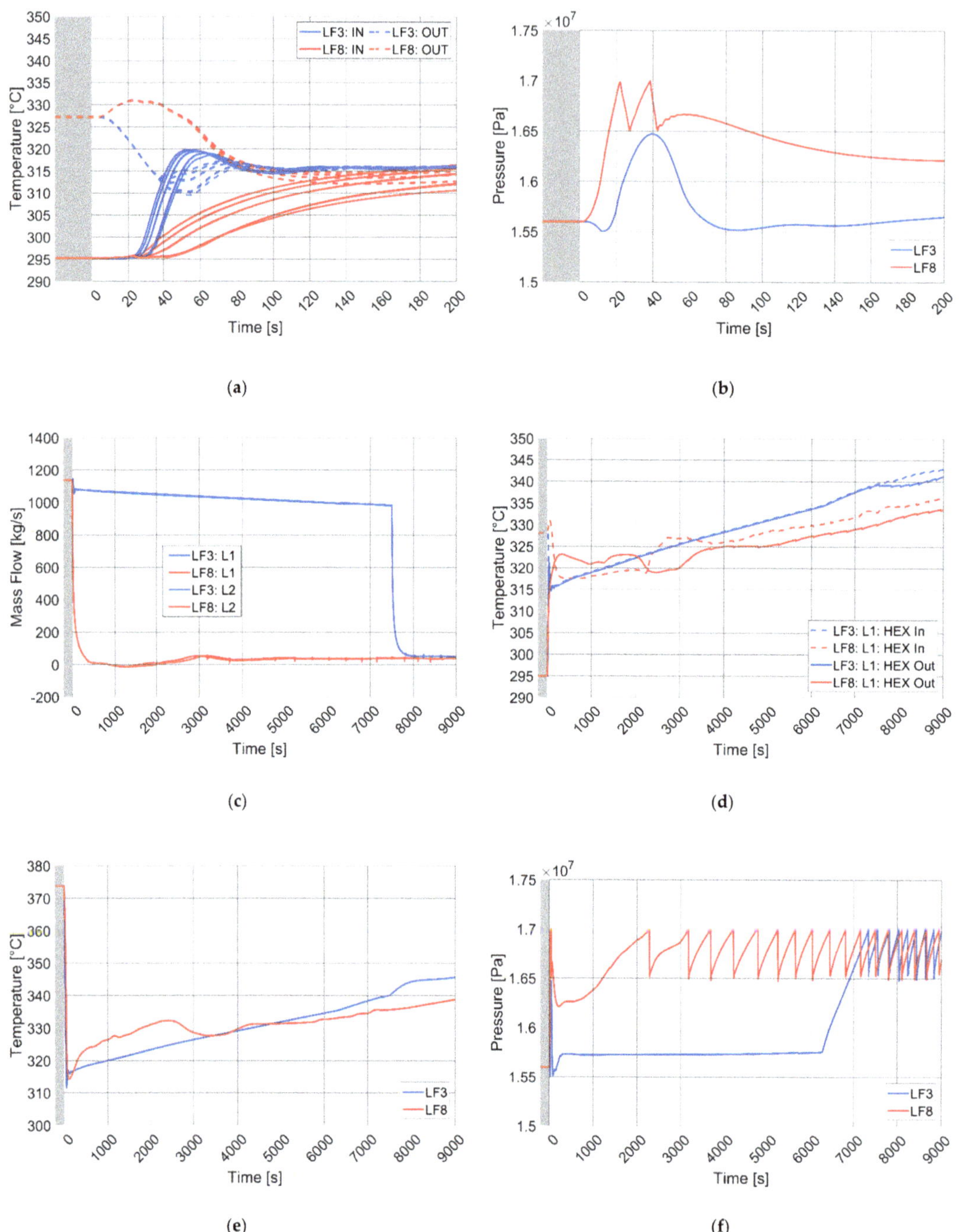

Figure 13. Comparison between LF3 and LF8 transients: (**a**) FW PHTS water temperatures at BB inlet & outlet (all sectors, early time, COB segment); (**b**) Pressure in FW PHTS (early time); (**c**) Mass flow elaborated by FW MCPs (full range); (**d**) FW PHTS water temperatures at HEX inlet & outlet (loop 1, full range); (**e**) Maximum Eurofer temperature in FW component (full range); (**f**) Pressure in FW PHTS (full range).

5. Discussion

Results presented in the previous section highlight how the type of circulation (natural or forced) characterizing each cooling system is the main element influencing its TH performances. According to the considered case, BZ and FW systems can have the same kind of circulation or not. However, as a general rule, for the suitability of the forced circulation in a primary cooling circuit is mandatory the presence of the off-site power. If its loss is assumed in combination with the initiating event, at the occurrence of turbine trip forced circulation is lost in both systems, if not already missing in one of them according to the specific PIE considered. In fact, the turbine generator set is the only element ensuring the AC power needed for the pumps operation and it is disconnected after the TT signal. If forced circulation is available, the following TH behavior can be observed in BZ and FW systems.

- Few seconds after the SOT, the temperature spikes at blanket outlet characterizing the trend of both BZ and FW PHTS water are significantly smoothed.
- In FW system, the availability of forced circulation in both primary and secondary (only for the first 10 s) circuits limits the pressure increase and avoids the intervention of the pressurizer PORV in the short term.
- The OTSGs cooling capability lasts less. The presence of forced circulation in the primary cooling system enhances the steam generator HTC, increasing the thermal power transferred to the PCS. This reduces the time between two subsequent steam line SRVs openings and speeds up the evacuation of the water mass present in the OTSGs secondary side. Once terminated, the steam generators are no more able to provide any cooling function to the BZ PHTS.
- For more or less two hours from PIE occurrence, the system pressure is controlled by the pressurizer sprays. The first PORV intervention in the long term is significantly delayed.
- The temperature slope characterizing both BZ and FW systems (thermally coupled) is higher since pumping power is added to the power balance. This is valid until the MCPs trip is triggered in each system.

Summarizing, forced circulation improves the BZ and FW TH performances in the short term, smoothing the temperature spikes, but reduces the ones in the mid-long term. In fact, it shortens the cooling interval provided to the BZ PHTS by the steam generators and increases the temperature slope experienced by BZ and FW systems, reducing the reactor grace time. The best management strategy for PHTS pumps is to use, at the SOT, the forced circulation they provide, in order to avoid excessive temperatures in the blanket, and then stop them, to increase the reactor grace time. To prove the effectiveness of this control logic, case LF3 was run again adding a new trip signal to BB MCPs. The level in the BZ OTSGs lower downcomer is monitored and when it reaches the 1% of the rated value in full plasma power state, both BZ and FW pumps are stopped. LF3 (partial LOFA in BZ PHTS without loss of off-site power) was selected as reference case since it is one of the two (together with LF1) where forced circulation is available for both primary cooling systems, even if reduced in the one involved in the PIE. The PHTS water temperatures at loop 1 OTSG/HEX inlet/outlet are reported in Figure 14. As shown, this new pump management strategy combines the benefits of forced circulation in the short term and of natural circulation in the long term.

In all the transient simulations, included the one discussed in this section, BZ and FW systems experience a positive temperature drift in the mid-long term. It is due to the unbalance between decay heat produced in the blanket and system heat losses, with the former overwhelming the latter. The temperature slope is higher if the forced circulation is still active. In these cases, it must be added another source term to the power balance, represented by the pumping power. In the calculations performed, no Decay Heat Removal (DHR) system was implemented in the input deck and the power surplus is managed by the pressurizer PORV. Power in excess produces a pressure increase and when this parameter reaches the PORV opening setpoint, PHTS water mass is discharged with its associated enthalpy content. This is the way adopted by BZ and FW system to dissipate the power surplus. However, a DHR system is foreseen for DEMO reactor in accidental conditions, as discussed in [5].

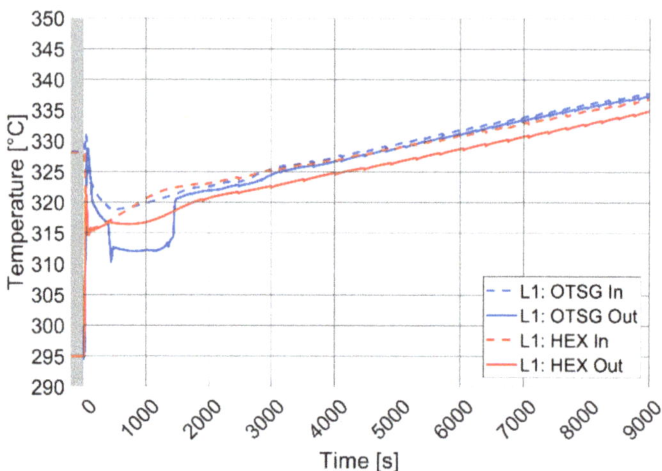

Figure 14. Partial LOFA on BZ PHTS without loss of off-site power, new BB MCPs management strategy: BZ PHTS water temperatures at loop 1 OTSG inlet & outlet and FW PHTS water temperatures at loop 1 HEX inlet & outlet (full range).

6. Conclusions

The analysis was performed with the aim of preliminary evaluating the WCLL BB PHTS behavior during anticipated transients and accidental conditions. A best-estimate system code, RELAP5/Mod3.3, was used to achieve this goal. A modified version was developed at DIAEE with the purpose of increasing the predictive capabilities of the code with respect to fusion reactors. Implemented features include new HTC correlations, new fluids, etc. A full RELAP5 TH model was prepared. Blanket was simulated with equivalent pipes, maintaining the overall thermal inertia. The PHTS cooling circuits were modelled in detail adopting one-dimensional hydrodynamic components. All the system equipment (pumps, heat exchangers, pressurizer) and piping were included in the model. The input deck was initially used to simulate the DEMO full plasma power state, that is the most challenging scenario during the reactor normal operations. This state was chosen as initial condition for the transient analysis. The selected initiating events consist in partial and complete LOFA. Simulations were run considering the PIEs occurring in both BZ or FW system and they were repeated also assuming the loss of off-site power. A matrix of interesting scenarios was individuated. A preliminary actuation logic, based on the consolidated PWR experience and the innovations related to GEN III+ nuclear reactor design, was proposed and implemented for some reactor components. Simulation outcomes highlight the appropriateness of the current PHTS design. BB temperatures do not experience excessive excursions during the plasma shutdown. Pressure transients in BZ PHTS, FW PHTS and PCS are effectively managed by the related relief systems. The results underline a strong dependence of the PHTS TH performances on the type of circulation characterizing each primary cooling circuit. The forced circulation is of great importance in the management of the initial power transient, while the natural circulation is advisable in the long term to increase the reactor grace time. On the basis of the calculation outcomes, a revised BB MCPs management strategy was defined for the cases where the off-site power is available. It combines the short term benefits of forced circulation and the long term advantages of natural circulation. In the long term, BZ and FW systems are heated up by the BB decay heat, overwhelming the system heat losses. In the current simulations, the power surplus is dissipated by the pressurizer PORV that opens and discharges PHTS water mass and related enthalpy. In the future developments of the activity, the DHR system foreseen for DEMO reactor will be implemented in the input deck to evaluate the effectiveness of its mitigation action.

Author Contributions: Conceptualization, C.C., F.G., A.D.N. and G.C.; methodology, C.C. and F.G.; software, C.C.; validation, C.C. and F.G.; formal analysis, C.C.; writing—original draft preparation, C.C.; writing—review and editing, F.G., A.D.N. and G.C.; supervision, F.G., A.D.N. and G.C.; project administration, A.D.N. and G.C. All authors have read and agreed to the published version of the manuscript.

Funding: This research was funded by Euratom research and training programme 2014-2018 and 2019-2020, grant number 633053.

Institutional Review Board Statement: Not applicable.

Informed Consent Statement: Not applicable.

Data Availability Statement: No new data were created or analyzed in this study. Data sharing is not applicable to this article.

Acknowledgments: This work has been carried out within the framework of the EUROfusion Consortium and has received funding from the Euratom research and training programme 2014–2018 and 2019–2020 under grant agreement No 633053. The views and opinions expressed herein do not necessarily reflect those of the European Commission.

Conflicts of Interest: The authors declare no conflict of interest.

Abbreviations

AC	Alternating Current
BB	Breeding Blanket
BC	Boundary Condition
BRC	Breeding Cell
BOL	Beginning Of Life
BZ	Breeder Zone
CAD	Computer Aided Design
CFETR	China Fusion Engineering Test Reactor
COB	Central Outboard Blanket
CV	Control Volume
DHR	Decay Heat Removal System
DIAEE	Dipartimento di Ingegneria Astronautica, Elettrica ed Energetica
DWT	Double Walled Tube
EOL	End Of Life
EOT	End Of Transient
ESS	Energy Storage System
EU-DEMO	European Demonstration Power Plant
FP	Feeding Pipe
FW	First Wall
HCPB	Helium Cooled Pebble Bed
HCSG	Helicoidal Coil Steam Generator
HEX	Heat EXchanger
HS	Heat Structure
HTC	Heat Transfer Coefficient
IB	Inboard Blanket
IHTS	Intermediate Heat Transfer System
K-DEMO	Korean Demonstration Power Plant
LIB	Left Inboard Blanket
LiPb	Lithium Lead
LOB	Left Outboard Blanket

LOCA	Loss of coolant accident
LOFA	Loss of flow accident
LOHS	Loss of Heat Sink
MARS-KS	Multi-dimensional Analysis of Reactor Safety
MCP	Main Coolant Pump
MS	Molten Salt
OB	Outboard Blanket
OTSG	Once-Through Steam Generator
PCS	Power Conversion System
PHTS	Primary Heat Transfer System
PIE	Postulated Initiating Event
PORV	Pilot (Power)-Operated Relief Valve
PRZ	Pressurizer
PT	Plasma Termination
PWR	Pressurized Water Reactor
RELAP5	Reactor Excursion Leak Analysis Program
RIB	Right Inboard Blanket
ROB	Right Outboard Blanket
SMS	Single Module Segment
SOT	Start Of Transient
SRV	Safety Relief Valve
TH	Thermal-Hydraulics
TSV	Turbine Stop Valve
TT	Turbine Trip
WCCB	Water-Cooled Ceramic Breeder
WCLL	Water-Cooled Lithium-Lead

References

1. Federici, G.; Boccaccini, L.; Cismondi, F.; Gasparotto, M.; Poitevin, Y.; Ricapito, I. An overview of the EU breeding blanket design strategy as an integral part of the DEMO design effort. *Fusion Eng. Des.* **2019**, *141*, 30–42. [CrossRef]
2. Del Nevo, A.; Arena, P.; Caruso, G.; Chiovaro, P.; Di Maio, P.A.; Eboli, M.; Edemetti, F.; Forgione, N.; Forte, R.; Froio, A.; et al. Recent progress in developing a feasible and integrated conceptual design of the WCLL BB in EUROfusion project. *Fusion Eng. Des.* **2019**, *146*, 1805–1809. [CrossRef]
3. Ricapito, I.; Cismondi, F.; Federici, G.; Poitevin, Y.; Zmitko, M. European TBM programme: First elements of RoX and technical performance assessment for DEMO breeding blankets. *Fusion Eng. Des.* **2020**, *156*, 111584. [CrossRef]
4. Martelli, E.; Giannetti, F.; Caruso, G.; Tarallo, A.; Polidori, M.; Barucca, L.; Del Nevo, A. A Study of EU DEMO WCLL breeding blanket and primary heat transfer system integration. *Fusion Eng. Des.* **2018**, *136*, 828–833. [CrossRef]
5. Barucca, L.; Bubelis, E.; Ciattaglia, S.; D'Alessandro, A.; Del Nevo, A.; Giannetti, F.; Hering, W.; Lorusso, P.; Martelli, E.; Moscato, I.; et al. Pre-conceptual design of EU DEMO balance of plant systems: Objectives and challenges. *Fusion Eng. Des.* **2017**. under review.
6. Martelli, E.; Giannetti, F.; Ciurluini, C.; Caruso, G.; Del Nevo, A. Thermal-hydraulic modeling and analyses of the water-cooled EU DEMO using RELAP5 system code. *Fusion Eng. Des.* **2019**, *146*, 1121–1125. [CrossRef]
7. Idaho National Laboratory; The RELAP5-3D© Code Development Team. *RELAP5-3D© Code Manual: Code Structure, System Models, and Solution Methods*; Revision 4.3, INL-MIS-15-36723; Idaho National Laboratory: Idaho Falls, ID, USA, 2015; Volume I.
8. Emonot, P.; Souyri, A.; Gandrille, J.L.; Barré, F. CATHARE-3: A new system code for thermal-hydraulics in the context of the NEPTUNE project. *Nucl. Eng. Des.* **2011**, *241*, 4476–4481. [CrossRef]
9. Hu, R. A fully-implicit high-order system thermal-hydraulics model for advanced non-LWR safety analyses. *Ann. Nucl. Energy* **2017**, *101*, 174–181. [CrossRef]
10. D'Onorio, M.; Giannetti, F.; Porfiri, M.T.; Caruso, G. Preliminary safety analysis of an in-vessel LOCA for the EU-DEMO WCLL blanket concept. *Fusion Eng. Des.* **2020**, *155*, 111560. [CrossRef]
11. D'Onorio, M.; Giannetti, F.; Porfiri, M.T.; Caruso, G. Preliminary sensitivity analysis for an ex-vessel LOCA without plasma shutdown for the EU DEMO WCLL blanket concept. *Fusion Eng. Des.* **2020**, *158*, 111745. [CrossRef]
12. Gauntt, R.O.; Cash, J.E.; Cole, R.K.; Erickson, C.M.; Humphries, L.L.; Rodriguez, S.B.; Young, M.F. *MELCOR Computer Code Manuals: Primer and Users*; Guide Version 1.8.6., Revision 3, NUREG/CR-6119; Sandia National Laboratory: Albuquerque, NM, USA, 2005; Volume 1.
13. D'Onorio, M.; Caruso, G. Pressure suppression system influence on vacuum vessel thermal-hydraulics and on source term mobilization during a multiple First Wall-Blanket pipe break. *Fusion Eng. Des.* **2021**, *164*, 112224. [CrossRef]

14. Ciurluini, C.; Giannetti, F.; Martelli, E.; Del Nevo, A.; Barucca, L.; Caruso, G. Analysis of the thermal-hydraulic behavior of the EU-DEMO WCLL Breeding Blanket cooling systems during a Loss Of Flow Accident. *Fusion Eng. Des.* **2021**, *164*, 112206. [CrossRef]
15. The US Nuclear Regulatory Commission (USNRC). *RELAP5/MOD3 Code Manual: Code Structure, System Models, and Solution Methods*; NUREG/CR-5535; USNRC: Washington, DC, USA, 1998; Volume I.
16. D'Amico, S.; Di Maio, P.A.; Jin, X.Z.; Hernández-Gonzalez, F.A.; Moscato, I.; Zhou, G. Preliminary thermal-hydraulic analysis of the EU-DEMO Helium-Cooled Pebble Bed fusion reactor by using the RELAP5-3D system code. *Fusion Eng. Des.* **2021**, *162*, 112111. [CrossRef]
17. Jin, X.Z.; Chen, Y.; Ghidersa, B.E. LOFA analysis for the FW of DEMO HCPB blanket concept. In Proceedings of the 1st IAEA Technical Meeting on the Safety, Design and Technology of Fusion Power Plants, Vienna, Austria, 3–5 May 2016.
18. Jin, X.Z. BB LOCA analysis for the reference design of the EU DEMO HCPB blanket concept. *Fusion Eng. Des.* **2018**, *136 Pt B*, 958–963. [CrossRef]
19. Cheng, X.; Ma, X.; Li, X.; Wang, W.; Liu, S. Steady states and LOFA analyses of the updated WCCB blanket for multiple fusion power modes of CFETR. *Fusion Eng. Des.* **2019**, *144*, 23–28. [CrossRef]
20. Cheng, X.; Lin, S.; Liu, S. Loss of flow accident and loss of heat sink accident analyses of the WCCB primary heat transfer system for CFETR. *Fusion Eng. Des.* **2019**, *147*, 111247. [CrossRef]
21. Moon, S.B.; Lim, S.M.; Bang, I.C. Analysis of hydrogen and dust explosion after vacuum vessel rupture: Preliminary safety analysis of Korean fusion demonstration reactor using MELCOR. *Int. J. Energy Res.* **2018**, *42*, 104–116. [CrossRef]
22. Cheng, X.; Ma, X.; Lu, P.; Wang, W.; Liu, S. Thermal dynamic analyses of the primary heat transfer system for the WCCB blanket of CFETR. *Fusion Eng. Des.* **2020**, *161*, 112067. [CrossRef]
23. Kim, G.-W.; Lee, J.-H.; Cho, H.-K.; Park, G.-C.; Im, K. Development of thermal-hydraulic analysis methodology for multiple modules of water-cooled breeder blanket in fusion DEMO reactor. *Fusion Eng. Des.* **2016**, *103*, 98–109. [CrossRef]
24. Jeong, J.-J.; Ha, K.S.; Chung, B.D.; Lee, W.J. Development of a multi-dimensional thermal-hydraulic system code, MARS 1.3.1. *Ann. Nucl. Energy* **1999**, *26*, 1611–1642. [CrossRef]
25. Giannetti, F.; D'Alessandro, T.; Ciurluini, C. *Development of a RELAP5 Mod3.3 Version for FUSION Applications*; DIAEE Sapienza Technical Report No. D1902_ENBR_T01, Revision 1; DIAEE, Sapienza University of Rome: Rome, Italy, 2019.
26. Edemetti, F.; Micheli, P.; Del Nevo, A.; Caruso, G. Optimization of the first wall cooling system for the DEMO WCLL blanket. *Fusion Eng. Des.* **2020**, *161*, 111903. [CrossRef]
27. Edemetti, F.; Di Piazza, I.; Del Nevo, A.; Caruso, G. Thermal-hydraulic analysis of the DEMO WCLL elementary cell: BZ tubes layout optimization. *Fusion Eng. Des.* **2020**, *160*, 111956. [CrossRef]
28. Idelchik, I.E. *Handbook of Hydraulic Resistance*, 2nd ed.; Hemisphere Publishing Corporation: Washington, DC, USA, 1986.
29. Sieder, E.N.; Tate, G.E. Heat transfer and pressure drop of liquids in tubes. *J. Ind. Eng. Chem.* **1936**, *28*, 1429–1435. [CrossRef]
30. U.S. Nuclear Regulatory Commission. Section 3.2—Reactor Coolant System. In Westinghouse Technology Systems Manual. 2012. Available online: https://www.nrc.gov/docs/ML1122/ML11223A213 (accessed on 10 March 2021).
31. U.S. Nuclear Regulatory Commission. Section 10.2—Pressurizer Pressure Control System. In Westinghouse Technology Systems Manual. 2012. Available online: https://www.nrc.gov/docs/ML1122/ML11223A287 (accessed on 10 March 2021).
32. Spagnuolo, A. *BB-9.2.1-T010-D005: Breeding Blanket Load Specifications Document*; EUROfusion Internal Deliverable, EFDA_D_2NLL6N v1.1; EUROfusion: Garching, Germany, 2019.

Article

Experimental Investigation of EU-DEMO Breeding Blanket First Wall Mock-Ups in Support of the Manufacturing and Material Development Programmes

Bradut-Eugen Ghidersa [1,*], Ali Abou Sena [1], Michael Rieth [2], Thomas Emmerich [2], Martin Lux [1] and Jarir Aktaa [2]

1   Karlsruhe Institute of Technology, Institute for Neutron Physics and Reactor Technology, 76131 Karlsruhe, Germany; ali.abou-sena@kit.edu (A.A.S.); Martin.Lux@kit.edu (M.L.)
2   Karlsruhe Institute of Technology, Institute for Applied Materials, 76131 Karlsruhe, Germany; michael.rieth@kit.edu (M.R.); thomas.emmerich@partner.kit.edu (T.E.); jarir.aktaa@kit.edu (J.A.)
*   Correspondence: bradut-eugen.ghidersa@kit.edu

Abstract: This paper presents the testing campaign of the two First Wall mock-ups in the HELOKA facility, one mock-up having a 3 mm thick Oxide Dispersion Strengthened (ODS) steel layer on its surface and the other featuring a tungsten functionally graded cover. Special consideration is given to the diagnostics used for these tests, in particular, the measurement of the surface temperature of the tungsten functionally graded layer with an infrared camera. Additionally, the paper looks into the uncertainty associated with the calorimetric evaluation of the applied heating power for these experiments.

Keywords: DEMO blanket; first wall; ODS steel layer; tungsten functionally graded coating; experimental investigation

Citation: Ghidersa, B.-E.; Abou Sena, A.; Rieth, M.; Emmerich, T.; Lux, M.; Aktaa, J. Experimental Investigation of EU-DEMO Breeding Blanket First Wall Mock-Ups in Support of the Manufacturing and Material Development Programmes. *Energies* **2021**, *14*, 7580. https://doi.org/10.3390/en14227580

Academic Editor: Alessandro Del Nevo

Received: 29 September 2021
Accepted: 8 November 2021
Published: 12 November 2021

Publisher's Note: MDPI stays neutral with regard to jurisdictional claims in published maps and institutional affiliations.

1. Introduction

The presently estimated surface loadings of the EU-DEMO blanket are pushing the design of the Helium Cooled Pebble Bed breeding blanket closer to the upper limits of the operation window for the Eurofer97 (European reduced activation ferritic martensitic steel). To mitigate this issue, various approaches have been investigated. To increase the temperature upper limits and improve the neutron irradiation performance, novel nanostructured Oxide Dispersion Strengthened (ODS) steels have been developed [1]. Additionally, to reduce the sputtering due to particle loading of the first wall (FW), the application of a tungsten coating has been considered [2]. Given the difference in the thermal expansion coefficient between the two materials, the experimental demonstration of the good adhesion and stability of such a coating under high heat flux loading is required. To evaluate experimentally the behavior of an ODS first wall under high heat loadings as well as the qualification of the tungsten coating procedures, two FW mock-ups have been manufactured and tested in the Helium Loop Karlsruhe (HELOKA) at KIT: (i) an FW mock-up with 3 mm ODS plate joined to Eurofer97 plate by diffusion welding [3], and (ii) an FW mock-up with Functionally Graded Tungsten/Eurofer97 coating (about 1.4 mm) by vacuum plasma spraying [4]. Both mock-ups have cooling channels of shapes and sizes used for the design of the DEMO blanket and have been subject to high heat flux cyclic loading while being cooled by helium at blanket relevant operating conditions of 8 MPa and 300 °C (mock-up inlet conditions). The FW mock-ups were installed inside the HELOKA vacuum vessel (VV), and heated using an electron beam gun (EBG) attached to the top of the VV. During testing the typical vacuum level was 10^{-4} mbar, a vacuum level that was found to be optimal for the EBG operation. More details and information about HELOKA may be found in reference [5].

The objectives of this paper are to cover the following: (i) testing of the FW mock-ups under blanket FW operating conditions, namely, cyclic high heat flux and cooling with helium at 80 bar and 300–350 °C, (ii) describing the diagnostics used in the testing campaign, especially the surface temperature measurement with an IR camera and the evaluation of the applied heating power, and (iii) presenting the relevant measurements obtained during the experiments. Regarding the paper structure, the FW mock-ups are introduced in Section 2. Then, in Section 3, the measurement sensors and diagnostic tools are presented, followed by Section 4, which contains the testing parameters for both mock-ups. Section 5 has two subsections presenting the experimental results of each mock-up. At the end of the paper, Section 6 is dedicated to the conclusions.

2. Mock-Ups and Experimental Set-Up Description

The two mock-ups, the Functional Graded Tungsten/Eurofer FW (hereafter designated by FG-FW) and the combined ODS/Eurofer97 FW (ODS-FW), have three and, correspondingly, five cooling channels of rectangular cross-section, 10×15 mm^2, with 2 mm filet radius at the corners (see Figures 1b and 2b). The only exception is the middle channel (3rd channel) in the ODS mock-up, which has a height of 11 mm, the wall thickness towards the heat-loaded surface being 2.5 mm, with 1 mm thinner than this for the rest of the channels. The channels are separated one from the other by a 5 mm thick wall. For the FG-FW, a thermocouple is inserted in a 1.5 mm hole made in the wall between the 2nd and 3rd channel at mid distance between the channel's inlet and outlet. The hole, drilled from the back-side of the mock-up, has a depth of 18 mm allowing monitoring of the temperature field at 3.4 mm below the heat-loaded surface (1.4 mm functional graded coating and 2 mm base material), as can be seen in Figure 1b.

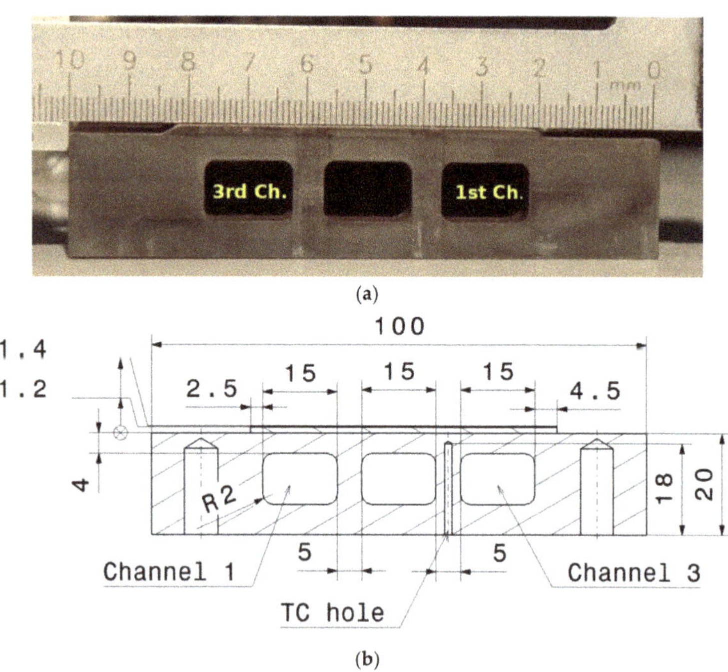

Figure 1. FG mock-up cross-section: (**a**) Picture of the mock-up cross-section as manufactured; (**b**) Drawing of the cross-section with dimensions including the final dimensions of the coating (1.2 mm FG layer +0.2 mm tungsten cover). Note that, during the final manufacturing steps, the mock-up lost about 0.6 mm from the initial 0.8 mm of the tungsten layer, as described in [4].

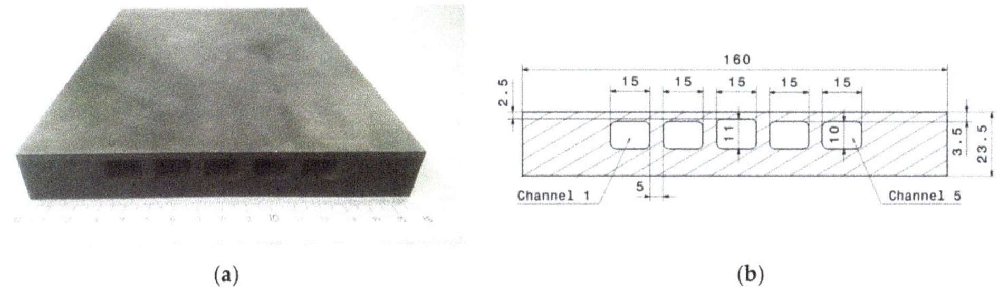

(a)   **(b)**

Figure 2. ODS mock-up: (**a**) the cooling plate with the cooling channels before the installation of the inlet/outlet manifolds; (**b**) cross-section through the mock-up with channels dimensions.

The FG mock-up was made out of a $300 \times 100 \times 20$ mm^3 Eurofer97 plate, the coating being applied over a 270×62 mm^2 surface. As can be seen in Figure 1, the coating is applied slightly asymmetric with respect to the plate width, one side being at about 20 mm away from the edge and at around 2.5 mm from the 1st channel side wall, the other side extending 4.5 mm from the 3rd channel wall.

For the ODS mock-up, the cooling channels were cut in a $208 \times 160 \times 24$ mm^3 plate through electro-discharge wire cutting, as can be seen in Figure 2a. From Figure 2b it can be seen that the cooling channels 1, 2, 4 and 5 have a wall thickness towards the heat-loaded side of 3.5 mm. From this wall thickness, 3 mm is ODS steel and 0.5 mm is Eurofer97. Channel 4's height was intentionally increased to 11 mm so that the channel wall contained only ODS steel. Thus, the wall thickness was reduced to 2.5 mm and the zone where the joining between ODS and Eurofer97 occurred is exposed to the coolant.

More details about the design and manufacture of the FG and ODS mock-ups are reported in references [3,4], respectively.

The helium distribution in the cooling channels was conducted using specially designed inlet/outlet manifolds. The design of these flow distributors was optimized for achieving a uniform mass flow distribution in the cooling channels of the ODS mock-up, assuming a nominal flow rate per channel of 40 g/s (200 g/s in total). As can be seen in Figure 3a, the manifolds have a large cylindrical chamber, 32 mm in diameter, where the helium either coming from the loop or from the channels gathers. For the inlet, the helium flows from this chamber towards the cooling channels through 11 mm in diameter and 30 mm long holes. These round channels have the same axis as the cooling channels themselves and have the function of stabilizing the flow reducing the transverse component of the flow. The Computational Fluid Dynamics (CFD) analyses performed with ANSYS CFX show that the expected deviation from the nominal value ranges from around 4% below the nominal flow rate, in the case of channel 1, and up to 3.5% excess flow in the case of channel 3 (see Table 1), values that are considered as acceptable for the foreseen testing conditions.

The inlet/outlet manifolds for the FG mock-up (Figure 3b) were manufactured without any prior CFD analysis by taking the same cross-section (in the flow direction) as for the ODS mock-up manifolds and simply reducing the flow distribution length from 100 mm (5 channels mock-up) to 60 mm (3 channels mock-up).

During testing, the mock-ups were cooled in parallel but, since the set-up does not allow controlling the flow independently for each mock-up, during a testing session, the surface heat loading was applied only on one mock-up, while the flow rate through that particular item was adjusted to the required value.

The two mock-ups were installed, one next to each other, on a joint fixation structure inside the VV, see Figure 4. This fixation structure was designed in a way that allows free thermal expansion of the mock-ups in the two directions parallel to the mock-ups surface but preventing an overall dislocation (movement) of the mock-ups. Each mock-up was fixed to the holding plate by four stainless steel bolts, two at the manifolds and two on the massive parts of the plates that were free to move in slotted holes. Several ceramic discs were placed between the mock-ups and the holding plate to minimize heat loss by conduction.

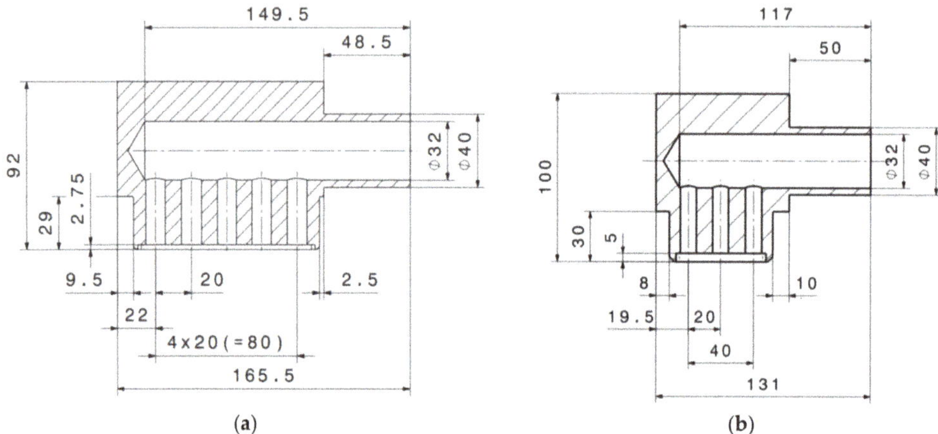

Figure 3. Geometry of the inlet/outlet manifolds: (**a**) ODS mock-up; (**b**) FG mock-up.

Table 1. Calculated channel flow rates under steady-state conditions and deviation from the nominal values (positive values represent a deficit in the flow and the negative values correspond to an excess with regard to nominal values).

Flow Channel mn1	Flow Calculated at the Channel Inlet (g/s)	Deviation from Nominal Flow (%)
Flow CH1	38.4	4
Flow CH2	39.5	1.25
Flow CH3	41.4	−3.5
Flow CH4	40.2	−0.5
Flow CH5	40.4	−1
Total	200.2	−0.1

Figure 4. Experimental assembly featuring the two mock-ups.

3. Measurements & Diagnostics

The two mock-ups were installed in parallel, as can be seen from Figure 5, the coolant parameters being monitored. Thus, the flow rate through each mock-up and their inlet and outlet temperatures were monitored, as indicated in Tables 2 and 3.

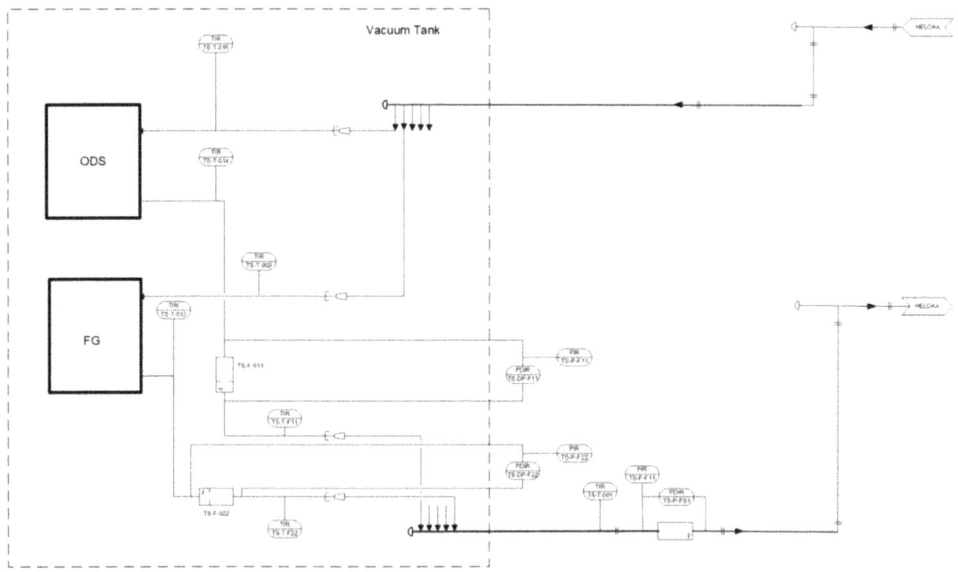

Figure 5. Piping and Instrumentation Diagram for the experimental set-up.

Table 2. List of helium parameters monitored for the ODS mock-up.

Parameter	Sensor/Signal Label	Sensor Type	Range Accuracy
Flow rate	TS-F-011	Orifice (d = 14.365 mm, β = 0.53921)	-
Differential pressure over the flowmeter	TS-DP-F11	Membrane differential pressure sensor (type SIEMENS Sitrans P DSIII,)	0 to 5 bar 0.065% from range
Pressure before the flowmeter (+tap)	TS-P-F11	Membrane pressure sensor (type DMP 320 from BD Sensors GmbH)	0 to 100 bar 0.1% from range
Temperature after flowmeter	TS-T-F11	Thermocouple (type K, class 1)	<1000 °C 0.4% × temperature (in °C)
Inlet temperature	TS-T-015	Thermocouple (type K, class 1)	<1000 °C 0.4% × temperature (in °C)
Outlet temperature	TS-T-014	Thermocouple (type K, class 1)	<1000 °C 0.4% × temperature (in °C)

Table 3. List of helium parameters monitored for the FG mock-up.

Parameter	Sensor/Signal Label	Sensor Type	Range Accuracy
Flow rate	TS-F-022	Orifice (d = 14.365 mm, β = 0.53921)	-
Differential pressure over the flowmeter	TS-DP-F22	Membrane differential pressure sensor (type Sitrans P DSIII from SIEMENS)	0 to 5 bar 0.065% from range
Pressure before the flowmeter (+tap)	TS-P-F22	Membrane pressure sensor (type DMP 320 from BD Sensors GmbH)	0 to 100 bar 0.1% from range
Temperature after flowmeter	TS-T-F22	Thermocouple (type K, class 1)	<1000 °C 0.4% × temperature (in °C)
Inlet temperature	TS-T-003	Thermocouple (type K, class 1)	<1000 °C 0.4% × temperature (in °C)
Outlet temperature	TS-T-013	Thermocouple (type K, class 1)	<1000 °C 0.4% × temperature (in °C)

The HELOKA facility has a dedicated data acquisition for the experimental test sections based on National Instruments CompactRIO NI cRIO-9025. This system runs in parallel with the HELOKA loop own control system, which is based on Siemens SIMATIC PCS7 (see reference [6] for further details on HELOKA control). The two systems are synchronized by obtaining the time stamp from the same radio clock (Siemens SICLOCK TC 400), otherwise they can be independently developed and maintained.

3.1. Mass Flow and Pressure Measurements

The helium mass flow rate is measured downstream of each mock-up using orifice flowmeters designated in Figure 5 as TS-F-011 for the ODS mock-up and TS-F-022 for the FG mock-up. The two flowmeters can operate at high pressure, up to 110 bar, and temperatures up to 550 °C. The flowmeter and their associated transducers were designed initially for another experiment operating at 400 °C, 8 MPa and 50 g/s. Since the two experiments required higher flow rates (170 g/s for FG mock-up and 200 g/s for the ODS mock-up), the original transducers were replaced by new ones of the same type (SIEMENS Sitrans P DSIII) but having a larger measuring range (0 to 5000 mbar). Differently from the original transducers that were calibrated and set to deliver the mass flow directly, the new transducers were used to measure the pressure loss over the orifice, the mass flow being calculated in the cRIO according to the provisions of DIN-EN-ISO 5167-1 and 2:2004-01 standards. The calculation uses the helium pressure and temperature measured at the level of the flow meter using DMP 320 sensors from BD Sensors GmbH (<0.1% uncertainty at a measuring range of 100 bar) and thermocouples type K (3 mm sheath; tolerance class 1), respectively.

Using the existing orifice flow meters means that we needed to cope with large pressure drops at the flow meters' level: around 1.5 bar for the FG mock-up and 2.9 bar for the ODS mock-up. However, the HELOKA loop can easily cope with such loads, while the systematic uncertainty for the mass flow measurement stays constant at around 1.31% for all of the measuring range of interest. This value, calculated according to DIN-EN-ISO 5167-2:2004 and ISO/TR-15377:2007, already takes into account the fact that, for the pressure and the differential pressure, there is an additional uncertainty of 0.54% associated with the National Instruments card (NI 9203) that was used during the testing. In the calculation, this uncertainty is added to the sensor's systematic uncertainty (0.065% from the measuring range).

3.2. Coolant Temperature Measurements

Four thermocouples (TCs) of type K (tolerance class 1) with a 3 mm sheath diameter measure the helium temperatures at the inlet and outlet pipes of the two mock-ups. The sensors are installed on the piping 100 mm downstream from the mock-ups inlet/outlet connecting points. These thermocouples are inserted directly in the helium stream to reduce the response time.

During testing, the inlet sensors are used to adjust the power of the HELOKA heater to match the temperature set-point for the loop control. In addition to this, these sensors, together with the corresponding outlet temperature sensors, are used for the calorimetric power evaluation.

3.3. Mock-Up Temperature Measurement

The experiments were focused mainly on phenomena occurring on the surface of the mock-up, therefore dedicated measurements of the mock-up steel are not included. The only exception is the thermocouple inserted in the wall between channel 2 and 3 of the FG mock-up. As indicated in [4], this sensor was used during manufacturing to monitor the surface temperature when the coating step was performed. During testing in HELOKA, the sensor, a type K sensor with class 1 accuracy, was included in the experimental data acquisition as an additional way to monitor the temperature of the steel near the heat-loaded surface. For the ODS mock-up, such a sensor was not needed during manufacturing and applying a sensor afterwards up to 2 mm below the surface was considered too intrusive, the sensor would be penetrating into the ODS layer.

Measuring the heated surface temperature using thermocouples was difficult to implement for the present mock-ups. Having relatively thin walls (3.5 and 4 mm), placing even thin thermocouples on the surface would impact on the local heat transfer, the thermocouples having a stainless-steel sheath with lower conductivity than the mock-up material. Additionally, since the experiments were conducted in a vacuum at high heat fluxes, the thermocouple wires needed to be brazed to the mock-up surface, the presence of the brazing increasing further the impact on the local heat transfer. This brazing needed to be performed on the whole path the thermocouple followed on the heated surface (not only on the tip), to avoid the (relatively fast) destruction of the thermocouple by the electron beam.

For these reasons, a thermal imaging infrared (IR) camera of type X6580 sc, from FLIR Systems, Inc., (Wilsonville, OR, USA), was used to measure the surface temperatures of the tested mock-ups. The camera images were also used to define and adjust the electron beam gun heating pattern by looking at the thermal response of the mock-up.

The camera is installed outside the vacuum vessel, behind an ZnSn window, and the images are recorded through a mirror system, the same as used in reference [7]. A detailed description of the camera set-up can be found in reference [7] and will not be repeated here. The only difference to the experiments presented in [7] is the use of a different camera filter (NA 3.90–4.01 μm 60%) calibrated for measurements between 300 °C and 1500 °C, the temperature range to be observed exceeding the limits of the filter used in [7], which was calibrated for temperatures between 100 °C and 600 °C.

The treatment of the images was carried out using FLIR Systems, Inc., (Wilsonville, OR, USA) proprietary software, the FLIR ResearchIR Max Version 4.40.9.30 (7 February 2019). In order to provide reliable values for the surface temperature, this software requires setting two parameters: the transmission coefficient and the surface emissivity. For the HELOKA test stand, similarly to the tests presented in [7], the transmission coefficient is always set to 0.33. The surface emissivity depends on the material and the status of the mock-up surface; therefore, its value is set (calibrated) at the beginning of each (daily) run and checked again at the end of the day. The calibration is conducted by selecting a Region of Interest (ROI) on the surface of the mock-up and bringing the mock-up at 300 °C using the coolant temperature only. In the absence of external heating and due to the fact that the mock-ups are in vacuum, there is practically no temperature gradient across the mock-up walls, the surface having the same temperature as the coolant. Thus, by providing the FLIR ResearchIR the coolant temperature value, the software is able to calculate the corresponding emissivity.

As indicated in [7], in the case of EUROFER97 mock-ups or made out of similar ferritic martensitic steels, calibrating the emissivity parameter at a coolant temperature of 300 °C results in a reliable reading for temperatures in the range of 300 °C to 550 °C, the parameter value staying almost constant above 300 °C. Thus, for the ODS mock-up, the surface temperatures measured with the IR camera are assumed to have the same uncertainty as those mentioned in [7], namely, 2% uncertainty for (area) averaged temperatures and 4% for maximum values.

The situation is different when it comes to tungsten-covered surfaces due to its substantial change in emissivity with temperature. During previous experiments, in which massive tungsten blocks were heated up from 300 °C or 450 °C up to temperatures above 1000 °C, it was observed that the IR camera readings were higher than the real surface temperatures. In these experiments, similar to the first wall experiments presented here, the emissivity parameter was set at the beginning of the heating cycle when the mock-up was in thermal equilibrium with the environment.

A well-defined correction formula for such experiments is difficult to produce due to the multitude of factors coming into play for high heat flux testing, the most significant being the change in the surface roughness that occurs during the experiment. Nevertheless, in order to have a better idea of the temperature levels on the surface of the mock-up, we performed a complementary experiment to deduce a correction relation for the IR camera readings over a temperature range of 300 to 800 °C. The experiment used the central part of the FG mock-up, a part that remained available after the samples for the material studies were cut off. Thus, the test object is only 202 mm long, about 30 mm being cut away from each end. The mock-up is installed inside the HELOKA vacuum vessel at the same position as the original mock-up. For the purpose of this experiment, the heating of the mock-up was conducted using six strip heaters (two heaters per each cooling channel). Three thermocouples (TC12–14) are fixed on the steel surface and aligned transversely to contact the tungsten coating as shown in Figure 6. The fourth thermocouple has the same position as the thermocouple used to monitor the mock-up base material temperature during the high heat flux testing. The TCs are type K (accuracy class 1) with 1 mm sheath diameter and temperature limit of 1000 °C.

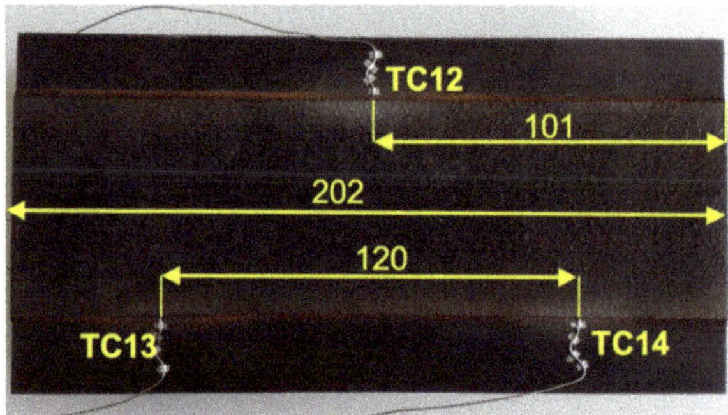

Figure 6. Experimental set-up for determining the correlation between IR camera temperature readings and the real surface temperature for tungsten surfaces.

The data for the correction formula were gathered during three runs. In each run, the six heaters were turned on to heat the mock-up top surface to 300 °C and the heaters' power supplies were adjusted so that the readings of the four thermocouples was stationary. This initial state was used to set the emissivity parameter of the IR camera the same way as was carried out in the FG mock-up testing campaign. Once the setting of the emissivity was performed, the temperature of the mock-up was increased in steps of 100 °C until reaching the final temperature of 800 °C. During the run, the temperature measurements of TC12–14 were saved while the IR camera software recorded images of the test object. Three ROI (Region of Interest), in the shape of 3 × 3-pixel regions, were placed on the tungsten coating area close to the thermocouples. Under stationary conditions, these regions should have the same temperatures (or very close ones) as the neighboring thermocouples on the steel surface. The temperature values obtained from FLIR ResearchIR for the three ROIs and the corresponding thermocouple readings were processed using OriginPro software (OriginPro, Version 2021, OriginLab Corporation, Northampton, MA, USA). Figure 7 shows the FV (Fasano and Vio [8]) linear fitting of the thermocouples' (TC12–14) measurements versus the IR camera's temperature measurements in the three runs. This fitting method was preferred to the OriginPro default method (York) since it allows us to take in account uncertainties in both coordinates/variables. All thermocouples measurements were plotted associated with both the systematic and random errors.

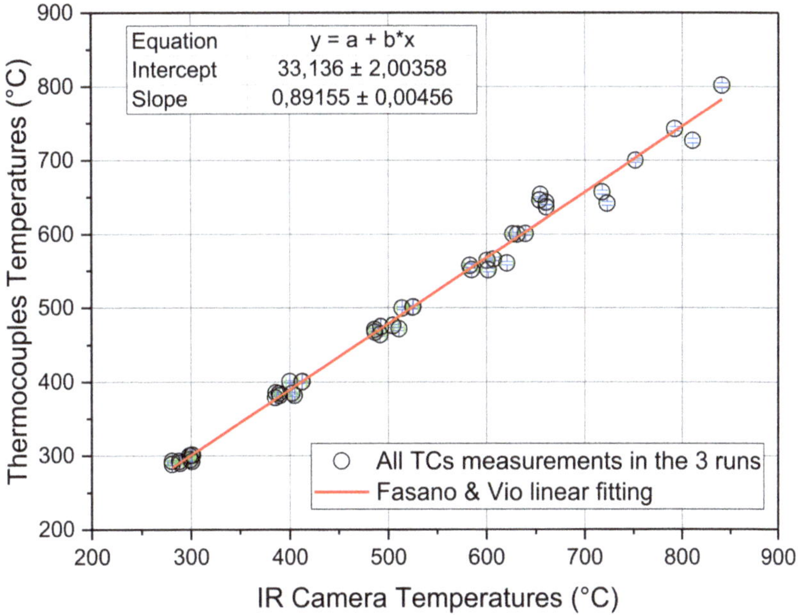

Figure 7. Fasano and Vio [8] fit of the IR camera measurements and the W-slab thermocouple.

4. Testing Parameters and Loading Conditions

The two FW mock-ups were subject to high heat cyclic loading, and were cooled by helium at 8 MPa. In both cases, the loading starts when the mock-ups have a temperature level around 300 °C. A predefined scanning pattern allows the application of a rather uniform surface heat load by means of an electron beam gun. The selected loading level is reached relatively fast, after 1 s, the beam power being already at the selected level. The duration of the beam pulse is mostly determined by the time needed by the helium outlet temperature to reach a stationary value so that the value of the power deposited on the mock-up can be evaluated calorimetrically. This duration was set at 150 s for the FG mock-up and 120 s for the ODS-mock-up. After the beam is shut-off, a cool-down period of similar length is needed so that the mock-ups return to the same thermal state as before the beam was applied, namely a uniform temperature of 300 °C.

The EB gun scanning pattern is defined in such a way that the loading of the weld seams between the inlet/outlet manifolds is avoided. Additionally, on the transverse direction to the channels, the heating patterns are limited to the area above the cooling channels. Both mock-ups have, left and right of the channel's region, massive steel parts that are of limited interest for the experiment, but they can influence the duration of a loading cycle, the more heat that is applied over these regions, the longer it will take for the mock-ups to cool-down. The shape of the electron beam pattern and the uniformity of the heating loads were estimated by looking at the mock-up's surface thermal images recorded by the IR camera.

For the FG mock-up, the testing parameters listed in Table 4 were selected so that the temperature of the steel substrate was maintained below the maximum working temperature of the material (550 °C), as indicated in reference [4]. Based on their calculations, for an applied heat flux of 750 kW/m^2, the steel temperature should be around 530 °C. However, estimating the heat flux is not trivial, an estimation directly from the beam power being far from straightforward due to the fact that the fraction of the power transferred as heat flux to the mock-up depends strongly on the surface material, and the FG mock-up has both tungsten and steel. For the experiment, the setting of the beam power was conducted at the beginning of the testing campaign by increasing, progressively, the beam power while maintaining the specified cooling conditions, until the thermocouple installed in the middle of the mock-up indicated a temperature around 500 °C. Using the simplified convection/conduction model from [4] for the wall temperature increase along the mock-up channel, having 500 °C half-distance along the channel would correspond to a heat flux of 625 kW/m^2. Under these conditions, the same model predicts that the substrate maximum temperature would stay below 550 °C. However, the evaluation of the experimental results indicate that the calorimetrically estimated heat flux was close to 730 kW/m^2, which means that the real heat transfer coefficients were higher than those from the model.

Table 4. Testing parameters of the FG mock-up (see [4] for details).

Helium Mass Flow Rate	170 g/s
Helium inlet temperature	300 °C
Helium pressure	8 MPa
Maximum heat flux	700 kW/m^2
Substrate temperature limit	<520 °C
Heating on/off time	150 s/150 s
Number of cycles	1000

Figure 8 shows an infrared image of the mock-up taken during testing. In this picture, one can see the thermal response to the applied heating pattern, indicating that the area of uniform heat flux is somewhat smaller than the coated area, both in the direction of the flow and in the transverse direction. This is mainly due to the fact that the beam spot has a diameter of about 15 mm and a Gaussian shape. This means that, in order to avoid heating the welds or the bulk steel on the sides, we have to restrict the scanning pattern inside the coated area. Nevertheless, there is an area sufficiently large towards the outlet of the mock-up that has suitable testing conditions.

In the case of the FG mock-up, the objective was to cycle for 1000 times the tungsten coating at the highest surface heat flux possible while keeping the substrate (steel) at temperatures below 550 °C. The flow rate was set to the maximum value allowed by the experimental set-up. For the ODS mock-up, the main objective was to apply 100 cycles while reaching a specified surface temperature under typical blanket First Wall cooling conditions, namely 40 g/s in each cooling channel (200 g/s in total). Table 5 presents the complete test matrix for the ODS mock-up. After the original testing objectives were reached, it was decided to apply the same load as that for the last fatigue tests (0.9 MW/m^2) and maintain it for a longer time in order to activate creep or other thermal effects in the ODS layer (see [4] for more details). The duration of these pulses was chosen having in mind a typical EU-DEMO pulse duration.

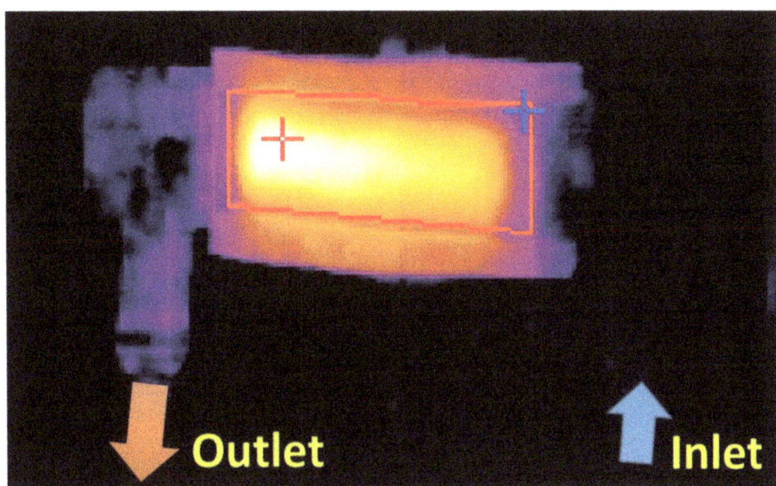

Figure 8. Thermal image of the FG mock-up by the IR camera. The red rectangle marks the coated area and the red and blue cross marks the position of the hottest and, respectively, the coldest point on the coating. The coolant at 300 °C enters the target from the right side and flows to the left side.

Table 5. Test matrix for the ODS mock-up.

Cycles	He Inlet T	Surface T	Heat on/off Time	Est. Heat Flux
100	300 °C	550 °C	120 s/120 s	700 kW/m^2
100	300 °C	600 °C	120 s/120 s	800 kW/m^2
100	350 °C	650 °C	120 s/120 s	900 kW/m^2
7	350 °C	650 °C	2 h/5 min	900 kW/m^2

For the ODS mock-up, the selection of the heating pattern was conducted following the same general guidelines and procedures as for the FG mock-up. Since the ODS layer covered the whole surface, the lateral extent of the electron beam scanning pattern was determined by looking at the two junctions between the inlet and outlet manifolds to the main plate. In the flow direction, the extent of the scanning pattern was decided based on a preliminary stress analysis carried out with ANSYS software. The reason for this was that the welds between the manifolds and the main plate contained ODS material and, at the time when the experiment was performed, we had no thorough investigation and qualification of these kind of welds (and the associated heat treatment). As such, in planning the experiment, it was decided to keep these zones at stress levels well below the allowable limits. Following these investigations, it was decided to use a heating pattern with a length of 11 mm in the flow direction.

The stress analysis also showed that, for the first two series of cycles (700 kW/m^2 and 800 kW/m^2), the stresses stay below the allowable limits defined by RCC-MRx code [9] for EUROFER97. When going to higher surface loadings (900 kW/m^2) and higher surface temperatures, the design code ratcheting criteria (3Sm) is no longer fulfilled. However, by increasing the inlet temperature to 350 °C and using a surface heat flux of 900 kW/m^2, the model predicts thermal stresses that exceed by only 10% the allowable values for EUROFER. These values occur within the mock-up wall towards the heated surface, which is mostly made out of ODS steel. While the characterization of the ODS steels is an on-going activity, it is expected that the mechanical properties of these steels will see an improvement as compared with the EUROFER in the range of 20%. As a consequence, it was decided that, for the tests aiming at 650 °C, the helium inlet temperature was increased from 300 °C to 350 °C.

5. Results and Discussion

5.1. Results for the FG Mock-Up

In this section, the results of the FW and FG mock-up experimental runs are presented. Testing of the FG mock-up was performed to investigate the durability of the Tungsten/Eurofer97 coating by exposing it to high heat loadings. Applying this loading repeatedly, the experiment aimed to demonstrate the robustness of the coating. Figure 9 shows the measured values of the inlet helium temperature (T-in), outlet helium temperature (T-out) and the internal mock-up temperature (T-TC) for a representative group of seven thermal cycles recorded during the experimental run of the 5 August 2019. The reproducibility of the three temperatures, measured in these cycles, shows that the experiment was producing similar heating cycles. As can be seen in Figure 9, when the heat load is applied to the mock-up, the outlet helium temperature reaches a maximum value of 314.5 °C, while the thermocouple embedded into the mock-up measures a temperature of 488 °C. During the beam-off period, both the outlet helium temperature and internal mock-up temperature go down to the inlet helium temperature level of 300 °C. The good reproducibility of the testing conditions can be seen also by compiling an "averaged" loading cycle. Thus, by taking 15 consecutive cycles (the 7 cycles shown in Figure 9 and the following 8 cycles) and treating them as individual realizations of a typical 300-s-long loading cycle, we can compute an averaged evolution of the various parameters. Figure 10 shows the average values over the 15 cycles of the temperature measured by the mock-up thermocouple and the associated standard deviations (random uncertainties). From this figure, one can see that the temperature near the surface of the mock-up reaches, rather quickly, a stationary value having a standard deviation in that region around 0.01 °C, which indicates a very good reproducibility of the testing conditions. The situation is slightly different during the ramp-up and ramp-down phases when the standard deviation can reach up to 14 °C. While the duration of the transient phases is the same for all of the 15 cycles, the large deviation could be due to the fact that during the heating-up phase, there might be slight variations in the electron beam gun operation. Moreover, the instant the shut-off signal for the beam is initiated could slightly differ from cycle to cycle, leading to high deviations in the cool-down phase.

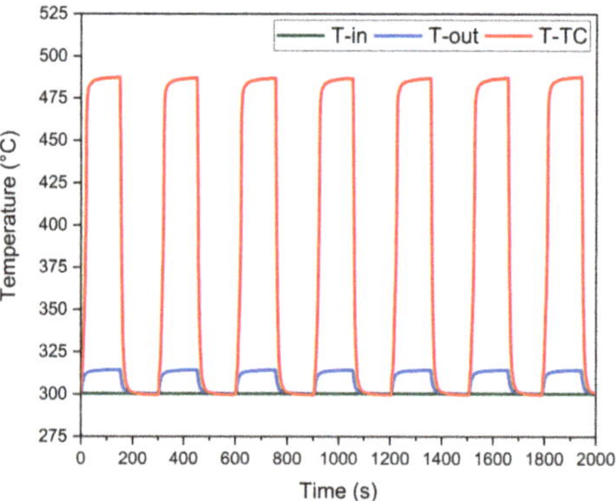

Figure 9. Temperatures (T-in, T-out, T-TC) in 7 cycles for FG mock-up during the run 5 August 2019.

From the available data we can evaluate the power increase in the helium stream. For this we apply a two-step procedure as follows:

1. From the measured data, we compile first typical evolution profiles for the helium mass flow rate $\dot{m}$, and the mock-up inlet and outlet temperatures, T_{in} and T_{out} in a similar way it was carried out for the mock-up thermocouple measurement. Together with the averaged values we also obtain the associated standard deviation,

which will give us the measure of the random uncertainty. The associated systematic uncertainties are calculated using the averaged values of the parameters.

2. From these averaged profiles, the evolution of the power rise over the mock-up is calculated using the formula

$$P = \dot{m}C_p(T_{out} - T_{in}) \tag{1}$$

where C_p is the helium specific heat capacity. Here, for the specific heat capacity (C_p) of helium, we consider a value of 5.195 kJ/kg·K with an uncertainty of 0.3% according to reference [10]. The associated systematic uncertainty is calculated as

$$u_P^2 = \left(C_p(T_{out} - T_{in})u_m\right)^2 + \left(\dot{m}(T_{out} - T_{in})u_{Cp}\right)^2 + \left(\dot{m}C_p\right)^2\left(u_{Tout}^2 + u_{Tin}^2\right) \tag{2}$$

where u_m, u_{Tout} and u_{Tin} are the systematic uncertainties calculated in the first step, while $u_{Cp} = 15.585$ W/kg·K. The random uncertainties are obtained using the same formula and taking the corresponding random uncertainties for the mass flow and temperatures, the random uncertainty for the specific heat capacity being taken as zero.

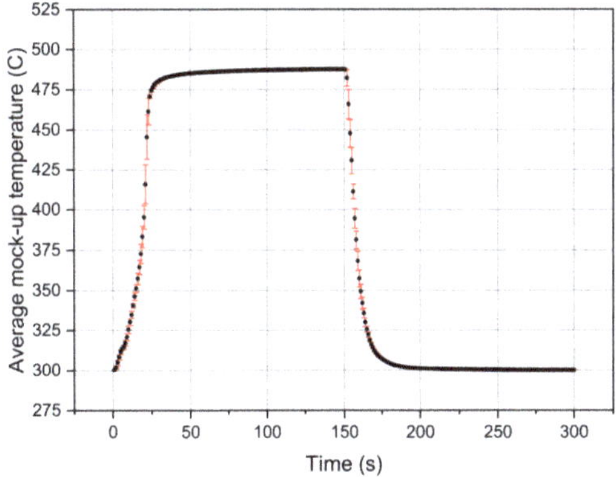

Figure 10. Temperature evolution at the mock-up thermocouple position averaged over 15 cycles during a run from 5 August 2019 for FG mock-up.

Similar to the thermocouple measurements shown in Figure 10, the power applied over the 15 cycles shows a good reproducibility, as can be seen in Figure 11a. In this figure, the mean values for the power and the associated random uncertainties are shown. From this figure, one can see that the time evolution is slightly different from that in Figure 10, the profile reaching a stationary value only shortly before the heating period stops. This behavior is mainly due to the impact of the outlet manifold thermal inertia to the helium power intake evolution, the hot helium exiting the mock-up channels losing a part of its enthalpy to the colder manifold walls. Once the manifold reaches the same temperature as the helium coming from the heated zone, the outlet temperature signal becomes stationary. Due to this delay in the outlet temperature measurement, the power applied by the electron beam gun to the mock-up can be correctly estimated only from the part where stationary conditions are achieved, and all of the input power exits the system as helium enthalpy. Taking the last 30 values before the end of the beam-on period, we obtain for the applied power a value of 12.3 kW with a standard deviation of ±0.04 kW. However, when taking into account the systematic uncertainties, the total uncertainty of the estimated power increases to 1.5 kW, which represents 12.3% from the estimated value. Assuming that the heat flux is applied uniformly over the whole coated area, the resulting applied average heat flux is then 734 kW/m² ± 90 kW/m², a value obtained by dividing the power by the area of the loaded zone ($1.674 \cdot 10^{-2}$ m²)

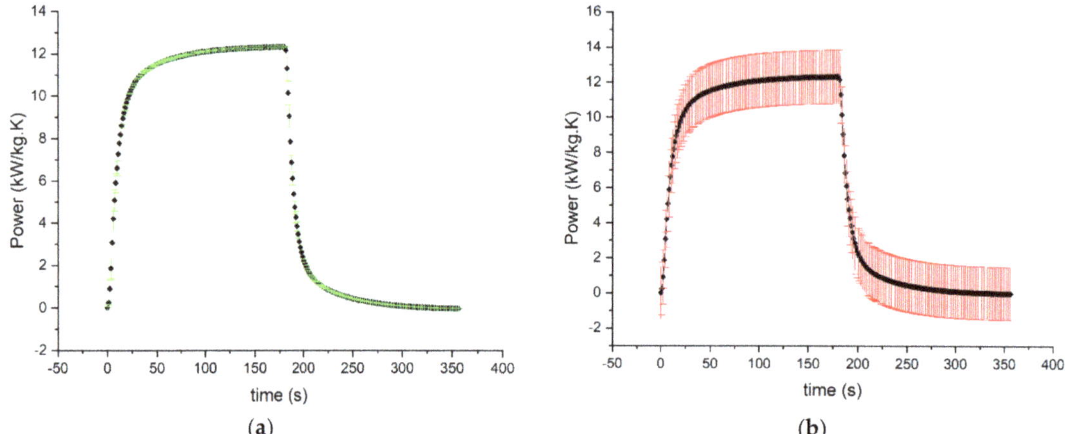

Figure 11. Helium power intake averaged over 15 pulses (run from 5 August 2019): (**a**) power with associated random uncertainty; (**b**) power with associated measurement uncertainty.

These results show that there is a relatively high uncertainty in the estimated power levels and, consequently, on the evaluated heat fluxes, despite using sensors with low uncertainty. Looking more in detail, one can see that this is due to having a low temperature increase over the mock-up: while the temperature sensors themselves have a low uncertainty (0.4% from the temperature, which corresponds roughly to 1.2 °C in our case), the temperature difference, which is close to 12 °C, has an associated uncertainty of 1.7 °C or 14.7% relative uncertainty. This high uncertainty dominates, in fact, the overall uncertainty of the power evaluation at any point during the pulse, as can be seen in Figure 11b.

A typical temperature distribution of the FG mock-up surface obtained by the IR camera can be seen in Figure 8. In this picture, the temperature scale was set from 300 °C (black) to 800 °C (white), and the image was recorded using a Plateau Equalization (PE) algorithm available in the FLIR ResearchIR Max Version 4.40.9.30 (7 February 2019) software from FLIR Systems, Inc., Wilsonville, OR, USA. In this picture, the maximum surface temperature recorded by the IR camera is close to 750 °C and is located at the white area near the helium outlet side of the mock-up. Using the correction formula deduced above (see Figure 7), the real surface temperature should be around 670 °C, as can be seen in Figure 12, where the axial and transverse temperature profiles through the point of maximum temperature are plotted.

For the last 200 cycles the beam power was increased, the power deposited into the mock-up being 13.25 ± 1.56 kW. This led to an increase in the surface temperature, the IR camera recorded value being around 800 °C, as reported in reference [4]. Using the correction formula introduced earlier, the estimated maximum temperature on the mock-up surface would be close to 750 °C (see the profiles using a star as a symbol in Figure 12).

Figure 13 shows the status of the heat-loaded coating area of the FG mock-up after finishing the testing campaign. The impact of the high heat flux on the surface can be observed clearly, however, the thermal stability of the coating successfully survived the thermal fatigue tests. The Tungsten/Eurofer97 coating layer did not fail or separate during (or after completing) the heating cycles.

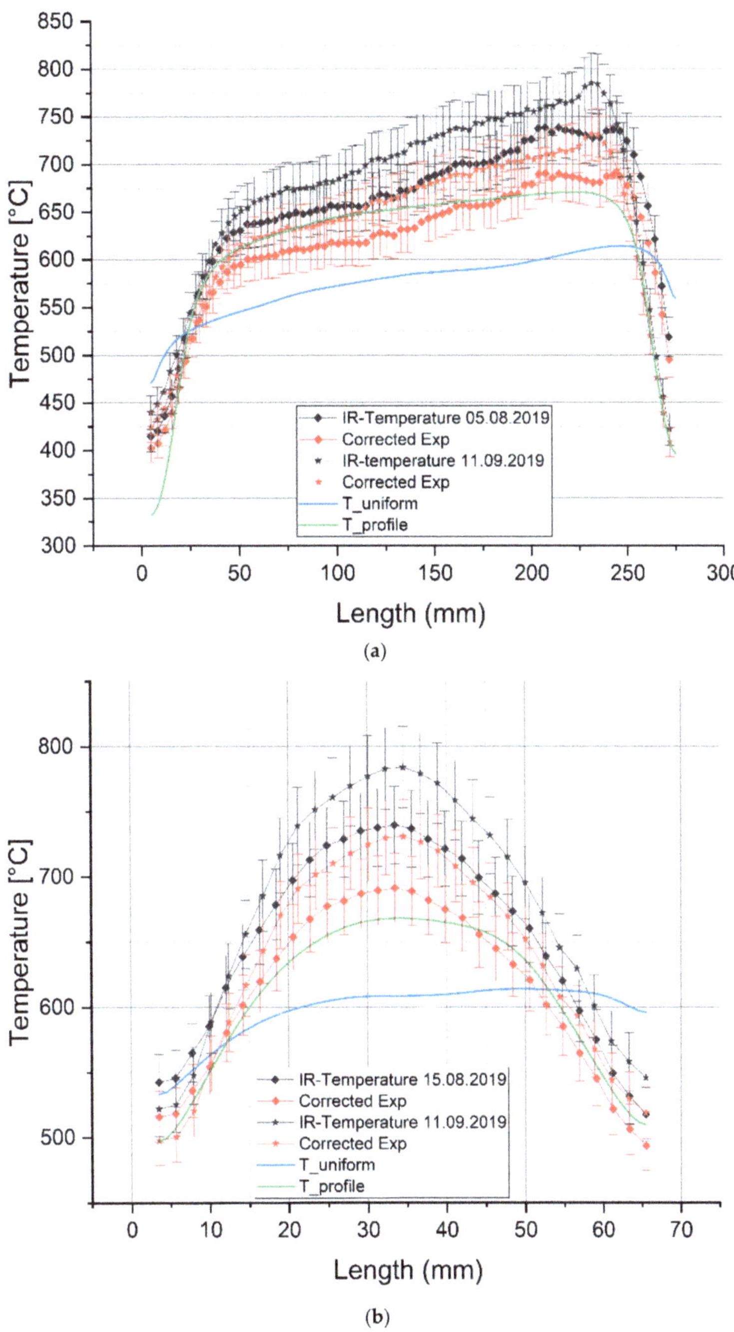

Figure 12. FG mock-up temperature profiles obtained from IR camera, the corresponding corrected profiles and (CFX) simulated temperature profiles: (**a**) axial temperature profile; (**b**) transversal temperature profile. The profiles are taken through the point of maximum temperature of the temperature field.

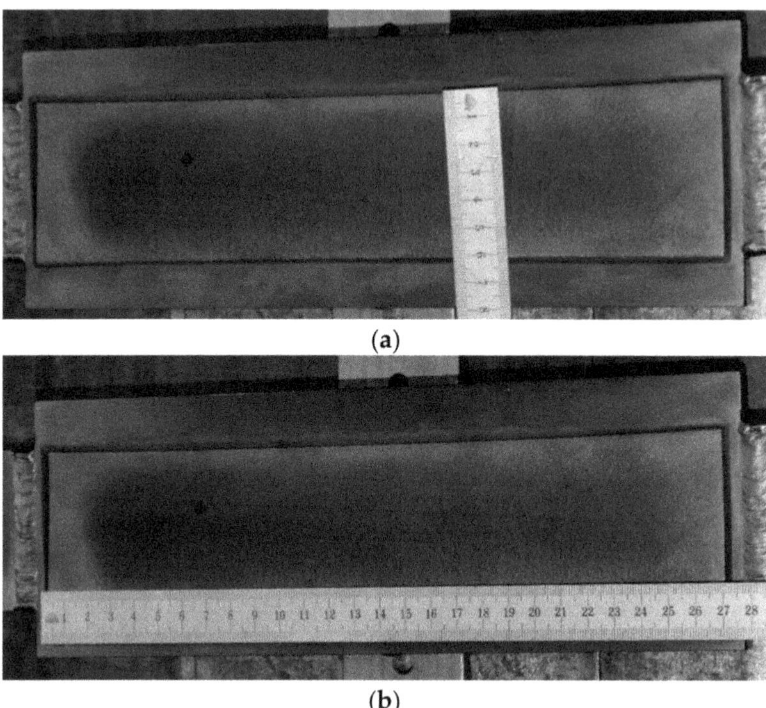

(a)

(b)

Figure 13. Status of the heat-loaded FG mock-up surface after the testing, the flow direction in the channels being from right to the left. The dark area in the center of the picture indicates the zone subject to high-heat flux loading: (**a**) transverse measurement of the FG-layer and of the loaded area; (**b**) measurement of the longitudinal dimensions of the FG-layer and the loaded surface.

From both the temperature profile measurements in Figure 12b and the shape of the darker zone in Figure 13, it can be seen that the heating profile is not uniform over the whole coated surface, meaning that the uniform heat flux level will be higher in the middle of the coating, dropping quickly towards the sides. Given the fact that the beam profile can be approximated with a Gaussian, the profile on the sides will follow the same law, the width being determined by the beam spot diameter, which has been estimated to 13.56 mm. To understand better what the real loading conditions on the mock-up were, we simulated the experiment with ANSYS CFX (2021 R1) both using a uniform heat flux and a profiled heat flux. In both cases the coolant enthalpy rise was the same, 12.3 kW, as in the experiments from 5 August 2019. For the profiled heat flux, the area where the applied heat flux is constant (red zone in Figure 14a) has been adjusted so that the resulting temperature field (Figure 14b) matches the darkened area observed on the surface of the mock-up from Figure 13. The resulting heat flux profile has a peak value close to 930 kW/m^2 for the same rise in helium enthalpy as for the case when the heat flux is applied uniformly over the whole coated area. The temperature profiles, both in the axial and transverse direction, are in good agreement with the corrected profiles obtained from the IR camera readings from the experiment on 5 August 2019.

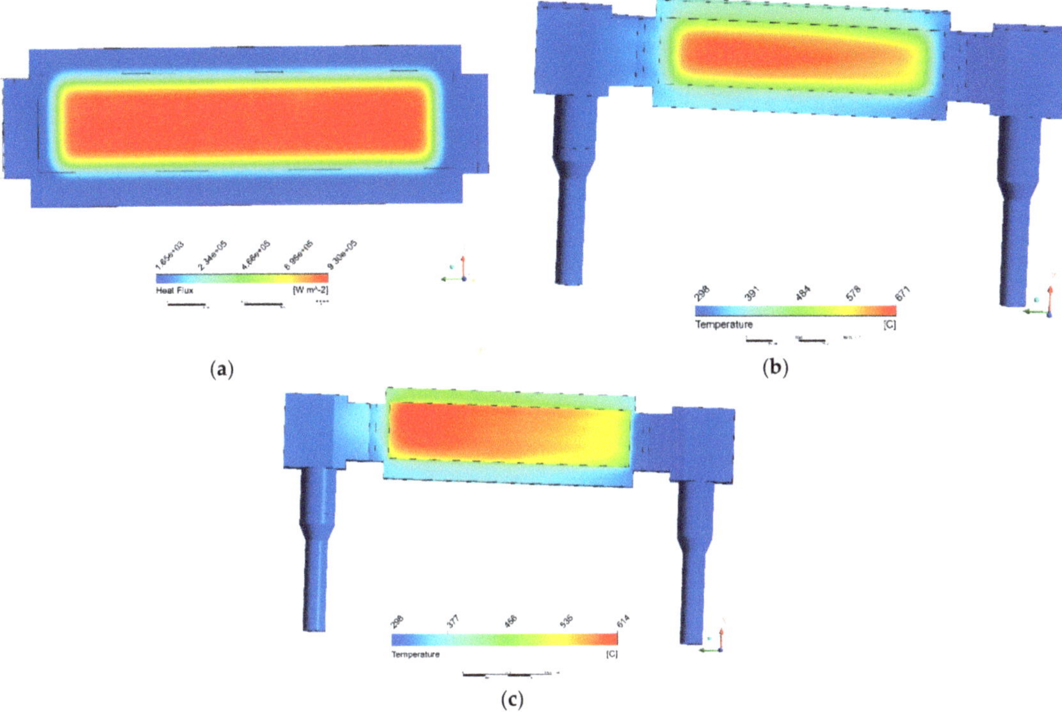

Figure 14. Surface temperature field simulated with CFX: (**a**) profiled heat flux distribution (930 kW/m^2 peak flux); (**b**) surface temperature using the profiled heat flux; (**c**) surface temperature obtained for a uniform heat flux of 734 kW/m^2.

5.2. Results for the ODS Mock-Up

Figure 15 shows the measured values of the inlet helium temperature (T-in) and outlet helium temperature (T-out) for seven short (4 min) heating cycles selected from the testing run of the 9 September 2019. Similar to the experimental sessions for the FG mock-up, in this case the experimental settings ensure a good reproducibility of the loading cycles and generate consistent results. For these tests, the outlet helium temperature reaches a maximum value of 307 °C during the heating-on phase and then decreases down to 301 °C at the end of the heating-off phase of the heating cycle. Additionally, in this run, the inlet helium temperature is about 300.4 °C, which is slightly higher than the set value of 300 °C.

After the completion of the specified number of short pulses, the mock-up was exposed to seven long pulses (one on 7 October, three on 8 October and three on 9 October, 2019), each one being 2 h long. Figure 16 shows the inlet helium temperature (T-in) and outlet helium temperature (T-out) versus the time during the testing run of 8 October 2019. In those runs, the helium mass flow rate was 201 g/s, and the inlet helium temperature was about 351.4 °C. The outlet helium temperature has a maximum value of 360.3 °C during the heating phase and a minimum value of about 352 °C at the end of the heating-off phase of the heating cycle. The helium temperature rise is about 8.3 °C.

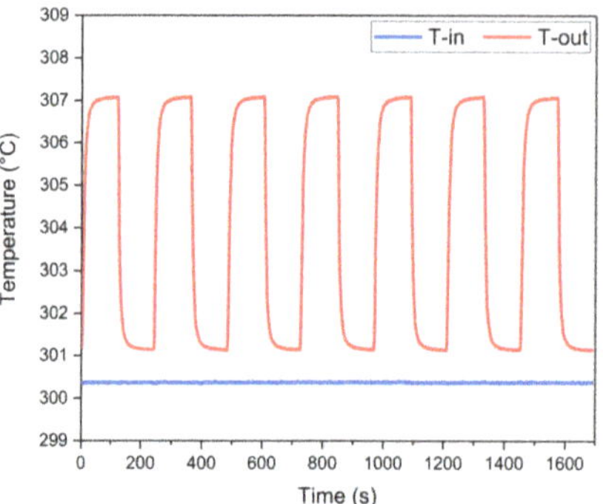

Figure 15. Temperatures (T-in and T-out) in 7 short cycles for ODS mock-up run 9 September 2019.

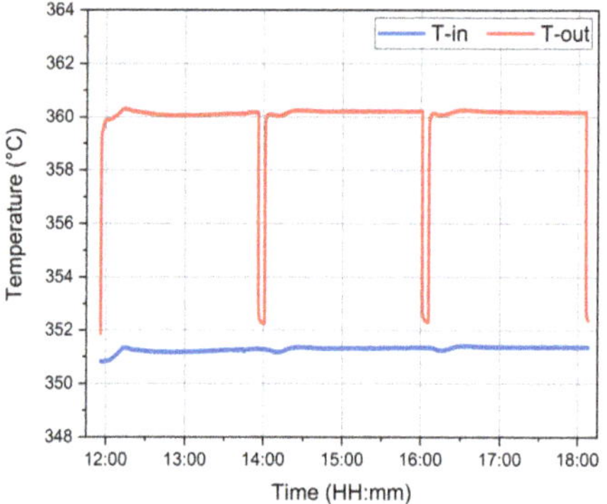

Figure 16. Temperatures (T-in and T-out) in the long cycles for ODS mock-up run from 8 October 2019.

Similar to the FG mock-up, Figure 17a shows the helium enthalpy rise over the mock-up averaged over 15 cycles as well as the associated random uncertainty. The low standard deviations associated with the data over most of the stationary side of the pulse are an indication of the good reproducibility of the loading cycles. When also taking into account the systematic uncertainties associated with the individual measurements, the same as for the FG mock-up, the total uncertainty for the calorimetrically estimated power intake (given in Figure 17b) is almost the same (minimum = ± 1.81 kW and maximum = ± 1.9 kW). At the top plateau where the power is about 7 kW, the total uncertainty is about 1.81 kW (i.e., the relative uncertainty is about $\pm 26\%$). The same as for the FG mock-up, the major contribution to the uncertainty associated with the power estimation comes from the uncertainty in evaluating the temperature difference, ΔT, which is relatively large because the temperature difference ΔT is relatively small (maximum $\Delta T = 6.7\ ^\circ$C). Since the uncertainties in the measurement of the temperatures are almost the same for the two mock-ups, having a lower temperature increase in the ODS experiment means that we have a larger power uncertainty compared to the FG mock-up.

For the long pulses we follow a different approach in evaluating the enthalpy rise in the mock-up. Having only seven pulses, it does not make sense in averaging over the pulses as we did for the short loading cycles. However, the long steady-state region provides us with enough measurement points to estimate the power input during each pulse or an overall value for this type of loading. If we consider the data from all three days, the heat load is estimated to be 9.11 kW/m^2 ± 2.1 kW/m^2 (relative total uncertainty 23.1%).

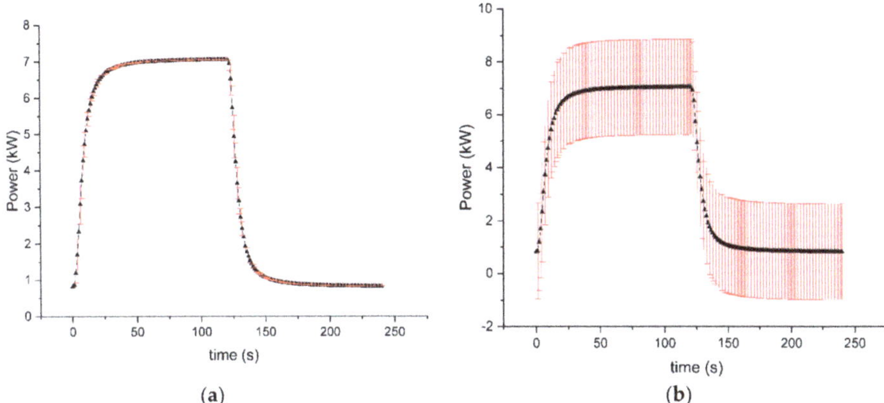

(a)

(b)

Figure 17. Power evolution for the run of 19 September 2019: (**a**) values averaged over 15 consecutive cycles with the associated standard deviation; (**b**) power evolution including the associated uncertainty.

The situation is even more complex when it comes to evaluating the applied heat flux because, during the experiment, there is no clear delimitation of the area on which the heat load is applied, for the FG mock-up, the limits of the coating provide a better reference. Looking at the surface of the mock-up after the experiment (Figure 18), one can see that there is a zone 110 mm long and 90 mm wide that has a rather uniform change in color (light yellow) surrounded by what seems a transition area with a thickness of 5 mm. The first observation that can be made is that, in the flow direction the heated zone covers more than half of the channels' length (110 mm from 208 mm total length), while in the transverse direction only three channels are below the high heat-loaded zone.

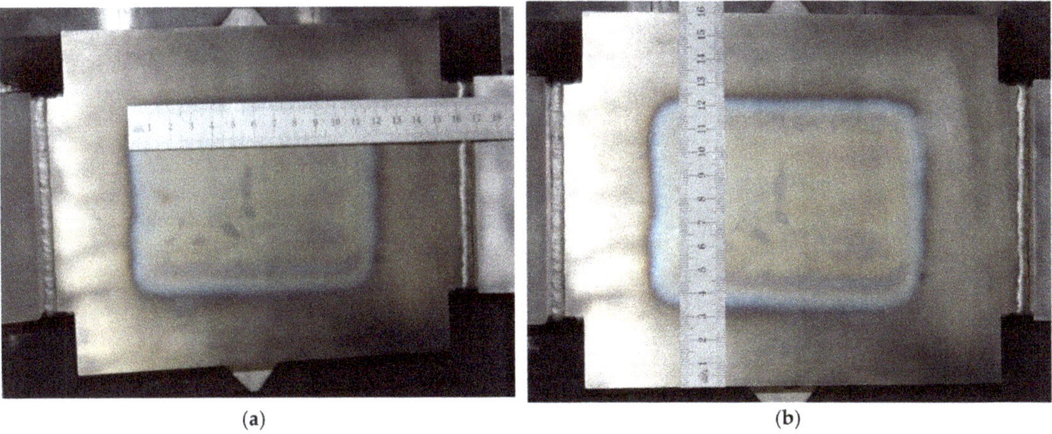

(a)

(b)

Figure 18. Status of the heat-loaded ODS mock-up surface after the testing, in both pictures the flow direction in the channels being from left to the right. The light-yellow area in the center of the picture indicates the zone subject to high-heat flux loading: (**a**) longitudinal measurement of the loaded area; (**b**) measurement of the mock-up transverse dimensions and of the loaded surface.

Figure 19 shows the ODS mock-up surface temperature distribution as recorded by the IR camera. The maximum surface temperature is about 656 °C and it occurs in the white colored area in the upper half of the mock-up. Using the channel numbering from Figure 2b, this maximum value occurs on top of channel number 2.

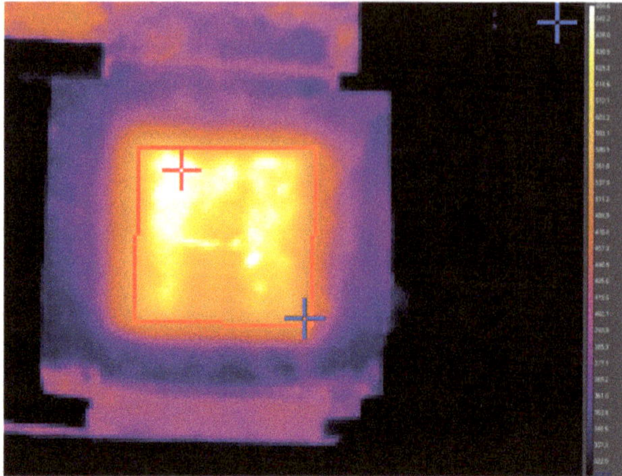

Figure 19. Thermal image of the ODS mock-up by the IR camera. The coolant inlet is on the bottom of the figure, the helium channels being oriented vertically in this picture. The red cross indicates the position of the maximum temperature (656 °C) and the blue cross the lowest temperature (504 °C) the averaged value being around 612 °C.

The longitudinal and transverse temperatures profiles (shown in Figure 20) across the ODS mock-up surface were also generated by the IR camera software for the same run from 7 October 2019, for which the picture in Figure 19 was taken.

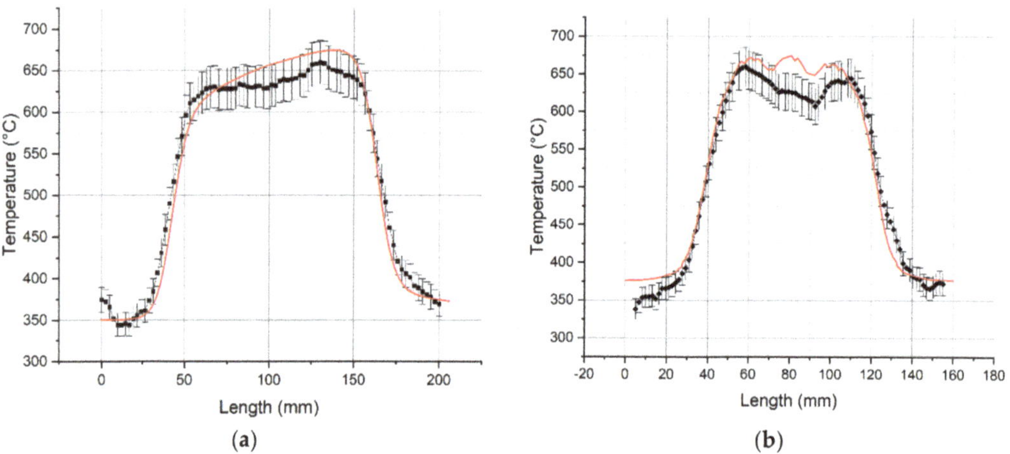

Figure 20. Temperatures profiles across the ODS mock-up in run of 7 October 2019: (**a**) longitudinal; (**b**) transversal. The numerically simulated profiles are drawn with red continuous line.

Both the IR image and the temperature profiles were taken during the long pulse when the highest surface temperature was obtained (close to 670 °C). For comparison, the tests were simulated using ANSYS CFX (2021 R1). The numerical simulation uses the same coolant conditions as in the experiment (350 °C and 201 g/s), the surface heating being applied uniformly over a 110×90 mm^2 area, similar with the yellow area measured on the mock-up itself. Similar to the simulation conducted for the FG mock-up, outside this zone the heating profile drops following a Gaussian with a full width at half maximum (FWHM) of 13.56 mm. From Figure 20 it can be seen that there is a relatively good agreement between the model and the experiment, in particular in what concerns the axial profile. On the transverse direction, the simulation gives a flatter profile in the top region than what is recorded by the IR camera. Moreover, the peak temperature values occur on top of the middle channel (number 3) and the second channel from the left, while the temperature slightly drops towards the right-hand side. The trend of having lower values on the right-hand side of the mock-up is also visible in the experiment but there the middle channel has lower temperatures as compared to the two neighboring channels. The fact that the middle channel has a thinner wall would explain, to some extent, the temperature profile observed experimentally; however, having a channel with a larger cross-section, the coolant velocity would be smaller than in the other channels, reducing the heat transfer performances. The latter would explain the profile obtained numerically. To clarify these aspects and to further analyze the mechanical stresses that occurred during the experiment, further numerical investigations are under way.

In general, it can be stated that the ODS mock-up was tested successfully (i.e., without any failure or damage) in three hundred short (4 min.) heating cycles and seven long (2 h) heating cycles, as planned. The surface area, which received the electron beam heating, has an obvious change in its color (see the blue-black square contours) as seen in Figure 18. However, besides this change in color, the material analysis conducted afterwards did not find any significant changes in the material structure, as reported in [3].

6. Conclusions

This paper describes the experimental campaign of two breeding blanket first wall mock-ups with special consideration of the diagnostics used, in particular, the measurement of the surface temperature with an IR camera and the evaluation of the applied heating power. Both mock-ups completed the specified testing program without any signs of deterioration. The data acquired during these experiments show a good reproducibility of the loading cycles and a stable loop operation at helium-cooled breeding blanket relevant conditions. The calorimetric evaluation of the power deposited on the mock-ups shows relatively large uncertainties, despite using sensors with the lowest commercially available uncertainty. The reason for this is mainly due to the uncertainty associated with the estimation of the temperature rise over the mock-up, which is calculated from the inlet and outlet temperature measurements. Since these experiments have a modest temperature difference between the inlet and outlet (typically around 10 °C), the associated uncertainty (around 1.7 °C at temperatures around 300 °C) represents about 17% of the calculated value. For the experiments presented in this paper, the value of the power and the associated heat fluxes had only an informative value, the focus here being on the behavior of the materials, therefore, a lower uncertainty on the power evaluation was not required. Nevertheless, numerical studies of these two experiments are underway to evaluate the associated uncertainty of the mechanical stresses.

The experimental investigation and obtained results for the two mock-ups will support the development activities of the advanced Eurofer97 steels as well as the manufacturing techniques for the fusion FW.

Author Contributions: Conceptualization, B.-E.G., M.R., T.E., A.A.S. and M.L.; writing—original draft preparation, B.-E.G.; writing—review and editing, A.A.S.; visualization, A.A.S., B.-E.G.; supervision, J.A., M.R. All authors have read and agreed to the published version of the manuscript.

Funding: This work has been carried out within the framework of the EUROfusion Consortium and has received funding from the Euratom research and training programme 2014–2018 and 2019–2020 under grant agreement No 633053. The views and opinions expressed herein do not necessarily reflect those of the European Commission.

Institutional Review Board Statement: Not applicable.

Informed Consent Statement: Has received funding from the Euratom research and training programme 2014–2018 and 2019–2020 under grant agreement No 633053. The views and opinions expressed herein do not necessarily reflect those of the European Commission.

Data Availability Statement: Not applicable.

Conflicts of Interest: The authors declare no conflict of interest.

References

1. Gräning, T.; Rieth, M.; Hoffmann, J.; Möslang, A. Production, microstructure and mechanical properties of two different austenitic ODS steels. *J. Nucl. Mater.* **2017**, *487*, 348–361. [CrossRef]
2. Emmerich, T.; Qu, D.D.; Vaßen, R.; Aktaa, J. Development of W-coating with functionally graded W/EUROFER-layers for protection of First-Wall materials. *Fusion Eng. Des.* **2018**, *128*, 58–67. [CrossRef]
3. Rieth, M.; Pintsuk, G.; Aiello, G.; Henry, J.; de Carlan, Y.; Ghidersa, B.E.; Neuberger, H.; Rey, J.; Dürrschnabel, M.; Bolich, D.; et al. Impact of materials technology on the Breeding Blanket design—Recent progress and case studies in materials technology. *Fusion Eng. Des.* **2021**, *166*, 112275. [CrossRef]
4. Emmerich, T.; Qu, D.; Ghidersa, B.E.; Lux, M.; Rey, J.; Vaßen, R.; Aktaa, J. Development progress of coating First Wall components with functionally graded W/Eurofer layers on laboratory scale. *Nucl. Fusion* **2020**, *60*, 126004. [CrossRef]
5. Ghidersa, B.E.; Ionescu-Bujor, M.; Janeschitz, G. Helium Loop Karlsruhe (HELOKA): A valuable tool for testing and qualifying ITER components and their He cooling circuits. *Fusion Eng. Des.* **2006**, *81*, 1471–1476. [CrossRef]
6. Jianu, A.; Marchese, V.; Ghidersa, B.E.; Messemer, G.; Ihli, T. HELOKA data acquisition and control system: Current development status. *Fusion Eng. Des.* **2009**, *84*, 974–978. [CrossRef]
7. Ranjithkumar, S.; Yadav, B.K.; Saraswat, A.; Paritosh Chaudhuri, E.; Kumar, R.; Kunze, A.; Ghidersa, B.E. Performance assessment of the Helium cooled First Wall mock-up in HELOKA facility. *Fusion Eng. Des.* **2020**, *150*, 111319. [CrossRef]
8. Fasano, G.; Vio, R. Fitting a straight line with errors on both coordinates. *Bull. D'inf. Cent. Donnees Stellaires* **1988**, *35*, 191.
9. Association Française pour les Règles de Conception, de Construction et Surveillance en Exploitation des Matériels des Chaudières Électro-Nucléaires. *RCC-MRx, Design and Construction Rules for Mechanical Components of Nuclear Installations: High Temperature, Research and Fusion Reactors*; AFCEN: Courbevoie, France, 2015.
10. Peterson, H. *The Properties of Helium: Density, Specific Heats, Viscosity, and Thermal Conductivity at Pressures from 1 to 100 Bar and from Room Temperature to About 1800k*; Risö Report no. 224; Forskningscenter Risoe: Roskilde, Denmark, 1970.

Article

Thermal Hydraulic Analysis on the Water Lead Lithium Cooled Blanket for CFETR

Kecheng Jiang [1], Yi Yu [1,2], Xuebin Ma [1], Qiuran Wu [1,2], Lei Chen [1], Songlin Liu [1] and Kai Huang [1,*]

[1] Institute of Plasma Physics, Hefei Institutes of Physical Science, Chinese Academy of Sciences, Hefei 230031, China; jiangkecheng@ipp.ac.cn (K.J.); yi.yu@ipp.ac.cn (Y.Y.); maxuebin@ipp.ac.cn (X.M.); qiuran.wu@ipp.ac.cn (Q.W.); chlei@ipp.ac.cn (L.C.); slliu@ipp.ac.cn (S.L.)

[2] University of Science and Technology of China, Hefei 230026, China

* Correspondence: huangkai@ipp.ac.cn

Abstract: A new type of Water Lead Lithium Cooled (WLLC) blanket that adopts the modular design scheme, water cooling the structure components, liquid PbLi as breeder and coolant, and SiC as the thermal insulator between PbLi and structures is under development as a candidate blanket concept for the Chinese Fusion Engineering Test Reactor (CFETR). Based on a poloidal-radial slice model, thermal hydraulic analysis is performed for this blanket to validate the feasibility of design goals. Results show that the present design can achieve the outlet temperature in the range of 600–700 °C, with all the material temperatures safely below the upper limits. A series of sensitivity analyses are also carried out. It indicates that the thermal conductivity (TC) of SiC would have a significant influence on the temperature field, streamlines and pressure drop; that is, lower TC of SiC can maintain the temperature of PbLi at a high level, and induce an increased number of vortices in the liquid PbLi flow as well as a larger pressure drop. On this basis, the joint effects of the TC of SiC and inlet velocity on the performance of blanket thermal hydraulics are analyzed, then the so-called *"attainable region"* is proposed. Finally, optimization design studies are carried out by decreasing the width of the front channel. Comparison results show that the present design is the most reasonable.

Keywords: thermal hydraulic; WLLC blanket; CFETR

Citation: Jiang, K.; Yu, Y.; Ma, X.; Wu, Q.; Chen, L.; Liu, S.; Huang, K. Thermal Hydraulic Analysis on the Water Lead Lithium Cooled Blanket for CFETR. *Energies* **2021**, *14*, 6350. https://doi.org/10.3390/en14196350

Academic Editor: Hyungdae Kim

Received: 31 July 2021
Accepted: 17 September 2021
Published: 5 October 2021

Publisher's Note: MDPI stays neutral with regard to jurisdictional claims in published maps and institutional affiliations.

1. Introduction

The mission of the Chinese Fusion Engineering Test Reactor (CFETR) is to demonstrate fusion energy production on the basis of ITER's existing technologies [1]. The blanket surrounding the plasma is an essential component, which should achieve three main functions: reliable shielding performance, sufficient tritium breeding, and efficient heat removal for electricity production. Blanket concepts can be classified into solid and liquid blankets according to the physical form of functional materials [2–4]. The liquid blanket relies on the compound or alloy containing lithium as tritium breeder and neutron multiplier, which has inherent advantages, i.e., good geometric adaptability, possibility to bring the breeders out of the blanket for tritium extraction, and higher thermal conductivity, etc. [5]. The family of liquid breeders mainly comprises pure lithium, lead lithium (PbLi), and low melting point ternary Pb-Li-X. Recently, the PbLi eutectic alloy has been widely used because of its much lower chemical reactivity with air and water, which can be used for blanket structure cooling.

Researchers have explored a variety of feasible liquid blankets, aiming at obtaining higher coolant temperatures for efficient electricity generation through reasonable structure and thermal hydraulic design. In some of the advanced conceptual designs [6–8], i.e., ARIES-AT, Tauro, and the He-Li-V blanket, the structure uses high-temperature resistant materials, such as ceramic composite (SiC$_f$/SiC) or vanadium alloy. SiC is attractive as it allows higher operation temperature and removes decay heat, alleviating the loss of coolant accident (LOCA) and loss of flow accident(LOFA). However, this material shows apparent deterioration in thermal and mechanical properties under neutron irradiation conditions, and it is challenging to bear the static and disruption pressure from liquid metal due to the low ductility. Besides, the manufacturing technology for both SiC and vanadium alloy is immature, and this limits industrial-scale applications. These problems may be overcome by using the mature Reduced Activation Ferritic Martensite (RAFM) steel as the structural material, which is cooled below 550 °C by gas or water. This blanket design is divided into self-cooled and separately cooled concepts, respectively, depending on the velocity level (~10 mm/s or ~1 mm/s) of liquid PbLi. The self-cooled blanket uses PbLi as tritium carrier and coolant, which is circulated at a large velocity towards the external heat exchanger, thus the Magnetohydrodynamics (MHD) effect is inevitable. It needs flow channel inserts (FCI) acting as a thermal and electricity insulator between the PbLi and steel structures [9–11]. On the contrary, the separately cooled blanket adopts stagnant or slowly flowing PbLi only for tritium carrying, whereas the heat is removed by another coolant in the structural component, thus the MHD effects are not important. The WCLL and HCLL blankets [12,13] are such examples.

In the liquid blanket design, researchers usually neglect the MHD effect and adopt commercial computation fluid dynamic (CFD) software to optimize the structure design and obtain preliminary thermal hydraulic results, then the MHD is further investigated. For example, E. Martelli established the sliced three-dimensional (3D) model based on the outboard equatorial plane, then adopted ANSYS CFX to evaluate the heat removal capacity of the first wall and breeder zone, and further optimize the cooling tube layout to simplify the structure [14]. Similarly, R. Boullon optimized the cooling channel design based on the 1/4 model of the HCLL blanket [15]. W. Li modified the system analysis code RELAP5/MOD3 by inserting into the governing equations the MHD pressure and heat transfer coefficient between PbLi and helium, then applied it to analyze the DFLL-TBM [16]. I. Fernández developed a 1D thermal hydraulic code, which can obtain the global design parameters for the outboard blanket. Besides, it is found that there appears to be buoyancy-driven flow in the front poloidal channel due to the radially decreasing power heating [17]. W. Ni further theoretically studied the effects of strong buoyancy on flow and heat transfer characteristics in the front channel [18].

The water lead lithium cooled blanket (WLLC) is being under development as a candidate blanket concept for CFETR. It adopts the self-cooled design, which uses pressurized water (15.5 MPa, 285/325 °C) removing heat from the structural components, and liquid PbLi at an inlet temperature of 460 °C flowing through the blanket to carry heat and tritium from the breeder zones. The temperature of the slowly moving PbLi finally approaches 600-700 °C for heat exchange. The present paper addresses the design feasibility of this WLLC blanket from the thermal hydraulic perspectives.

2. Structural Design

As shown in Figure 1, the CFETR blanket system consists of 16 sectors along the toroidal direction, and each sector includes three outboard and two inboard segments. To maintain the fusion machine conveniently by remote handling and reduce the electromagnetic force, the blanket layout adopts a multi-modular segment (MMS) design. The outboard and inboard segments have 5 and 6 blanket modules, respectively. The detailed structure design of outboard module 3# located on the equatorial plane is shown in Figure 2.

The blanket uses RAFM steel as the structural material, and the main components include the first wall (FW), cooling plates (CPs), stiffening plates (SPs), and cover plates. These components share the same manifolds, which makes water flowing into them in parallel. The FW is a U-shaped structure in which the coolant water passes through the channels along the radial-toroidal direction, and the flowing direction between adjacent channels is set opposite to alleviate the mechanical stress. Moreover, there is a layer of tungsten coating on the FW to protect it from plasma erosion and corrosion. The CPs are designed as an intersecting "7" shape, and it collaborates with the SPs to form the 5 (toroidal) × 3 (radial) PbLi channels. The PbLi flows upwards, then turns into two separate channels, finally converges before the manifolds. To reduce the MHD effects and increase the outlet temperature of PbLi, the SiC layer acting as a thermal and electrical insulator is inserted between PbLi and the structural steel. A gap filling with PbLi between the SiC and structural steel is preserved on purpose, which can avoid structural stress resulting from the direct contact between two different materials. Between the adjacent SPs, $Be_{12}Ti$ blocks are inserted to further enhance the capability of neutron multiplying.

The design feature must satisfy multiple requirements because there are strong coupling effects between neutronics and thermal hydraulics. For example, neutronic transportation creates nuclear heat on the structure, and thermal hydraulic analysis is based on the power distribution. On the contrary, thermal hydraulic optimization on the structure arrangement will change the neutronics performance, i.e., TBR and nuclear heating [19]. There are several design criteria for thermal hydraulic analysis, as shown in Table 1. In principle, under the condition that the material temperature is below the upper limit, the outlet temperature of PbLi should be as high as possible (600–700 °C) to achieve efficient power conversion, which is the key objective for blanket design. Besides, the interface temperature between RAFM steel and PbLi in the gap is controlled by coolant water, and it should be kept below 480 °C to prevent corrosion acceleration.

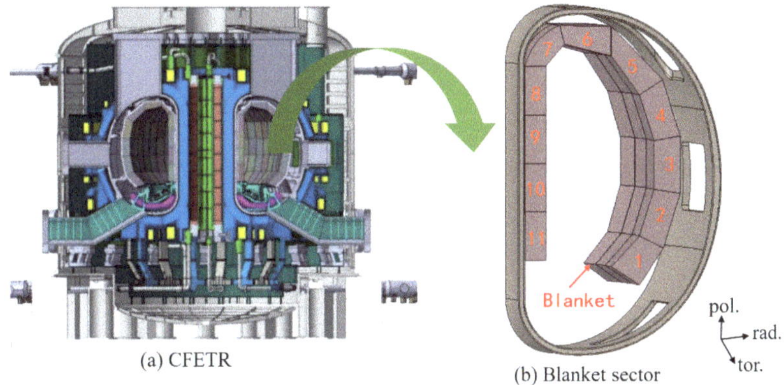

(a) CFETR

(b) Blanket sector

Figure 1. CFETR and blanket layout.

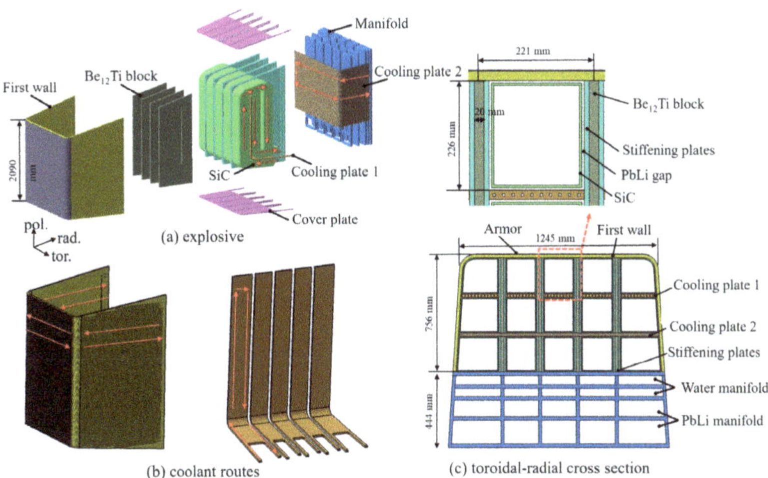

Figure 2. Structural design of the WLLC blanket (module 3#).

Table 1. Design criteria for thermal hydraulics.

Items	Temp. Limit ($^\circ$C)
Water	inlet: 285/outlet: 325
PbLi	inlet: 460/outlet: 600–700
$Be_{12}Ti$	900
RAFM	500
SiC	1000
PbLi-RAFM interface	480

3. Numerical Model

The commercial computational fluid dynamic (CFD) code FLUENT is adopted to perform the analyses. In all the analyses, the minimum Reynolds number under the velocity of 0.03 m/s is 41,574, and it is already within the turbulent region. The SST k-ω model is used to model turbulence as it is capable of simulating the internal channel flow separation against the pressure gradient, which is a common phenomenon in this WLLC model. Since we only consider the performance of the blanket steady-state operation, the steady-state model is employed. And the pressure-based solver is used. Besides, the PbLi flow has an uneven distribution of nuclear heating along the radial direction, and it is subjected to the gravity force, thus there are strong buoyancy-driven effects. We use the Boussinesq approximation to model the thermal convection equation, as given in

$$\rho_0 \frac{\partial u_i}{\partial t} + \rho_0 u_j \frac{u_i}{x_j} = -\frac{\partial p'}{\partial x_i} + u \frac{\partial^2 u_i}{\partial x_j^2} - \alpha \rho_0 g_i (T - T_0) \tag{1}$$

where α is the thermal expansion coefficient of the coolant, $1/^\circ$C; ρ_0 and T_0 is reference density (kg/m^3) and temperature ($^\circ$C), respectively. This approximation considers the density change caused by temperature change and applies it to adjust the momentum governing equation. By employing this approximation, the buoyancy-driven flow can be simulated.

In order to obtain the preliminary thermal hydraulic results for optimization with improved calculation efficiency, a 2D model is developed and adopted for the analysis (Figure 3). This model is a poloidal and radial slice, which is located on the toroidal central plane based on the 3D blanket model. The heat flux from plasma is set as 0.5 MW/m^2. The temperature on the backplate is fixed as 285 $^\circ$C since it directly contacts the inlet manifold containing water. A heat-transfer boundary condition of the third kind derived from Newton's law of cooling, which includes the bulk temperature of coolant and heat transfer coefficient (HTC), is applied. The bulk temperature in each component and the corresponding HTC values [20,21] used in the calculation are listed in Table 2. The nuclear power distribution along the radial direction is obtained using a 3D MCNP analysis, and the results are normalized to 1.5 GW fusion power. Meanwhile, the power distribution is fitted as a user defined function (UDF) implemented into the FLUENT code, which allows each mesh cell to have a more precise nuclear heating source term. The gravity force is set as the actual operation condition. Besides, the thermal properties of materials are listed in Table 3.

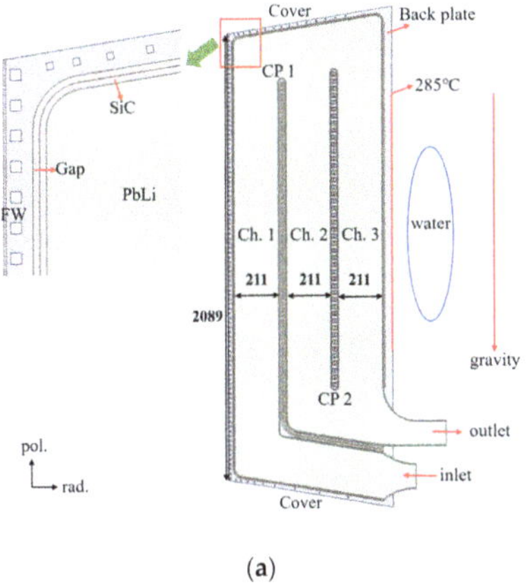

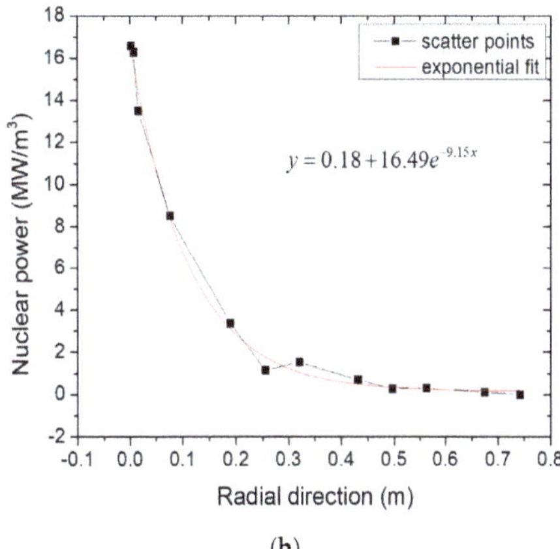

Figure 3. (**a**) Two dimensional numerical model; (**b**) radial power distribution.

Table 2. Boundary conditions for numerical analysis [20,21].

Components	Temperature (°C)	HTC ($\times 10^4$ W/m$^2\cdot$°C)
CP 1	290	1.0
CP 2	290	1.0
Cover	290	1.0
FW	300	1.9

Table 3. Thermal properties of materials.

Materials	Thermal Properties
Tungsten [22]	$\lambda = 207.98 - 0.136T + 5.469 \times 10^{-5}T^2 - 7.835 \times 10^{-9}T^3$
RAFM [23]	$\lambda = 32.5$
SiC [24]	$\lambda = 3.5$
PbLi [25]	$\rho = 10520 - 1.189T$ $\lambda = 1.9463 + 1.96 \times 10^{-2}T$ $\mu = 0.00914 - 1.77459 \times 10^{-5}T + 9.5521 \times 10^{-9}T^2$ $C_p = 194.74 - 9 \times 10^{-3}T$ $\alpha_v = \left(11.221 + 1.531 \times 10^{-3} \cdot T\right) \times 10^{-5}$ [26]

T: temperature, K; λ: thermal conductivity, W/(m·K); ρ: density, kg/m^3; μ: dynamic viscosity, Pa·s; C_p: specific heat, J/(kg·K).

The model uses the mixed structured and unstructured meshes, as shown in Figure 4. By adjusting the mesh size in the solid and fluid domain, mesh independence verification is performed based on the typical thermal hydraulic model (see Section 4.1), as shown in Figure 5. It indicates that the element number of 179,090 is already enough to obtain the precise results, and there is only a temperature increment of 0.35 °C if the element number increases to 332,139. Therefore, the mesh layout with 179,090 elements is adopted for the analyses in this paper. The mesh size of the solid and fluid domains is 2.5 mm and 5.0 mm, respectively. In the boundary layer, the growth rate is 1.4, and there are 18 layers.

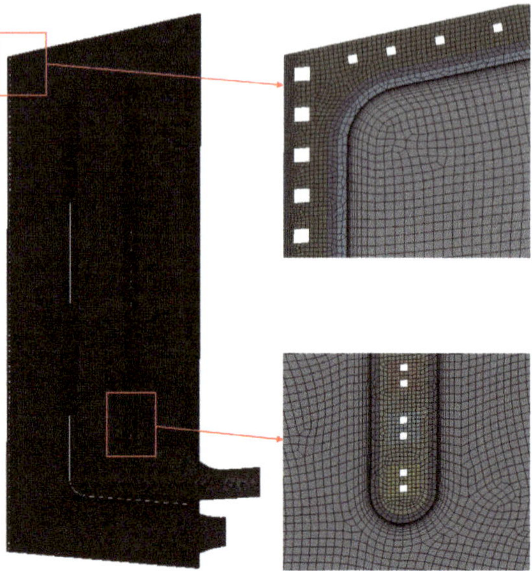

Figure 4. Mesh arrangement.

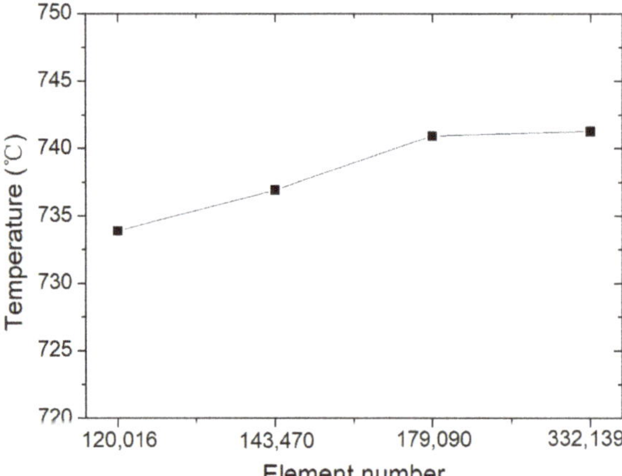

Figure 5. Mesh independence verification.

4. Results and Discussion

4.1. Typical Thermal Hydraulic Results

The thermal hydraulic results of a typical model which can achieve a satisfactory outlet temperature are summarized in Table 4. The thermal conductivity of SiC has considerable effects on heat transfer between PbLi and water, and former researchers optimized this parameter as 3.5 W/(m·°C) to have better performance of thermal and electrical insulation [24]. During the analysis, the inlet velocity of PbLi is iteratively adjusted. The velocity of 0.05 m/s is able to elevate the outlet PbLi temperature to as high as 612.85 °C that is within the targeting range of 600–700 °C. However, this is at the cost of a 1.79 kPa pressure drop. Moreover, all the material temperatures stay below the upper limits, as well as the interface in the gap between PbLi and the RAFM steel. Among the materials, PbLi has the highest temperature since it enters at a relatively high temperature aiming at better power conversion efficiency.

Table 4. Results of typical thermal hydraulic analysis cases.

λ_{SiC} (W/m·°C)	Velocity (m/s)	Max. Temperature (°C)						ΔP (kPa)
		Tungsten	RAFM	SiC	PbLi	Interface	Outlet (Ave.)	
3.5	0.05	468.83	468.83	704.89	740.95	468.83	612.85	1.79

4.1.1. Temperature Field

The temperature field of the entire model is shown in Figure 6a. As can be seen, the PbLi entered Channel 1 and is heated continuously to the upper location, where the peak temperature of 741 °C takes place. However, when it splits and then flows into Channel 2 and 3, the temperature decreases rapidly to the average outlet temperature of 612.85 °C. This is because the nuclear heat source in the back zone of the blanket is lower than the front, and CP2 further cools the PbLi. Besides, the inlet channel with a low temperature of 460 °C connects directly with the outlet channel. Heat conduction between the inlet and outlet channels brings additional cooling effects.

Along the radial direction through point A, as shown in Figure 6b, the temperature of PbLi in each channel remains at a high level, and the distribution in most areas is relatively uniform, but it decreases rapidly near the CPs because of the strong cooling effects. In Channel 1, the temperature in the front zone is slightly higher due to the larger heating source, but it is not apparent since the coolant velocity decreases along the radial direction (Figure 7). As shown in Figure 6d of the FW, although the tungsten armor faces high heat flux from the plasma, the peak temperature is located at the breeder side for the reason that the PbLi contains larger amount of heat. Because the FW is cooled by water flowing through the embedded channels, sources including the plasma and breeder deposit heat into the FW coolant from both directions. This phenomenon can be understood from the radial temperature distribution, in which it decreases from armor to the channel at first, then increases to the SiC side. In Channel 1 and 2, the temperature difference ($T_1 - T_0$) between the mainstream and channel wall is 50 °C and 125 °C, respectively. Thus, the Grashof number Gr can be calculated by Equation (2), as 3.3×10^{10} and 8.3×10^{10}.

$$Gr = \frac{\alpha g (T_1 - T_0) l^3}{\nu^2} \tag{2}$$

where α is the volumetric thermal expansion coefficient, 1/K; g is the gravity force, m/s^2; l is the characteristic length, m; ν is kinematic viscosity, m^2/s.

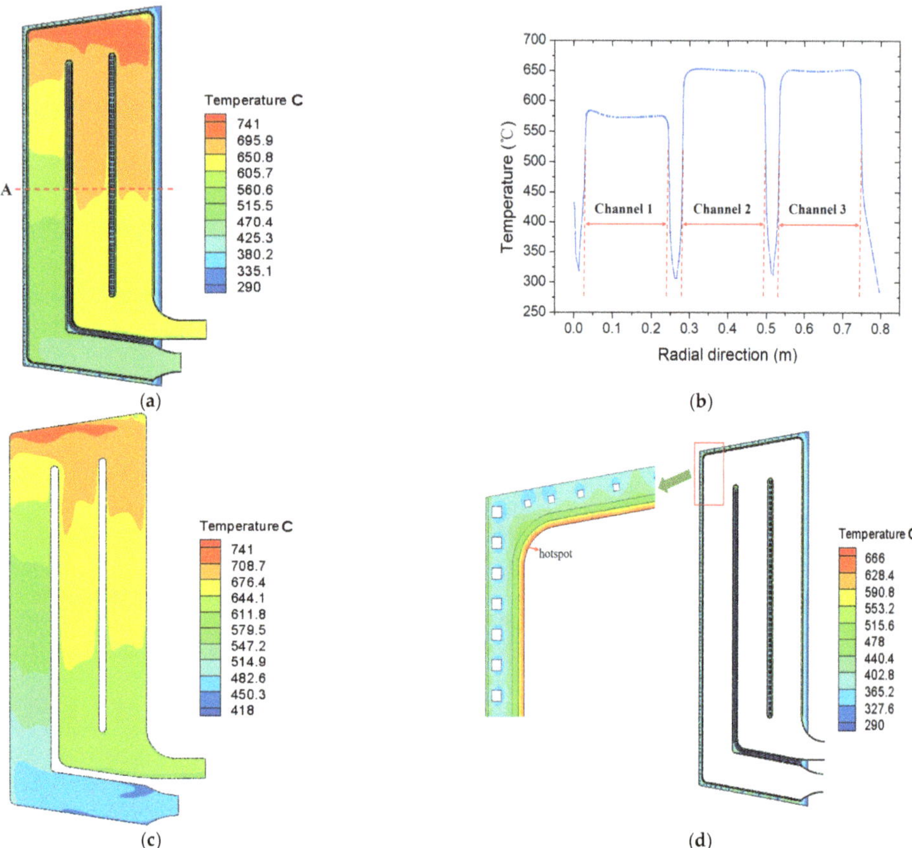

Figure 6. Temperature contour (**a**) entire model; (**b**) radial temperature distribution; (**c**) PbLi; (**d**) FW, gap and SiC.

4.1.2. Velocity Field

The flow velocity field is presented in Figure 7, in which the velocity magnitude is synthesized by x and y velocities. It shows that the velocity near the channel wall is relatively larger, and a considerable number of vortices emerge nearby. This is closely related to the geometry design, gravity effects, and radial distribution of heating power. Take Channel 1 for example, PbLi flows into a nozzle with a sudden enlargement that causes larger velocity at the bottom of the radial pipe, which induces vortices. When it turns into the poloidal channel, the velocity near the front wall becomes significantly larger due to the centrifugal force, and it reduces sharply along the radial direction yet increases again since there are vortices caused by the different velocity gradients in the same channel. In Channel 1 and 2, because the poloidal temperature difference is 130 °C, it brings about buoyancy lift due to the density difference. Therefore, vortices in the two channels show more complicated patterns. A vortex appears in the middle of the channel, which blocks the coolant flow, thus the velocity near the channel wall becomes larger. Furthermore, the velocity magnitude along the radial direction in Figure 7c shows that the mass flow in Channel 2 is smaller, which indicates that the natural convection dominates the flow field. This explains why the number of vortices in this channel is greater. It should be emphasized that vortices are to be avoided in liquid blankets because they may increase the momentum dissipation and further increase the pressure drop, which can reduce the power conversion efficiency.

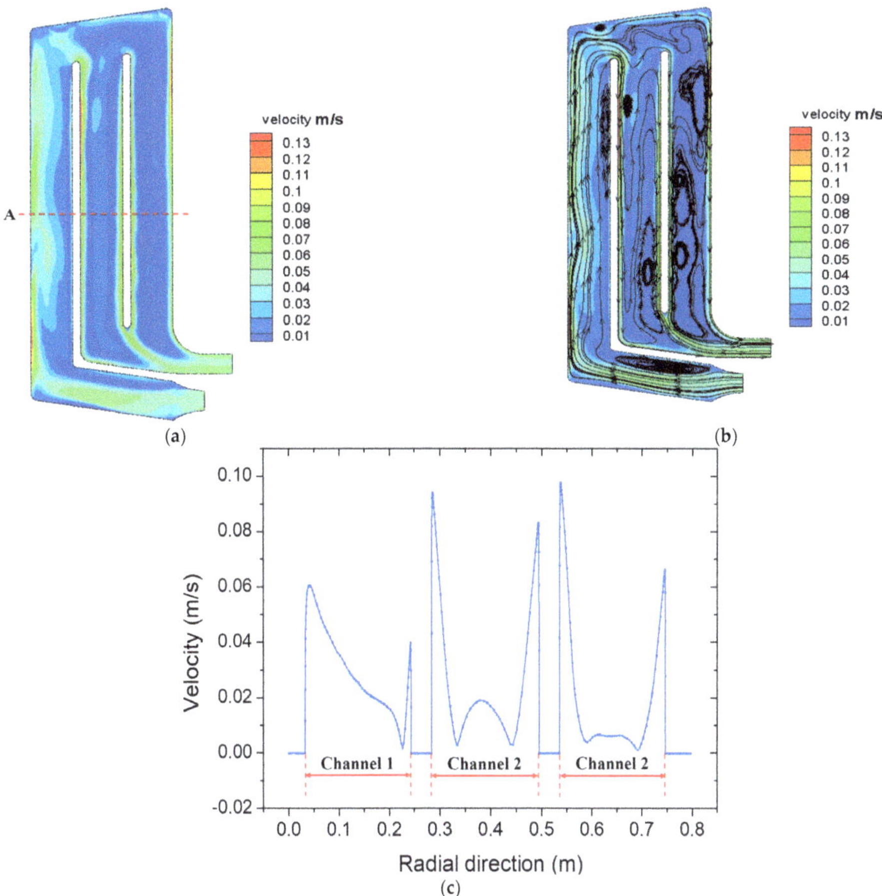

Figure 7. Velocity field (**a**) contour of magnitude; (**b**) streamline; (**c**) radial distribution of velocity magnitude.

4.2. Sensitivity Analysis

In the WLLC blanket design, the temperature of coolant water in the embedded channels inside the structural components is significantly lower than that of the flowing PbLi, in order to prevent the RAFM steel from exceeding the upper limit. Hence, the heat transfers from PbLi to water. The SiC is inserted between PbLi and water to insulate heat. Basically, the thermal conductivity of SiC determines the heat transfer ability of this material. As SiC is a type of synthetic composite material, its thermal properties can be changed. Therefore, it is necessary to study the effects of thermal conductivity and coolant inlet velocity on the performance of thermal hydraulics.

4.2.1. The Effects of SiC Thermal Conductivity

In this sensitivity analysis, the thermal conductivity of SiC ranges from 1.0 W/(m·K) to 10.0 W/(m·K), while the velocity of PbLi is kept constant. The variation of maximum temperature and coolant pressure drop for all materials is plotted in Figures 8 and 9. As the TC of SiC increases, the insulation performance of SiC deteriorates, and this causes more heat losses from PbLi to the structure materials, i.e., tungsten and RAFM. Therefore, the temperature of PbLi and SiC was decreasing, as well as the outlet temperature, while the temperature of tungsten and RAFM changes in the opposite way. When TC increases from

1.0 W/(m·K) to 10.0 W/(m·K), the temperature of PbLi reduces by 120 °C. As mentioned above, the temperature of the FW on the breeder side is higher, and it decreases along the reversed radial direction. Therefore, the hotspot appears at the interface between PbLi and RAFM, thus there is no difference in peak temperatures between RAFM and the interface. The temperature rise for the RAFM steel is 60 °C, but it is only 5 °C for tungsten because the distance to PbLi is farther. The outlet temperature for TC of SiC at 4.0 W/(m·K) is 603 °C, and it will not satisfy the requirement (outlet $T \geq 600$ °C) when the TC of SiC increases further.

Keeping the coolant velocity constant, the pressure drop is also significantly affected by the TC of SiC, as it decreases by almost 90% when the TC increases from 1.0 W/(m·K) to 10.0 W/(m·K). This can be clarified from the contour of streamlines, as shown in Figure 10. As the temperature at the upper location becomes larger when TC decreases, the temperature difference term ΔT in Equation (2) becomes larger since the inlet temperature is the same. Thus, the buoyancy effects will be apparent and induce more vortexes. This will cause large momentum loss and pressure drop. However, the outlet temperature will be lower. And a compromise should be made among these design parameters.

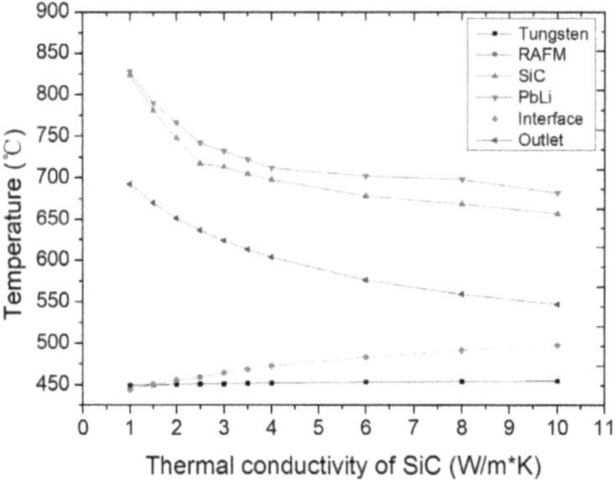

Figure 8. Variation of maximum temperature for different thermal conductivity of SiC.

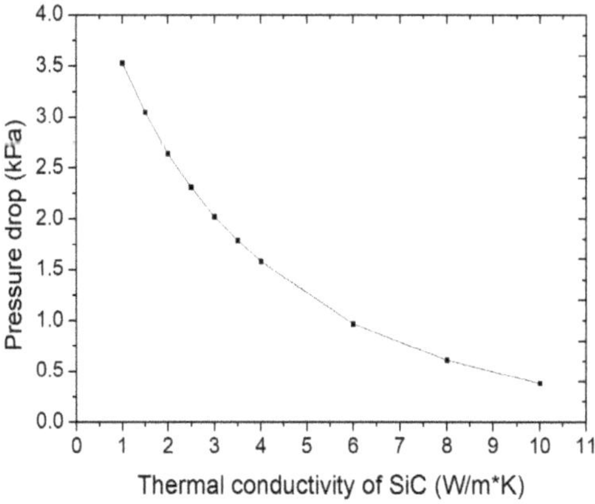

Figure 9. Variation of pressure drop for different thermal conductivity of SiC.

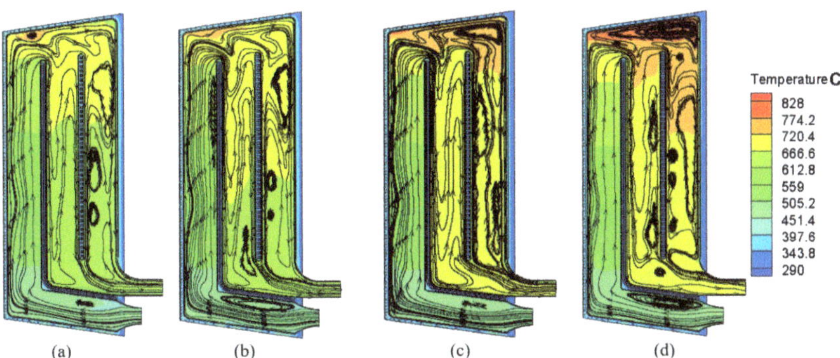

Figure 10. Contour of temperature and streamline for different thermal conductivity of SiC (**a**) 3.5 W/(m·K); (**b**) 2.5 W/(m·K); (**c**) 1.5 W/(m·K); (**d**) 1.0 W/(m·K).

As shown in Figures 11 and 12, the effects of the TC of SiC on thermal hydraulics can also be further understood through the contour of streamlines. For different TCs, the general distribution of velocity magnitude remains alike, in which the velocity near the channel wall is still larger. Along the radial direction, it can be seen that there is no profound difference of velocity distribution in Channel 1. However, as the TC of SiC decreases, the vortex at the upper location becames stronger, which blocks the coolant from flowing into Channel 3. Because the total mass flow rate is unchanged, this results in more PbLi entering Channel 2. When the TC of SiC decreases from 3.5 W/(m·K) to 1.0 W/(m·K), the velocity in Channel 2 increases by 40%. This trend is opposite for Channel 3.

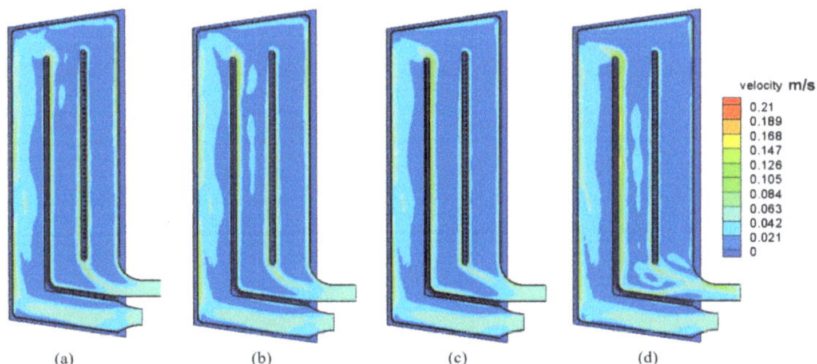

Figure 11. Contour of velocity magnitude for different thermal conductivity of SiC (**a**) 3.5 W/(m·K); (**b**) 2.5 W/(m·K); (**c**) 1.5W/(m·K); (**d**) 1.0 W/(m·K).

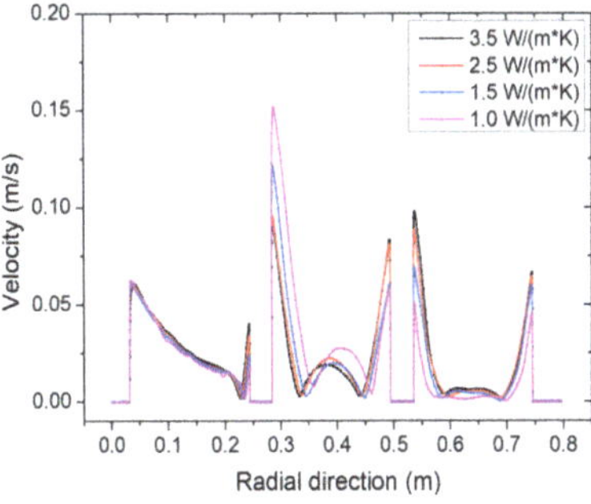

Figure 12. Radial distribution of velocity magnitude for different thermal conductivity of SiC.

4.2.2. The Effects of Coolant Velocity

In the sensitivity analysis regarding coolant velocity, the TC of SiC is kept constant as 3.5 W/(m·K), while the coolant inlet velocity ranges from 0.03 m/s to 0.12 m/s. As shown in Figures 13 and 14, increasing velocity can significantly reduce the material temperature, especially for PbLi and SiC. For the outlet temperature, when the velocity increases over 0.055 m/s, the PbLi outlet temperature will be lower than 600 °C, which cannot satisfy the requirement. A temperature decline of 4.02 °C and 51.51 °C takes place for tungsten and RAFM, respectively. For the pressure drop variation, there is a peak at the velocity of 0.05 m/s. Before this peak, the pressure drop increases as velocity increases. This is because the buoyancy-lift-induced vortices dominate in this area, and increasing velocity will enhance the vortices. However, this buoyancy effect will be replaced gradually by forced convection when the velocity is increasing beyond 0.05 m/s, where the number of vortices decreases, such that less momentum of PbLi flow is lost. The temperature field is shown in Figure 15. It is observed that low velocity will make the temperature distribution uneven, and this unevenness can be alleviated by increasing the velocity. For example, a poloidal temperature difference of 210 °C occurs in Channel 2 and 3 when the PbLi flow velocity is 0.03 m/s, and the resulting temperature gradient will bring stronger buoyancy lift that leads to a larger pressure drop. However, this will diminish as the velocity increases. For instance, the temperature difference for the velocity of 0.09 m/s is almost indistinguishable (Figure 15).

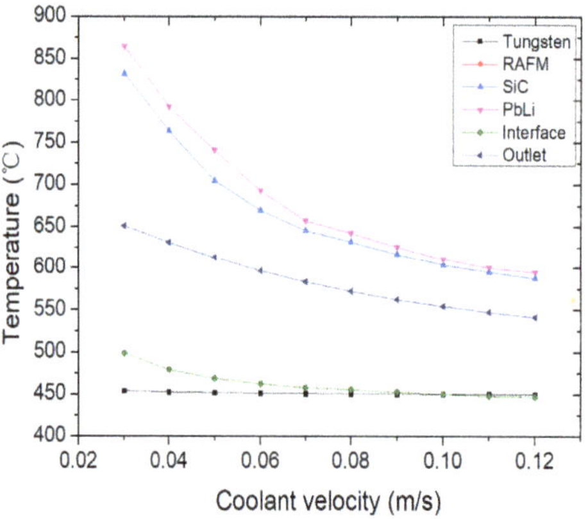

Figure 13. Variation of maximum temperature versus different coolant inlet velocity.

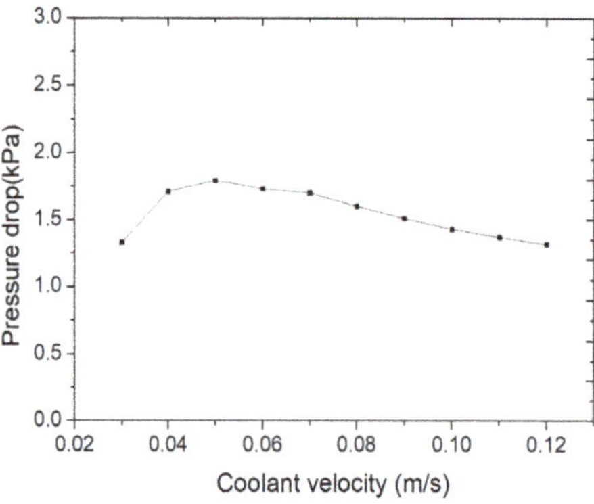

Figure 14. Variation of pressure drop versus different coolant inlet velocity.

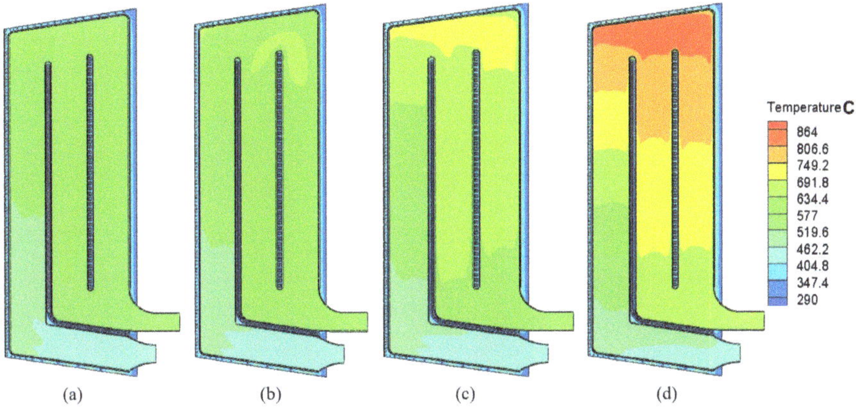

Figure 15. Temperature field under different coolant inlet velocity (**a**) 0.09 m/s; (**b**) 0.07 m/s; (**c**) 0.05 m/s; (**d**) 0.03 m/s.

4.2.3. Adjustment on the Outlet Temperature

As mentioned above, the thermal conductivity and coolant inlet velocity have significant effects on the thermal hydraulic performance of the WLLC blanket. The design target of the outlet temperature is 600–700 °C in order to achieve high power conversion efficiency. These two parameters are adjusted and optimized simultaneously to obtain outlet temperatures of 600 °C and 700 °C, respectively. As shown in Figure 16, it indicates that the relationship between the velocity of PbLi flow and the TC of SiC is almost linear. Higher TC of SiC means poorer insulation performance. Thus, a lower velocity is needed to compensate and obtain a satisfactory outlet temperature. Besides, it shows that the design target of the outlet temperature cannot be achieved by adjusting only one parameter. For example, when the TC of SiC is greater than 3.5 W/(m·K), decreasing the inlet velocity is impossible to sustain outlet temperature ≥600 °C, and it needs to improve the insulation of SiC further. In all scenarios, the TC of SiC = 1.0–2.5 W/(m·K) can realize the outlet temperature of 700 °C. And when the target outlet temperature is 600 °C, the corresponding TC of SiC is 1.0–3.5 W/(m·K). Using the TC of SiC-velocity value pairs corresponding to the outlet temperature of 600 °C and 700 °C as boundary points, an *"attainable region"* is defined, where it is easy to find the possible combination of the PbLi velocity and TC of SiC to have outlet temperature within the target range. The results of these scenarios are summarized in Table 5, in which all the material temperatures can satisfy the requirements.

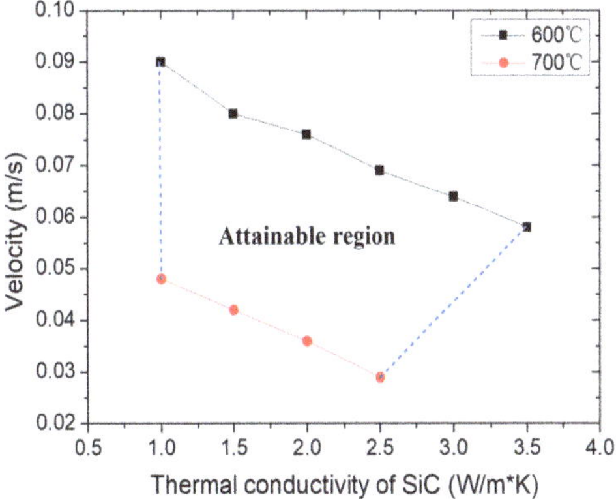

Figure 16. Attainable region for satisfactory outlet temperatures.

Table 5. Results summary for the selective scenarios.

#	TC (W/m·K)	Inlet Velocity (m/s)	Max. T (°C)					Outlet T. (°C)	ΔP (kPa)
			Tungsten	RAFM	SiC	PbLi	Interface		
1	1.0	0.09	448.67	441.96	637.02	637.23	432.03	600	2.29
2	1.0	0.048	449.43	444.35	833.24	838.22	444.35	700	3.61
3	1.5	0.08	449.06	442.30	649.03	650.86	438.38	603	2.27
4	1.5	0.042	450.33	455.37	825.51	836.06	455.37	700	3.35
5	2.0	0.076	449.62	444.46	651.21	656.02	444.46	600	2.15
6	2.0	0.036	451.50	468.01	844.71	861.78	468.01	700	2.92
7	2.5	0.069	450.22	450.55	656.34	665.31	450.55	600	2.05
8	2.5	0.029	453.02	484.61	877.48	902.16	485.57	700	2.20
9	3.0	0.064	450.78	457.06	665.47	681.77	457.06	600	1.90
10	3.5	0.058	451.37	463.90	677.09	702.97	463.90	600	1.74

4.3. Structure Optimization

In the present design of the WLLC blanket, the PbLi flow channels are formed by the two CPs, and the channel width is equally spaced. The typical results (see Section 4.1) indicate that the hotspot would occur at the upper location due to the existing vortices (Figure 6), and this has a negative effect on increasing the outlet temperature. Therefore, the structure optimization is investigated by adjusting the width ratio of PbLi channels in order to change the flow field, then further optimize the temperature field. As shown in Figure 17, the total radial length remains the same, and the other three structures are designed on the basis of the standard model using three ratios, namely, 0.5:1:1.5, 0.5:1.5:1, and 1:0.5:1. Thermal hydraulic analyses are carried out for these optimization designs, and all the boundary conditions remain the same as in the typical case discussed in the previous sections.

The temperature contour with streamlines is shown in Figure 18, and the detailed thermal hydraulic results are summarized in Table 6. In general, compared with Design 1#, the structural optimizations (Design 2# and 3#) have slight effects on the peak temperature of materials, but Design 1# presents larger poloidal temperature differences and lower outlet temperatures, which causes more vortices in the channels. On the contrary, because the flow velocity in Channel 1 is twice as large, the streamline is smoother, and there are no obvious vortices. However, the flow in Designs 2# and 3# will rotate around CP2, which causes a pressure drop of about 27.8% larger than Design 1. Meanwhile, keeping the width of Channel 1 unchanged, yet reducing the width of Channel 2 by half, there is no large difference of the material temperatures, except for the lower outlet temperature, which is adverse for achieving high power conversion efficiency. From this point of view, although Design 1 has a lower outlet temperature, it still presents better thermal hydraulic performances than the other designs due to the significantly reduced pressure drop.

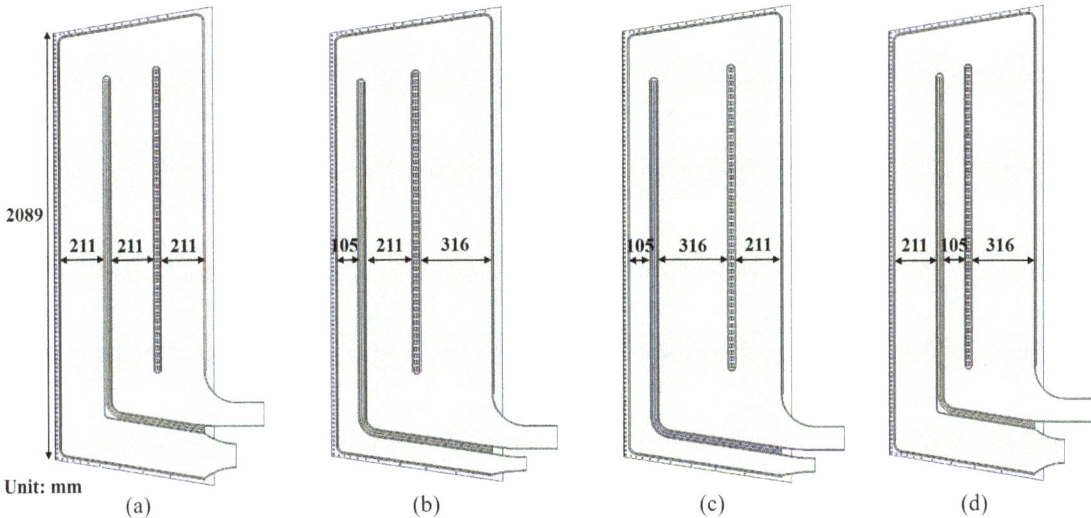

Figure 17. Three structure designs for optimization (**a**) 1:1:1; (**b**) 0.5:1:1.5; (**c**) 0.5:1.5:1; (**d**)1:0.5:1.

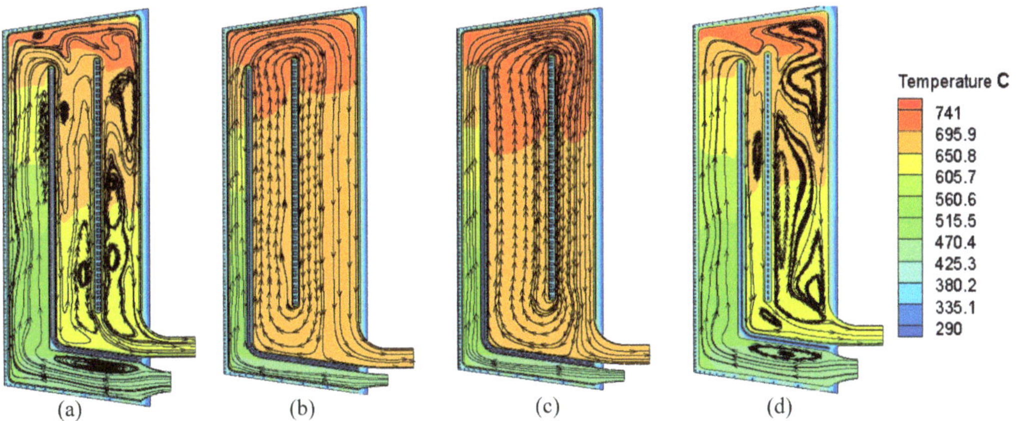

Figure 18. Temperature field combined with streamline (**a**) 1:1:1; (**b**) 0.5:1:1.5; (**c**) 0.5:1.5:1; (**d**)1:0.5:1.

Table 6. Results summary of the four designs.

#	Max. *T* (°C)					Outlet *T.* (°C)	ΔP (kPa)
	Tungsten	RAFM	SiC	PbLi	Interface		
1	451.82	468.83	704.89	740.95	468.83	612.85	1.79
2	452.27	479.96	723.76	733.7	485.75	662.94	2.31
3	452.25	481.65	722.40	732.52	487.50	663.73	2.34
4	449.82	472.95	712.94	740.85	472.95	609.74	1.70

5. Conclusions

The Water Lead Lithium Cooled (WLLC) blanket that adopts pressurized water (15.5 MPa, 285/325 °C) cooling the structural components, liquid PbLi as breeder and coolant, and SiC as thermal insulator between PbLi and the structures is being under development for the Chinese Fusion Engineering Test Reactor. In the present work, the thermal hydraulic analysis based on a poloidal-radial slice model is performed for this WLLC blanket design. The analyses results show that the present design can achieve high outlet temperature in the range of 600–700 °C without material temperatures exceeding the upper limits, which is beneficial for high power conversion efficiency.

A series of sensitivity analyses are conducted with regard to the flow inlet velocity and thermal conductivity of the SiC insulator. Results indicate that the thermal conductivity of SiC has a significant influence on the temperature field, streamlines, and pressure drop. It shows that lower TC of SiC can maintain the temperature of PbLi at high level, yet induce unfavorably an increased number of vortices in the liquid PbLi flow as well as a larger pressure drop. On this basis, the joint effects of the TC of SiC and inlet velocity on the performance of blanket thermal hydraulics are analyzed. Then, using the TC of SiC-velocity value pairs corresponding to the outlet temperature of upper and lower target values as boundary points, a so-called "attainable region" is proposed, where the possible combination of the PbLi velocity and TC of SiC are easily identified to ensure the outlet temperature within the design target range.

The structure optimization is further performed by decreasing the width of the front channel to reduce the buoyancy effects at the upper location. Four cases with the width ratio of the three channels being 1:1:1, 0.5:1:1.5, 0.5:1.5:1, and 1:0.5:1.5, are studied. Compared with the 1:1:1 design, although Design 2# and 3# can make the temperature field more evenly distributed and reduce obvious local vortices, the flow rotation emerging in these two designs around CP2 would cause a pressure drop that is 27.8% larger than the 1:1:1 design. Therefore, the present design of the WLLC blanket is still the most reasonable.

Author Contributions: Conceptualization, K.J., L.C. and K.H.; methodology, K.J.; software, K.J., Y.Y. and X.M.; validation, Y.Y.; formal analysis, K.J. and Y.Y.; investigation, K.J.; resources, Q.W. and X.M.; visualization, K.J.; writing, K.J. and K.H.; project administration, S.L. and K.H.; funding acquisition, K.J., S.L. and K.H. All authors have read and agreed to the published version of the manuscript.

Funding: This work is supported by the Comprehensive Research Facility for Fusion Technology (CRAFT) Program of China [Contract No. 2018-000052-73-01-001228], and the Science Foundation of ASIPP [No. DSJJ-2020-10].

Conflicts of Interest: The authors declare no conflict of interest.

References

1. Wan, Y.; Li, J.; Liu, Y.; Yong, L.; Wang, X.; Chan, V.; Chen, C.; Duan, X.; Fu, P.; Gao, X.; et al. Overview of the present progress and activities on the CFETR. *Nucl. Fusion* **2017**, *57*, 102009. [CrossRef]
2. Leys, O.; Leys, J.M.; Knitter, R. Current status and future perspectives of EU ceramic breeder development. *Fusion Eng. Des.* **2021**, *164*, 112171. [CrossRef]
3. Frano, R.L.; Puccini, M. Preliminary investigation of Li_4SiO_4 pebbles structural performance. *Fusion Eng. Des.* **2021**, *167*, 112388. [CrossRef]
4. Smolentsev, S.; Moreau, R.; Abdou, M. Characterization of key magnetohydrodynamic phenomena in PbLi flows for the US DCLL blanket. *Fusion Eng. Des.* **2008**, *83*, 771–783. [CrossRef]
5. John, H.; Malang, S.; Sebening, H. *KfK Contribution to the Development of DEMO-Relevant Test Blankets for NET/ITER. I: Self-Cooled Liquid Metal Breeder Blanket. Volume 1: Summary*; Kernforschungszentrum Karlsruhe: Karlsruhe, Germany, 1991; pp. 1–67.
6. Wong, C.; Mcquillan, B.W.; Schleicher, R.W. Evaluation of US demo helium-cooled blanket options. In Proceedings of the 16th International Symposium on Fusion Engineering, Champaign, IL, USA, 30 September–5 October 1995.
7. Ramirez, A.P.; Caso, A.; Giancarli, L. Tauro: A ceramic composite structural material self-cooled Pb-17Li breeder blanket concept. *J. Nucl. Mater.* **1996**, *233–237*, 1257–1261. [CrossRef]
8. Raffray, A.R.; El-Guebaly, L.; Malang, S. Advanced power core system for the ARIES-AT power plant. *Fusion Eng. Des.* **2007**, *82*, 217–236. [CrossRef]
9. Smolentsev, S.; Morley, N.; Abdou, M.; Malang, S. Dual-coolant lead–lithium (DCLL) blanket status and R&D needs. *Fusion Eng. Des.* **2015**, *100*, 44–54.
10. Zeng, Z.; Jiang, J.; Chen, S.; Wang, F. Preliminary neutronics analyses of China Dual-Functional Lithium-Lead (DFLL) test blanket module for CFETR. *Fusion Eng. Des.* **2020**, *152*, 111414. [CrossRef]
11. Fernández, I.; Palermo, I.; Urgorri, F.R. Alternatives for upgrading the EU DCLL breeding blanket from MMS to SMS. *Fusion Eng. Des.* **2021**, *167*, 112380. [CrossRef]

12. Del Nevo, A.; Arena, P.; Caruso, G.; Chiovaro, P. Recent progress in developing a feasible and integrated conceptual design of the WCLL BB in EUROfusion project. *Fusion Eng. Des.* **2019**, *146*, 1805–1809. [CrossRef]

13. Auber, J.; Aiello, G.; Arena, P. Status of the EU DEMO HCLL breeding blanket design development. *Fusion Eng. Des.* **2018**, *136*, 1428–1432. [CrossRef]

14. Martelli, E.; Caruso, G.; Giannetti, F.; Del Nevo, A. Thermo-hydraulic analysis of EU DEMO WCLL breeding blanket. *Fusion Eng. Des.* **2018**, *130*, 48–55. [CrossRef]

15. Boullon, R.; Aubert, J.; Aiello, G.; Jaboulay, J.-C.; Morin, A. The DEMO Helium Cooled Lithium Lead "Advanced-Plus" Breeding Blanket: Design Improvement and FEM Studies. *Fusion Eng. Des.* **2019**, *146*, 2026–2030. [CrossRef]

16. Li, W.; Tian, W.; Qiu, S.; Su, G.; Jiao, H.; Bai, Y.; Chen, H.; Wu, Y. Preliminary thermal-hydraulic and safety analysis of China DFLLTBM system. *Fusion Eng. Des.* **2013**, *88*, 286–294. [CrossRef]

17. Fernández-Berceruelo, I.; Rapisarda, D.; Palermo, I.; Maqueda, L.; Alonso, D.; Melichar, T.; Frybort, O.; Vala, L.; Ibarra, A. Thermal-hydraulic design of a DCLL breeding blanket for the EU DEMO. *Fusion Eng. Des.* **2017**, *124*, 822–826. [CrossRef]

18. Ni, W.F.; Qiu, S.Z.; Su, G.H.; Tian, W.X.; Wu, Y.W. Numerical investigation of buoyant effect on flow and heat transfer of Lithium-Lead Eutectic in DFLL-TBM. *Prog. Nucl. Energy* **2012**, *58*, 108–115. [CrossRef]

19. Jiang, K.; Ding, W.; Zhang, X. Development of neutronic-thermal hydraulic-mechanic-coupled platform for WCCB blanket design for CFETR. *Fusion Eng. Des.* **2018**, *137*, 312–324. [CrossRef]

20. Jiang, K.; Ma, X.; Cheng, X.; Liu, S. Thermal hydraulic analysis on the whole module of water cooled ceramic breeder blanket for CFETR. *Fusion Eng. Des.* **2016**, *112*, 81–88. [CrossRef]

21. Jiang, K.; Martelli, E.; Agostini, P.; Del Nevo, A. Investigation on cooling performance of WCLL breeding blanket first wall for EU DEMO. *Fusion Eng. Des.* **2019**, *146*, 2748–2756. [CrossRef]

22. Anonymous. *Report on the Mechanical and Thermal Properties of Tungsten and TZM Sheet Produced in the Refractory Metal Sheet Rolling Program*; Southern Research Institute Report 7563-1479-XII to the U.S. Bureau of Naval Weapons (638631 CE); Southern Research Institute: Birmingham, AL, USA, 31 August 1966.

23. Hirose, T.; Nozawa, T.; Stoller, R.E.; Hamaguchi, D.; Sakasegawa, H.; Tanigawa, H.; Eneoda, M.; Katoh, Y.; Snead, L.L. Physical properties of F82H for fusion blanket design. *Fusion Eng. Des.* **2014**, *89*, 1595–1599. [CrossRef]

24. Wong, C.P.C.; Malang, S.; Sawan, M.; Dagher, M.; Smolentsev, S.; Merrill, B.; Youssef, M.; Reyes, S.; Sze, D.K.; Morley, N.B.; et al. An overview of dual coolant Pb–17Li breeder first wall and blanket concept development for the US ITER-TBM design. *Fusion Eng. Des.* **2006**, *81*, 461–467. [CrossRef]

25. Martelli, E.; Del Nevo, A.; Arena, P.; Bongiovi, G.; Caruso, G.; Di Maio, P.A.; Eboli, M.; Mariano, G.; Marinari, R.; Moro, F.; et al. Advancements in DEMO WCLL breeding blanket design and integration. *Int. J. Energy Res.* **2018**, *42*, 27–52. [CrossRef]

26. Stankus, S.V.; Khairulin, R.A.; Mozgovoi, A.G. An experimental investigation of the density and thermal expansion of advanced materials and heat–transfer agents of liquid–metal systems of fusion reactor: Lead–lithium eutectic. *High Temp.* **2006**, *44*, 829–837. [CrossRef]

Article

A Water Loop Design for the CRAFT Project towards the Testing of CFETR Water-Cooled Blanket and Divertor

Xiaoman Cheng [1], Zihan Liu [2], Songlin Liu [3,*], Changhong Peng [2], Wenjia Wang [2,3] and Qixin Ling [2,3]

[1] Hefei Comprehensive National Science Center, Institute of Energy, Hefei 230000, China; cxm@ie.ah.cn
[2] School of Nuclear Science and Technology, University of Science and Technology of China, Hefei 230026, China; zihanliu@mail.ustc.edu.cn (Z.L.); pengch@ustc.edu.cn (C.P.); wangwenjia@ipp.ac.cn (W.W.); qixin.ling@ipp.ac.cn (Q.L.)
[3] Institute of Plasma Physics, Hefei Institutes of Physical Science, Chinese Academy of Sciences, Hefei 230031, China
*   Correspondence: slliu@ipp.ac.cn

Abstract: As one of the tasks of the Comprehensive Research Facility for Fusion Technology (CRAFT), a High Heat Flux (HHF) testing device will be built to test the blanket and divertor of Chinese Fusion Engineering Testing Reactor (CFETR). The water loop is a key system of the HHF testing device. The main objective of the water loop is to provide deionized water at specific temperature, pressure, and flow rate for different testing experiments of the water-cooled blanket and water-cooled divertor components. The design of the water loop has been through three major steps. Firstly, the water cooled blanket and divertor were designed and analyzed, in detail, for CFETR. Secondly, thermal hydraulic features of the prototypes were abstracted from the analyses results. Then, the experiment plan was made so that the preliminary design of the water loop was carried out. The third step was the engineering design, which was conducted through cooperation with an industrial enterprise with certifications. At present, the water loop is ready for fabrication and construction. The water loop will be completed, for commissioning operation, by August 2022, as scheduled. After that, the experiments will be carried out step by step and provide solid technical base to CFETR.

Keywords: CFETR; CRAFT; blanket and divertor; experiment plan; water loop design

Citation: Cheng, X.; Liu, Z.; Liu, S.; Peng, C.; Wang, W.; Ling, Q. A Water Loop Design for the CRAFT Project towards the Testing of CFETR Water-Cooled Blanket and Divertor. *Energies* **2021**, *14*, 7354. https://doi.org/10.3390/en14217354

Academic Editors: Dimitris Drikakis, Hiroshi Sekimoto and Dan Gabriel Cacuci

Received: 30 July 2021
Accepted: 2 November 2021
Published: 4 November 2021

Publisher's Note: MDPI stays neutral with regard to jurisdictional claims in published maps and institutional affiliations.

1. Introduction

Fusion energy is an environment-friendly new energy, which is a promising way to solve the increasingly serious energy crisis and global warming. According to the roadmap for Chinese magnetic confinement fusion development, China will independently design and build the Chinese Fusion Engineering Testing Reactor (CFETR). The goal of CFETR is to achieve stable operation, tritium self-sufficiency, and, finally, to realize commercial operation and power generation [1].

Blanket and divertor are two crucial components in CFETR and future fusion DEMO. The main functions of the blanket are to realize tritium self-sufficiency, convert fusion energy for electricity production, and provide shielding. The divertor is used for removing the impurities and helium ashes in the plasma. Besides, the divertor needs to resist extremely high heat flux from the plasma. For CFETR, the blanket and the divertor both have water-cooled [2,3] and helium-cooled concepts [4,5]. Experiments are indispensable to verify and validate the design and performance of the blanket and divertor.

Comprehensive Research Facility for Fusion Technology (CRAFT) is one of the national big science and technology facilities in China. Its objectives are to explore and master crucial technologies of key components and systems, to establish standards of manufacture, to build key prototype systems, and to validate the technologies for the successful construction of CFETR. The construction of CRAFT started on 20 September 2019 in Hefei, Anhui

Province, and it will last for 5 years and 8 months with joint funds from central and local governments.

CRAFT consists of 20 different facilities that address most of the key technologies and systems of CFETR. Thereinto, a High Heat Flux (HHF) testing facility will be built to test the blanket and divertor of CFETR. This test device is equipped with two Electron Beam Guns (EBGs), water loop, helium loop, and vacuum chamber. It is constructed to:

- provide a testing environment of high thermal radiation loads, thermal hydraulics for water-cooled and helium-cooled divertor, and blanket of CFETR;
- test prototype components to verify their heat-resisting and thermal hydraulic performances;
- evaluate the manufacturing technologies to determine the standards and criteria for design and manufacture.

The main objective of the water loop is to provide deionized water at specific temperature, pressure and flow rate for testing water-cooled blanket, and water-cooled divertor components towards CFETR. Based on the design and analyses of the CFETR prototype, the main parameters and functions of the water loop were determined. In favor of the experiment performance, the water loop consists of high pressure water loop and low pressure water loop. The operation condition of the high pressure water loop is 4~16.5 MPa/70~340 °C. The low pressure water loop is capable of two-phase flow experiments under atmospheric pressure.

In this paper, Section 2 introduces the design features and analyses results of the CFETR water-cooled blanket and divertor. Based on that, the experiment plan is introduced in Section 3. Then, the water loop design is indicated in Sections 4 and 5. At last, is the discussion and conclusion.

2. Water Cooled Blanket and Water-Cooled Divertor of CFETR

The engineering design of CFETR started in December 2017. Last year, the water-cooled blanket and water-cooled divertor basically completed the design. The water-cooled blanket uses the design scheme of multi-module segment [2]. The water coolant operates under the pressure of 15.5 MPa, and the inlet/outlet temperature of the blanket module is 285 °C/325 °C. Square channels with a cross section of 8×8 mm^2 are embedded in the First Wall (FW) in which the coolant flows along the radial-toroidal-radial direction. The breeding zones are filled with a mixed pebble bed of Li_2TiO_3 and $Be_{12}Ti$, which is cooled by the cooling tubes with multiple bends. The water-cooled blanket is able to withstand steady state heat flux of 0.5 MW/m^2 and transient heat flux of 1 MW/m^2 [6,7]. As for the divertor, the maximum inlet pressure of water coolant is 5 MPa. Inlet temperature is 140 °C, and the temperature rise is below 40 °C [8]. From the perspective of thermal hydraulics, the divertor is required to resist steady state heat flux of 10 MW/m^2 and transient heat flux of 20 MW/m^2. The prototype of the water cooled blanket and the water-cooled divertor for CFETR is shown in Figure 1. Although the water-cooled blanket and the water-cooled divertor are different in functions, they still have some common grounds:

- facing one-side high heat flux
- using high pressure and high temperature water coolant
- having complex coolant channels
- operating in high vacuum environment

The operation events of CFETR can be classified into five categories referring to ITER, as shown in Table 1 [9]. Events in Category I and II are taken into account for the test condition of water coolant due to the high frequency. Besides, the capability and safety of the experimental water loop should also be considered. In Category I, the plasma pulse operation is defined as normal operating condition due to the pulsed nature of fusion plasma so far. As for Category II, the event description still refers to ITER for lack of detailed frequency data sources for CFETR [10]. The events of the blanket and divertor system in Category II are listed in Table 2.

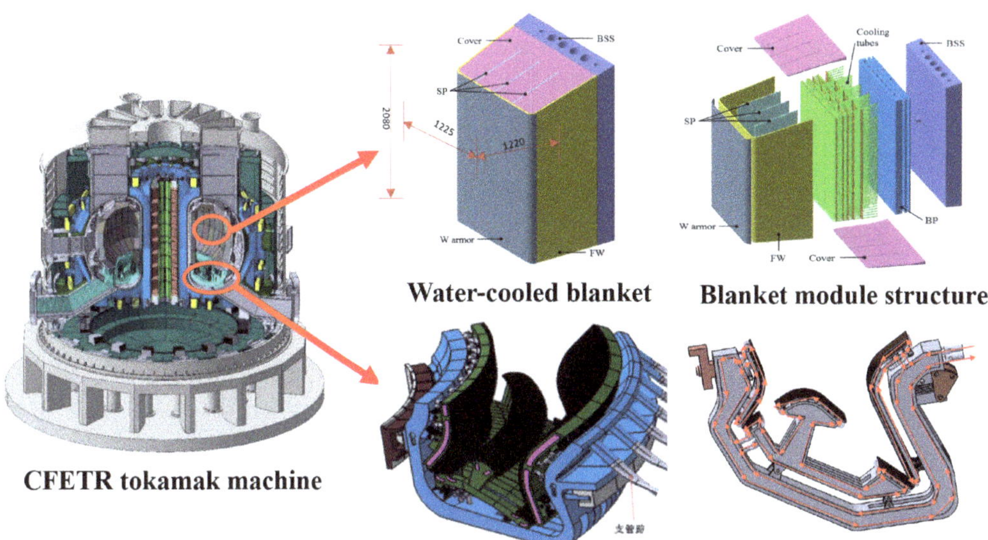

Figure 1. Water-cooled blanket and water-cooled divertor of CFETR.

Table 1. Event categories of CFETR.

Category	Event Frequency/yr	CFETR Plant Condition
I	>1	Normal operation
II	10^{-2}–1	Incident
III	10^{-4}–10^{-2}	Accident
IV	10^{-6}–10^{-4}	Accident
V	$<10^{-6}$	Hypothetical

Table 2. Events in Category II for the blanket and divertor system.

No.	Description of Event
1	Loss of flow in a blanket coolant circuit due to pump trip
2	Loss of flow in a blanket coolant channel
3	Ex-vessel coolant leakage due to small rupture of blanket coolant circuit pipe inside TCWS [1] vault
4	Small blanket in-vessel coolant leakage-equivalent break size: a few cm^2
5	Loss of flow in a divertor coolant circuit due to pump trip
6	Ex-vessel coolant leakage due to small rupture of divertor cooling circuit inside TCWS vault

TCWS [1]: Tokamak Cooling Water System.

Those selected events were analyzed to obtain the enveloping test condition for the water loop. Moreover, key phenomena were captured, and will be validated, in experiments. For normal operation in Category I, the thermal hydraulic parameters of the coolant also have pulsed characteristics due to pulsed heating loads. Taking the base case of the blanket system as an example [6], the pressure range is 15.38–16.0 MPa, and the inlet temperature range is 273–285 °C during the pulse operation. The two-phase flow does not occur during the whole process.

In category II, the ex-vessel small rupture of the blanket coolant circuit pipe is explained as an example. The RELAP5 model can be found in a previous report [11]. Main results are shown in Figure 2. In this case, no mitigation method is applied. Therefore, the coolant continues to leak until all the coolant runs out. Then, the coolant pressure declines continuously. As the decay heat decreases with time, after plasma shutdown, the coolant temperature decreases too. However, the two-phase flow occurs for lack of cooling. Due to the parallel structure of blanket modules and sectors, the flow instability is observed. From the results, it is clear that the thermal hydraulic parameters of the coolant can decrease to a much lower level, even without human intervention, at 1000 s after the event. Therefore, we pay more attention to the period of 0–1000 s. During this period, the pressure is in the range of 8–15.5 MPa, and the temperature is in the range of 292–325 °C. The maximum void fraction is 0.68.

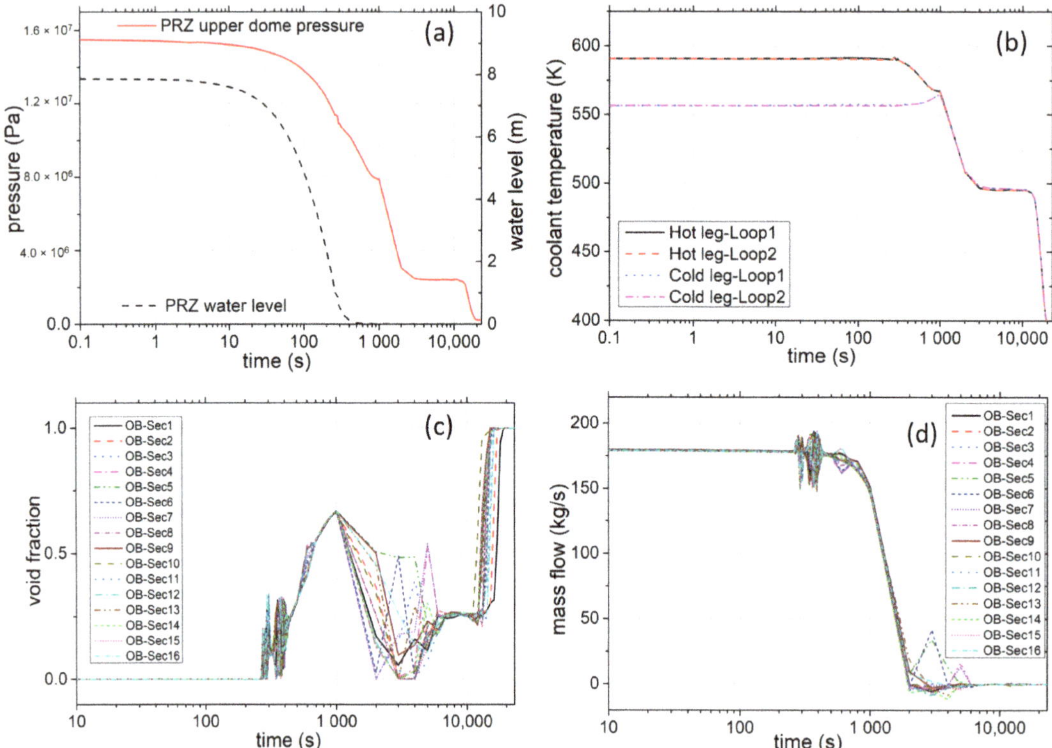

Figure 2. Ex-vessel small rupture of water-cooled blanket coolant circuit pipe: (**a**) pressurizer pressure and water level; (**b**) coolant temperature; (**c**) void fraction of coolant; (**d**) mass flow of coolant.

It should be noted that all events herein are caused by single fault, and only the most serious period of each event is considered. For the water-cooled divertor, most attention is focused on the manufacturing technology and the heat-resisting capability for one-side high heat flux. Besides, the operating temperature and pressure of the water-cooled divertor is below that of the water-cooled blanket [3,8]. Therefore, the operation condition of the water-cooled blanket is considered as enveloping test condition. According to the analyses of events in Category I and II [6,11,12], it can be concluded that:

- the mass flow is 40–110% of the steady state level;
- the temperature is in the range of 70–180 °C for divertor and 227–337 °C for blanket;
- the pressure is in the range of 4.0–5.0 MPa for divertor and 8.0–16.5 MPa for blanket;
- the maximum void fraction is 68%.

3. Experiment Plan

The experiment plan for the water loop of CRAFT was made according to the design and analyses results of the water-cooled blanket and divertor of CFETR. Primary experiments are introduced below.

3.1. Experiment of Critical Heat Flux

The FW of the blanket faces the one-side heat flux from plasma. The average steady state heat flux is assumed to be 0.5 MW/m^2. However, the heat flux might be uneven and much higher at certain positions [13,14], which may damage the structures. At the same time, there is nuclear heat induced by fast neutrons. To remove the heat, 95 channels, with the cross section of 8×8 mm^2, are embedded in the FW. Every two channels are set up as one group and flows in the opposite direction. The purposes of this experiment are to:

- verify the heat removal capability of the FW;
- obtain the Critical Heat Flux (CHF) to prevent burn up of the FW.

There are 4 parallel channels abstracted as typical experiment units, namely 2 groups of square channels with opposite flow directions. The schematic view of the test section is shown in Figure 3. The manifolds and inlet/outlet channels provide similar fully-developed flow conditions. Since the nuclear heat of the manifold and inlet/outlet channel occupies only ~6% of the FW total heat load, only the plasma facing section is going to be heated in the experiment [15,16]. Specifically, the heated section will be heated by two separate Direct-Current (DC) power supplies, which represent the nuclear heat and the heat flux, respectively. The maximum heat flux required is ~5 MW/m^2. Inconel 625 is selected as the structural material to facilitate the DC heating. The flow area remains 8×8 mm^2 in each channel, and the flow path keeps the same as the prototype. Since the nuclear heat decreases along the radial direction, the heated Inconel structure needs equivalent cross section design to reproduce that feature [17]. The detailed structure design of the test section is under development.

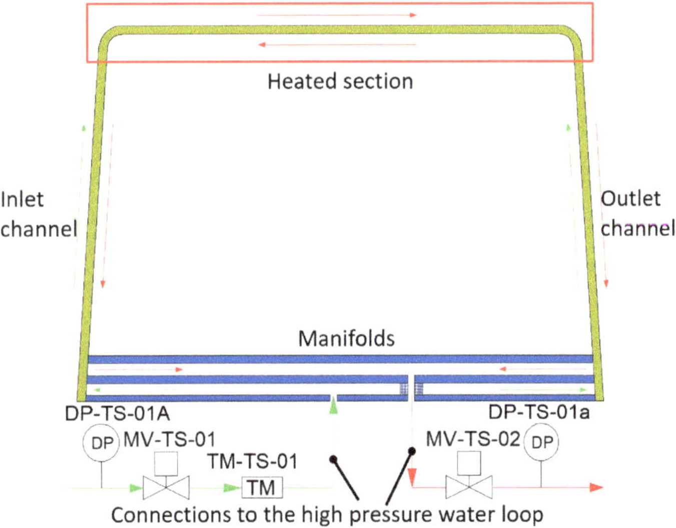

Figure 3. Schematic view of the CHF test section.

3.2. Experiment of One-Side Thermal Radiation by EBGs

The water-cooled blanket and divertor will use Reduced Activated Ferritic Martensitic (RAFM) steel. Tungsten is coated on the plasma facing side. In this experiment, the test sections will use the exact same structure and material with the prototype. The test section will be put into the vacuum chamber, and the two EBGs will be employed to generate one-side steady state and transient heat flux [18], as shown in Figure 4. The purposes of this experiment are to:

- verify the steady state and transient heat-resisting capability of the water-cooled blanket and divertor;
- verify the material performance and manufacturing techniques.

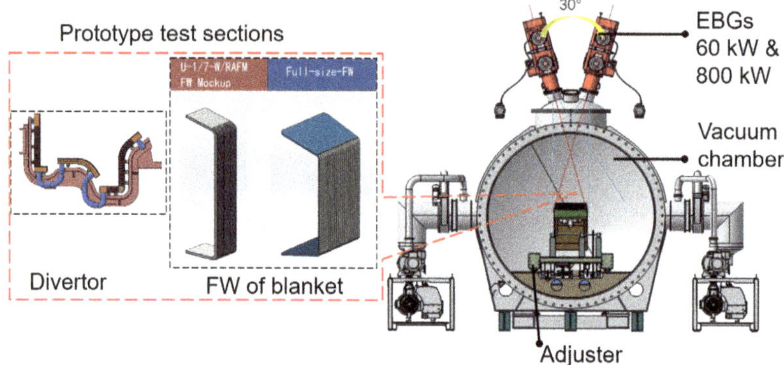

Figure 4. Schematic view of the one-side thermal radiation by EBGs.

3.3. Experiment of Flow Instability

The blanket system has many parallel structures. Specifically, the 95 channels in the FW of one blanket module, the 27 blanket modules in one blanket sector, and 16 blanket sectors in the tokamak machine are all in parallel from different levels. Uneven heat flux or hot spots may result in two-phase flow in the FW. Besides, two-phase flow may also appear in the coolant system during abnormal operation events, such as small rupture of cooling circuit pipes. Then, the flow instability may occur [19], which will damage the system and is not expected to happen. Therefore, the purposes of this experiment are to:

- clarify the mechanism of flow instability from different parallel levels;
- obtain the critical void fraction for flow instability to occur and prevent damage of the system.

Flow instability between different blanket modules in one sub-sector will use equivalent modules in the experiment. Each equivalent module shall keep the structure characteristics, including inclination, altitude, and heat source distribution, as shown in Figure 5. Detailed structure design of the test section is ongoing.

3.4. Experiment of Flow Distribution

As mentioned in Section 3.3, the blanket system has many parallel structures from different levels. The visualization of the flow distribution, among parallel channels/modules/ sectors, is going to be carried out under atmospheric pressure and temperature for the safety concern in the experiment. The test section will use Acrylic in favor of observation. The air will be used as the gas phase instead of steam. Apart from the flow distribution, many other important phenomena can also be observed through this experiment, including the void fraction distribution, the two-phase flow pattern, the pressure drop etc. The detailed structure design of the test section is underway.

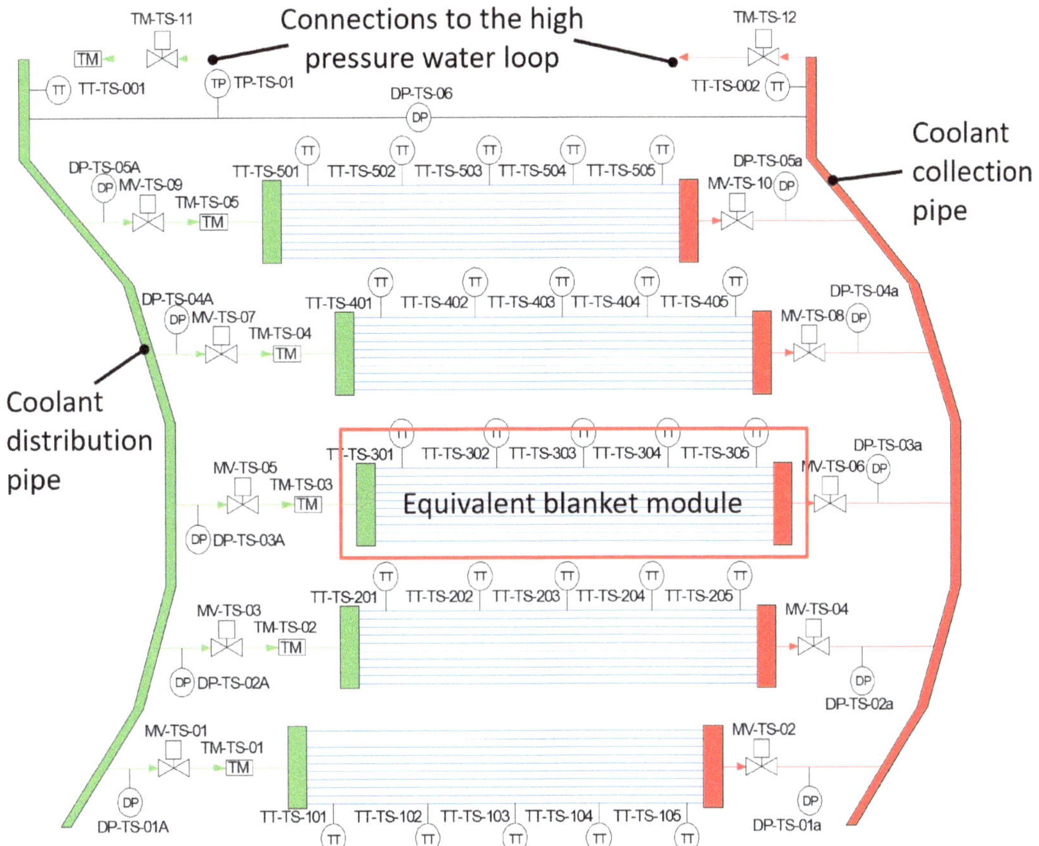

Figure 5. Schematic view of the flow instability test section.

4. Water Loop Design

The functions and main parameters of the water loop are determined based on the operating condition of each experiment. The water loop mainly consists of four subsystems, namely the high pressure water loop, the low pressure water loop, the component cooling water system, and the water charging system.

4.1. High Pressure Water Loop

The function of the high pressure water loop is to support high pressure experiments in Sections 3.1–3.3, while keeping stable operation. The high pressure water loop is composed of high temperature and high pressure canned pump, electric heating pressurizer (PZR), preheater, test section, high pressure mixer, 1# heat exchanger (HX), 2# HX, flow meters, control valves, related pipelines, and valve components. Besides, there is a discharge tank related to the PZR. The main function of the discharge tank is to collect water and steam, discharged through the Pressure Safety Valve (PSV) of the PZR and the test section. The scheme of the high pressure water loop is shown in Figure 6.

During the normal operation of the high pressure water loop, the canned pump provides a stable flow rate. Then, the deionized water enters the preheater after passing through the control valve and flow meter. After that, the water is heated further in the test section. The high temperature liquid, or two-phase flow at the outlet of the test section, is mixed with the lower temperature fluid from the bypass. Subsequently, the water is cooled by the 1# HX or the 2# HX and finally returns to the inlet of the canned pump.

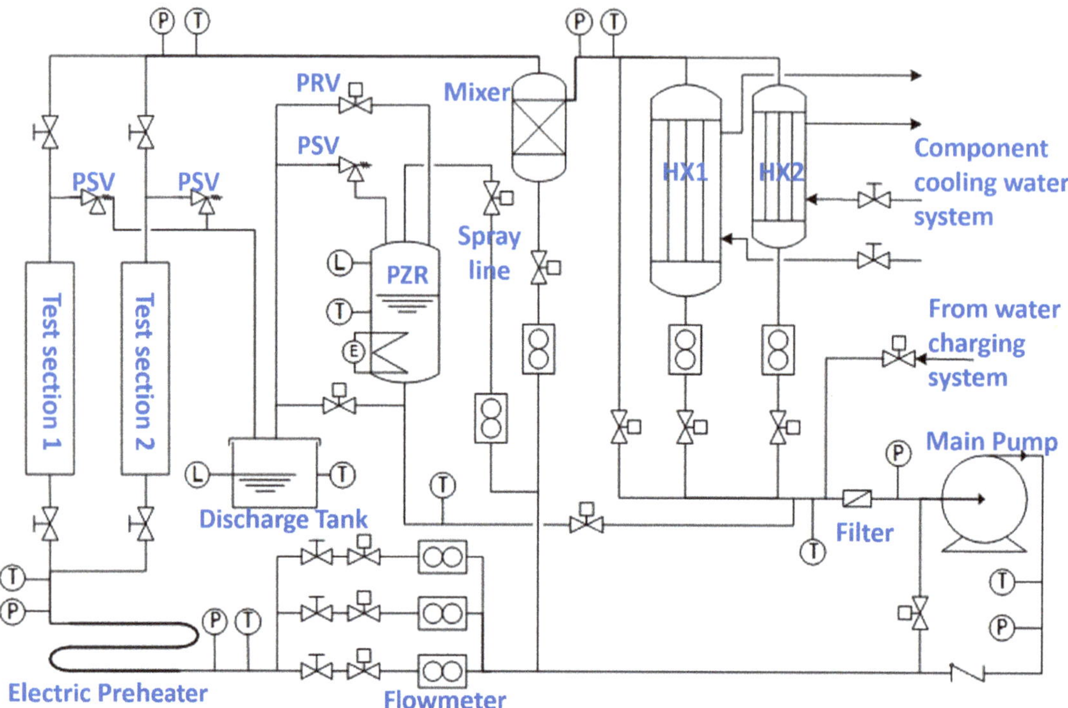

Figure 6. Scheme of the high pressure water loop.

The PZR surge line is connected to the canned pump inlet pipe for the pressure stability. The internal pressure of the PZR can be set manually, so the inlet pressure of the canned pump can be stabilized at any value between 3.5 MPa and 16.5 MPa, according to experiment requirements. The pressure control method refers to that of the primary heat transfer system of the water-cooled blanket [11]. Three sets of control valve and flow meters are arranged in front of the preheater, which can accurately adjust the inlet flow of the test section. Only one set is put into service at each experiment condition. The power of the preheater can be adjusted precisely to ensure that the inlet temperature of the test section is stable at the set value. The use of the mixer is to prevent the two-phase flow at the outlet of the test section from directly entering the HXs and causing damage. Thus, 1# HX and 2# HX are in parallel. When the total heating power of the test section and the preheater is less than 400 kW, the 2# HX is used for heat transfer. When the total heating power of the test section and the preheater is higher than 400 kW, the 1# HX is used. The HX power can be adjusted by regulating the control valves at the outlets of the two HXs and the control valve on the bypass of the HXs to maintain the stability of the inlet temperature of the canned pump. Main design parameters of the high pressure water loop are listed in Table 3.

4.2. Low Pressure Water Loop

The low pressure water loop consists of three parts, namely the air supply line, the water supply line, and the air-water mixing and separation line as shown in Figure 7. The air supply line mainly includes air compression, air tank, Pressure Reducing Valve (PRV), air flow meters, control valves, as well as related pipelines and valves. The main function of the air supply line is to provide air with stable pressure and adjustable flow rate for the low pressure mixer. At the same time, the compressed air stored in the air tank is used as the driving air source of pneumatic valves at the inlets and outlets of the high pressure test sections for emergency isolation.

Table 3. Design parameters of the high pressure water loop.

Item	Value	Unit
Design pressure	17.5	MPa
Design temperature	355	°C
Volume of the PZR	0.96	m^3
PZR electric heater power	60	kW
1# heat exchanger capacity	2600	kW
2# heat exchanger capacity	415	kW
Flow rate of the main pump	25	m^3/h
Electric preheater power	200	kW
Volume of discharge tank	3.8	m^3

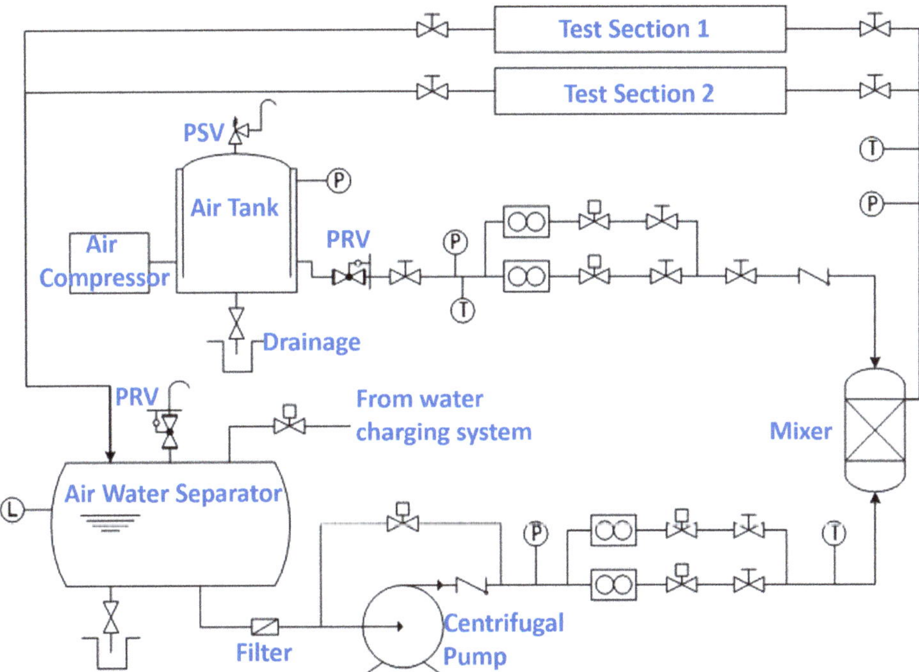

Figure 7. Scheme of the low pressure water loop.

The water supply line mainly includes the centrifugal pump, filters, water flow meters, control valves, related pipelines and valves. The main function of the water supply line is to provide water with stable pressure and adjustable flow rate for the low pressure mixer.

The air-water mixing and separation line is composed of low pressure mixer, air-water separator, test sections, related pipelines, and valves. The volume of the air-water separator is big enough and it is also used as the water tank for the centrifugal pump. The main function is to mix air and deionized water to provide a two-phase flow for the low-pressure test section, then separate air and water through the air-water separator, so the deionized water can be reused. Main parameters of the low pressure loop are listed in Table 4.

4.3. Component Cooling Water System

The component cooling water system is composed of circulating cooling water pump, plate heat exchanger, cooling water tank, filter, related pipeline, and valves, as shown in Figure 8. The main function is to provide cooling water for 1# HX, 2# HX, canned pump, air compression, DC power supply and vacuum chamber with stable temperature and flow rate. Main parameters of the component cooling water system are listed in Table 5.

Table 4. Design parameters of the low pressure water loop.

Item	Value	Unit
Design pressure of air supply line	1.0	MPa
Design pressure of water supply line	1.2	MPa
Design temperature	80	°C
Flow rate of air compressor	2–90	m^3/h
Flow rate of centrifugal pump	10–160	m^3/h
Volume of air tank	3.3	m^3
Volume of air water separator	16.2	m^3

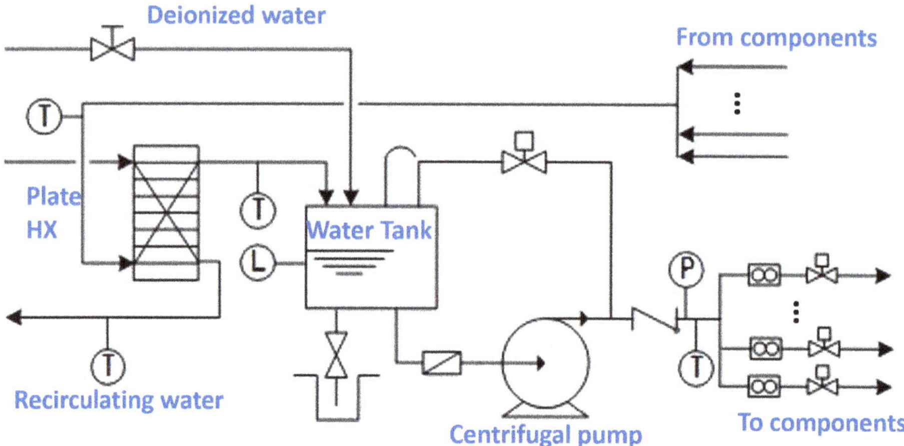

Figure 8. Scheme of the component cooling water system.

Table 5. Design parameters of the component cooling water system.

Item	Value	Unit
Design pressure	0.8	MPa
Design temperature	80	°C
Design flow rate	280	m^3/h
Cooling capacity	3500	kW

4.4. Water Charging System

The water charging system mainly includes deionized water tank, plunger pump, metering pump, filter, related pipelines and valves as shown in Figure 9. The main function is to supply water for the high pressure water loop and the low pressure water loop using the plunger pump. Then, the metering pump is used for increasing the pressure of the high pressure water loop. The deionized water tank is equipped with an electric heater and a nitrogen sealing system, which can deoxygenate the water and prevent corrosion of the system pipelines and the test sections. Main parameters are listed in Table 6.

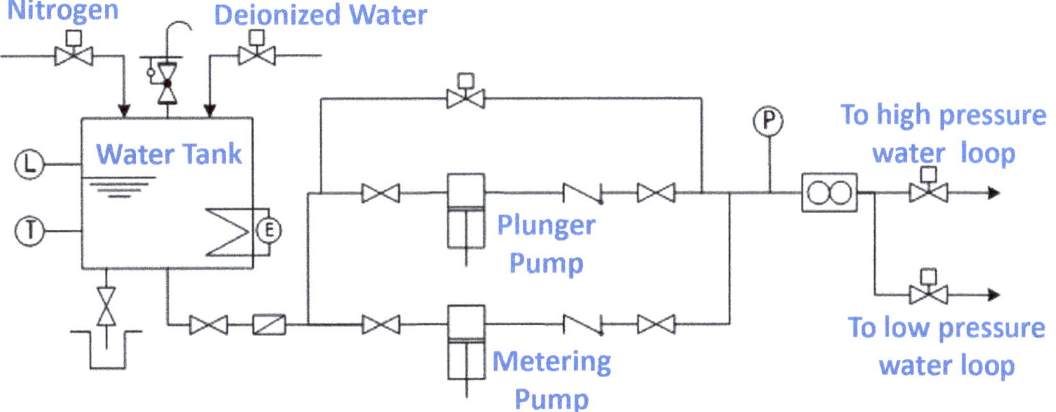

Figure 9. Scheme of the water charging system.

Table 6. Design parameters of the water charging system.

Item	Value	Unit
Outlet pressure of plunger pump	6	MPa
Flow rate of plunger pump	5	m^3/h
Outlet pressure of metering pump	20	MPa
Flow rate of metering pump	0.4	m^3/h
Volume of air tank	2.4	m^3

5. Layout of the Water Loop

Based on the process design in Section 4, the engineering design of the water loop was completed by cooperation with an industrial enterprise with certifications. The 3D layout of the water loop in the CRAFT plant is shown in Figure 10. The water loop covers an area of 10×12 m^2. To accelerate the construction and ensure the quality, the water loop adopts modular design, consisting of 9 blocks. The 9 blocks can be fabricated at the same time in the factory so as to save time and reduce onsite assembling and inspection work. Block 1–4 are located on the first floor at the elevation of 0.0 m. Block 5–8 are located on the second floor at the elevation of +5.0 m. Block 9 and the control cabinet are located on the third flow at the elevation of +10.0 m. Since the high pressure water loop and the low pressure loop are independent, they are arranged separately on different blocks. Block 1, 5, 7, and 9 belong to the high pressure water loop, which are located on the west side of the steel platform. Block 2 and 6 belong to the low pressure water loop, which are located at the south-east side of the steel platform. The other blocks are for auxiliary components. Except the test section for "experiment of one-side thermal radiation by EBGs" which will be put inside a vacuum chamber, test sections for other experiments will be connected to the water loop from the north side. On every block, there are respective instrument junction boxes and electrical junction boxes, so each block can realize independent functions. Then, the control signals are collected to the control cabinet through cable. After that, the signals are transmitted from the onsite cabinet to the control room though optical fiber. Main information of each block is summarized in Table 7.

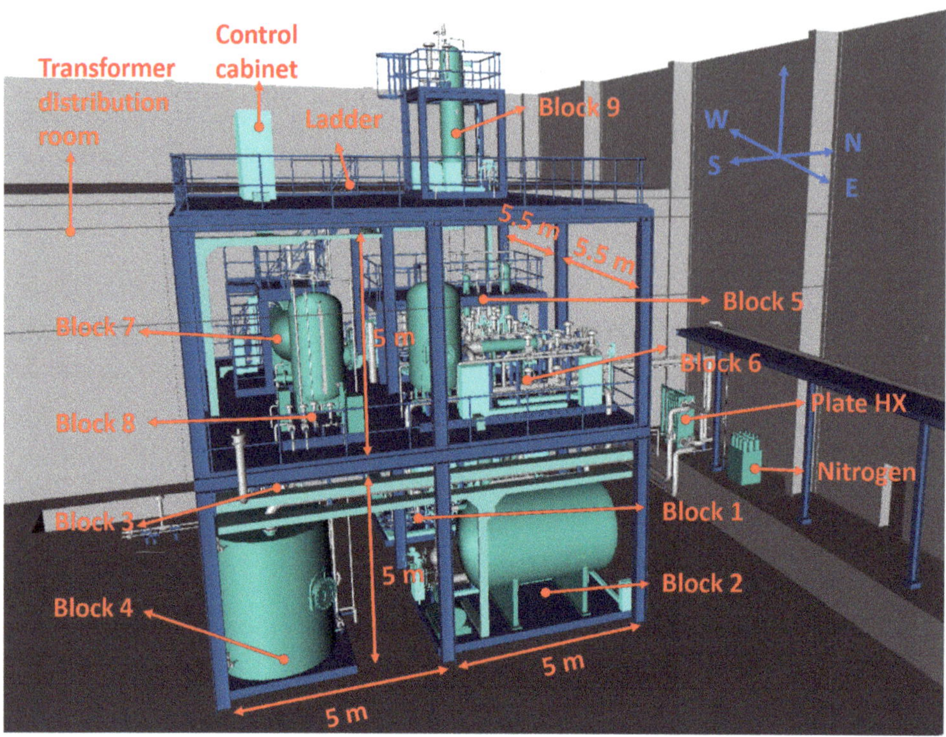

Figure 10. A 3D layout of the water loop in plant of CRAFT.

Table 7. Information of Block 1–9 of the water loop.

No.	Components	Functions
Block 1	Canned pump, preheater, high pressure water flow meters and control valves	High pressure water pumping, flow control and measuring, preheating
Block 2	Centrifugal pump of low pressure water loop, air-water separator	Air-water separating, low pressure water pumping
Block 3	Centrifugal pump of component cooling water system, plunger pump and metering pump of water charging system	Water pumping
Block 4	Water tank of component cooling water system, air compressor	Component cooling water storage, air compressing
Block 5	High pressure mixer, 1# HX, 2# HX	Mixing, heat exchanging
Block 6	Air tank, air flow meters and control valves, water flow meters and control valves, low pressure mixer	Air and water supply and mixing
Block 7	Discharge tank	Water and steam discharge
Block 8	Water tank of water charging system, air dryer	Deionized water storage, air drying
Block 9	PZR	Pressure control

6. Discussion and Conclusions

With the financial support from the CRAFT project, a water loop will be constructed for testing the water-cooled blanket and water-cooled divertor components of CFETR in the frame of a HHF testing facility. This paper introduces the process of the water loop design. It started from the design and analyses of the water-cooled blanket and the water-cooled divertor prototypes of CFETR. Events in Category I and II were selected to obtain enveloping thermal hydraulic parameters for the water loop design. The experiment plan was set out by abstracting features of the prototypes from analyses results. Then, the function of the water loop was determined to fulfill experiment requirements. The conceptual design, preliminary design, and engineering design of the water loop were carried out step by step. Until now, the design of the water loop has been completed. The fabrication and construction of the water loop is about to start. According to the schedule, the water loop will be ready for commissioning by August 2022. The experiments will provide an important technical base for CFETR.

Author Contributions: Conceptualization, X.C. and Z.L.; methodology, S.L. and C.P.; software, X.C.; validation, W.W. and Q.L.; formal analysis, X.C. and Z.L.; investigation, X.C. and Z.L.; writing—original draft preparation, X.C.; writing—review and editing, S.L.; supervision, S.L. and C.P.; project administration, X.C. and S.L. All authors have read and agreed to the published version of the manuscript.

Funding: This work was funded by Comprehensive Research Facility for Fusion Technology Program of China under Contract No. 2018-000052-73-01-001228.

Data Availability Statement: Not applicable.

Acknowledgments: This work has been carried out in the framework of strategic cooperation of Institute of Energy with ASIPP and USTC. The project receives funding from the CRAFT program. The views and opinions expressed herein do not necessarily reflect those of ASIPP, USTC and the CRAFT program committee.

Conflicts of Interest: The authors declare no conflict of interest.

References

1. Li, J.; Ni, M.; Lu, Y. The frontier and perspective for tokamak development. *Natl. Sci. Rev.* **2019**, *6*, 382–383. [CrossRef]
2. Liu, S.; Cheng, X.; Ma, X.; Chen, L.; Jiang, K.; Xia, L.; Bao, H.; Wang, J.; Wang, W.; Peng, C.; et al. Progress on design and related R&D activities for the water-cooled breeder blanket for CFETR. *Theor. Appl. Mech. Lett.* **2019**, *9*, 161–172.
3. Qin, S.; Yao, D.; Wang, Q.; Mao, X.; Liu, P.; Qian, X.; Xu, T.; Li, L.; Peng, X.; Lu, K.; et al. Preliminary design progress of the CFETR water-cooled divertor. *IEEE Trans. Plasma Sci.* **2020**, *48*, 1733–1742. [CrossRef]
4. Wang, X. Helium-cooled blanket design. In Proceedings of the CFETR Engineering Design Annual Conference, Langfang, China, 26–29 October 2020. (In Chinese)
5. Huang, W.; Lu, Y.; Zhang, L.; Liu, K.; Jin, Y.; Zheng, G.; Cai, L.; Zhu, Y.; Xue, M. Preliminary design and analysis of CFETR He-cooled divertor. *Nucl. Fusion Plasma Phys.* **2021**, *41*, 37–44. (In Chinese)
6. Cheng, X.; Ma, X.; Lu, P.; Wang, W.; Liu, S. Thermal dynamic analyses of the primary heat transfer system for the WCCB blanket of CFETR. *Fusion Eng. Des.* **2020**, *161*, 112067. [CrossRef]
7. Abdou, M.; Morley, N.B.; Smolentsev, S.; Ying, A.; Malang, S.; Rowcliffe, A.; Ulrickson, M. Blanket/first wall challenges and required R&D on the pathway to DEMO. *Fusion Eng. Des.* **2015**, *100*, 2–43.
8. Yao, D. Water-cooled divertor design. In Proceedings of the CFETR Engineering Design Annual Conference, Langfang, China, 26–29 October 2020. (In Chinese)
9. Taylor, N. Accident Analysis Guidelines 4. IDM No. ITER_D_24TDZ8 v2.4. Available online: https://www.iter.org/ (accessed on 24 May 2021).
10. Taylor, N. Accident Analysis Report (AAR) Volume 1-Event Identification and Selection. IDM No. ITER_D_2DPVGT v1.4. Available online: https://www.iter.org/ (accessed on 24 May 2021).
11. Cheng, X.; Ma, X.; Wang, W.; Chen, L.; Liu, S.; Xu, Y. Primary heat transfer system design of the WCCB blanket for multiple operation modes of CFETR. *Fusion Eng. Des.* **2020**, *153*, 111489. [CrossRef]
12. Cheng, X. Water-cooled blanket design. In Proceedings of the CFETR Engineering Design Annual Conference, Langfang, China, 26–29 October 2020. (In Chinese)
13. Mitteau, R.; Stangeby, P.; Lowry, C.; Merola, M.; The ITER Blanket Section. Heat loads and shape design of the ITER first wall. *Fusion Eng. Des.* **2010**, *85*, 2049–2053. [CrossRef]
14. Igitkhanov, Y.; Fetzer, R.; Bazylev, B. Effect of heat loads on the plasma facing components of demo. *Fusion Eng. Des.* **2016**, *109–111*, 768–772. [CrossRef]
15. Ilic, M.; Messemer, G.; Zinn, K.; Kiss, B. HETRA experiment for investigation of heat removal from the first wall of helium-cooled-pebble-bed test blanket module. *Fusion Eng. Des.* **2011**, *86*, 2250–2253. [CrossRef]
16. Ilic, M.; Messemer, G.; Zinn, K.; Meyder, R.; Kecskés, S.; Kiss, B. Experimental and numerical investigation of heat transfer in the first wall of Helium-Cooled-Pebble-Bed Test Blanket Module-Part 1: Presentation of test section and 3D CFD model. *Fusion Eng. Des.* **2015**, *90*, 29–36. [CrossRef]

17. Peng, J.; Huang, Y.; Xu, J.; Xiao, Z.; Duan, S. Method of the modeling and abstraction of typical thermal experiment unit for sub-critical fuel components. *Nucl. Power Eng.* **2015**, *36*, 9–13.
18. Kim, S.K.; Park, S.D.; Jin, H.G.; Lee, E.H.; Yoon, J.S.; Lee, D.W. Current status of Korea heat load test facility KoHLT-EB for fusion reactor materials. *IEEE Trans. Plasma Sci.* **2017**, *45*, 1820–1823. [CrossRef]
19. Lian, Q.; Tian, W.; Gao, X.; Qiu, S.; Su, G. Study of flow instability in parallel channel system for water cooled blanket. *Fusion Eng. Des.* **2019**, *148*, 111291. [CrossRef]

Article

MHD R&D Activities for Liquid Metal Blankets

Chiara Mistrangelo [1,*], Leo Bühler [1], Ciro Alberghi [2], Serena Bassini [3], Luigi Candido [2], Cyril Courtessole [1], Alessandro Tassone [4], Fernando R. Urgorri [5] and Oleg Zikanov [6]

[1] Karlsruhe Institute of Technology (KIT), 76131 Karlsruhe, Germany; leo.buehler@kit.edu (L.B.); cyril.courtessole@kit.edu (C.C.)
[2] ESSENTIAL Group, Politecnico di Torino, Corso Duca degli Abruzzi 24, 10129 Torino, Italy; ciro.alberghi@polito.it (C.A.); luigi.candido@polito.it (L.C.)
[3] ENEA FSN-ING Division, C.R. Brasimone, Bacino del Brasimone, 40032 Camugnano, Italy; serena.bassini@enea.it
[4] DIAEE Nuclear Section, Sapienza University of Rome, 00186 Rome, Italy; alessandro.tassone@uniroma1.it
[5] National Fusion Laboratory, CIEMAT, 28040 Madrid, Spain; fernando.roca@ciemat.es
[6] College of Engineering and Computer Science, University of Michigan-Dearborn, Dearborn, MI 48128-1491, USA; zikanov@umich.edu
* Correspondence: chiara.mistrangelo@kit.edu

Abstract: According to the most recently revised European design strategy for DEMO breeding blankets, mature concepts have been identified that require a reduced technological extrapolation towards DEMO and will be tested in ITER. In order to optimize and finalize the design of test blanket modules, a number of issues have to be better understood that are related to the magnetohydrodynamic (MHD) interactions of the liquid breeder with the strong magnetic field that confines the fusion plasma. The aim of the present paper is to describe the state of the art of the study of MHD effects coupled with other physical phenomena, such as tritium transport, corrosion and heat transfer. Both numerical and experimental approaches are discussed, as well as future requirements to achieve a reliable prediction of these processes in liquid metal blankets.

Keywords: liquid metal blankets; magnetohydrodynamics (MHD); tritium; corrosion; convection; turbulence; WCLL blanket; DCLL blanket

Citation: Mistrangelo, C.; Bühler, L.; Alberghi, C.; Bassini, S.; Candido, L.; Courtessole, C.; Tassone, A.; Urgorri, F.R.; Zikanov, O. MHD R&D Activities for Liquid Metal Blankets. *Energies* **2021**, *14*, 6640. https://doi.org/10.3390/en14206640

Academic Editor: Hyungdae Kim

Received: 31 August 2021
Accepted: 22 September 2021
Published: 14 October 2021

Publisher's Note: MDPI stays neutral with regard to jurisdictional claims in published maps and institutional affiliations.

1. Introduction

As stated in the most recent European roadmap for a DEMO reactor [1,2], one of the candidate driver blankets that will be investigated in ITER is the Water Cooled Lead Lithium (WCLL) blanket concept. It uses water as a coolant at typical pressurized water reactor conditions (290–325 °C, 15.5 MPa) and lead lithium PbLi in eutectic composition as tritium breeder, neutron multiplier and tritium carrier. Other concepts, such as the Helium Cooled Lead Lithium (HCLL) and the Dual Coolant Lead Lithium (DCLL) blankets, are still being considered, in limited R&D activities, as potential long-term options.

In liquid metal (LM) blankets, the electrically conducting PbLi moves in the system under the action of the intense magnetic field that confines the fusion plasma.

It is well known that electromagnetic forces induced by the interaction of magnetic field and velocity significantly modify liquid metal flow behavior compared to hydrodynamic conditions [3]. The peculiar magnetohydrodynamic (MHD) flow distribution affects all phenomena that depend on the near-wall velocity profile, such as transport of tritium and corrosion products, and heat transfer. In addition, significant temperature gradients are present in breeding blankets, giving rise to buoyancy forces.

Recently, a review paper has been published aiming to identify the main MHD issues related to the design of liquid metal blankets and to present the state of the art of the study of some fundamental MHD phenomena that occur in these complex systems [4]. In the present paper, we want to provide a complementary description of multiphysics

Energies **2021**, *14*, 6640. https://doi.org/10.3390/en14206640

MHD phenomena, focusing on recent progress and fusion-relevant applications. For instance, tritium transport and corrosion in hydrodynamic LM systems have been studied extensively for different cases and operating conditions. However, those processes in MHD flows have been investigated only in a few publications. Since transport processes depend on the velocity distribution, the MHD flow that results from the interaction of the moving LM with the external magnetic field has to be used as input for their analysis. Features to be taken into account are, e.g., increased velocity at electrically conducting walls aligned with the magnetic field, turbulence anisotropy, reversed flows caused by magneto-convection and electromagnetic coupling.

Models have been proposed to integrate into the analysis of MHD flows corrosion and tritium transfer across material interfaces, permeation through structural materials and into the coolant. In order to identify the fundamental steps for a long-term strategy aimed at implementing and validating accurate and robust modeling tools for application to fusion technology, it is first required to outline the actual state of the art of the different R&D topics. Therefore, a review of modelling approaches, validation experiments and future requirements for analyses of coupled MHD phenomena is presented in this paper. Another important step towards a better understanding of multiphysics MHD flows in complex systems is the development of system codes. The main features, advantages and limitations of these predictive tools are discussed as well.

2. Problem Description

We consider in the following the general equations governing the magneto-convective flow of an electrically conducting fluid in a magnetic field B. They describe the conservation of momentum and mass:

$$\rho\left(\frac{\partial v}{\partial t} + (v \cdot \nabla)v\right) = -\nabla p + \rho \nu \nabla^2 v - \rho \beta (T - T_0)g + j \times B, \nabla \cdot v = 0, \tag{1}$$

where v denotes the velocity, p is the deviation of pressure from isothermal hydrostatic conditions at the reference temperature T_0, g stands for gravitational acceleration, and $j \times B$ is the electromagnetic Lorentz force induced by the interaction of imposed magnetic field B and electric current density j. The latter is determined via Ohm's law $j = \sigma(-\nabla\phi + v \times B)$ that, combined with the condition for charge conservation, $\nabla \cdot j = 0$, results in a Poisson equation for the electric potential ϕ:

$$\nabla \cdot \nabla\phi = \nabla \cdot (v \times B). \tag{2}$$

Density changes due to temperature variations in the liquid metal are described by the Boussinesq approximation. The physical properties of the fluid, density ρ, kinematic viscosity ν, volumetric thermal expansion coefficient β, and electric conductivity σ are taken at the reference temperature T_0.

The temperature distribution in the fluid is given by the energy balance equation

$$\rho c_p \left(\frac{\partial T}{\partial t} + (v \cdot \nabla)T\right) = k\nabla^2 T + Q \tag{3}$$

where c_p is the specific heat of the fluid at constant pressure, k the thermal conductivity and Q a volumetric thermal source.

The flow can be characterized by three dimensionless groups, the Hartmann number Ha, the Reynolds number Re and the Grashof number Gr,

$$Ha = LB\sqrt{\frac{\sigma}{\rho \nu}}, \ Re = \frac{u_0 L}{\nu}, \ Gr = \frac{g\beta L^3 \Delta T}{\nu^2}. \tag{4}$$

The former one gives a nondimensional measure for the strength B of the imposed magnetic field and its square quantifies the ratio between electromagnetic and viscous forces. The second expresses the ratio of inertia to viscous forces. Alternatively, the interaction parameter $N = Ha^2/Re$ can be used to weigh the relative importance of electromagnetic and inertia forces. The Grashof number characterizes the intensity of buoyancy. The quantities L and u_0 are a typical size and mean velocity in the considered geometry. A characteristic temperature difference ΔT can be defined for instance through the volumetric heat as $\Delta T = QL^2/k$.

Equations (1)–(3) are solved in the fluid and Equations (2) and (3) also in electrically and thermally conducting solid structures, by applying appropriate boundary conditions. The fluid velocity vanishes at walls (no-slip condition) and continuity of wall-normal currents and electric potential is assumed at interfaces (perfect electrical contact between fluid and solid material). At external surfaces towards an electrically insulating environment, wall-normal currents vanish. At the entrance to the computational domain, the flow is prescribed and often assumed as being fully developed, while at the exit an advective outflow condition can be applied.

3. Theoretical Description of MHD Flows

The design of test blanket modules (TBMs) for ITER and breeding blankets for DEMO requires accurate and reliable predictive tools. Depending on the application, the type of information required, and the level of detail, different methods can be employed to study theoretically MHD flows under fusion-relevant conditions.

In the following, we discuss the advantages, limitations, development progress, and application examples of various approaches, such as Computational Fluid Dynamics (CFD) simulations, asymptotic analysis and system codes.

3.1. Numerical Simulations

CFD codes provide a detailed description of flow features, since the domain discretization is highly refined. Moreover, they may be used to simulate turbulent flows, heat transfer and other coupled MHD phenomena.

Preliminary numerical simulations prior to the start of an experimental campaign can serve to determine which system's parameters need the highest measurement accuracy. This type of information is very useful during the experiment's conceptual phase, since it helps to select the type of sensors based on their accuracy when recording the desired quantity and hence it increases the consistency of experimental and numerical results. Simulations can provide indications for a suitable arrangement and for the needed number of sensors in specific zones of interest in the system. Numerical simulations can also be used to analyze the sensitivity of a model to the variation of given parameters.

The simplicity of the electromagnetic part of the governing Equations (1)–(3) can be misleading. In fact, accurate numerical simulation of flows at blanket-relevant high values of Ha is a computationally challenging task, significantly more than in the case of hydrodynamic flows at the same Re and Gr. The key reason is the numerical stiffness of the problem, which manifests itself as thin boundary and internal shear layers and the very large ratio between the largest and smallest typical time and length scales (see Section 3.4 as well as [3,5] for a more detailed discussion). One significant consequence of the stiffness is that high-Ha flows, especially those with unsteady behavior, are not directly amenable to analysis by general-purpose CFD codes based on finite-volume or finite-element discretization on unstructured grids. A review of numerical methods and outstanding questions can be found, e.g., in [3].

Another important factor in analyses of unsteady flows is the availability of effective, accurate and well parallelizable solvers for elliptic equations. A time steep in a numerical solution of the system (1)–(3) inevitably requires solving two such equations, one for electric potential (2) and one for pressure (here we assume that a projection method is used to satisfy incompressibility). The typically very small Prandtl number, $Pr = \rho v c_p/k$, of liquid metals means that the heat conduction term in the energy Equation (3) has to be treated implicitly in order to avoid severe limitations on the size of the time step, so another elliptic equation for temperature must be solved. Implicit treatment of viscous terms in (1) is also often applied, which implies another three elliptic equations for velocity components. It is not uncommon that over 90% of the computational time in an analysis of an unsteady high-*Ha* flow is spent on the solution of elliptic equations. High-performance computing based on massive parallelization becomes necessary in many cases (see, e.g., [6]).

In recent years, considerable advances have been made in the numerical prediction of coupled MHD-heat transfer phenomena in complex blanket-relevant geometries and under fusion conditions [7,8]. However, even when relatively small computational domains are considered, large computing resources are needed, which results in long simulation time. The latter can be reduced by introducing some assumptions in the flow modelling, as described in the following section for the asymptotic numerical approach.

3.2. Asymptotic Analysis

While the study of MHD flows under very strong magnetic fields represents a considerable challenge for numerical simulations, asymptotic analyses benefit from flow conditions that are typical for fusion applications. From a mathematical point of view, steady-state flows in strong magnetic fields constitute a singular perturbation problem. One may neglect inertia and viscous forces in comparison with the strong Lorentz and pressure forces (which largely balance each other) in the core of the flow domain if $N \to \infty$ and $Ha \gg 1$. Viscous effects at walls can be taken into account by a boundary layer analysis, which allows satisfying the no-slip condition at fluid–solid interfaces. Asymptotic analyses can be used to study fully developed flows in long straight pipes and ducts (see, e.g., [9,10]). However, the method has been also applied to some 3D flows such as those in ducts with variable cross-section [11] or in non-uniform magnetic fields [12]. For the latter case a good agreement with experiments has been achieved for $Ha \gg 1$.

The initial idea of an asymptotic description of general 3D flows, which dates back to Kulikovskii [13], has been reformulated in tensor notation, extended by a boundary layer analysis, and developed to treat walls of arbitrary electric conductivity [14]. The latter approach takes advantage of the fact that variations of all flow quantities along magnetic field lines are analytically known, which allows for a kind of "projection" of the 3D problem onto the duct walls, where the remaining 2D equations for pressure and electric potential are numerically solved using boundary-fitted coordinates. In a final step, the entire 3D flow is reconstructed by analytical means. The method allows for fast computations (seconds to minutes) on standard PCs, even for relatively complex geometries. On very coarse grids, one already achieves a good approximation of flow quantities, as shown by the example of a 3D flow in a fringing magnetic field displayed in Figure 1. The validity of the method has been verified by comparison with analytical solutions, full 3D simulations, and experiment [12,15]. The general formulation of the code allows calculations for quite complex domains, as illustrated by the examples in Figure 2, which shows the entrance and exit flows of a DCLL blanket sector, or in a model geometry of a DCLL blanket module.

While asymptotic methods give accurate results with almost negligible numerical effort, they are unable to predict the influence of inertia on MHD flows. Even if the core flow remains almost inertialess for $N \gg 1$, inertia at sharp corners or in high-velocity boundary layers may lead to local flow separation or instabilities that cannot be predicted by the latter method. If such phenomena are expected to occur, one has to perform full 3D time-dependent numerical simulations.

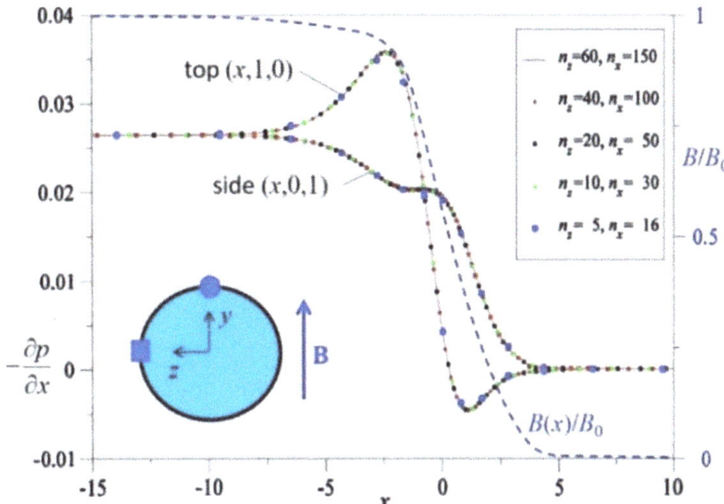

Figure 1. MHD flow in a circular pipe in a fringing magnetic field $B(x)$ as used in [12]. In the figure, $B(x)$ is normalized by the value B_0 in the center of the magnet. Axial pressure gradients $\partial p/\partial x$ are calculated at the top and side of the circular pipe for $Ha = 6600$. Results displayed for various numerical resolutions n_x and n_z in x and z directions highlight the rapid convergence of the asymptotic method.

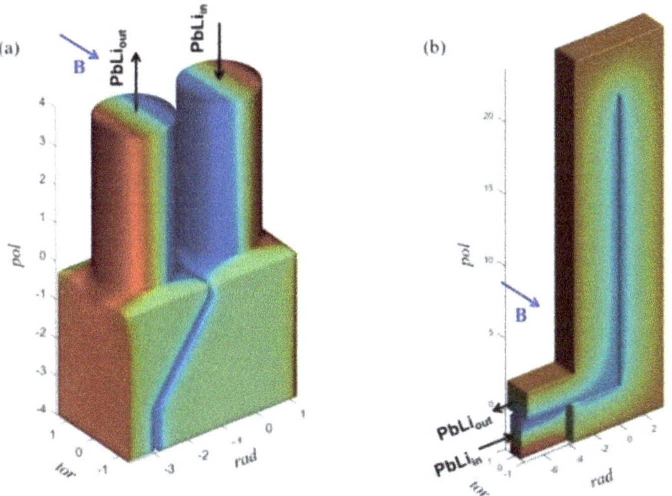

Figure 2. Examples of application of asymptotic methods for the study of pressure-driven MHD flows in (**a**) a DCLL geometry proposed in [16] and in (**b**) a model geometry for a modular DCLL, as shown in [17]; contours of electric potential are plotted on the fluid–wall interfaces.

3.3. System Codes

System thermal-hydraulics (SYS-TH) codes are a class of numerical tools developed since the 1970s in the framework of nuclear reactor safety technology for fission reactors [18]. SYS-TH codes aim to represent with sufficient accuracy all the main phenomena occurring in a nuclear reactor during operational and accidental transients and, as such, are used to perform deterministic safety analyses for licensing purposes. A SYS-TH model is created through a process called "nodalization" that subdivides the reactor into control volumes in which a set of linearized algebraic equations, derived by variants of the finite volume and difference methods, are solved to obtain the distribution of physical properties (temperature, velocity, pressure, etc.) in a set of nodes. The key features of SYS-TH codes are the quite coarse spatial resolution and the reliance on closure laws derived empirically to model complex phenomena. Modelling choices are available to represent reactor components with an increasing level of fidelity from lumped (0D) or 1D modules to simulate flow in primary and secondary loops, up to fully 3D to recreate the phenomena occurring in the reactor core. Notable examples of SYS-TH codes are, among others, RELAP5, TRAC/TRACE, ATHLET and CATHARE.

SYS-TH codes are often described as multiphysics, best-estimate, safety and industrial computational tools [18]:

- Multiphysics: to include all relevant phenomena encountered in a nuclear reactor from two-phase thermal-hydraulics to neutron kinetics diffusion equations and fuel thermomechanics.
- Best-estimate: to provide an accurate prediction of selected figures of merit and to be as close as possible to the physical reality that one aims to simulate.
- Safety: suitable to demonstrate the reactor safety for licensing purposes, namely developed according to a clearly defined quality assurance (QA) methodology, accompanied by uncertainty quantification (UQ), extensively verified and validated to prove the code scalability.
- Industrial: i.e., robust and computationally inexpensive to allow simulations on a wide test matrix for sensitivity analysis and UQ in a reasonable timeframe.

In the past two decades, SYS-TH codes have been further developed to make them able to represent dominant characteristic phenomena in fast breeder reactors (FBR) and other innovative Generation IV fission reactor concepts [19]. The development of a DEMO design has also motivated lines of research aimed at improving the capability of SYS-TH codes to deal with fusion reactor environment for both liquid metal and solid breeder blankets [20,21]. Despite the progress achieved, SYS-TH codes for fusion applications are still far from an acceptable maturity. In particular, the prediction of MHD effects in liquid metal blanket systems is still very challenging due to the limited availability of closure laws to describe them [22,23]. This is in stark contrast with the advancements made in recent years in the field of CFD analysis, where both commercial and open-source codes are nowadays able to simulate liquid metal MHD flows at reactor-relevant magnetic field intensity and at blanket scale [24].

Relying on closure laws to ensure accurate representation of physical reality is not necessarily associated with a lack of accuracy, and, in fact, the experience of the nuclear industry has clearly demonstrated that SYS-TH codes can provide reliable results for reactor safety and design if closure laws of sufficient quality are available [18]. The possibility of performing system-scale coupled thermal-hydraulics/MHD analyses at largely reduced computational cost is attractive for performing sensitivity and uncertainty analyses even when mature CFD codes for fusion modelling are available. The critical point is to develop closure laws for MHD effects, such as MHD pressure drop, electromagnetic coupling and heat transfer, based on current knowledge, as discussed in the next section, and to define future R&D steps for their improvement.

3.3.1. MHD Pressure Loss, Electromagnetic Coupling and Heat Transfer

Among the various MHD effects that influence liquid metal blanket performance, pressure losses and modified heat transfer coefficients are often considered the most critical ones [25]. Therefore, it is reasonable that they should be prioritized for implementation in SYS-TH codes.

Regarding MHD pressure losses, SYS-TH codes must be able to predict distributed (2D) and concentrated (3D) losses and how they are impacted by electromagnetic coupling, buoyancy and Q2D turbulence. Pressure losses in fully developed flows at fusion-relevant parameters are well characterized by theoretical and experimental works and scaling laws are available for both circular pipes and rectangular ducts [4,26,27].

The prediction of 3D MHD losses is more demanding due to the wider range of governing parameters involved and the need to take into account the effects of inertia and viscous forces. Moreover, they are caused by many scenarios, such as the presence of fringing magnetic fields of general orientation, discontinuous wall conductivity, and complex geometry (bends, cross-section variation, etc.) [26]. Therefore, it is difficult to develop a scaling law that has universal validity. The needed coefficients have to be chosen by experienced analysts, from detailed numerical simulations or dedicated experiments in representative and specific conditions. A recent review of theoretical and experimental works on 3D MHD losses may be found in [4].

While 2D and many 3D phenomena in MHD duct flows are reasonably well understood and ready for implementation in SYS-TH codes by adaption of friction factors, other phenomena that do not occur in hydrodynamic flows make their implementation in SYS-TH codes challenging. One example is the electromagnetic coupling caused by leakage currents between parallel channels that share electrically conductive walls. Its effect depends on the number of coupled channels, their orientation with respect to the magnetic field, the conductivity of the walls, etc. [4]. Analytical solutions exist only for very special electrical boundary conditions [28] so that for applications in complex blanket geometries only coupled 2D or 3D simulations can provide the relevant friction coefficients when general scaling laws do not exist [29–31]. The problem becomes even more complex for flows in multiple coupled bends, where the coupling can be responsible for strongly increased flow in external ducts, while the flow in the central channels is significantly reduced, as predicted by Madarame et al. [32] and shown in experiments by Stieglitz et al. [33].

Another phenomenon that is difficult to implement in SYS-TH codes is the buoyancy force that aid or hamper the forced flow circulation if directed upward or downward, respectively [26]. While it is straightforward to take changes of cross-section averaged density into account in modified correlations for friction coefficients, strong temperature gradients across channels may lead to flow separation and the formation of vortices (preferentially Q2D in strong magnetic fields) with unexpected impact on pressure distribution and heat transfer [34].

The development of closure laws for heat transfer modelling in liquid metals is still a topic under active research for advanced SYS-TH codes for fusion reactor applications. Since these fluids are characterized by a small Prandtl number, $Pr \ll 1$, the Nusselt number, Nu, cannot be predicted by correlations originally developed for air and water. For forced convection, we have in general $Nu = Nu(Pe, Pr)$, where $Pe = \rho u_0 c_p L / k$ is the Peclet number [35]. For applications in fusion blankets, in which the flow is subject to strong magnetic fields, heat transfer in terms of the Nusselt number depends in addition on MHD parameters, $Nu = Nu(Pe, Pr, Ha, N)$. For $N \gg 1$ and $Ha \gg 1$, the magnetic field is expected to dampen velocity oscillations in the flow and revert it to a laminar state, thus degrading the heat transfer [27,36]. Nevertheless, experiments have demonstrated that MHD velocity profiles can promote a higher heat transfer than hydrodynamic flows. This is partially attributable to thin MHD boundary layers. Likewise important are the high-amplitude large-scale flow structures (vortices and jets) that develop in flows with intense magnetic field and other strong forcing. Such structures often significantly increase rates of mixing and local heat transfer (see the recent review [5] and references therein). An important example of this general phenomenon is flow in ducts with electrically conducting walls, where high-velocity jets at the sidewalls may become unstable and locally enhance the heat transfer [37,38]. Heat transfer in turbulent pipe flow at moderate Hartmann numbers has been studied in [39] and summarized in a correlation $Nu(Pe, Ha)$. Despite these findings, the knowledge accrued has yet to be consolidated in scaling laws that can be then implemented in SYS-TH codes.

3.3.2. State-of-the-Art of MHD Modelling in SYS-TH Codes

To the best of our knowledge, there are five SYS-TH codes that feature some MHD modelling capabilities for liquid metal blanket systems: RELAP5 [40], MARS-FR [41], MELCOR [42], MHD-SYS [43] and GETTHEM [44]. They are listed in Table 1 with regard to their capacity to predict MHD pressure losses, electromagnetic coupling, and heat transfer. RELAP5-3D is the only code that provides MHD modelling capabilities out-of-the-box, whereas all the others have been modified with custom models for this purpose. MHD-SYS has been developed specifically to simulate MHD effects.

Table 1. SYS-TH capabilities for MHD modelling.

Code	2D Loss	3D Loss	Coupling	Heat Transfer
RELAP5-3D [40]	Yes	Only for fringing magnetic field	No	No
RELAP5/MOD3.3 [23]	Yes	Yes	No	No
MARS-FR [45]	Yes	No	No	No
MELCOR 1.8.5/1.8.6 [46]	Yes	No	No	No
MHD-SYS [43]	Yes	Estimated through coupling	Based on analytical relations	Based on analytical relations
GETTHEM [47]	Yes	Yes	No	No

Prediction of MHD pressure losses for fully developed flows (2D loss) is implemented in all the codes with a similar approach, i.e., the modification of the friction loss coefficient. Different implementations are instead adopted for 3D losses, and it should be highlighted that no SYS-TH code is currently able to cover all the scenarios that can cause these additional pressure drops. MARS-FR and MELCOR do not provide any support for this feature, whereas RELAP5-3D uses an ad hoc treatment only for the special case of a fringing magnetic field. MHD-SYS is unable to directly calculate 3D losses, but it supplies boundary conditions to a coupled CFD tool (the mhdFoam solver of OpenFOAM) to estimate them. A more general modelling of 3D losses is used by RELAP5/MOD3.3 that supports the automatic calculation of 3D loss coefficients through the specification of geometrical parameters in reserved words within the input deck, e.g., due to the presence of bends, cross-section variation and discontinuous electrical insulation. GETTHEM follows a similar approach for bends, whereas cross-section changes are represented with a fixed loss coefficient. No support is provided for insulation discontinuities, while a preliminary model for pressure penalty due to obstacles is available.

Electromagnetic coupling and heat transfer modelling are features present only in MHD-SYS. The used models rely exclusively on analytical solutions for very special combinations of wall conductivity and orientation of the magnetic field so that their generalization to blanket relevant applications is not possible.

3.3.3. A Validation Exercise for SYS-TH Code MHD Modelling: HCLL TBM Mock-Up

To illustrate the capability of a SYS-TH code including MHD effects, the custom module for RELAP5/MOD3.3 developed at Sapienza University of Rome is used to reproduce pressure distribution measured in a scaled mock-up of a HCLL TBM [48]. The experimental campaign predicts the global isothermal flow distribution in the considered blanket concept. This experimental data can be regarded as a good approximation of an integral effect test (IET) to validate the developed module. A full description of the validation exercise and the models implemented can be found in [23].

Figure 3a depicts a poloidal–radial view of the mock-up, pressure taps (A–E), and the liquid metal (NaK) flow path: feeding pipe (A), inlet manifold (B,C), inflow in BU1 (D,E) and outflow in BU2 (F,G), outlet manifold (H,I). The liquid metal flow path consists of numerous bends and cross-section variations, which introduce significant 3D MHD pressure losses, and features electromagnetic coupling of neighboring fluid domains, since the walls are electrically conducting.

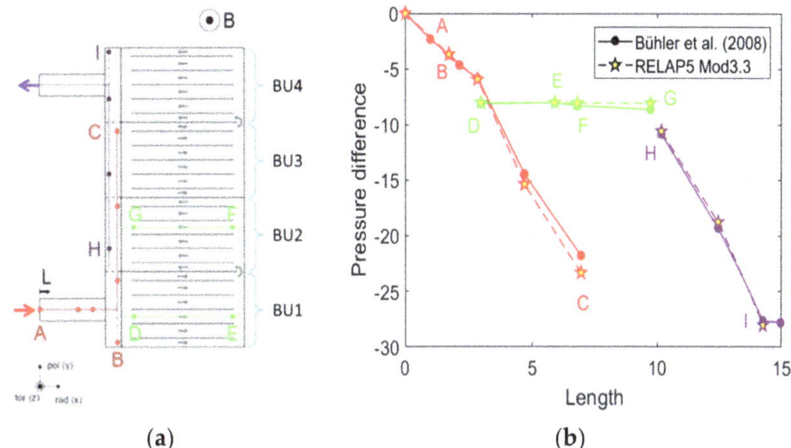

Figure 3. (**a**) Poloidal–radial sketch of the TBM geometry. The flow path (A–I) is depicted with colored lines; (**b**) Numerical results (stars) and experimental data (dots) for nondimensional pressure drop vs. normalized length of the flow path (Ha = 3000, Re = 3360).

In Figure 3b, the comparison between the experimental and RELAP5/MOD3.3 results for nondimensional pressure drop is shown for Ha = 3000 and Re = 3360. As scaling parameters we choose the toroidal half-length of a BU, L = 0.045 m, and the pressure scale $p_0 = \sigma u_0 L B^2$. The reference pressure value is imposed at location A. The code shows an excellent agreement with the experimental data in terms of total pressure loss and local values ($-1\% < \varepsilon < +14\%$), even without a model to represent electromagnetic coupling. The latter does not affect the overall pressure loss estimate due to the low contribution from BUs. On the contrary, the system code is not able to reproduce the flow distribution in the BUs, since it is significantly influenced by coupling [49]. Further model development is required to provide accurate predictions of flow distribution in Bus, in particular when heat transfer has to be quantified.

3.4. Requirements for Theoretical Analysis under Fusion Conditions

Common to all analysis methods discussed in Section 3, is the need for a database containing experimental data for validation of implemented models. Some of the experimental results that can be used for such a purpose are discussed in Section 4.

Various types of benchmark experiments have to be planned depending on validation purposes, e.g., tests to validate specific models or to improve the accuracy of parameters that are required for calculations. Both single effect and integrated multiple effect experiments have to be foreseen. The latter ones are also referred to as mock-up experiments. They include all essential features of a blanket concept with the aim to improve and finalize the design.

In the following section, we present some of the requirements for a proper theoretical investigation of fusion-relevant problems both by means of CFD and by using system codes.

3.4.1. Grid Generation

One of the most challenging aspects of numerical simulation of MHD flows in a strong magnetic field is the resolution of very thin boundary and internal layers, which form along walls and in the fluid at electrical or geometrical discontinuities of the wall. In order to properly resolve these layers, a minimum number of grid points is required [3]. For this purpose, non-uniform meshes that are coarser in the core regions and refined towards the layers have to be employed. MHD boundary layers are best resolved by using a prism layer. In general, structured grids, i.e., orthogonal and with non-skewed cells, lead to better performance. However, when studying MHD flows in complex geometries related to fusion blankets, hybrid unstructured meshes seem more suitable to meet the resolution requirements without excessive computational costs. This type of grid has some advantages compared to structured meshes, since it facilitates re-meshing and local refinement due to the possibility of clustering nodes in selected zones. For the study of MHD flows in complex geometries, as present in fusion reactors, the meshing tool should allow an automatic generation of the computational grid starting from a CAD model and permit effective control of local refinement in shear layers and regions of interest. In Section 3.4.2 an example of the influence of grid topology on the accuracy of the solution is discussed.

An option to be taken into account is the use of an adaptive mesh refinement (AMR) technique to adjust the accuracy of the solution within certain sensitive regions of the simulation domain by refining the mesh during runtime dynamically and locally according to given criteria [50].

Due to the very large number of nodes expected to be required to discretize a blanket related geometry, high-performance parallel computing is essential to reduce the computational time. However, simulation of MHD flows in an intense magnetic field, coupled with mass and heat transport phenomena, in realistic blanket models, remains a challenging time- and resource-consuming procedure.

3.4.2. Numerical Schemes

When simulating MHD flows at large Hartmann numbers, there exists a strong dependency of numerical errors in the solution on the grid topology. If unstructured meshes are used, discretization corrections are required, which account for non-orthogonality and skewness of the cells. A significant source of error can also originate from the numerical diffusion that arises from the truncation error. By using second-order discretization schemes and mesh refinement, the effects of numerical diffusion on the solution are reduced.

Discretization schemes with good conservation properties (conservation of mass, momentum, internal and kinetic energy, and, critically, electric charge) appear to be a preferable choice for flows at high Ha [6–8]. For instance, Ni et al. [51,52] developed an algorithm, which retains the conservative properties of the current density field, by using a proper interpolation procedure of current fluxes from cell-faces to cell-centers. Numerical errors can also derive from the computation of the electric current via Ohm's law in flow regions where j is very small. This is because the two terms $\nabla \phi$ and $v \times B$ are of nearly the same magnitude but with opposite signs. The computation of gradients of electric potential requires appropriate numerical schemes, such as Least-Squares (LS) or skew-corrected Green–Gauss (GG_{corr}) [53].

As an example to show the influence of grid topology on the prediction of MHD flows and the need for suitable corrections for the discretization of the potential gradient when using unstructured meshes, the MHD flow in an electrically insulating pipe at $Ha = 1000$ has been calculated. Two types of grids have been considered. The former one represents a block-structured mesh, a so-called O-grid, which consists of a central square block surrounded by four blocks that adapt to the curved pipe wall (see Figure 4a). The grid corners of the inner block, highlighted by red circles, represent discontinuities in the mesh structure that can introduce numerical errors in the solution. The second mesh, created by using the software STAR-CCM+, consists of polygonal cells in the duct cross-section (Figure 4b). In both grids, boundary layers are resolved by hexahedral elements with adequate grading in wall-normal direction. In Figure 4 only the core grid is visualized.

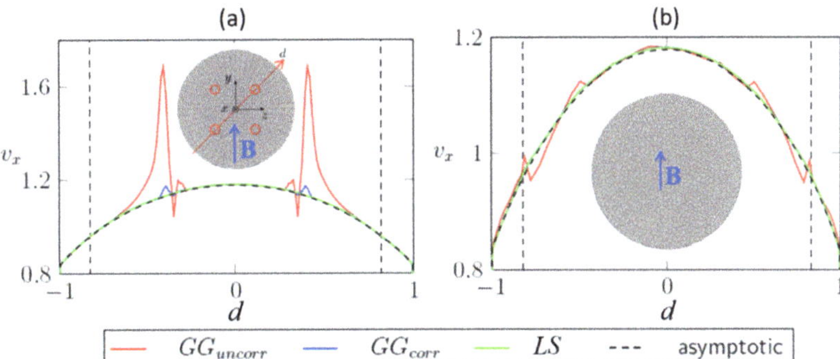

Figure 4. Influence of discretization schemes on MHD flow in an insulating pipe at *Ha* = 1000 [54]. Axial velocity profile along the diagonal marked in red in the sketch in (**a**), for two grid types, O-grid (**a**) and polyhedral mesh (**b**), and various discretization schemes for the electric potential gradient: Least-Squares (*LS*), skew-corrected Green–Gauss (GG_{corr}), uncorrected Green–Gauss (GG_{uncorr}). Vertical dashed lines indicate the boundary of the core grid. The black dashed curve shows the asymptotic solution according to Chang and Lundgren [55].

Simulations have been performed by changing the discretization scheme used for the calculation of the electric potential gradient. The results displayed in red in Figure 4 have been obtained by using the Green–Gauss scheme without skewness correction (GG_{uncorr}), the blue curve by Green–Gauss combined with a linear interpolation for the correction (GG_{corr}), and the green profile by a least square (*LS*) scheme. For insulating walls, the determination of the electric potential gradient by means of GG_{uncorr} introduces an error whose magnitude depends on the mesh topology, and this scheme is very sensitive to grid irregularities. Applying either corrected GG_{corr} or *LS* for the electric potential gradient leads to a significant improvement of the solution. Skewness-related perturbations are minimized so that errors are caused mainly by the grid cell size and not by the mesh type. However, for very large *Ha* (fusion conditions) the most accurate results for MHD flow in circular pipe have been obtained by using a grid with a structured core, a layer of hexahedral elements to resolve the boundary layers, and an unstructured mesh to connect both of them, together with *LS* scheme for electric potential gradient discretization [56].

3.4.3. Suitable Closing Laws for SYS-TH Codes

In SYS-TH codes the main issue is the absence of closure laws able to describe the effects of electromagnetic coupling and MHD velocity modifications which affect heat and mass transfer. In addition, the coefficients to be used for scaling laws to estimate 3D MHD losses have to be determined. Experimental and numerical data available in the literature must be expanded for these purposes. On the same note, it is challenging to demonstrate the code's scalability since no past experimental work qualifies as IET and among planned ones, only the ITER campaign will satisfy all requirements (cf. Section 4 and [57]). Therefore, it would be highly desirable to plan separate effect tests to obtain experimental data for the development of SYS-TH models, which can be integrated with numerical activities performed with validated CFD tools. Moreover, there is the need for at least one IET to provide support for SYS-TH code validation before the experimental campaigns in ITER.

4. Experiments

Despite the great progress of computational tools, made possible by the increased availability of high-performance computing, numerical codes are not yet able to simulate all types of 3D MHD flows at fusion-relevant parameters in blanket geometries with enough reliability and accuracy. Empirical research remains, therefore, the principal method for advancing the knowledge of complex blanket flows, and experiments are necessary for the validation of computational tools.

MHD flow in an outboard blanket module in ITER for a typical magnetic field $B = 4$ T, length $L = 0.1$ m, $u_0 = 0.1$ m/s and for PbLi properties as shown in Table 2, is characterized by a Hartmann number close to $Ha = 9000$. Experimental reproduction of such numbers is challenging for several reasons. In laboratory experiments, the strength of the magnetic field is limited to about $B_{exp} \lesssim 2$ T when normal conducting copper magnets are used. Moreover, the available space in the magnets reduces the typical length scale for mock-up experiments by at least a factor of two (see, e.g., [58]), i.e., $L_{exp} \lesssim 0.05$ m. If thermal insulation is required, L_{exp} could become even smaller. Therefore, Hartmann numbers in PbLi mock-up experiments are limited to $Ha < 1000$–2000. Higher values of Ha are possible only in superconducting solenoids, where, however, the length of channels with transverse magnetic field is limited by the size of the magnet bore [59]. In addition to this, MHD experiments with PbLi have to be performed at high temperatures. This results in thermoelectric disturbances of signals of the measured induced electric potential and difficulties in operating the flow meters and pressure transducers.

Table 2. Thermophysical properties of PbLi at 400 °C and examples of possible model fluids that allow experimentation at room temperature. Nondimensional parameters have been evaluated for experimental conditions with $B_{exp} = 2$ T, $L_{exp} = 0.05$ m and $u_0 = 0.1$ m/s. For the Grashof number Gr a wall heat flux of 10 W/cm^2 is assumed.

	ρ kg/m^3	$\nu \times 10^6$ m^2/s	$\sigma \times 10^{-6}$ 1/Ω/m	k W/m/K	$\beta \times 10^3$ 1/K	Re	Ha	Gr
PbLi $_{400\,°C}$	9719	0.161	0.849	22.4	0.122	29,585	2273	1.2×10^9
Hg $_{20\,°C}$	13,546	0.115	1.04	8.72	0.181	43,478	2584	9.6×10^9
GaInSn $_{20\,°C}$	6353	0.340	3.32	24	0.122	14,706	3920	2.7×10^8
NaK $_{20\,°C}$	868	1.06	2.87	21.8	0.29	4717	5585	7.3×10^7

Given the problems associated with the high-temperature operation of PbLi loops, one may consider using model fluids, such as mercury, Hg [60], alloys such as gallium indium tin, GaInSn [61,62], or sodium potassium, NaK [63–65], which allow for MHD experiments at room temperature. As shown in Table 2, Hg is a preferred choice for studying mixed magneto-convection since its usage leads to Gr values one or two orders of magnitude higher than other liquid metals in the same conditions. The highest Hartmann numbers may be reached by using NaK. The drawback of using Hg is toxicity and for NaK is its chemical reactivity with oxygen or water, so that special precautions are required during preparation and conduction of experiments. NaK has the advantage that good electrical contact between fluid and wall is established after wetting, and corrosion issues do not exist in the temperature range of operation. Reasonable Hartmann numbers can be reached also with GaInSn, but electrical contact at the interfaces suffers from poor wetting capability. This requires special manual treatment of all surfaces exposed to the alloy in MHD experiments, since contact resistance by residual oxide layers influences the flow behavior in terms of pressure drop, velocity and potential distribution [66].

MHD experiments with model fluids fall into two major categories: fundamental or applied research. The first group includes experiments in generic geometries such as pipes, ducts, expansions and bends, for investigations of stability of laminar flows, transition to turbulence, heat transfer and buoyant flows. The results from these experiments constitute a valuable database for the validation of numerical tools [3,4].

The other type of experiment aims at demonstrating the feasibility of technological aspects, such as a reduction in pressure drop by electrically insulating coatings or flow channel inserts [67,68], or at studying complex electrically coupled MHD flows in scaled mock-ups of entire blanket modules.

Before discussing some examples of MHD experiments, it is worth mentioning a few aspects concerning measuring techniques for MHD flows at high Hartmann numbers.

Measurements of flow rate, pressure differences, and electric potential are straightforward, at least for model fluids. In experiments performed, e.g., in the MEKKA facility at KIT [64], it is possible to determine flow rates up to 25 m^3/h, differential pressure between 30 pressure taps, and distribution of potential by up to 600 sensors on the wall or at the fluid–wall interface. The latter ones are of particular importance, since for $Ha \gg 1$, potential is constant along magnetic field lines and hence, the wall data give a good picture of the values inside the core of the fluid. Moreover, it is possible to calculate from Ohm's law the components of velocity in the plane perpendicular to the magnetic field according to

$$ v_\perp = \frac{1}{B^2} \left(-\nabla\phi \times B - \underbrace{\frac{\nabla p}{\sigma}}_{\approx 0} \right), $$

where for flows in insulating or thin-wall ducts, the current density (or pressure gradient) is negligible compared to potential gradients. Therefore, the potential ϕ may be interpreted as an approximate stream function of the flow and we can determine the velocity field from wall potential measurements, i.e., $u \approx B^{-1}\partial\phi/\partial z$ and $w \approx -B^{-1}\partial\phi/\partial x$. As an example, Figure 5 shows the distribution of axial velocity determined from wall potential measurements in a circular pipe flow in a spatially varying magnetic field $B(x) = B(x)\,\hat{y}$ for $Ha = 5485$ and $Re = 10{,}043$. The strength of the transverse magnetic field is displayed in nondimensional form as $Ha(x)$. For further details, see [69].

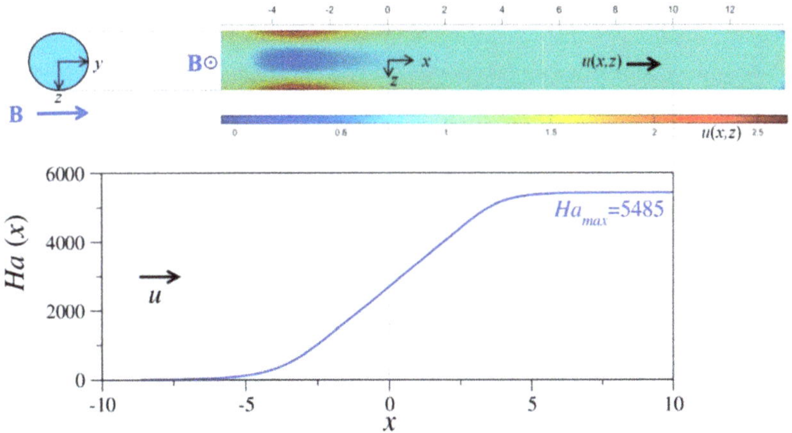

Figure 5. Pipe flow in an axially increasing magnetic field ($Ha = 5485$ and $Re = 10{,}043$). Colored contours of axial velocity on the pipe symmetry plane, displayed in the upper plot, have been obtained from potential measurements at the electrically conducting wall.

Using the same principles, i.e., measuring potential differences $\Delta\phi$ between tips of local traversable probes at distance Δz, allows for the investigation of time-resolved velocity profiles $u \approx B^{-1}\Delta\phi/\Delta z$ in the core of duct flows and even across field-aligned parallel layers. These probes, called Liquid metal Electromagnetic Velocity Instruments (LEVI) [70] or conduction anemometers, can be used to study the stability of laminar flows or to quantify turbulent properties. Slight asymmetries in measured velocity profiles, observed among others, e.g., in [70], can be explained by the formation of internal field-aligned layers tangential to the probe shaft, which partially blocks a fraction of the duct cross-section [71]. The latter reference further shows that higher accuracy may be obtained by proper calibration of the probe.

Another example of MHD experiments is the study of magneto-convection in a geometry relevant to the WCLL blanket concept. It addresses the topic of magneto-buoyant flow in a liquid metal filled box with internal obstacles. Due to the presence of cooling pipes generating large thermal gradients in the breeding zone and the slow circulation of the liquid metal, the major flow in a WCLL blanket is expected to result from the balance of buoyancy and electromagnetic forces. To examine generic buoyancy-driven MHD flows, a simplified rectangular model geometry has been considered in which two parallel pipes are inserted (Figure 6b). Both pipes are kept at constant temperature T_1 and $T_2 > T_1$ to generate the horizontal temperature gradient driving the flow. To provide clear boundary conditions, the pipes were made of copper to ensure that their temperature is as uniform and constant as possible. Their outer surface is coated with a very thin electrically insulating layer to prohibit induced currents from closing into their walls and to avoid parasitic thermoelectric effects that could occur in contact with the model fluid GaInSn. The box is made of PEEK plastic and thermally insulated to provide adiabatic conditions before being inserted in the MEKKA magnet that produces a vertical magnetic field (Figure 6a). Temperature and electric potential were recorded with high-accuracy instrumentation at the most pertinent locations identified by preliminary simulations [72]. As an example of the data collected, nondimensional temperature profiles measured at the center of the cavity for hydrodynamic flows ($Ha = 0$) and various Gr are presented in Figure 6c. A convection cell forms in the center of the cavity, between the two pipes, and the buoyant flow results in a thermal stratification with the hot fluid staying on the top and the cold fluid on the bottom. Experiments have been performed for a variety of temperature differences (Gr) and magnetic field strengths up to $Ha = 3000$.

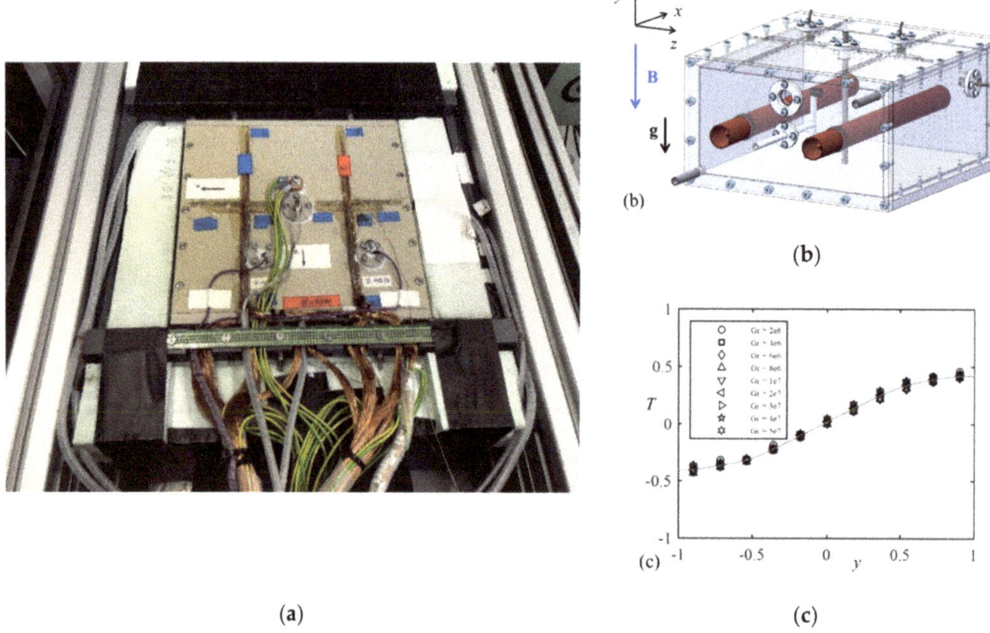

(a)   **(b)**   **(c)**

Figure 6. (**a**) Magneto-convection test section in front of the magnet during installation. For thermal insulation the box is surrounded from all sides by Styrofoam. The top insulation has been removed to obtain a view of the box and of the instrumentation. (**b**) Design; (**c**) nondimensional temperature profiles measured at the center of the cavity for several Gr and $Ha = 0$.

As mentioned above, a second class of experiments exists for the investigation of MHD flows in scaled mock-ups of entire TBMs. As an example, the results of potential measured on the surface of a mock-up for a HCLL TBM are shown in Figure 7. The test section consists of eight breeder units (BUs) which are fed and drained by a system of manifolds. From the experimental results (symbols) of the potential measurements in the middle of the module, it can be seen that higher gradients exist in external BU pairs, i.e., BU1-BU2 and BU7-BU8, while the values in the central ones are considerably smaller. Since potential gradients may be interpreted as an approximation of core velocities, it is obvious that there is a flow imbalance between external and central pairs of BUs. A numerical simulation of eight electrically coupled BUs agrees well with the experimental data on the middle plane of the mock-up and additionally yields details of the velocity profile, including flow in field-aligned boundary layers that cannot be detected from potential measurements on the Hartmann wall [73]. In this sense, numerical simulations and experiments are complementary.

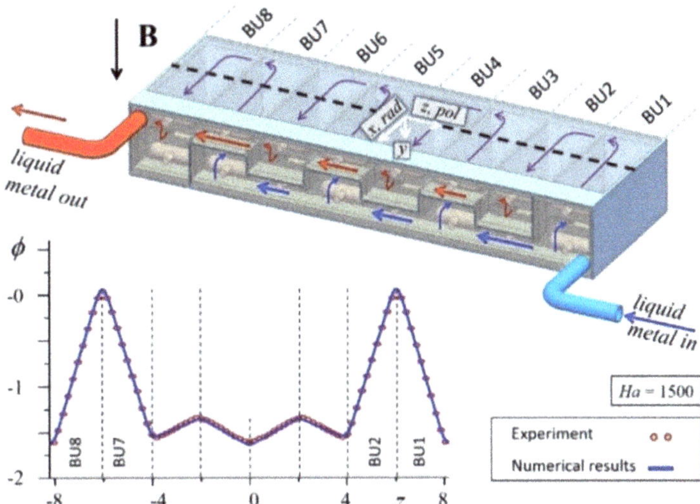

Figure 7. Sketch of liquid metal flow paths in an experimental MHD mock-up of a HCLL TBM for ITER. Nondimensional results for measured electric potential ϕ (symbols) along the upper wall (dashed line) are compared with numerical simulations (solid line).

The observation that external BUs receive more flow than the central ones is confirmed by measurements of pressure drop along typical flow paths in manifolds and BUs. This behavior originates from the design of manifolds whose cross-sections do not adjust to the changing flow rates when the fluid is distributed and collected along the poloidal direction.

More recently, the trend towards the development of PbLi MHD facilities gathers speed with the purpose of promoting integrated experiments and supporting TBM development. Related PbLi activities have been reported from China [74,75], India [76], Japan [77], Korea [78], Russia [79], Europe [80] and the USA, which operated the MaPLE facility [81]. The latter, initially built by the University of California Los Angeles, has been upgraded by a joint EU–US collaboration. It has been transferred to Europe and it is currently being reassembled at KIT, where it will contribute to the EUROfusion blanket program. MaPLE will provide experimental data on MHD heat transfer in blanket-typical geometries allowing different inclinations of the test section with respect to gravity and various orientations of the magnetic gap (horizontal, inclined, vertical). The ability to lift and tilt the 20-ton magnet to any desired position is a unique feature of this installation compared to other existing liquid metal facilities (see Figure 8). In MaPLE it is further planned to test measuring techniques to record pressure, flow rate and electric potential at reactor-relevant temperature and to gain experience in long-term operation of a PbLi MHD facility.

Figure 8. MaPLE PbLi loop and magnet during installation at KIT.

5. MHD Phenomena and Coupling with Heat and Mass Transfer

The occurrence of MHD effects can give rise to counterintuitive phenomena that can affect blanket performance. This is the case for flows in parallel electrically conducting ducts, in which electromagnetic forces induced by currents leaking across common walls modify flows in individual channels. Other examples are so-called coupled MHD phenomena, i.e., heat and mass transport in MHD flows. When multiple factors, such as induced currents and thermal gradients, determine the velocity distribution, the action of the resulting flow on material corrosion and tritium transfer is significantly different than in case of isothermal hydrodynamic conditions.

5.1. Electromagnetic Flow Coupling

When a LM moves in electrically coupled channels flow imbalance, reversal and recirculation may occur, which can cause the formation of regions of stagnant flow [25]. The latter is a primary concern for reactor safety, since it can lead to the accumulation of tritium and increased permeation towards coolant and structures, or to the formation of hotspots, in which the temperature exceeds the maximum allowable value of the structural materials.

5.1.1. Flow Distribution in Electromagnetically Coupled Parallel Channels

In Figure 9 an example of velocity distribution in five stacked electrically coupled ducts is depicted, where α is the angle between the horizontal coordinate x and the imposed magnetic field. In general, any inclination angle is possible between the two following limiting cases where:

- Channels are stacked along magnetic field direction, $\alpha = 0°$ (Hartmann wall coupling);
- Channels are stacked transverse to the magnetic field, $\alpha = 90°$ (sidewall coupling).

Let us consider the case of an array of channels where the flow in each duct is driven by the same imposed pressure head.

For $\alpha = 0°$, mean velocity is predicted to increase in central channels [28]. Figure 9a shows qualitatively the variation of the flow rate depending on the duct number. Experiments conducted at the Efremov Institute [65] confirm the predicted effect and find a 13% increase in flow rate for the central duct in an array of three subchannels coupled at Hartmann walls.

For $\alpha = 90°$, the velocity in each core is almost the same and sidewall jets in the boundary layers along the internal walls are suppressed. Because high-velocity layers, which are still present at external sidewalls, are able to carry a significant amount of flux, the flow rates in these external channels are increased compared to the inner ducts [82]. This effect is also qualitatively shown in Figure 9a.

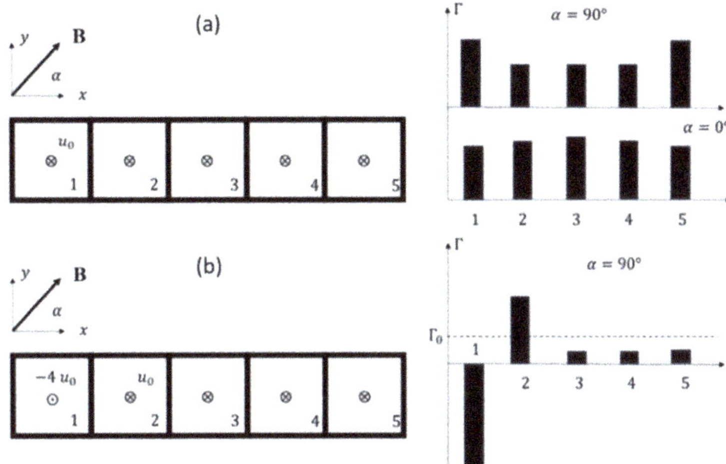

Figure 9. Qualitative representation of flow rate imbalance in parallel ducts due to electromagnetic coupling (Ha = 10,000). (**a**) Coupling between channels with equal mean velocity through Hartmann wall (α = 0°) and side walls (α = 90°). (**b**) Coupling between channels with large difference in mean velocity and α = 90°. Channel no. 1 collects and returns all the flow rate incoming from no. 2 to no. 5 and Γ_0 represents the mean flux in ducts 2–5 assuming uniform flow distribution.

Flow imbalance can also be observed when counter-flowing channels with different mean velocity are coupled through sidewalls (α = 90°). An example is found in the Lead Lithium Ceramic Breeder (LLCB) in which a number of ascending poloidal channels are electrically coupled with a descending return duct that carries their combined flow rates [83]. A qualitative representation of a similar case is displayed in Figure 9b. The high mean velocity in the return channel (no. 1) induces strong electric currents that leak in the adjacent duct and generate Lorentz forces. The latter ones draw more than double the flow rate into that channel from the manifold compared with the other ducts.

In general, in the breeding zone of a blanket, none of both limiting coupling modes exists. Solutions for 0° < α < 90° cannot be obtained just by superimposing the ideal cases described above. Only numerical simulations with a correct inclination of the magnetic field may reveal the complex physics determined by the electromagnetic flow coupling at common conducting walls, characterized by the spreading of internal layers along magnetic field lines [84]. When internal layers originating from singularities (e.g., corners) in one duct hit a common wall, the electromagnetic disturbance in potential and currents is transferred to the neighboring channel, in which the layer continues developing along field lines. This effect may even cause unexpected local flow reversal in parts of the neighboring channels depending on the driving pressure gradients. An example of flows coupled at common conducting walls and exposed to an inclined magnetic field (α = 67.5°) is displayed in Figure 10. Here, velocity contours are plotted on the cross-section of the channels together with the velocity profile along the central line of the duct array. Internal layers divide fluid regions inti cores with a uniform velocity that spans across the walls.

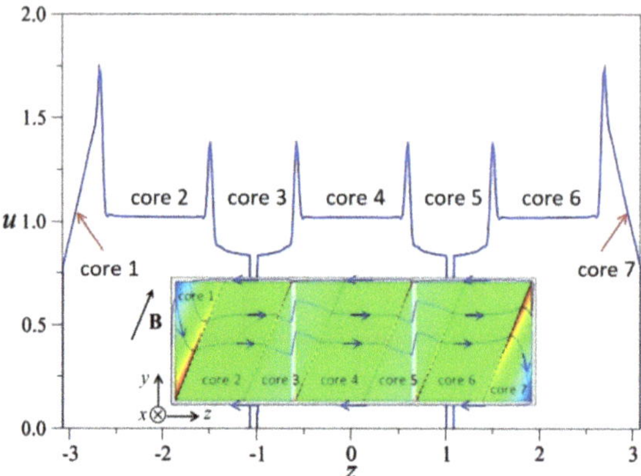

Figure 10. MHD flow in three electrically coupled ducts exposed to an inclined magnetic field, $\alpha = 67.5°$, for $Ha = 2500$ and conductance parameter $c = \sigma L/(\sigma t_w) = 0.038$. Velocity is plotted along the central line of the duct array and contours on the horizontal midplane.

5.1.2. Coupled Flow in Manifolds of LM Blankets

Due to the complex geometry of fusion blankets and the fact that the major pressure drop arises from manifolds, it is not guaranteed that the driving pressure heads in all parallel channels of BUs are equal. The influence of manifolds on the flow in eight electrically coupled BUs of a HCLL blanket has been investigated experimentally and the electric potential measurements show that the BUs at the mock-up extremities are characterized by higher flow rates compared with central BUs (see Figure 7). The velocity profiles deduced from electric potential data are consistent with coupling through sidewalls and the strong difference in flow rates is linked to significantly different pressure heads driving the flow in individual BUs [73] (see also Section 4).

Electromagnetic coupling can also cause the appearance of flow reversals. This phenomenon is an issue in the breeding zone but particularly in the outflow manifold, where one wishes to avoid stagnation of the tritium-rich breeder. Coupling-mediated flow reversals may appear in co-flowing manifold channels of HCLL [85] or WCLL [30] blankets. They are associated with different flow rates or pressure heads between adjacent channels that, in turn, lead to an imbalance in mean flow velocity and leaking currents. For instance, the co-flowing feeding and draining manifolds in the WCLL TBM, in which flow rates reduce and increase, respectively, along the poloidal direction, are coupled through common conducting Hartmann walls [30]. At the top and bottom of the TBM, the mean velocity difference between manifold channels is the highest and currents leaking into the duct with the lower velocity tend to drive its core. The pressure build-up there leads to a reversed flow in the side layers that can reach a significant fraction of the flow rate.

As an example of a different manifold concept, we discuss the case of the co-axial manifold proposed for the WCLL blanket [31,86], where a reverse flow was observed even for equal imposed flow rate in the two concentric channels due to the higher mean velocity in the internal duct [31]. The flows in the two channels are along z, isothermal, steady and fully developed, and a uniform toroidal magnetic field is imposed in x-direction. The liquid metal enters the blanket through the external duct and exits it through the internal one. As a result, the external channel flow rate (Γ_{in}) decreases, moving towards the top of the blanket, while the internal duct flow rate (Γ_{out}) increases. Local flow reversals caused by electromagnetic coupling can be observed in Figure 11. In general, it is advisable to tailor the manifold configuration to ensure that all the channels have a similar mean velocity along their length in order to avoid coupling-mediated flow reversals.

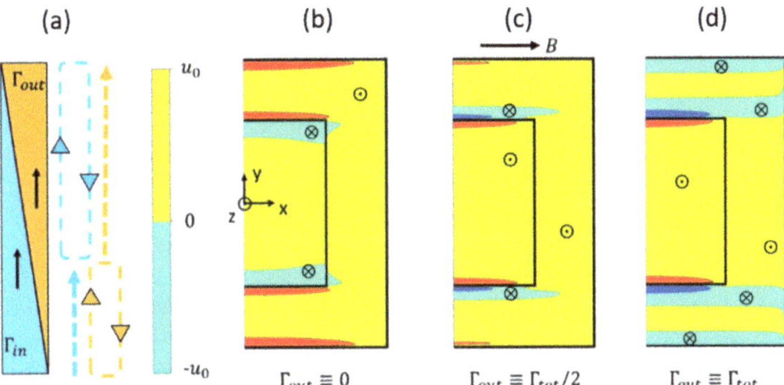

Figure 11. Coupling-mediated flow reversal in a concentric co-axial manifold. Numerical results for $Ha = 2000$ for half channel (symmetry with respect to $x = 0$). (**a**) Sketch of flow rate distribution along poloidal coordinate ($\Gamma_{in} + \Gamma_{out} = \Gamma_{tot}$) and qualitative representation of regions where the flow is expected to recirculate in inflow and outflow manifolds. Velocity distribution in the manifold channel (**b**) at the first cell ($\Gamma_{out} = 0$, $\Gamma_{in} = \Gamma_{tot}$), (**c**) at the equatorial plane ($\Gamma_{out} = \Gamma_{in} = \Gamma_{tot}/2$), and (**d**) at the last cell ($\Gamma_{out} = \Gamma_{tot}$). Crossed circles $\otimes$ mark flow reversal regions.

5.2. Turbulence and Heat Transfer

Turbulent fluctuations in a flow of an electrically conducting fluid are suppressed by an imposed magnetic field via conversion of their kinetic energy into heat by Joule dissipation of induced electric currents. As a result, MHD flows are found in a laminar or transitional state at much higher Reynolds and Grashof numbers than their hydrodynamic counterparts. For example, transition to turbulence in isothermal flows in ducts and pipes with electrically insulating walls and an imposed transverse magnetic field occurs at Re growing with Ha so that Re/Ha remains in the range between 200 and 400 [87]. Even when turbulence occurs, it is transformed into an anisotropic state with suppressed flow gradients in the direction of the magnetic field lines [88–90].

Practically, the suppression of turbulent fluctuations by a magnetic field implies that conventional (three-dimensional, albeit anisotropic) turbulence can occur in typical blanket conditions only if poorly electrically conducting fluids, such as the molten salts FLiNaK or FLiBe, are used as working liquids. In that case, turbulent flows with high Re and moderate Hartmann number, $Ha{\sim}100$, are anticipated. The properties of such flows include decreased fluctuation amplitude, anisotropy, reduced rate of energy transfer to small length scales and the corresponding (steeper slope) transformation of the energy power spectrum. Moreover, reduced rates of turbulent transport, and development of flow regimes with localized turbulent zones and large-scale spatio-temporal intermittency are observed (see, e.g., [5,6,90–94]). Modeling of such flows can be accomplished by LES or RANS methods, but models have to be adapted for the peculiar nature of MHD turbulence (see, e.g., [5] and references therein for a discussion of this interesting topic). The models commonly applied in conventional hydrodynamics (for example, the $k - \varepsilon$ model in RANS or the classical Smagorinsky model in LES) do not produce accurate results due to their excessive dissipation and inability to reproduce anisotropy and intermittency effects.

In the case of strongly electrically conducting fluids, such as PbLi, $Ha{\sim}10^4$ typical for blanket conditions implies that 3D turbulence cannot occur. This does not mean that the flow necessarily becomes laminar and steady-state. From the general physical perspective, it is evident that even a very strong magnetic field cannot prevent the growth of hydrodynamically unstable perturbations, if the perturbations are quasi-two-dimensional (with nearly zero velocity gradients along the magnetic field outside the boundary layers) and, therefore, do not generate significant Joule dissipation. The physical nature of such instability may vary. For example, growth of perturbations has been demonstrated for shear-flow instability of the Kelvin–Helmholtz type, thermal convection, or a combination of the two (see, e.g., [95–102] or the review [5] for further examples).

The growth occurs in the absence of conventional turbulent mixing, in the conditions of quasi-two-dimensionality, and, typically, at high values of control parameters, such as *Re* or *Gr*. The often-observed outcome is an ultimate flow state in the form of unsteady regimes with strong inertia effects. The flows can be characterized as experiencing quasi-two-dimensional turbulence with friction imposed on the flow at walls perpendicular to the magnetic field. More important than this classification is that the properties of such flows are very different from the ones predicted at the same parameters by models based on assumptions of a steady-state and laminar or inertialess behavior. The flows are dominated by large-scale quasi-two-dimensional vortices and jets. Temperature oscillations of high amplitude (up to 50 K) and low frequency have been observed in experiments and obtained in calculations at moderately high *Gr* and *Ha* achievable in a laboratory [5]. The state of the flows at higher *Gr* and *Ha* typical for blanket conditions is debatable, but there are strong indications that fluctuations of even higher amplitude are to be expected [98,102]. Should such high-amplitude low-frequency fluctuations occur in components of a liquid metal blanket, potentially serious consequences include a substantial increase in the rates of transport of heat and tritium, the modification of balances controlling corrosion, and strong unsteady thermal stresses in the walls.

5.3. Tritium Transport

Tritium consumption for a 2000 MW fusion power reactor is about 112 kg per full power year, which is an enormous amount when compared with the actual worldwide availability, estimated as 20–30 kg [103]. Even if tritium can be produced from lithium, the bred tritium has to be almost completely recovered for subsequent use as fuel. This means that tritium self-sufficiency represents one of the most challenging issues for future deployment of fusion electricity. Therefore, an efficient characterization of processes and technical solutions to manage and control tritium transfer and release is mandatory.

For the development of liquid metal breeding blankets, the assessment of fusion reactor safety and the ability to breed tritium, it is necessary to accurately predict how tritium transport is affected by complex phenomena such as diffusion, surface reactions, MHD effects and heat transfer. Therefore, mathematical models and computational codes have to be developed to quantify tritium distribution, inventory, retention in PbLi and structural materials, and permeation losses from PbLi to the coolant, to the reactor building and the environment. Among the different phenomena affecting tritium inventories and losses, one should mention Sieverts' law describing solubility in metals, but also surface reactions and the possibility of tritium becoming trapped in the structure.

Hydrogen isotope solubility in PbLi (Sieverts' solubility constants), including that of tritium, suffers from uncertainties of up to three orders of magnitude depending on the adopted measurement technique (absorption or desorption). The huge difference between the two techniques needs to be better understood. For these reasons, experimental campaigns are being conducted in Europe in order to re-evaluate PbLi properties including crosschecking of data from different laboratories under controlled conditions of PbLi manufacturing and chemical composition.

For the determination of tritium inventory and flux in different materials, which are in contact, it is necessary to define suitable interface conditions. Tritium concentration c_T (mol m^{-3}) at a certain interface is linked to tritium partial pressure p_T through Sieverts' law [104]

$$c_T = k_S \sqrt{p_T},\qquad(5)$$

where k_S (mol m^{-3} Pa$^{-0.5}$) is the Sieverts' constant. At the interface between liquid metal and steel, the continuity of p_T is assumed,

$$p_{T,ls} = p_{T,sl},\qquad(6)$$

where the subscripts ls and sl indicate the liquid phase at a steel interface and steel at an interface to a liquid, respectively. Substituting Equation (5) into Equation (6) yields a concentration discontinuity related to the ratio of Sieverts' constants of liquid metal and steel, $k_{s,l}/k_{s,s}$, called the partition coefficient

$$\frac{c_{T,ls}}{c_{T,sl}} = \frac{k_{s,l}}{k_{s,s}}.$$

(7)

As a result, the uncertainty in the PbLi solubility directly translates into a lack of confidence in tritium permeation rate and inventories. To deal with these uncertainties in a conservative way, most of the tritium models use the lowest available solubility values, which provide the highest permeation rates from PbLi to the coolant.

Tritium permeation rate across interfaces of different materials is affected by surface conditions. Oxidized or clean walls have different properties, e.g., absorption, desorption, and recombination constants, with consequences on tritium permeation fluxes, especially at lower tritium partial pressures, i.e., when the limiting process characterizing the permeation is due to surface effects. The recombination constant, $\sigma k_2 \left[\mathrm{m}^4 \, \mathrm{mol}^{-1} \, \mathrm{s}^{-1} \right]$ is related to the Sieverts' constant as $k_s^2 = \sigma k_1 / \sigma k_2$, where $\sigma k_1 \left[\mathrm{mol} \, \mathrm{m}^{-2} \mathrm{s}^{-1} \mathrm{Pa}^{-1} \right]$ is the dissociation constant and σ the sticking probability [105]. The recombination constant is strictly dependent on the condition of the material surface and therefore its value suffers from huge uncertainties. In addition, up to now consolidate reference values for transport properties in Eurofer are still lacking and properties of Optifer-IV steel are typically used instead.

In addition to the aforementioned uncertainty of physical parameters, tritium trapping in solid structures, which can result from impurities and material defects or neutron irradiation damages, can considerably affect tritium inventory and losses. The density of trap sites increases with the damage induced by neutron irradiation. Tritium implantation depends on the ion flux density and energy, and the peak implantation usually occurs at few nanometers under the surface of the metal. Permeation, therefore, depends on surface conditions of plasma-facing components and on defect traps, hence it can increase on contaminated surfaces.

Recent calculations with system-level codes for WCLL blankets estimate a total tritium release to the coolant without any mitigation strategy of 38 g/d, which is about 11.7% of the total tritium generated. This implies high tritium inventories retained in the water circuits, surpassing the safety limit in few days of operation [106]. Therefore, the reduction of tritium permeation is a primary issue for the licensing of a DEMO reactor [107].

In order to mitigate tritium release to the coolant two main strategies are foreseen. The first one concerns the adoption of efficient tritium extraction systems able to guarantee low inventories reducing the gradients of concentration, which are responsible for permeation phenomena. Among the different technologies, the most promising ones [12] are the Gas-Liquid Contactor (GLC), which is the reference technology for ITER, the Permeator Against Vacuum (PAV) and the Vacuum-Liquid Contactor (VLC). It must be kept in mind that the scalability of these tritium extraction units from ITER to DEMO has to be addressed from the point of view of technological feasibility and costs. A second approach to reduce tritium losses is the use of anti-permeation barriers (APB) [108]. This strategy appears to be required when considering practical and economic limits on the size and hence the efficiency of tritium extraction systems. The APB performance in terms of permeation flux reduction is quantitatively assessed by the Permeation Reduction Factor (PRF), which quantifies the ability to reduce the permeation of hydrogen isotopes. The PRF is defined as the ratio of tritium permeation fluxes through an uncoated wall ($j_{T_2}^{uncoated}$) to that across a coated one ($j_{T_2}^{coated}$),

$$PRF = j_{T_2}^{uncoated} / j_{T_2}^{coated}.$$

(8)

Several coatings have been proposed over the years [109], among which the most promising ones in terms of PRF are Er_2O_3, ZrN, Al-Cr-O, Al_2O_3. These coatings may guarantee PRF higher than 1000 in the range 400–700 °C. Alumina Al_2O_3 serves also to mitigate corrosion of Eurofer steel and represents the reference coating for DEMO [107]. Additional research is necessary in order to ensure high adhesive strength of coatings, compatibility with complex shapes, and good performance in a radiation environment.

5.3.1. Tritium Analysis Methods, Transport Modeling and Coupling with MHD

Tritium transport in blankets is complex, since it includes many different phenomena and involves a large number of physical properties and parameters, many of which have not yet been determined with an adequate level of accuracy. Moreover, simplified models, which use empirical coefficients and are employed to obtain a first estimate of tritium inventory, have often limited applicability in terms of operating conditions. Therefore, they are less relevant for DEMO applications. The physics included in modelling tools for investigation of tritium cycle in fusion blankets should have a sufficient complexity, in order to support the design of TBMs and the definition of the experimental program for ITER. Additionally, models should allow exploiting experimental data from ITER and to extrapolate them to DEMO conditions. A four-step structure of the development strategy for tritium transport modeling tool is presented in [110]. It is proposed to keep a modular structure of the predictive tool, starting from the description of single components and by including progressively subsystems, such as TBM and ancillary systems, up to the complete test blanket system.

Tritium analyses can be performed either by employing system-level models for computing the global performance of blankets and ancillary systems, or by detailed 3D models and numerical simulations for spatially limited domains and critical regions. System-level models, which are discussed in more detail in Section 3.3, consider at most 1D equations [111]. Flowing PbLi is taken into account via imposing a mass transfer coefficient at interfaces between liquid metal and walls. This is the case of TMAP7 [112], FUS-TPC [113], EcosimPro or mHIT [114], which have been used for estimating tritium transport and inventory in different blanket concepts and divertors, and performing parametric sweeps. Exercises of verification of these codes against analytical solutions, validation against experimental data and benchmarking between them have been performed [112,114]. Most of the validation procedures are performed in solid systems and are devoted to validating processes such as diffusion, surface recombination, chemistry reactions, trapping, etc. System-level codes present important limitations when dealing with MHD coupling, since correlations between the Sherwood, Reynolds and Hartmann numbers are not known in blanket conditions. The Sherwood number *Sh* describes the ratio of effective mass transfer to the rate of diffusive mass transport. Therefore, these codes use either classical hydrodynamic correlations or an analogy with heat transfer. However, the latter type of correlations has been derived for relatively low Hartmann numbers ($Ha < 375$ [39]) and they are not applicable to blanket conditions.

Regarding detailed models, various finite volume and finite elements models have been developed for simulating tritium transport coupled with MHD effects. Several academic codes typically use the outputs of available MHD solvers as input for the velocity field in the calculations of mass transfer. This includes, for example, the case of CATRYS [115], which solves the equations for tritium transport and uses the velocity field from the MHD HIMAG code as input. Other tritium transport codes have been developed with an integrated MHD solver (e.g., [116]).

Another category comprises 3D simulations with commercial multiphysics codes, such as ANSYS-Fluent and COMSOL-Multiphysics. These simulation platforms include customization capabilities, where MHD modules can be integrated into available fluid dynamic solvers. For validation of non-MHD mass transfer in these codes, there are few facilities available where hydrogen isotopes are dissolved in flowing PbLi, as described in [117] or the recently constructed CLIPPER loop [118] at CIEMAT. At ENEA Brasimone, the TRIEX-II facility is able to qualify GLC, PAV and VLC technologies at different temperatures, PbLi mass flow rates and hydrogen isotopes concentrations [119]. Unfortunately, in none of the existing facilities can a magnetic field be applied on a section of the flow path.

Because of this experimental limitation, tritium transport codes have been validated using experiments under hydrodynamic conditions or verified against analytical solutions [120–122]. These exercises do not ensure code validation in MHD flows, but once the hydrodynamic transport of dissolved species has been verified, there is high confidence that the transport equations for passive scalars are implemented in a correct way, taking into account all relevant physical phenomena and applying them to MHD velocity distributions as well. The degree of uncertainty in numerical modeling, due to a lack of validation against MHD experiments, seems far less compared to the poor knowledge of physical parameters (orders of magnitude differences) required for the simulations.

5.3.2. Tritium Transport under Fusion-Relevant Conditions

WCLL DEMO blanket. A number of studies have been performed to address the effect of MHD on tritium transport for the WCLL blanket concept for DEMO reactor [123,124]. In particular, flow in a portion of the breeding unit, as shown in Figure 12 on the top, has been simulated adopting a novel coupling strategy for the physics involved, and differences between pure hydrodynamic ($Gr = 4.78 \times 10^{10}$, $Ha = 0$) and magnetohydrodynamic ($Gr = 4.78 \times 10^{10}$, $B = 4$ T, $Ha \approx 11{,}000$) conditions have been highlighted. In both models, the system has been assumed to be operated at steady-state conditions. Buoyancy effects have been introduced using the Boussinesq approximation. The radial profile of the volumetric nuclear heating on the equatorial midplane has been determined by means of the MCNP Monte Carlo code.

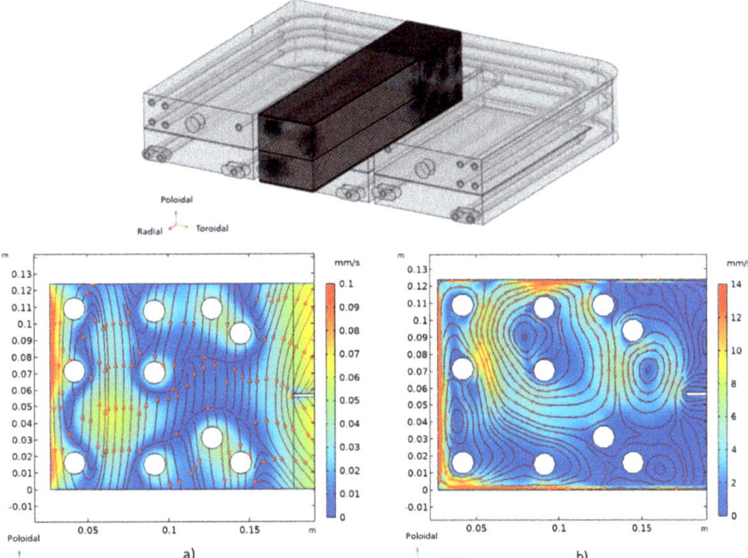

Figure 12. On the top the investigated geometry and the used mesh are shown. Contours of PbLi velocity for (**a**) hydrodynamic ($Ha = 0$, $Gr = 4.78 \times 10^{10}$) and (**b**) magnetohydrodynamic ($Ha = 10{,}830$, $Gr = 4.78 \times 10^{10}$) cases [124,125] are displayed on a radial poloidal plane in the middle of the central submodule.

Figure 12 shows the comparison between the velocity field on a radial–poloidal plane in the middle of the central submodule of the BU for the two cases. The presence of the magnetic field changes completely the velocity distribution and its magnitude. Near the piping region, the PbLi is quasi-stagnant in the hydrodynamic case in Figure 12a, whereas in the MHD case (Figure 12b), jets are found near the electrically conducting walls parallel to the toroidal magnetic field and a fast recirculating zone, driven by buoyancy forces, occurs between the pipes.

The MHD velocity distribution has, as expected, a significant effect on tritium transport. In the performed analysis, the Sieverts' constant of tritium in PbLi is the one proposed by Reiter [126]. In Figure 13, the steady-state tritium concentrations in the PbLi and Eurofer domains are shown for the hydrodynamic and MHD cases. In the former one, most of the tritium is found in the zone between the first wall and the first pipe column, reaching values greater than $0.7 \ mol/m^3$. By applying a toroidal magnetic field, the concentration is much more evenly distributed between the first wall and the edge of the stiffening plate, and the maximum value is smaller than $0.4 \ mol/m^3$. Nevertheless, the presence of recirculation zones increments the tritium mean permanence time in the breeder unit, and the tritium concentration decreases rapidly with the radial coordinate.

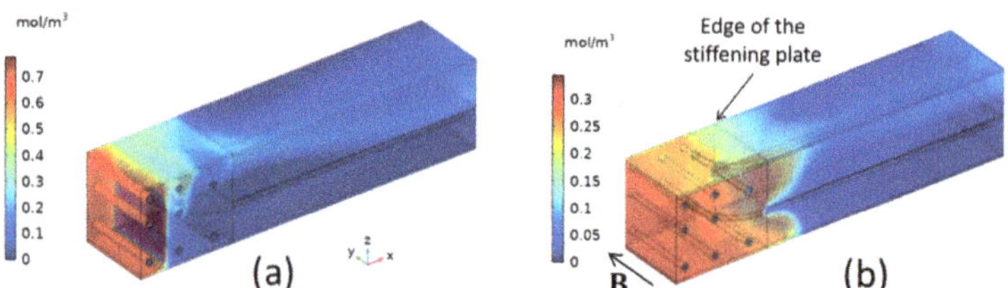

Figure 13. Tritium concentration in LiPb and Eurofer for (**a**) hydrodynamic ($Ha = 0$, $Gr = 4.78 \times 10^{10}$) and (**b**) MHD ($Ha = 10830$, $Gr = 4.78 \times 10^{10}$) cases.

The major effect caused by the different velocity fields is on the distribution of the tritium fluxes out of the blanket. In Figure 14, the comparison between the tritium that leaves the BU in the PbLi (Out) and the tritium that permeates through the piping system into the water (Permeation) is shown. As evident, MHD has a beneficial effect, and the permeation rate is reduced by around 60%.

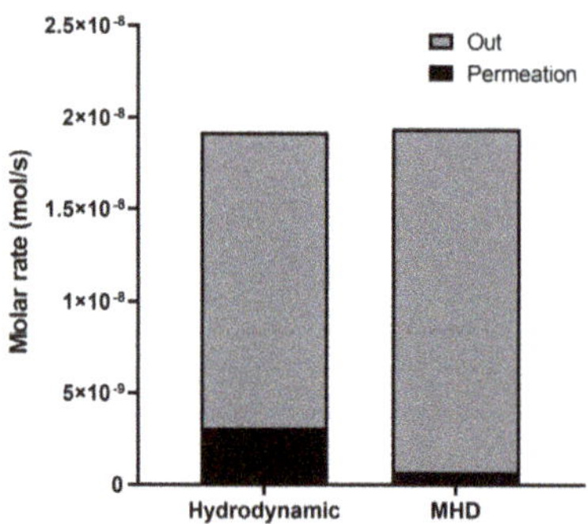

Figure 14. Steady-state molar flux exiting from the BU in PbLi (Out) and permeated through the piping system into the water (permeation) for the two cases.

Concerning tritium inventories in PbLi, Eurofer and water, there is no significant difference between hydrodynamic and MHD cases. Only a slight reduction in water has been found in case of MHD flow.

DCLL breeding blanket. The DCLL blanket concept is characterized by relatively high PbLi velocities compared to the WCLL blanket. The MHD pressure drop is reduced by electrically decoupling the PbLi flow from the metallic structure by using insulating Flow Channel Inserts (FCI) [127]. In the case of the European DCLL blanket design [128], a sandwich-type insert is proposed, which consists of a thin alumina layer protected by two

Eurofer sheets. The FCI divides the flow into two regions: the core flow and the gap flow between insert and wall.

From the tritium transport perspective, the latter is probably the most critical region in the DCLL blanket. This is due to the fact that, since alumina is a very effective permeation barrier, only the tritium generated in the gap is suitable to permeate into the coolant. Assuming that the alumina is a perfect electrical insulator, fully developed models of the gap flow can be applied independently from the flow in the core. This kind of simulation has been launched using the ANSYS-Fluent MHD solver.

In Figure 15 (left) the velocity profile in the gap is shown for the pressure-driven MHD flow at $Ha = 7630$ and $\partial p / \partial x = 1740$ Pa/m representative for the frontal channel of the central outboard module. The resulting flow is characterized by quasi-stagnant PbLi regions in the gaps perpendicular to the magnetic field (Hartmann gaps), while most of the flow goes through the gaps parallel to the magnetic field (side gaps). The obtained velocity profile has been used as input for a 3D tritium transport model of the complete annular channel between external wall and FCI. This model considers an exponentially decreasing volumetric tritium generation along the radial direction and constant transport properties [129].

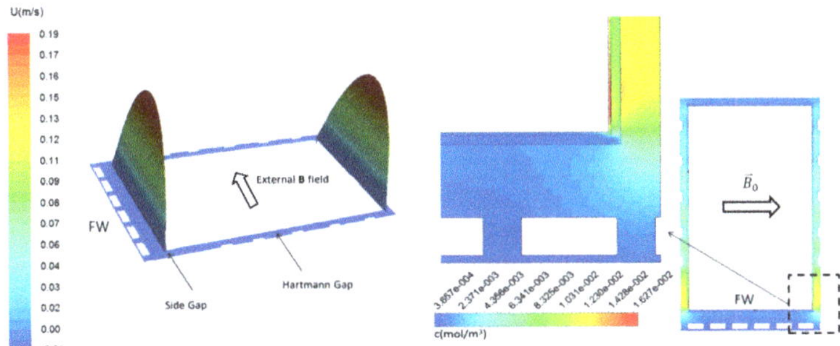

Figure 15. Left: fully developed velocity profile in the gap of a central outboard poloidal channel of a DCLL blanket for $Ha = 7630$ and $\partial p / \partial x = 1740$ Pa/m. Right: tritium concentration contours in the middle section of the PbLi channel, including PbLi gap, Eurofer walls and external Eurofer layer of the FCI.

Figure 15 (right) displays concentration contours in the midsection of the channel. In spite of the volumetric generation being maximal next to the FW, the high-velocity jets in the side gaps are able to effectively transport most of the tritium to the channel outlet. In steady-state, this results in low tritium concentrations in the parallel gaps and larger concentrations in the quasi-stagnant Hartmann gaps. Inside these gaps, the concentration decreases with the radial direction. This is a consequence of the exponential shape of the tritium source. Analogous conclusions are described in [115], showing the causal connection between very low velocities in the Hartmann gaps and larger tritium concentration.

In Figure 15 (right) the concentration discontinuity due to the different solubilities of PbLi and Eurofer can be observed at both PbLi/wall and PbLi/FCI interfaces. There is one order of magnitude of difference in tritium concentration in side and Hartmann gaps. The steady-state permeation rate, from the gap flow into the helium channels, is around six times higher than the value predicted by a system-level model that considers an evenly distributed flow rate through the gaps [129]. This comparison highlights the need for 3D transport models to predict tritium distribution in liquid metal blankets in order to reduce uncertainties and inaccuracy caused by assumptions, geometrical approximations and simplifications of the physics, as used in system models. In particular, the MHD velocity distribution and the complexity of the blanket configuration have a noticeable influence on the results.

In [130] parametric studies have been carried out to quantify the major factors that govern tritium transport and permeation in the DCLL blanket. Tritium solubility significantly affects these phenomena, since it directly influences tritium partial pressure. The parameters that determine the velocity distribution in the liquid metal gap between FCI and wall, such as Hartmann number, FCI and wall electrical conductivity, also have a strong impact on the tritium permeation rate.

5.4. Corrosion

In the frame of blanket engineering for DEMO and ITER, the chemical compatibility of Reduced Activation Ferritic-Martensitic (RAFM) steel Eurofer in contact with the corrosive flowing liquid PbLi represents a serious problem for fusion blanket development. Along with the possible wall thinning in the hotter part of the liquid metal loop that could lead to a deterioration of the mechanical integrity of the blanket structure, critical safety issues are related to the transport of corrosion products in the PbLi loop [131]. They can be activated by the intense neutron flux giving rise to local accumulation of activated materials in the liquid metal system. Moreover, their precipitation in its cold section results in the potential plugging of the loop.

Corrosion in heavy liquid metals originates from physical-chemical phenomena involving the dissolution of the alloying elements, their transport, chemical reactions with dissolved non-metallic impurities (oxygen, nitrogen, etc.), and the formation of intermetallic compounds with the liquid metal and/or other dissolved metal impurities [132].

In PbLi environment, RAFM steels experience degradation phenomena mainly related to solution-based mechanisms [133–135]. In particular, almost-uniform dissolution of the main alloying elements (Fe, Cr) is observed on the exposed surface, and subsequent penetration of PbLi along grain boundaries. The higher diffusion rate of Cr in PbLi and its slower diffusivity in the steel matrix result in a selective leaching up to some μm of depth at the interface, with consequent surface enrichment of the low-solubility elements (e.g., W in Eurofer) [134,135]. Porous Cr-poor ferrite with PbLi penetration is then formed at the surface. Interactions with dissolved impurities such as O and N have not been reported.

Corrosion experiments carried out over the last two decades on RAFM steels have shown that corrosion occurs in two steps [136]. The first one consists of the non-uniform dissolution of the native oxide layer on the steel surface. This process occurs during the "incubation time", whose duration depends on PbLi temperature and flow. The second step is the dissolution of the bulk via leaching of Fe and Cr. According to some experiments, the incubation time for RAFM steels in flowing PbLi is shorter than under stagnant conditions and is reported to be 500 h at 550 °C [137] and it can last up to 1000 h at 480 °C [138].

Besides the effect on the incubation time, PbLi temperature and fluid-dynamic conditions are the main parameters that influence the rate of corrosion of RAFM materials [133,137,138]. Larger temperature enhances the dissolution effect because of the increased solubility and diffusivity of the alloying elements in the liquid metal. The increased flow rate and the transition from laminar to turbulent regime raise the mass transfer in boundary layers. Finally, erosive effects, such as the spall-off of the porous Cr-poor ferrite layer, are inherent to flowing conditions. This is shown by experimental studies that found a corrosion rate for Eurofer steel at 550 °C of about 400 μm/yr at 22 cm/s [137], 220 μm at 10 cm/s [133], and 18 μm in static PbLi [135]. The strong impact of elevated temperature becomes obvious when corrosion rates are compared with results for lower temperatures for which, e.g., at 480 °C, rates of only 90 μm/yr at 22 cm/s are observed [138].

Dissolved corrosion products and impurities are reported to have an influence on the corrosion of the materials in PbLi in isothermal systems, since they reduce the corrosion rate of steel in comparison with "fresh" melt [139]. However, dissolved corrosion products are expected to participate in re-crystallization and deposition phenomena in colder sections of non-isothermal dynamic systems, such as PbLi loops. Crystallized particles may move with the melt flow or be deposited on colder surfaces where plugging may occur in the narrow sections, affecting PbLi circulation.

5.4.1. Modelling of Steel Corrosion in PbLi

Although the PbLi loop piping will be internally coated by alumina ceramics to face the high corrosiveness of PbLi and to reduce tritium permeation to the coolant and environment [109], theoretical predictions are needed to assess the feasibility and safe operation of breeding blankets for DEMO and ITER. Modelling tools have to be validated against experimental data [140]. Several experimental campaigns have been performed to study the corrosion of RAFM steels in flowing PbLi. Some of the most relevant recent results are reported in Table 3 and briefly described in the following.

The PICOLO loop at KIT (Germany) was extensively used to perform long-term exposure of different RAFM steels (Manet I, F82H-mod, Optifer IVa, CLAM), Eurofer and ODS-Eurofer at 480 and 550 °C in turbulent flowing conditions (10 and 22 cm/s) up to 12,000 h [133,137,141,142]. The loop has a "figure-of-eight" configuration, with a cold leg equipped with magnetic trap, electromagnetic pump, and magnetic flow meter.

LIFUS-2 loop at ENEA (Italy) was used to test Eurofer steel at 480 °C in laminar flowing conditions (1 cm/s) up to 4500 h [143]. In the "figure-of-eight" loop PbLi flow rate was ensured by a mechanical pump in the cold leg. An upgraded configuration was recently proposed to study the effect of the dissolved impurities, including cold-trap, sampling system for PbLi chemical analysis and the use of a coating on the internal piping of the loop [144].

DRAGON PbLi loop series at INEST-CAS (China) are foreseen to study PbLi corrosion in both thermal and forced convection conditions [75]. The DRAGON-IV loop was also constructed to investigate the corrosion behavior at high temperature (480–800 °C) with flow up to 1 m/s under magnetic field (B = 2 T) and stress-applied conditions in PbLi.

An Indian loop was also recently put in operation at IPR (India) to study PbLi corrosion. An experimental campaign on IN-RAFM steel was conducted in flowing PbLi at 465 °C, 10 cm/s up to 5000 h [136].

Table 3. Most relevant corrosion data of RAFM steels in flowing PbLi.

Material	Loop	T_{hot} °C	T_{cold} °C	Velocity m/s	Corrosion μm/year	Reference
IN-RAFM	Indian	465	400	0.10	31–44	[136]
Eurofer	LIFUS 2	480	400	0.01	40	[143]
Manet I, F82H-mod, Optifer IVa, Eurofer	PICOLO	480	350–400	0.22	90	[138]
CLAM, Eurofer, ODS-Eurofer	PICOLO	550	400	0.10	200–220	[141,142]
Eurofer	PICOLO	550	400	0.22	400	[137]

During experimental campaigns, correlations have been identified in order to explain the experimental results from loop facilities and to extrapolate data to other operating conditions. For instance, Sannier's correlation serves to calculate the metal loss of steels at a given temperature and PbLi flow rate in turbulent/mixed regime [145]. Good agreement between experimental results was found for data obtained from PICOLO and the Indian loop on Eurofer and RAFM steels [133,136], but no consistency was found for data in the laminar regime, which exhibit some scattering.

Since corrosion in fusion blankets is a complex phenomenon that includes kinetics of solution and deposition mass transfer and interaction with impurities, the development and validation of modelling tools is necessary for predictive design input data. To date, the code MATLIM found validation against corrosion data of RAFM steels in turbulent/mixed regime [140,146], but validation still needs to be performed in laminar conditions, as expected in BUs of WCLL blankets, where the PbLi velocity is few mm/s. Hence, experimental campaign in relevant conditions of flow and thermal gradients should be performed for code validation, also considering the effect of magnetic field and applied stress.

For reliable application of predictive tools, values for solubility and diffusivity of dissolved species have to be known with better accuracy. These properties are very important for the implementation of models as they reflect the mass transfer coefficients [140,145]. Solubility values of Fe in PbLi were obtained experimentally by various authors, but they differ by orders of magnitude [147]. The solubility value of Cr is completely missing. In addition, in most of the experimental campaigns no monitoring of impurity levels, e.g., of dissolved metals and non-metal impurities, has been performed [140]. The presence of dissolved corrosion products and the formation of precipitates affect the evaluation of corrosion rates during the tests. Moreover, the influence of impurities, such as H, O, N, on the corrosion of structural materials has never been investigated experimentally.

5.4.2. Corrosion with Magnetic Field

A review of possible effects of a magnetic field on corrosion and deposition in flowing PbLi is given in [148]. The major influence of the magnetic field on corrosion processes is related to the change in the velocity profile due to the action of the induced electromagnetic forces, which results in steeper velocity gradients in the near-wall region and modified transport properties in the flow. Depending on the electric characteristics of the wall material, the flow rate inside the boundary layers at the walls parallel to the magnetic field can reach very large values. The thickness of the boundary layers reduces with increasing magnetic field strength; hence, the diffusion layer becomes thinner. Moreover, MHD flow in electrically coupled ducts can be characterized by flow reversals (cf. Section 5.1), and turbulent MHD flow is typically anisotropic with formation of Q2D turbulent structures (see Section 5.2). All these aspects determine the velocity distribution close to the fluid–solid interface, which controls mass transfer phenomena. In addition to that, the effect of the magnetic field on chemical reactions at the material surface should be taken into account, together with an electrodissolution mechanism due to induced currents entering the walls, as discussed in [149] for the corrosion of Hartmann walls.

Only a few experimental investigations are available in the literature in which the influence of a magnetic field on steel corrosion is studied, as required for fusion applications.

The compatibility of PbLi with austenitic (316 L) and martensitic (1.4914) steel under the influence of an imposed magnetic field (1.4 T) perpendicular to the module axis has been analyzed in the devices CELIMENE and ALCESTE [150,151]. The liquid metal (PbLi) was filled in a 30 mm annular space between a hot and a cold tube. In the former loop, the weak flow was due only to natural convection; in the second one, equipped with an electromagnetic pump for PbLi circulation, the mean velocity in a section was about 1 mm/s. It was observed that by applying a magnetic field, the corrosion rates of the two types of steel in semi-stagnant LM increase by about 50% for 316 L steel and 30% for 1.4914 steel compared to hydrodynamic conditions. The corrosion process is mainly due to dissolution of iron in the liquid metal. Concerning the deposition rate, it was found to be larger in the direction parallel to the magnetic field.

At IPUL (Latvia) experiments with samples of both P-91 steel [152,153] and Eurofer [154] have been carried out to study the influence of a magnetic field on corrosion phenomena in flowing PbLi. The loop consists of a cold and a hot part. The test section with samples is located partly in the magnetic field and partly outside, so that corrosion rates in hydrodynamic and MHD flows can be compared.

In this loop, non-isothermal corrosion of Eurofer steel has been investigated in a PbLi flow with a mean velocity of (5 ± 0.5) cm/s, a maximum operating temperature of 550 °C and a magnetic field of 1.7 T [154]. During a 2000 h experimental session, an intensification of corrosion processes was observed due to the presence of the magnetic field. Mass losses for the samples in the magnetic field were $1.5 \div 2$ times higher than for the ones in the region with $B = 0$. It was also observed that samples in hydrodynamic conditions had a rather smooth surface, while those in MHD flow had a regular wavy pattern on the Hartmann walls, in the form of grooves oriented in the PbLi flow direction [155,156]. A more detailed observation of the sample surfaces revealed that, differently than in hydrodynamic flow, where corrosion takes place predominantly at grain boundaries, when applying a magnetic field, the bulk is also involved in the process due to a deeper dissolution. The diffusion of elements (e.g., Fe and Cr) from Eurofer into the liquid metal is enhanced by the magnetic field (faster erosion) [156]. Experimental results have also shown that the Eurofer corrosion rate is significantly affected by the temperature of the melt.

Experimental corrosion data of FM steels in flowing PbLi, in the presence of a magnetic field, are summarized in Table 4. When available, the results for hydrodynamic conditions with the same parameters are also indicated for a direct comparison.

Table 4. Experimental corrosion data of FM steels in flowing PbLi with magnetic field.

Material	Loop	T_{hot} °C	T_{cold} °C	Velocity m/s	B T	Exposure h	Corrosion µm/year	Ref.
P-91 steel	IPUL/Riga	550	370–430	0.15–0.3	1.7	1000–2700	320–360	[152,153]
				0.15–0.3	Entrance	1000–2700	200–226	[152,153]
				0.15–0.3	Exit	1000–2700	100–218	[152,153]
Eurofer				0.05	1.7	1000		[156]
				0.05	1.7	2000	$+50 \div 100\%$	[154]
		550	350	0.05	1.8	1000	$+80 \div 140\%$	[155]
		515	350	0.05	1.8	2500	$+80 \div 180\%$	[155]
316 L 1.4914	CELIMENE ALCESTE	thermal gradient DT = 40K, convection		0.001 and 0	1.4		$\div 50\%$ $+30\%$	[151]
316 L	CELIMENE annular geometry	thermal gradient		Natural convection movements	1.4		Corrosion rate in quasi-stagnant conditions is larger on average than with $B = 0$ Magnetic field results in dissymmetry between dissolution and deposition rates in directions $\perp$ or θ to B	[150]

A small number of theoretical investigations have been carried out to predict corrosion of Eurofer in PbLi exposed to a magnetic field. Data related to the diffusion coefficients of metallic elements in liquid PbLi are very limited; for instance, values of diffusivity and saturation concentrations of iron in PbLi exhibit a large scattering. The latter parameter is one of the most important in corrosion models, such as the one developed by Smolentsev et al. [147]. The code called TRANSMAG uses a fully developed 2D MHD flow model and solves in 3D mass and heat transfer equations. In [147], in order to improve the available data on the saturation concentration of iron, the values for this property are reconstructed from experimental results in turbulent hydrodynamic flow by solving an inverse problem. The obtained data are approximated by a new correlation, which is employed to predict corrosion in laminar MHD flow in rectangular channels. The results confirm the experimental observations: the corrosion rate increases when imposing a magnetic field, the appearance of the corroded surface depends on the orientation of the wall with respect to B, and a larger mass loss occurs at the sidewalls (2–3 times stronger corrosion rates if compared to the Hartmann wall). These studies highlight the need for further experimental campaigns to increase the amount and the accuracy of corrosion data in MHD flows. Moreover, the method used to achieve improved correlations for the properties needed for corrosion modelling by matching calculated and experimental data, represents a valid procedure to obtain reliable predictions of corrosion rates in MHD flows.

Numerical investigations of corrosion have been also carried out in MHD turbulent flows [157,158]. Since the magnetic field tends to suppress turbulent motions, smaller corrosion rates were found compared to the case $B = 0$, in contrast with observations in MHD laminar regime. The reduction of the corrosion rate depends on the orientation of the magnetic field with respect to the flow direction, due to the anisotropic action of Joule dissipation, and on the intensity of the field. A magnetic field perpendicular to duct walls leads to the largest reduction in the corrosion rate [157].

6. Conclusions and Future R&D

The development of engineering designs for liquid metal TBMs for ITER and blankets for a DEMO reactor requires profound knowledge about magnetohydrodynamics coupled with multiphysics phenomena such as heat transfer, neutron physics, tritium breeding and transport, and the corrosion of wall material.

The progress in the numerical modelling of MHD flows combined with the availability of high-performance computing makes it possible to attain detailed insights into flow properties (pressure, velocity, temperature, electric potential) for various types of geometries and parameters close to those in fusion applications. There exists good confidence in the validity and accuracy of predictions, since numerical codes have been carefully validated against analytical solutions and model experiments. Often experiments and theoretical analyses are complementary since measured data are required for code validation or derivation of design correlations, and numerical results may support the interpretation of experimental observations.

The complexity of engineering designs represents a challenge for the numerical prediction of MHD flows in blanket-relevant geometries, since it implies special requirements for grid generation and for the employed numerical schemes. The computational meshes have to provide reasonable resolution of boundary layers, but, unlike in hydrodynamics, they also have to account for thin internal layers that spread along field lines from wall singularities and electrical discontinuities. Future research should find strategies for automatic grid generation that consider these particular needs. Discretization algorithms have to preserve conservation of mass, momentum, internal and kinetic energy, and electric charge.

Numerical investigation of coupled MHD effects in liquid metal blankets by means of CFD codes has achieved remarkable progress in recent years in terms of prediction accuracy and the complexity of the geometry and physics that can be modelled. However, even by using HPCs, simulations of large problems are still very time-consuming. Therefore, the use of SYS-TH codes for fusion applications represents a valuable option that should be significantly promoted. As pointed out, the prediction of MHD effects in liquid metal blanket systems by means of SYS-TH codes is still very challenging and far from suitably mature. Pressure drops in blanket components and heat transfer can be well described by friction coefficients and Nusselt number correlations. On the other hand, further effort should be put into the determination of reliable closure laws for the description of coupled MHD phenomena, such as electromagnetic coupling and mass transport, and for the prediction of blanket accidental transients. A number of studies are available in the literature on coupled system/CFD codes for thermal-hydraulics and safety analyses for nuclear reactors. Methods of coupling, of data transfer processing between CFD and system codes, as well as validation procedures have been reviewed in [159] and they may also be applied in fusion engineering.

Apart from heat transfer, where buoyancy may directly affect the flow, all other coupled phenomena may be considered as convective diffusive transport problems of passive scalars, i.e., dissolved tritium or corrosion products. The inventory and flux of dissolved species can be described by a convection–diffusion equation and the solution is straightforward as in hydrodynamics, once the PbLi velocity is known from MHD analysis. The major drawback in these analyses is the low precision of thermophysical properties such as solubility and diffusivity, since their reported values in the literature differ by orders of magnitude, as well as unknown kinetics of reactions at interfaces. As a result, the major uncertainties in theoretical predictions derive from scattering in the temperature-dependent thermophysical properties of PbLi required for the analyses. Smolentsev et al. [147] showed the importance of exploiting the synergy between experiments and simulations in order to obtain more reliable correlations for parameters used, for instance, in the numerical studies of corrosion in MHD flows. As long as precise input data for numerical simulations are not available, sensitivity studies by varying the uncertain quantities in a wide range may be used to achieve preliminary conservative estimates for corrosion rates or tritium permeation losses.

Once a precise measurement of material properties is available, target-oriented benchmark experiments have to be performed to validate models for tritium and corrosion transport in MHD flows. For instance, models for tritium transfer analysis are validated only against hydrodynamic experiments, and validation of the implemented coupled phenomena is missing. Therefore, experimental data are essential to increase the reliability of these models.

Author Contributions: Conceptualization, writing—original draft preparation, editing and review, C.M.; all authors have contributed to individual subsections by providing drafts and by reviewing the final manuscript (writing—review and editing). All authors have read and agreed to the published version of the manuscript.

Funding: This work has been carried out within the framework of the EUROfusion Consortium and has received funding from the Euratom research and training programme 2014–2018 and 2019–2020 under grant agreement No 633053. The views and opinions expressed herein do not necessarily reflect those of the European Commission. O.Z. was funded by the US NSF (grant CBET 1803730 "Extreme magnetoconvection").

Acknowledgments: The authors appreciate the contribution of Camillo Sartorio and Massimo Angiolini to the preparation of Section 5.4.1.

Conflicts of Interest: The authors declare no conflict of interest.

References

1. Federici, G.; Boccaccini, L.; Cismondi, F.; Gasparotto, M.; Poitevin, Y.; Ricapito, I. An overview of the EU breeding blanket design strategy as an integral part of the DEMO design effort. *Fusion Eng. Des.* **2019**, *141*, 30–42. [CrossRef]
2. Cismondi, F.; Spagnuolo, G.; Boccaccini, L.; Chiovaro, P.; Ciattaglia, S.; Cristescu, I.; Day, C.; del Nevo, A.; di Maio, P.; Federici, G.; et al. Progress of the conceptual design of the European DEMO breeding blanket, tritium extraction and coolant purification systems. *Fusion Eng. Des.* **2020**, *157*, 111640. [CrossRef]
3. Smolentsev, S.; Badia, S.; Bhattacharyay, R.; Bühler, L.; Chen, L.; Huang, Q.; Jin, H.-G.; Krasnov, D.; Lee, D.-W.; Valls, E.M.d.; et al. An approach to verification and validation of MHD codes for fusion applications. *Fusion Eng. Des.* **2015**, *100*, 65–72. [CrossRef]
4. Mistrangelo, C.; Bühler, L.; Smolentsev, S.; Klüber, V.; Maione, I.A.; Aubert, J. MHD flow in liquid metal blankets: Major design issues, MHD guidelines and numerical analysis. *Fusion Eng. Des.* **2021**, *173*, 112795. [CrossRef]
5. Zikanov, O.; Belyaev, I.; Listratov, Y.; Frick, P.; Razuvanov, N.; Sviridov, V. Mixed convection in pipe and duct flows with strong magnetic fields. *Appl. Mech. Rev.* **2021**, *73*, 010801. [CrossRef]
6. Krasnov, D.; Zikanov, O.; Boeck, T. Numerical study of magnetohydrodynamic duct flow at high Reynolds and Hartmann numbers. *J. Fluid Mech.* **2012**, *704*, 421–446. [CrossRef]
7. Chen, L.; Smolentsev, S.; Ni, M.-J. Toward full simulations for a liquid metal blanket: MHD flow computations for a PbLi blanket prototype at Ha $\sim$ 104. *Nucl. Fusion* **2020**, *60*, 076003. [CrossRef]
8. Yan, Y.; Ying, A.; Abdou, M. Numerical study of magneto-convection flows in a complex prototypical liquid-metal fusion blanket geometry. *Fusion Eng. Des.* **2020**, *159*, 111688. [CrossRef]
9. Shercliff, J.A. Steady motion of conducting fluids in pipes under transverse magnetic fields. *Math. Proc. Camb. Philos. Soc.* **1953**, *49*, 136–144. [CrossRef]
10. Walker, J.S. Magnetohydrodynamic flows in rectangular ducts with thin conducting walls. *J. Mécanique* **1981**, *20*, 79–112.
11. Walker, J.S.; Ludford, G.S.S.; Hunt, J.C.R. Three-dimensional MHD duct flows with strong transverse magnetic fields. Part 2. Variable-area rectangular ducts with conducting sides. *J. Fluid Mech.* **1971**, *46*, 657–684. [CrossRef]
12. Picologlou, B.F.; Reed, C.B.; Dauzvardis, P.V.; Walker, J.S. Experimental and Analytical Investigations of Magnetohydrodynamic Flows Near the Entrance to a Strong Magnetic Field. *Fusion Technol.* **1986**, *10*, 860–865. [CrossRef]
13. Kulikovskii, A.G. Slow steady flows of a conducting fluid at large Hartmann numbers. *Fluid Dyn.* **1968**, *3*, 1–5. [CrossRef]
14. Bühler, L. Magnetohydrodynamic flows in abitrary geometries in strong, nonuniform magnetic fields. *Fusion Technol.* **1995**, *27*, 3–24. [CrossRef]
15. Bühler, L. *Three-Dimensional Liquid Metal Flows in Strong Magnetic Fields*; FZKA-7412; Forschungszentrum Karlsruhe GmbH Technik und Umwelt: Berlin, Germany, 2008.

16. Malang, S.; Bojarsky, E.; Bühler, L.; Deckers, H.; Fischer, U.; Norajitra, P.; Reiser, H. Dual coolant liquid metal breeder blanket. In *Fusion Technology 1992, Proceedings of the 17th Symposium on Fusion Technology, Rome, Italy, 14–18 September 1992*; Ferro, C., Gasparotto, M., Knoepfel, H., Eds.; Elsevier: Amsterdam, The Netherlands, 1992; pp. 1424–1428.

17. Fernandez-Berceruelo, I.; Rapisarda, D.; Palermo, I.; Urgorri, F.R.; Agostinetti, P.; Cismondi, F.; De-Esch, H.P.L.; Ibarra, A. Integration of the neutral beam injector system Into the DCLL breeding blanket for the EU DEMO. *IEEE Trans. Plasma Sci.* **2018**, *46*, 2708. [CrossRef]

18. Bestion, D. The structure of system thermal-hydraulic (SYS-TH) code for nuclear energy applications. In *Thermal-Hydraulics of Water Cooled Nuclear Reactors*; Elsevier: Amsterdam, The Netherlands, 2017; pp. 639–727.

19. Liu, L.; Zhang, D.; Yan, Q.; Xu, R.; Wang, C.; Qiu, S.; Su, G.H. RELAP5 MOD3.2 modification and application to the transient analysis of a fluoride-salt-cooled high-temperature reactor. *Ann. Nucl. Energy* **2017**, *101*, 504–515. [CrossRef]

20. Eboli, M.; Forgione, N.; Nevo, A.D. Assessment of SIMMER-III code in predicting Water Cooled Lithium Lead Breeding Blanket\textquotedblleftin-box-Loss of Coolant Accident\textquotedblright. *Fusion Eng. Des.* **2021**, *163*, 112127. [CrossRef]

21. Panayotov, D.; Grief, A.; Merrill, B.J.; Humrickhouse, P.; Trow, M.; Dillistone, M.; Murgatroyd, J.T.; Owen, S.; Poitevin, Y.; Peers, K.; et al. Methodology for accident analyses of fusion breeder blankets and its application to helium-cooled pebble bed blanket. *Fusion Eng. Des.* **2016**, *109–111*, 1574–1580. [CrossRef]

22. Seo, S.B.; Hernandez, R.; O'Neal, M.; Meehan, N.; Novais, F.S.; Rizk, M.; Maldonado, G.I.; Brown, N.R. A review of thermal hydraulics systems analysis for breeding blanket design and future needs for fusion engineering demonstration facility design and licensing. *Fusion Eng. Des.* **2021**, *172*, 112769. [CrossRef]

23. Melchiorri, L.; Narcisi, V.; Giannetti, F.; Caruso, G.; Tassone, A. Development of a RELAP5/MOD3.3 module for MHD pressure drop analysis in liquid metals loops: Verification and Validation. *Energies* **2021**, *14*, 5538. [CrossRef]

24. Smolentsev, S.; Spagnuolo, G.A.; Serikov, A.; Rasmussen, J.J.; Nielsen, A.H.; Naulin, V.; Marian, J.; Coleman, M.; Malerba, L. On the role of integrated computer modelling in fusion technology. *Fusion Eng. Des.* **2020**, *157*, 111671. [CrossRef]

25. Smolentsev, S.; Moreau, R.; Bühler, L.; Mistrangelo, C. MHD thermofluid issues of liquid-metal blankets: Phenomena and advances. *Fusion Eng. Des.* **2010**, *85*, 1196–1205. [CrossRef]

26. Smolentsev, S. Physical Background, Computations and Practical Issues of the Magnetohydrodynamic Pressure Drop in a Fusion Liquid Metal Blanket. *Fluids* **2021**, *6*, 110. [CrossRef]

27. Kirillov, I.R.; Reed, C.B.; Barleon, L.; Miyazaki, K. Present understanding of MHD and heat transfer phenomena for liquid metal blankets. *Fusion Eng. Des.* **1995**, *27*, 553–569. [CrossRef]

28. Bluck, M.J.; Wolfendale, M.J. An analytical solution to electromagnetically coupled duct flow in MHD. *J. Fluid Mech.* **2015**, *771*, 595–623. [CrossRef]

29. Zhang, X.; Wang, L.; Pan, C. Effects of inclined transversal magnetic fields on magnetohydrodynamic coupling duct flow states in liquid metal blankets: Under uniform magnetic fields. *Nucl. Fusion* **2021**, *61*, 016005. [CrossRef]

30. Mistrangelo, C.; Bühler, L.; Koehly, C.; Ricapito, I. Magnetohydrodynamic velocity and pressure drop in WCLL TBM. *Nucl. Fusion* **2021**. in print. [CrossRef]

31. Siriano, S.; Tassone, A.; Caruso, G.; del Nevo, A. Electromagnetic coupling phenomena in co-axial rectangular channels. *Fusion Eng. Des.* **2020**, *160*, 111854. [CrossRef]

32. Madarame, H.; Taghavi, K.; Tillack, M.S. The influence of leakage currents on MHD pressure drop. *Fusion Technol.* **1985**, *8*, 264–269. [CrossRef]

33. Stieglitz, R.; Molokov, S. Experimental study of magnetohydrodynamic flows in electrically coupled bends. *J. Fluid Mech.* **1997**, *343*, 1–28. [CrossRef]

34. Smolentsev, S.; Rhodes, T.; Yan, Y.; Tassone, A.; Mistrangelo, C.; Bühler, L.; Urgorri, F.R. Code-to-code comparison for a PbLi mixed-convection MHD flow. *Fusion Sci. Technol.* **2020**, *76*, 653–669. [CrossRef]

35. Pacio, J.; Marocco, L.; Wetzel, T. Review of data and correlations for turbulent forced convective heat transfer of liquid metals in pipes. *Heat Mass Transf.* **2015**, *51*, 153–164. [CrossRef]

36. Barleon, L.; Burr, U.; Mack, K.-J.; Stieglitz, R. Heat transfer in liquid metal cooled fusion blankets. *Fusion Eng. Des.* **2000**, *51*, 723–733. [CrossRef]

37. Miyazaki, K.; Inoue, H.; Kimoto, T.; Yamashita, S.; Inoue, S.; Yamaoka, N. Heat transfer and temperature fluctuation of lithium flowing under transverse magnetic field. *J. Nucl. Sci. Technol.* **1986**, *23*, 582–593. [CrossRef]

38. Burr, U.; Barleon, L.; Müller, U.; Tsinober, A. Turbulent transport of momentum and heat in magnetohydrodynamic rectangular duct flow with strong sidewall jets. *J. Fluid Mech.* **2000**, *406*, 247–279. [CrossRef]

39. Ji, H.-C.; Gardner, R.A. Numerical analysis of turbulent pipe flow in a transverse magnetic field. *Int. J. Heat Mass Transf.* **1997**, *40*, 1839–1851. [CrossRef]

40. RELAP5-3D Code Development Team. *RELAP5-3D Code Manual Volume I: Code Structure, System Models and Solution Methods*; Idaho National Laboratory: Idaho Falls, ID, USA, 2015.

41. Chung, B.D.; Kim, K.D.; Bae, S.W.; Jeong, J.J.; Lee, S.W.; Hwang, M.K.; Yoon, C. *MARS Code Manual Volume I: Code Structure, System Models, and Solution Methods*; No. KAERI/TR–2812/2004; Korea Atomic Energy Research Institute: Daejeon, Korea, 2010.

42. Gauntt, R.O.; Cole, R.K.; Erickson, C.M.; Gido, R.G.; Gasser, R.D.; Rodriguez, S.B.; Young, M.F. MELCOR computer code manuals. *Sandia Natl. Lab. NUREG/CR-6119* **2000**, *2*, 1–247.

43. Wolfendale, M.J.; Bluck, M.J. A coupled systems code-CFD MHD solver for fusion blanket design. *Fusion Eng. Des.* **2015**, *98*, 1902–1906. [CrossRef]

44. Froio, A.; Bertinetti, A.; Savoldi, L.; Zanino, R.; Cismondi, F.; Ciattaglia, S. Benchmark of the GETTHEM Vacuum Vessel Pressure Suppression System (VVPSS) model for a helium-cooled EU DEMO blanket. *Saf. Reliab. Theory Appl.* **2017**, *11*, 59–66.

45. Kim, S.H.; Kim, M.H.; Lee, D.W.; Choi, C. Code validation and development for MHD analysis of liquid metal flow in Korean TBM. *Fusion Eng. Des.* **2012**, *87*, 951–955. [CrossRef]

46. Panayotov, D.; Poitevin, Y.; Grief, A.; Trow, M.; Dillistone, M.; Murgatroyd, J.T.; Owen, S.; Peers, K.; Lyons, A.; Heaton, A.; et al. A Methodology for Accident Analysis of Fusion Breeder Blankets and Its Application to Helium-Cooled Lead–Lithium Blanket. *IEEE Trans. Plasma Sci.* **2016**, *44*, 2511–2522. [CrossRef]

47. Froio, A.; Batti, A.; del Nevo, A.; Savoldi, L.; Spagnuolo, A.; Zanino, R. Implementation of a system-level magnetohydrodynamic model in the GETTHEM code for the analysis of the EU DEMO WCLL Breeding Blanket. In Proceedings of the TOFE 2020, Charleston, SC, USA, 15–19 November 2020.

48. Bühler, L.; Brinkmann, H.-J.; Horanyi, S.; Starke, K. *Magnetohydrodynamic Flow in a Mock-Up of a HCLL Blanket*; Forschungszentrum Karlsruhe: Karlsruhe, Germany, 2008.

49. Bühler, L.; Mistrangelo, C. Determination of flow distribution in a HCLL blanket mock-up through electric potential measurements. *Fusion Eng. Des.* **2011**, *86*, 2301–2303. [CrossRef]

50. Zhang, J.; Ni, M.-J. Direct simulation of multi-phase MHD flows on an unstructured Cartesian adaptive system. *J. Comput. Phys.* **2014**, *270*, 345–365. [CrossRef]

51. Ni, M.-J.; Munipalli, R.; Morley, N.B.; Huang, P.; Abdou, M.A. A current density conservative scheme for incompressible MHD flows at a low magnetic Reynolds number. Part I: On a rectangular collocated grid system. *J. Comput. Phys.* **2007**, *227*, 174–204. [CrossRef]

52. Ni, M.-J.; Munipalli, R.; Huang, P.; Morley, N.B.; Abdou, M.A. A current density conservative scheme for incompressible MHD flows at a low magnetic Reynolds number. Part II: On an arbitrary collocated mesh. *J. Comput. Phys.* **2007**, *227*, 205–228. [CrossRef]

53. Syrakos, A.; Varchanis, S.; Dimakopoulos, Y.; Goulas, A.; Tsamopoulos, J. A critical analysis of some popular methods for the discretisation of the gradient operator in finite volume methods. *Phys. Fluids* **2017**, *29*, 127103. [CrossRef]

54. Klüber, V. Three-Dimensional Magnetohydrodynamic Phenomena in Circular Pipe Flow. Ph.D. Thesis, Karlsruhe Institute of Technology (KIT), Karlsruhe, Germany, 2021.

55. Chang, C.; Lundgren, S. Duct flow in Magnetohydrodynamics. In *Zeitschrift für angewandte Mathematik und Physik ZAMP*; Birkhäuser: Basel, Switzerland, 1961; Volume 12, pp. 100–114.

56. Klüber, V.; Bühler, L.; Mistrangelo, C. Numerical simulation of 3D magnetohydrodynamic liquid metal flow in a spatially varying solenoidal magnetic field. *Fusion Eng. Des.* **2020**, *156*, 111659. [CrossRef]

57. Bühler, L.; Mistrangelo, C.; Konys, J.; Bhattacharyay, R.; Huang, Q.; Obukhov, D.; Smolentsev, S.; Utili, M. Facilities, testing program and modeling needs for studying liquid metal magnetohydrodynamic flows in fusion blankets. *Fusion Eng. Des.* **2015**, *100*, 55–64. [CrossRef]

58. Bühler, L.; Mistrangelo, C.; Brinkmann, H.-J.; Koehly, C. Pressure distribution in MHD flows in an experimental test-section for a HCLL blanket. *Fusion Eng. Des.* **2018**, *127*, 168–172. [CrossRef]

59. Swain, P.; Shishko, A.; Mukherjee, P.; Tiwari, V.; Ghorui, S.; Bhattacharyay, R.; Patel, A.; Satyamurthy, P.; Ivanov, S.; Platacis, E.; et al. Numerical and experimental MHD studies of lead-lithium liquid metal flows in multichannel test-section at high magnetic fields. *Fusion Eng. Des.* **2018**, *132*, 73–85. [CrossRef]

60. Kirillov, I.; Obukhov, D.; Sviridov, V.; Razuvanov, N.; Belyaev, I.; Poddubnyi, I.; Kostichev, P. Buoyancy effects in vertical rectangular duct with coplanar magnetic field and single sided heat load—Downward and upward flow. *Fusion Eng. Des.* **2018**, *127*, 226–233. [CrossRef]

61. Morley, N.B.; Burris, J.; Cadwallader, L.C.; Nornberg, M.D. GaInSn Usage in the Research Laboratory. *Rev. Sci. Instrum.* **2008**, *79*, 056107. [CrossRef]

62. Chowdhury, V.; Bühler, L.; Mistrangelo, C.; Brinkmann, H.-J. Experimental study of instabilities in magnetohydrodynamic boundary layers. *Fusion Eng. Des.* **2015**, *98–99*, 1751–1754. [CrossRef]

63. Reed, C.B.; Picologlou, B.F.; Hua, T.Q.; Walker, J.S. Alex results—A comparison of measurements from a round and a rectangular duct with 3-D code predictions. In Proceedings of the 12th Symposium on Fusion Engineering, Monterey, CA, USA, 13–16 October 1987; IEEE: Piscataway, NJ, USA, 1987; pp. 1267–1270.

64. Barleon, L.; Mack, K.-J.; Stieglitz, R. *The MEKKA-Facility a Flexible Tool to Investigate MHD-Flow Phenomena*; FZKA 5821; Forschungszentrum Karlsruhe: Karlsruhe, Germany, 1996.

65. Satyamurthy, P.; Swain, P.; Tiwari, V.; Kirillov, I.; Obukhov, D.; Pertsev, D. Experiments and numerical MHD analysis of LLCB TBM test-section with NaK at 1 T magnetic field. *Fusion Eng. Des.* **2015**, *91*, 44–51. [CrossRef]

66. Chowdhury, V.; Bühler, L.; Mistrangelo, C. Influence of surface oxidation on electric potential measurements in MHD liquid metal flows. *Fusion Eng. Des.* **2014**, *89*, 1299–1303. [CrossRef]

67. Smolentsev, S.; Courtessole, C.; Abdou, M.; Sharafat, S.; Sahu, S.; Sketchley, T. Numerical modeling of first experiments on PbLi MHD flows in a rectangular duct with foam-based SiC flow channel insert. *Fusion Eng. Des.* **2016**, *108*, 7–20. [CrossRef]

68. Bühler, L.; Brinkmann, H.-J.; Koehly, C. Experimental study of liquid metal magnetohydrodynamic flows near gaps between flow channel inserts. *Fusion Eng. Des.* **2019**, *146*, 1399–1402. [CrossRef]

69. Bühler, L.; Lyu, B.; Brinkmann, H.-J.; Mistrangelo, C. Reconstruction of 3D MHD liquid metal velocity from measurements of electric potential on the external surface of a thick-walled pipe. *Fusion Eng. Des.* **2021**, *168*, 112590. [CrossRef]

70. Reed, C.B.; Picologlou, B.F.; Dauzvardis, P.V.; Bailey, J.L. Techniques for measurement of velocity in liquid-metal MHD flows. *Fusion Technol.* **1986**, *10*, 813–821. [CrossRef]

71. Mistrangelo, C.; Bühler, L. Perturbing Effects of Electric Potential Probes on MHD Duct Flows. *Exp. Fluids* **2010**, *48*, 157–165. [CrossRef]

72. Koehly, C.; Bühler, L.; Mistrangelo, C. Design of a test section to analyze magneto-convection effects in WCLL blankets. *Fusion Sci. Technol.* **2019**, *75*, 1010–1015. [CrossRef]

73. Mistrangelo, C.; Bühler, L. Determination of multichannel MHD velocity profiles from wall-potential measurements and numerical simulations. *Fusion Eng. Des.* **2018**, *130*, 137–141. [CrossRef]

74. Wu, Y.; Huang, Q.; Zhu, Z.; Gao, S.; Song, Y. R&D of DRAGON Series Lithium-Lead Loops for Material and Blanket Technology Testing. *Fusion Sci. Technol.* **2012**, *62*, 272–275.

75. Huang, Q.; Zhu, Z.; Team, F. Development and experiments of LiPb coolant technologies for fusion blanket in China. *Int. J. Energy Res.* **2021**, *45*, 11384–11398. [CrossRef]

76. Kumar, M.; Patel, A.; Jaiswal, A.; Ranjan, A.; Mohanta, D.; Sahu, S.; Saraswat, A.; Rao, P.; Rao, T.; Mehta, V.; et al. Engineering design and development of lead lithium loop for thermo-fluid MHD studies. *Fusion Eng. Des.* **2019**, *138*, 1–5. [CrossRef]

77. Tanaka, T.; Sagara, A.; Yagi, J.; Muroga, T. Liquid blanket collaboration platform Oroshhi-2 at NIFS with FLiNaK/LiPb twin loops. *Fusion Sci. Technol.* **2019**, *75*, 1002–1009. [CrossRef]

78. Yoon, J.S.; Lee, D.W.; Bae, Y.-D.; Kim, S.K.; Jung, K.S.; Cho, S. Development of an experimental facility for a liquid breeder in Korea. *Fusion Eng. Des.* **2011**, *86*, 2212–2215. [CrossRef]

79. Hon, A.; Kirillov, I.; Komov, K.; Kovalchuk, O.; Lancetov, A.; Obukhov, D.; Pertsev, D.; Pugachev, A.; Rodin, I.; Zapretilina, E. Lead-lithium facility with superconducting magnet for MHD/HT tests of liquid metal breeder blanket. *Fusion Eng. Des.* **2017**, *124*, 832–836. [CrossRef]

80. Ivanov, S.; Shishko, A.; Flerov, A.; Platacis, E.; Romanchuks, A.; Zik, A. MHD PbLi loop at IPUL. In Proceedings of the 9th PAMIR International Conference on Fundamental and Applied MHD, Riga, Latvia, 16–20 June 2014; pp. 76–80.

81. Smolentsev, S.; Li, F.-C.; Morley, N.; Ueki, Y.; Abdou, M.; Sketchley, T. Construction and initial operation of MHD PbLi facility at UCLA. *Fusion Eng. Des.* **2013**, *88*, 317–326. [CrossRef]

82. Mistrangelo, C.; Bühler, L. MHD phenomena related to electromagnetic flow coupling. *Magnetohydrodynamics* **2017**, *53*, 141–148. [CrossRef]

83. Swain, P.K.; Koli, P.; Ghorui, S.; Mukherjee, P.; Deshpande, A.V. Thermofluid MHD studies in a model of Indian LLCB TBM at high magnetic field relevant to ITER. *Fusion Eng. Des.* **2020**, *150*, 111374. [CrossRef]

84. Mistrangelo, C.; Bühler, L. Electric flow coupling in the HCLL blanket concept. *Fusion Eng. Des.* **2008**, *83*, 1232–1237. [CrossRef]

85. Bühler, L.; Mistrangelo, C. Theoretical studies of MHD flows in support to HCLL design activities. *Fusion Eng. Des.* **2016**, *109–111*, 1609–1613. [CrossRef]

86. Siriano, S.; Tassone, A.; Caruso, G.; del Nevo, A. MHD forced convection flow in dielectric and electro-conductive rectangular annuli. *Fusion Eng. Des.* **2020**, *159*, 111773. [CrossRef]

87. Zikanov, O.; Krasnov, D.; Boeck, T.; Thess, A.; Rossi, M. Laminar-Turbulent Transition in Magnetohydrodynamic Duct, Pipe, and Channel Flows. *Appl. Mech. Rev.* **2014**, *66*, 030802. [CrossRef]

88. Moffatt, K. On the Suppression of Turbulence by a Uniform Magnetic Field. *J. Fluid Mech.* **1967**, *28*, 571–592. [CrossRef]

89. Davidson, P.A. *Introduction to Magnetohydrodynamics*; Cambridge University Press: Cambridge, UK, 2016.

90. Vorobev, A.; Zikanov, O.; Davidson, P.; Knaepen, B. Anisotropy of magnetohydrodynamic turbulence at low magnetic Reynolds number. *Phys. Fluids* **2005**, *17*, 125105. [CrossRef]

91. Krasnov, D.; Zikanov, O.; Schumacher, J.; Boeck, T. Magnetohydrodynamic turbulence in a channel with spanwise magnetic field. *Phys. Fluids* **2008**, *20*, 095105. [CrossRef]

92. Boeck, T.; Krasnov, D.; Thess, A.; Zikanov, O. Large-scale intermittency of liquid-metal channel flow in a magnetic field. *Phys. Rev. Lett.* **2008**, *101*, 244501. [CrossRef]

93. Krasnov, D.; Thess, A.; Boeck, T.; Zhao, Y.; Zikanov, O. Patterned turbulence in liquid metal flow: Computational reconstruction of the Hartmann experiment. *Phys. Rev. Lett.* **2013**, *110*, 084501. [CrossRef]

94. Knaepen, B.; Moreau, R. Magnetohydrodynamic turbulence at low magnetic Reynolds number. *Annu. Rev. Fluid Mech.* **2008**, *40*, 25–45. [CrossRef]

95. Smolentsev, S.; Vetcha, N.; Moreau, R. Study of instabilities and transitions for a family of quasi-two-dimensional magnetohydro-dynamic flows based on a parametrical model. *Phys. Fluids* **2012**, *24*, 024101. [CrossRef]

96. Zikanov, O.; Listratov, Y.; Sviridov, V.G. Natural convection in horizontal pipe flow with a strong transverse magnetic field. *J. Fluid Mech.* **2013**, *720*, 486–516. [CrossRef]

97. Zikanov, O.; Listratov, Y. Numerical investigation of MHD heat transfer in a vertical round tube affected by transverse magnetic field. *Fusion Eng. Des.* **2016**, *113*, 151–161. [CrossRef]

98. Zhang, X.; Zikanov, O. Convection instability in a downward flow in a vertical duct with strong transverse magnetic field. *Phys. Fluids* **2018**, *30*, 117101. [CrossRef]

99. Sahu, S.; Courtessole, C.; Ranjan, A.; Bhattacharyay, R.; Sketchley, T.; Smolentsev, S. Thermal convection studies in liquid metal flow inside a horizontal duct under the influence of transverse magnetic field. *Phys. Fluids* **2020**, *32*, 067107. [CrossRef]
100. Liu, L.; Zikanov, O. Elevator Mode Convection in Flows With Strong Magnetic Fields. *Phys. Fluids* **2015**, *27*, 044103. [CrossRef]
101. Zhang, X.; Zikanov, O. Two-dimensional turbulent convection in a toroidal duct of a liquid metal blanket of a fusion reactor. *J. Fluid Mech.* **2015**, *779*, 36–52. [CrossRef]
102. Akhmedagaev, R.; Zikanov, O.; Listratov, Y. Magnetoconvection in a horizontal duct flow at very high Hartmann and Grashof numbers. *arXiv* **2021**, arXiv:2106.04231.
103. Ni, M.; Wang, Y.; Yuan, B.; Jiang, J.; Wu, Y. Tritium supply assessment for ITER and DEMOnstration power plant. *Fusion Eng. Des.* **2013**, *88*, 2422–2426. [CrossRef]
104. Sieverts, A. The absorption of gases by metals. *Z. Für Met.* **1929**, *21*, 37–46.
105. Ali-Khan, I.; Dietz, K.; Waelbroeck, F.; Wienhold, P. The rate of hydrogen release out of clean metallic surfaces. *J. Nucl. Mater.* **1978**, *76–77*, 337–343. [CrossRef]
106. Spagnuolo, G.A.; Arredondo, R.; Boccaccini, L.V.; Coleman, M.; Cristescu, I.; Federici, G.; Franza, F.; Garcinuño, B.; Moreno, C.; Rapisarda, D.; et al. Integration issues on tritium management of the European DEMO Breeding Blanket and ancillary systems. *Fusion Eng. Des.* **2021**, *171*, 112573. [CrossRef]
107. Santucci, A.; Incelli, M.; Noschese, L.; Moreno, C.; di Fonzo, F.; Utili, M.; Tosti, S.; Day, C. The issue of Tritium in DEMO coolant and mitigation strategies. *Fusion Eng. Des.* **2020**, *158*, 111759. [CrossRef]
108. Demange, D.; Boccaccini, L.; Franza, F.; Santucci, A.; Tosti, S.; Wagner, R. Tritium management and anti-permeation strategies for three different breeding blanket options foreseen for the European Power Plant Physics and Technology Demonstration reactor study. *Fusion Eng. Des.* **2014**, *89*, 1219–1222. [CrossRef]
109. Utili, M.; Bassini, S.; Cataldo, S.; Di Fonzo, F.; Kordac, M.; Hernandez, T.; Kunzova, K.; Lorenz, J.; Martelli, D.; Padino, B.; et al. Development of anti-permeation and corrosion barrier coatings for the WCLL breeding blanket of the European DEMO. *Fusion Eng. Des.* **2021**, *170*, 112453. [CrossRef]
110. Ricapito, I.; Calderoni, P.; Poitevin, Y.; Sedano, L. Tritium transport modeling for breeding blanket: State of the art and strategy for future development in the EU fusion program. *Fusion Eng. Des.* **2012**, *87*, 793–797. [CrossRef]
111. Ricapito, I.; Aiello, A.; Bükki-Deme, A.; Galabert, J.; Moreno, C.; Poitevin, Y.; Radloff, D.; Rueda, A.; Tincani, A.; Utili, M. Tritium technologies and transport modelling: Main outcomes from the European TBM Project. *Fusion Eng. Des.* **2018**, *136*, 128–134. [CrossRef]
112. Longhurst, G.R.; Ambrosek, J. Verification and Validaton of the Tritium Transport Code TMAP7. *Fusion Sci. Technol.* **2005**, *48*, 468–471. [CrossRef]
113. Franza, F.; Ciampichetti, A.; Ricapito, I.; Zucchetti, M. A model for tritium transport in fusion reactor components: The FUS-TPC code. *Fusion Eng. Des.* **2012**, *87*, 299–302. [CrossRef]
114. Candido, L.; Alberghi, C. Verification and validation of mHIT code over TMAP for hydrogen isotopes transport studies in fusion-relevant enviroments. *Fusion Eng. Des.* **2021**, *172*, 112740. [CrossRef]
115. Pattison, M.J.; Smolentsev, S.; Munipalli, R.; Abdou, M.A. Tritium Transport in Poloidal Flows of a DCLL Blanket. *Fusion Sci. Technol.* **2011**, *60*, 809–813. [CrossRef]
116. Gabriel, F.; Escuriol, Y.; Dabbene, F.; Gastaldi, O.; Salavy, J.; Giancarli, L. A 2D finite element modelling of tritium permeation for the HCLL DEMO blanket module. *Fusion Eng. Des.* **2007**, *82*, 2204–2211. [CrossRef]
117. Fukada, S.; Muneoka, T.; Kinjyo, M.; Yoshimura, R.; Katayama, K. Hydrogen transfer in PbLi forced convection flow with permeable wall. *Fusion Eng. Des.* **2015**, *96–97*, 95–100. [CrossRef]
118. Garcinuno, B.; Rapisarda, D.; Fernandez-Berceruelo, I.; Carella, E.; Sanz, J. The CIEMAT LiPb Loop Permeation Experiment. *Fusion Eng. Des.* **2019**, *146*, 1228–1232. [CrossRef]
119. Candido, L.; Cantore, M.; Galli, E.; Testoni, R.; Utili, M.; Zucchetti, M. An integrated hydrogen isotopes transport model for the TRIEX-II facility. *Fusion Eng. Des.* **2020**, *155*, 111585. [CrossRef]
120. Humrickhouse, P.W.; Calderoni, P.; Merrill, B.J. Implementation of Tritium Permeation Models in the CFD Code Fluent. *Fusion Sci. Technol.* **2011**, *60*, 1564–1567. [CrossRef]
121. Hendricks, S.; Carella, E.; Moreno, C.; Molla, J. Numerical investigation of hydrogen isotope retention by an yttrium pebble-bed from flowing liquid lithium. *Nucl. Fusion* **2020**, *60*, 106017. [CrossRef]
122. Candido, L.; Cantore, M.; Galli, E.; Testoni, R.; Zucchetti, M.; Utili, M.; Ciampichetti, A. Characterization of Pb-15.7Li Hydrogen Isotopes Permeation Sensors and Upgrade of Hyper- Quarch Experimental Device. *IEEE Trans. Plasma Sci.* **2020**, *48*, 1505–1511. [CrossRef]
123. Candido, L.; Testoni, R.; Utili, M.; Zucchetti, M. Tritium transport model at breeder unit level for WCLL breeding blanket. *Fusion Eng. Des.* **2019**, *146*, 1207–1210. [CrossRef]
124. Alberghi, C.; Candido, L.; Testoni, R.; Utili, M.; Zucchetti, M. Magneto-convective effect on tritium transport at breeder unit level for the WCLL breeding blanket of DEMO. *Fusion Eng. Des.* **2020**, *160*, 111996. [CrossRef]
125. Candido, L.; Alberghi, C.; Moro, F.; Testoni, R.; Utili, M.; Zucchetti, M. A novel approach to the study of magneto-hydro-dynamics effect on tritium transport in WCLL breeding blanket of DEMO. *Fusion Eng. Des.* **2021**, *167*, 112334. [CrossRef]
126. Reiter, F. Solubility and diffusivity of hydrogen isotopes in liquid Pb-17Li. *Fusion Eng. Des.* **1991**, *14*, 207–211. [CrossRef]

127. Norajitra, P.; Basuki, W.W.; Gonzalez, M.; Rapisarda, D.; Rohde, M.; Spatafora, L. Development of Sandwich Flow Channel Inserts for an EU DEMO Dual Coolant Blanket Concept. *Fusion Sci. Technol.* **2015**, *68*, 501–506. [CrossRef]

128. Fernández-Berceruelo, I.; Palermo, I.; Urgorri, F.R.; Rapisarda, D.; Garcinuño, B.; Ibarra, A. Remarks on the performance of the EU DCLL breeding blanket adapted to DEMO 2017. *Nucl. Fusion* **2020**, *155*, 111559. [CrossRef]

129. Urgorri, F.R.; Moreno, C.; Carella, E.; Rapisarda, D.; Fernandez-Berceruelo, I.; Palermo, I.; Ibarra, A. Tritium transport modeling at system level for the EUROfusion dual coolant lithium-lead breeding blanket. *Nucl. Fusion* **2017**, *57*, 116045. [CrossRef]

130. Zhang, H.; Ying, A.; Abdou, M. Quantification of Dominating Factors in Tritium Permeation in PbLi Blankets. *Fusion Sci. Technol.* **2015**, *68*, 362–367. [CrossRef]

131. Konys, J.; Krauss, W. Corrosion and precipitation effects in a forced-convection Pb–15.7Li loop. *J. Nucl. Mater.* **2013**, *442*, S576–S579. [CrossRef]

132. Draley, J.E.; Weeks, J.R. Corrosion by Liquid Metals. In Proceedings of the Sessions on Corrosion by Liquid Metals of the 1969 Fall Meeting of the Metallurgical Society of AIME, Philadelphia, PA, USA, 13–16 October 1969.

133. Konys, J.; Krauss, W.; Steiner, H.; Novotny, J.; Skrypnik, A. Flow rate dependent corrosion behavior of Eurofer steel in Pb–15.7Li. *J. Nucl. Mater.* **2011**, *417*, 1191–1194. [CrossRef]

134. Chakraborty, P.; Singh, V.; Bysakh, S.; Tewari, R.; Kain, V. Short-term corrosion behavior of Indian RAFM steel in liquid Pb-Li: Corrosion mechanism and effect of alloying elements. *J. Nucl. Mater.* **2019**, *520*, 208–217. [CrossRef]

135. Bassini, S.; Cuzzola, V.; Antonelli, A.; Utili, M. Long-term corrosion behavior of EUROFER RAFM steel in static liquid Pb-16Li at 550 °C. *Fusion Eng. Des.* **2020**, *160*, 111829. [CrossRef]

136. Atchutuni, S.S.; Saraswat, A.; Sasmal, C.S.; Verma, S.; Prajapati, A.K.; Jaiswal, A.; Gupta, S.; Chauhan, J.; Pandya, K.B.; Makwana, M.; et al. Corrosion experiments on IN-RAFM steel in flowing lead-lithium for Indian LLCB TBM. *Fusion Eng. Des.* **2018**, *132*, 52–59. [CrossRef]

137. Konys, J.; Krauss, W.; Novotny, J.; Steiner, H.; Voss, Z.; Wedemeyer, O. Compatibility behavior of EUROFER steel in flowing Pb–17Li. *J. Nucl. Mater.* **2009**, *386–388*, 678–681. [CrossRef]

138. Konys, J.; Krauss, W.; Voss, Z.; Wedemeyer, O. Corrosion behavior of EUROFER steel in flowing eutectic Pb–17Li alloy. *J. Nucl. Mater.* **2004**, *329–333*, 1379–1383. [CrossRef]

139. Chakraborty, P.; Pradhan, P.K.; Fotedar, R.K.; Krishnamurthy, N. Corrosion prevention of type 316L stainless steel in Pb-17Li through Nickel addition. *Fusion Sci. Technol.* **2014**, *65*, 332–337. [CrossRef]

140. Krauss, W.; Konys, J.; Li-Puma, A. TBM Testing in ITER: Requirements for the Development of Predictive Tools to Describe Corrosion-Related Phenomena in HCLL Blankets Towards DEMO. *Fusion Eng. Des.* **2012**, *87*, 403–406. [CrossRef]

141. Konys, J.; Krauss, W.; Zhu, Z.; Huang, Q. Comparison of corrosion behavior of EUROFER and CLAM steels in flowing Pb–15.7Li. *J. Nucl. Mater.* **2014**, *455*, 491–495. [CrossRef]

142. Krauss, W.; Wulf, S.-E.; Konys, J. Long-term corrosion behavior of ODS-Eurofer in flowing Pb-15.7Li at 550 °C. *Nucl. Mater. Energy* **2016**, *9*, 512–518. [CrossRef]

143. Benamati, G.; Fazio, C.; Ricapito, I. Mechanical and Corrosion Behaviour of EUROFER 97 Steel Exposed to Pb–17Li. *J. Nucl. Mater.* **2002**, *307–311*, 1391–1395. [CrossRef]

144. Martelli, D.; Bassini, S.; Utili, M.; Tarantino, M.; Lionetti, S.; Zanin, E. LIFUS II corrosion loop final design and screening of an Al based diffusion coating in stagnant LLE environment. *Fusion Eng. Des.* **2020**, *160*, 112034. [CrossRef]

145. Sannier, J.; Flamment, T.; Terlain, A. Corrosion of Martensitic Steel in Flowing Pb17Li. In *Fusion Technology 1990*; Elsevier: Amsterdam, The Netherlands, 1991; pp. 881–885.

146. Steiner, H.; Krauss, W.; Konys, J. Calculation of dissolution/deposition rates in flowing eutectic Pb-17Li with the MATLIM code. *J. Nucl. Mater.* **2009**, *386–388*, 675–677. [CrossRef]

147. Smolentsev, S.; Saedi, S.; Malang, S.; Abdou, M. Numerical study of corrosion of ferritic/martensitic steels in the flowing PbLi with and without a magnetic field. *J. Nucl. Mater.* **2013**, *432*, 294–304. [CrossRef]

148. Barbier, F.; Alemany, A.; Martemianov, S. On the Influence of a High Magnetic Field on the Corrosion and Deposition Processes in the Liquid Pb-17Li Alloy. *Fusion Eng. Des.* **1998**, *43*, 199–208. [CrossRef]

149. Moreau, R.; Brechet, Y.; Maniguet, L. Eurofer corrosion by the flow of the eutectic alloy Pb-Li in the presence of a strong magnetic field. *Fusion Eng. Des.* **2011**, *86*, 106–120. [CrossRef]

150. Flament, T.; Terlain, A.; Sannier, J.; Labbé, P. Influence of magnetic field on thermohydraulic and corrosion in the case of the water-cooled blanket concept. In Proceedings of the 16th Symposium on Fusion Technology, London, UK, 3–7 September 1991; pp. 911–915.

151. Terlain, A.; Dufrenoy, T. Influence of a magnetic field on the corrosion of austenitic and martensitic steels by semi-stagnant Pb17Li. *J. Nucl. Mater.* **1994**, *212–215*, 1504–1508. [CrossRef]

152. Platacis, E.; Ziks, A.; Poznjak, A.; Muktepavela, F.; Shisko, A.; Sarada, S.; Chakraborty, P.; Sanjay, K.; Vrushank, M.; Fotedar, R.; et al. Investigation of the Li-Pb flow corrosion attack on the surface of P91 steel in the presence of magnetic field. *Magnetohydrodynamics* **2012**, *48*, 343–350.

153. Sree, A.S.; Tanaji, K.; Poulami, C.; Fotedar, R.; Kumar, E.R.; Suri, A.; Platacis, E.; Ziks, A.; Bucenieks, I.E.; Poznjaks, A.; et al. Preliminary corrosion studies of P-91 in flowing lead lithium with and without magnetic field for Indian lead lithium ceramic breeder test blanket module. *Nucl. Fusion* **2014**, *54*, 083029. [CrossRef]

154. Bucenieks, I.; Krishbergs, R.; Platacis, E.; Lipsbergs, G.; Shishko, A.; Zik, A.; Muktepavela, F. Investigation of corrosion phenomena in Eurofer steel in Pb-17Li stationary flow exposed to a magnetic field. *Magnetohydrodynamics* **2006**, *42*, 237–251.
155. Krishbergs, R.; Ligere, E.; Muktepavela, F.; Platacis, E.; Shishko, A.; Zik, A. Experimental studies of the strong magnetic field action on the corrosion of RAFM steels in Pb17Li melt flows. *Magnetohydrodynamics* **2009**, *45*, 289–296. [CrossRef]
156. Gazquez, M.; Hernandez, T.; Muktepavela, F.; Platacis, E.; Shishko, A. Magnetic field effect on the corrosion processes at the Eurofer-Pb-17Li flow interface. *J. Nucl. Mater.* **2015**, *465*, 633–639. [CrossRef]
157. Saeidi, S.; Smolentsev, S. Numerical study of the effect of a magnetic field on corrosion of ferritic/martensitic steel in a turbulent PbLi flow. *Magnetohydrodynamics* **2014**, *50*, 109–120.
158. Saeidi, S.; Smolentsev, S.; Abdou, M. Study of MHD Corrosion of RAFM Steel in Laminar and Turbulent PbLi Flows in a Wall-Normal Magnetic Field. *Fusion Sci. Technol.* **2015**, *68*, 282–287. [CrossRef]
159. Long, J.; Zhang, B.; Yang, B.-W.; Wang, S. Review of researches on coupled system and CFD codes. *Nucl. Eng. Technol.* **2021**, *53*, 2775–2787. [CrossRef]

Article

Magneto-Convective Analyses of the PbLi Flow for the EU-WCLL Fusion Breeding Blanket

Fernando R. Urgorri *, Ivan Fernández-Berceruelo and David Rapisarda

Centre for Energy Environment and Technology Research, National Fusion Laboratory, 28040 Madrid, Spain; ivan.fernandez@ciemat.es (I.F.-B.); david.rapisarda@ciemat.es (D.R.)
* Correspondence: fernando.roca@ciemat.es

Abstract: The Water Cooled Lithium Lead (WCLL) breeding blanket is one of the driver blanket concepts under development for the European Demonstration Reactor (DEMO). The majority of the blanket volume is occupied by flowing PbLi at eutectic composition. This liquid metal flow is subdued to high fluxes of particles coming from the plasma which are translated into a high non-homogeneous heat volumetric source inside the fluid. The heat is removed from the PbLi thanks to several water tubes immersed in the metal. The dynamics of the PbLi is heavily affected by the heat source and by the position of the tubes. Moreover, the conducting fluid is electrically coupled with the intense magnetic field used for the plasma confinement. As a result, the PbLi flow is strongly affected by the Magnetohydrodynamics (MHD) forces. In the WCLL, the MHD and convective interactions are expected to be comparable. Therefore, the PbLi dynamics and consequently the heat transfer between the liquid metal and the water coolant will be ruled by the magneto-convective phenomenon. This work presents 3D computational analyses of the PbLi flow in the frontal region of the WCLL design. The simulations include the combined effect of MHD forces caused by the magnetic field and the buoyancy interaction created by the temperature distribution. The latter is determined by the PbLi dynamics, the volumetric heat source and the position of the water tubes. Simulations have allowed computing the heat transfer between the PbLi and the water tubes. Nusselt and Grashof numbers have been obtained in the different regions of the system.

Keywords: magneto-convection; Magnetohydrodynamics; heat transfer; WCLL

Citation: Urgorri, F.R.; Fernández-Berceruelo, I.; Rapisarda, D. Magneto-Convective Analyses of the PbLi Flow for the EU-WCLL Fusion Breeding Blanket. *Energies* **2021**, *14*, 6192. https://doi.org/10.3390/en14196192

Academic Editors: Marica Eboli and Dan Gabriel Cacuci

Received: 19 July 2021
Accepted: 21 September 2021
Published: 28 September 2021

Publisher's Note: MDPI stays neutral with regard to jurisdictional claims in published maps and institutional affiliations.

1. Introduction

Breeding blankets are crucial systems projected in future nuclear fusion reactors by magnetic confinement whose main purpose is to regenerate the tritium burnt in the plasma. They are designed to absorb the energetic neutrons created in the fusion reactions ($D + T \rightarrow {}^4 He + n$). In the core of the blanket, the neutrons produce tritium by transmuting the lithium contained in the blanket. The amount of tritium bred this way has to be extracted from the reactor for being processed, stored and eventually injected in the plasma. When including a neutron multiplier in the blanket, the tritium bred in the blanket can be higher than the tritium burnt in the plasma which allows maintaining the desired tritium self-sufficiency of the plant.

Within the framework of the EUROfusion blanket project [1,2], the WCLL blanket concept [3] has been selected as one of the two candidates for driver blanket of the European DEMO. This concept is based on an eutectic alloy of PbLi (Pb acting as neutron multiplier and Li acting as breeder) flowing at low velocities. The blanket is directly exposed to highly energetic particle fluxes coming from the blanket. For this reason, liquid water pressurized at 155 bar is used for cooling both the PbLi and the steel structures. A network of water tubes immersed in the liquid metal is included in the design for this purpose.

The PbLi is a good electrical conductor immersed in the strong magnetic field used for the plasma confinement. Therefore, electric currents will be induced inside the flow

201

producing Lorentz forces that alter its dynamics. This interaction known as Magneto-hydrodynamics (MHD) is dominant in most of the WCLL PbLi flowpath. Nevertheless, in the regions closer to the first wall (FW), the high particle fluxes from the plasma are very energetic which are translated into a very high non-homogeneous heat source. This together with the arrangement of the cold water tubes immersed in the PbLi flow can produce temperature differences inside the bulk of the fluid of hundreds of degrees. In these areas, the buoyancy forces are expected to be comparable or even higher than the Lorentz ones.

In this work, magneto-convective simulations of the WCLL blanket are presented. Those are focused on a very particular region of the design: the frontal part of the central outboard elementary cell (Figure 1). This region is located very close to the plasma and it is characterized by a very high heat flux and a consequent dense arrangement of water tubes. This location is of the special interest from the blanket design perspective since it presents a very high tritium production. In its vicinity, it is expected that buoyancy forces play a very important role on PbLi dynamics.

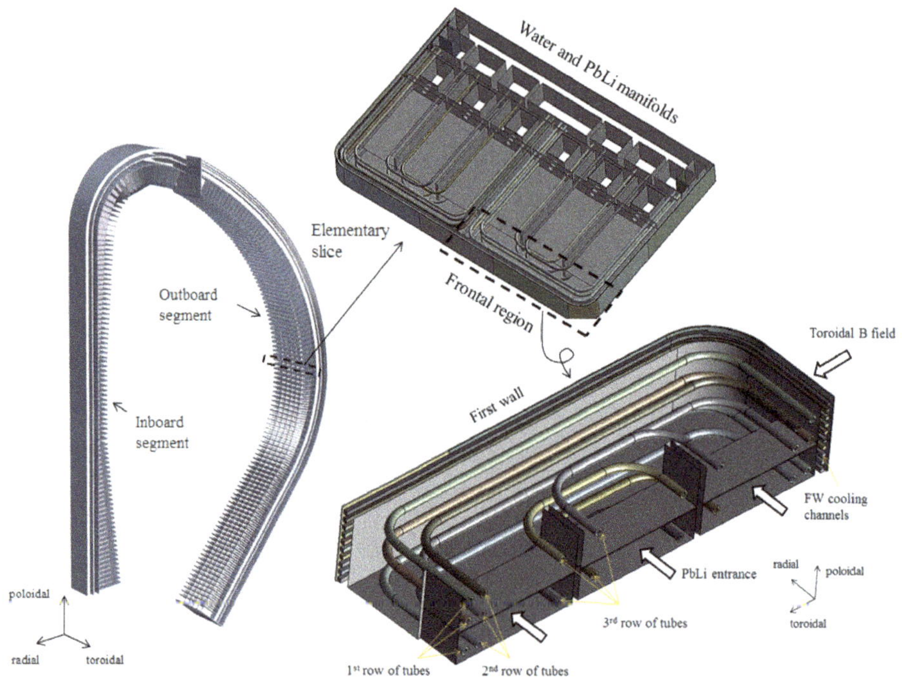

Figure 1. Detailed view of the WCLL blanket design: WCLL segments (**left**), elementary cell (**top-right**), frontal region (**bottom-right**).

The WCLL elementary cell is composed of six parallel circuits along the toroidal direction. Each of them is fed by a rear PbLi manifold and presents a U-shape flowpath along the radial direction. The analyses of this work are focused on the frontal part of one of the six parallel circuits; in the area where the flow describes a 180° turn, close to the FW of the blanket. It is worth noting that there is not a physical separation in between the frontal regions of the six parallel circuits. Therefore, in a central circuit, such as the one considered in this analysis, the lateral radial-poloidal walls only cover the rear part of the domain (the so-called radial channels Figure 2).

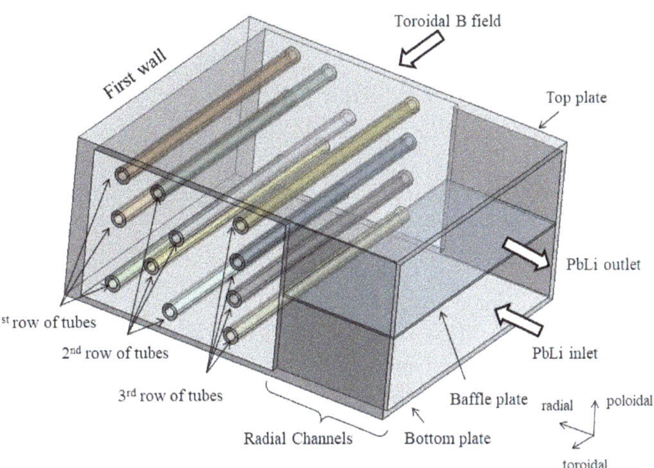

Figure 2. Geometry considered in the magneto-convective simulations.

The numerically analyzed geometry has been simplified. In particular, the shape of the third row of tubes has been straightened, making them completely toroidaly oriented . This way, the studied geometry presents a good symmetry along the toroidal direction which allows using structured meshes in the analyses. Indeed, structured meshes provide better stability in MHD simulations. In the ANSYS-Fluent MHD solver, unstructured meshes can lead to violations of charge conservation forcing extremely small time steps which would rise the computational cost.

The simplification performed does not affect the first and second rows of tubes which have the same shape and position than in the WCLL design. These tubes are closer to the FW where buoyancy forces are expected to be stronger. The square section FW cooling channels are not directly included in the computational domain. It only covers the part of the FW in between the cooling channels and the PbLi.

2. Numerical Model

In the present analyses, the dynamics of the PbLi flow is ruled by the Navier–Stokes equation in which the Lorentz and buoyancy terms are added as a volumetric sources:

$$\rho_0\left(\frac{\partial \vec{u}}{\partial t} + (\vec{u} \cdot \nabla)\vec{u}\right) = \eta \nabla^2 \vec{u} - \nabla p + \vec{j} \times \vec{B}_0 + \rho(T)\vec{g} \tag{1}$$

where the PbLi density and dynamic viscosity are denoted by ρ and η, respectively. The vector field $\vec{u}$ is the velocity of the fluid. The Lorentz force is the cross product of the current density $(\vec{j})$ and the external magnetic field $(\vec{B}_0)$ which is assumed to be constant. $\vec{g}$ represents the gravity vector field and T the temperature field.

Under the inductionless approximation (the magnetic field created by the induced currents is negligible in comparison with the external one), the electric potential (ϕ) is well defined and the generalized Ohm's law can be written as follows:

$$\vec{j} = \sigma(-\nabla\phi + \vec{u} \times \vec{B}_0) \tag{2}$$

where σ is the electrical conductivity of the media. The MHD problem can be closed by applying the divergence to (2) and assuming charge conservation $(\nabla\vec{j} = 0)$. Therefore, a Poisson's equation for the electric potential has to be solved coupled with (1). This equation is solved also in the solid domains. The electric potential and electric currents are continuous across the solid-fluid interfaces.

Concerning the buoyancy forces in the fluid, they are treated using the Boussinesq approximation:

$$\rho(T) \sim \rho_0(1 - \beta(T - T_0))$$ (3)

where T_0 is the characteristic temperature of the problem, ρ_0 is the PbLi density at that temperature and β is the thermal expansion coefficient. The rest of the fluid and solid properties are assumed to be constant.

The temperature field evolves following the energy conservation equation:

$$\rho_0 c_p\left(\frac{\partial T}{\partial t} + u^j\partial_j T\right) = \kappa\partial_j\partial^j T + Q$$ (4)

where c_p is the specific heat capacity of the material and κ its thermal conductivity. The volumetric heat source Q represents the effect of the neutrons and photons (nuclear heating). As a first approximation, this source only depends on the radial coordinate. The shape of this kind of functions are derived from the neutronics analyses performed for the WCLL (e.g., [4]). In this work, the following piecewise functions have been used. These functions provide a steeper power density in the region close to the FW in comparison with the exponential functions used in previous studies which are only accurate in the tail of the curve (e.g., [5,6]).

$$Q(r)_{PbLi}\left[\frac{MW}{m^3}\right] = \begin{cases} 98.962r + 9.5968 & r > -0.05 \\ 365.35r^4 + 44.3r^3 + 215r^2 + 54.376r + 6.8797 & r < -0.05 \end{cases}$$ (5)

$$Q(r)_{steel}\left[\frac{MW}{m^3}\right] = \begin{cases} 9185.7r - 221.25 & r > -0.025 \\ 483.1r^3 + 576.96r^2 + 85.61r + 5.88 & -0.025 > r > -0.075 \\ 253.5r^4 + 296.81r^3 + 142.72r^2 + 35.2r + 3.95 & r < -0.075 \end{cases}$$ (6)

The origin of the radial coordinate is located in the contact plane between the FW and the PbLi. The r coordinate decreases towards the PbLi inlet and grows towards the FW. The shape of the volumetric heat source is plotted in Figure 3.

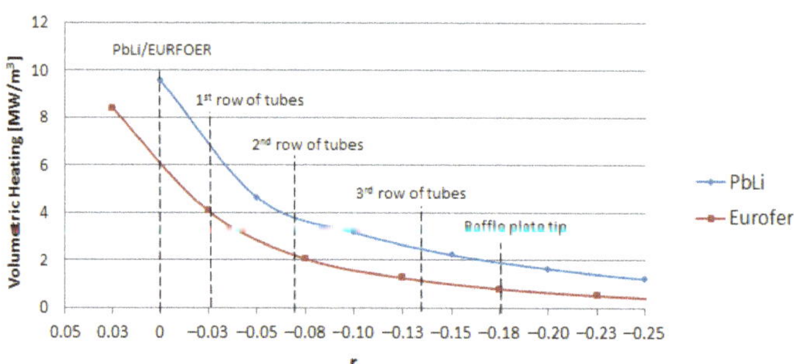

Figure 3. Volumetric heat source used in the computations.

The magnitude of the nuclear heating is of the order of MW/m³. The Ohmic or Joule heating ($\sim j^2/\sigma$) is expected to be several orders of magnitude smaller than the nuclear heating. Indeed, in the WCLL conditions: $u \sim$ mm/s, $B_0 \sim T$ and $\sigma \sim 10^5$ ($\Omega\cdot$ m)$^{-1}$, which means that: $j \sim \sigma u B_0 \sim 10^2$ A/m². Thus, the Joule heating ($\sim 10^{-1}$ W/m³) has been neglected in the present analysis.

In a dimensionless analysis of (1), the square of the Hartmann number (Ha) represents the ratio between Lorentz forces and Viscous forces:

$$Ha := aB_0\sqrt{\frac{\sigma}{\eta}}$$ (7)

where a is the channel semilength along the field direction (toroidal direction). In the central outboard WCLL conditions, the Hartmann number in the frontal part of the blanket is estimated to be approximately 8800. This means that viscous forces will be strongly suppressed by the magnetic field.

Complementary, the Grashof number (Gr) represents the ratio between the buoyancy and viscous forces:

$$Gr := \frac{g\beta\Delta T b^3 \rho^2}{\eta^2} \tag{8}$$

where b is the characteristic length of the heat transfer process, ΔT is the characteristic temperature difference and g is the gravitational acceleration. Estimating Gr in the different regions of the WCLL is not a straightforward task. Indeed, the heat source is extended along the complete computational domain (in a non-homogeneous way) and there are multiple cooling regions. Therefore, both the characteristic length and the temperature difference are not known before performing an specific thermal calculation. It is sometimes found in the literature (e.g., [7,8]) an estimation of the characteristic temperature as $\Delta T = \frac{Qb^2}{\kappa}$, using the semilength along the radial direction as the characteristic length. In the case of study, this strategy would provide an unrealistic temperature difference of several thousands of degrees and $Gr \sim 10^{12}$ or more. This assumption is only valid (although very conservative) when the cooling comes from the sides of the system (e.g., a cooling plate of the Dual Coolant Lithium Lead blanket (DCLL)). Indeed, the expression does not take into account the cooling effect of the internal tubes which will keep temperatures differences in more moderate values, reducing Gr as well. More realistic estimations can be made with this formula considering only the PbLi volume contained in between two rows of tubes. With this strategy, $\Delta T \sim 10^2$ K and $Gr \sim 10^8$.

In purely natural or free magneto-convection problems, buoyant forces are balanced with Lorentz forces. The characteristic velocity scale is given by the ratio between the Gr and Ha^2: $U_0 = \frac{\eta}{\rho_0 L}\frac{Gr}{Ha^2}$ [9]. Alternatively, in mixed-convection problems, the velocity scale is given by the characteristic velocity scale of the pressure driven flow (U_0). In these scenarios, the ratio between Gr and ($Re \cdot Ha^2$) weights the relation between buoyant and Lorentz forces [10]. Indeed, normalizing $\vec{B}$, $\vec{u}$, p, $\vec{j}$, $\vec{g}$ and T by B_0, U_0, $\sigma B_0^2 L U_0$, $\sigma B_0 U_0$, $|\vec{g}|$ and ΔT_0, respectively, Equation (1) can be written as follows:

$$\frac{Re}{Ha^2}\left(\frac{\partial \hat{u}}{\partial t} + (\hat{u} \cdot \nabla)\hat{u}\right) = \frac{1}{Ha^2}\nabla^2\hat{u} - \nabla\hat{p} + \hat{j} \times \hat{B}_0 + \frac{Gr}{ReHa^2}\hat{T}\hat{g} \tag{9}$$

where the symbol ^ represents a normalized variable (e.g., $\hat{u} = \vec{u}/U_0$). Inertial effects are weighted to the Lorentz forces with the usual interaction parameter ($N - \frac{Ha^2}{Re}$) or by the square of the so-called Lykoudis number (Ly) in natural convection problems: $Ly^2 = \frac{Ha^4}{Gr}$. In such problems, only for Gr of the order of Ha^4, inertial effects are significant.

In natural convection situations, the Péclet number ($Pe = Re \cdot Pr$) can be related also with Gr and Ha: $Pe = \frac{Gr}{Ha^2}Pr$. Therefore, the advective heat transfer is strongly suppressed by the magnetic field. This effect is also true on mixed-convection problems although a sufficiently high Re can as usual make advection a significant heat transfer mechanism.

Both natural and mixed-convection scenarios have been studied in the past for different combinations of Ha, Re and Gr under the influence of a transversal magnetic field. MHD flows in vertical channels with a non-homogeneous heat source and insulated walls have been studied within the framework of the DCLL blanket [11,12]. Above a critical value of Ha, the magnetic field is able to stabilize both the buoyancy-assisted (upward) flow [13] and buoyancy-opposed (downward) flow [14]. The latter requires more intense magnetic fields for the stabilization.

In the case of horizontal ducts, both natural convection [15] and mixed-convection scenarios [10] have been studied considering heated surfaces. Volumetric heat sources in horizontal channels have been studied for natural-convection regimes within the framework of the Helium Cooled Lithium Lead (HCLL) blanket concept analyses [16]. In this case, the cooling plates and the spatially varying heat source originate the flow movement. More recently, studies dedicated to the WCLL blanket show that the presence of the cooling tubes immersed in the PbLi also plays an important role on determining the buoyant recirculation patterns [17,18].

The strong magnetic field and heat source present in the frontal part of the WCLL blanket implies a very challenging problem from the computational point of view. Being able to resolve the MHD boundary layers while capturing the dynamics of the buoyant vortexes requires very small mesh sizes and very small time steps in a complex geometry. To relax the computational requirements, this work considers reduced Ha values. The final objective would be to gradually increase Ha until reproducing the real blanket conditions. In this paper, two different situations are compared: $Ha = 1000$ and $Ha = 2000$. In each case, the external magnetic field has been tuned accordingly.

To keep the ratio between Lorentz and buoyant forces as similar as possible to the WCLL conditions, the gravity field has been reduced as well keeping the ratio between Gr and Ha^2 constant. For this purpose, the reduced gravity field (g_r) and reduced magnetic field (B_r) follow the following scaling rule:

$$\frac{g_r}{g_0} = \left(\frac{B_r}{B_0}\right)^2 \tag{10}$$

Equation (10) will only keep the ratio $\frac{Gr}{Ha^2}$ constant if the characteristic temperature difference of the reduced case (ΔT_r) is equal to real ΔT. Since the heat source is the same, this is expected to be approximately true. Results obtained in Section 3 for different values of B_r and g_r are consistent with this approximation.

Simulation Conditions

The material properties of the PbLi [19] and the steel (EUROFER [20]) at $T_0 = 600$ K are exposed in Table 1.

Table 1. PbLi and EUROFER properties at $T_0 = 600$ K.

	PbLi	**EUROFER**
ρ (kg/m^3)	9806	7674
c_p (J/kg· K)	189.5	565
κ (W/m· K)	20.93	29.74
σ (Ω^{-1}· m^{-1})	7.82×10^5	1.07×10^6
β (K^{-1})	1.21×10^{-4}	—
η (Pa· s)	1.93×10^{-3}	—

Additionally, the geometrical inputs taken from the WCLL frontal region are exposed in Figure 4. The toroidal length of the domain is 243 mm.

Regarding the boundary conditions, convective ones ($q = h(T - T_f)$) have been used in the internal surfaces of the tubes and in the external side of the FW to represent the cooling effect of the water. The water stream temperatures (T_f) of both circuits are assumed at 311.5 °C. This value is the average temperature of the water thermal cycle [18]. The heat transfer coefficients (h) are 11,175 W/m^2 K for the tubes and 22,012 W/m^2 K for the FW. The values are obtained using the Dittus-Boelter correlation taking into account the design water velocities of both circuits. Electrically insulated boundary conditions ($\partial_n \phi = 0$) are applied in these surfaces as well. In the inlet channel, the PbLi enters at a constant temperature $T = 650$ K and with a constant velocity $U_0 = 0.2$ mm/s.

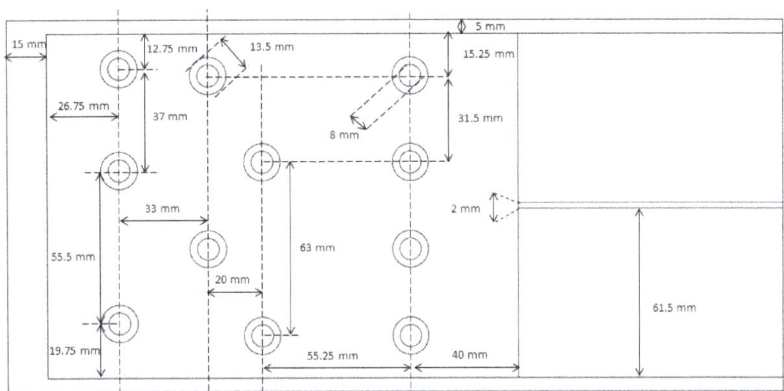

Figure 4. Geometrical input data.

Periodic boundary conditions are applied in the top and bottom steel plates to reproduce the presence of an analogous WCLL cell at the top and at the bottom of the studied one (translational symmetry). Likewise, periodic boundary conditions are applied in the lateral radial-poloidal surfaces since the PbLi parallel circuits stacked along the toroidal direction are supposed to be thermally similar and circulations between parallel circuits are expected to be negligible.

In the internal PbLi/steel surfaces, continuity of the temperature, heat flux, electric potential and electric currents are considered.

Concerning the initial conditions, a pure conductive heat transfer model (treating the PbLi as a solid material) has been used to obtain an initial temperature map. Using this map as the initial condition reduces the time needed to reach relevant conditions, which reduces significantly the computational time. This map is exposed in Figure 5. In agreement with previous studies [21], the hottest region is located at the end of the radial channels while the cooling tubes keep the frontal region at moderate temperatures (600–700 K).

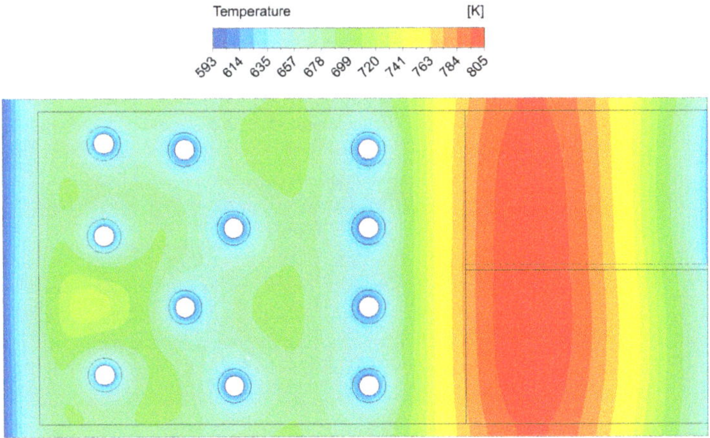

Figure 5. Initial temperature field obtained with a pure conductive model.

As mentioned in Section 1, the geometry has been simplified in order to employ a structured mesh. A multi-block structured mesh has been designed for this purpose using ICEM-CFD. O-grid structures are included in the vicinity of the tubes. After some testing, it was preferred to extend the radial elements of the tube walls towards the fluid domain in order to ensure good resolution in the boundary layers close to the tubes. For the same reasons, a hyperbolic clustering of cells has been applied towards the walls. Figure 6 depicts a zoomed view of the computational mesh used for the $Ha = 1000$ case.

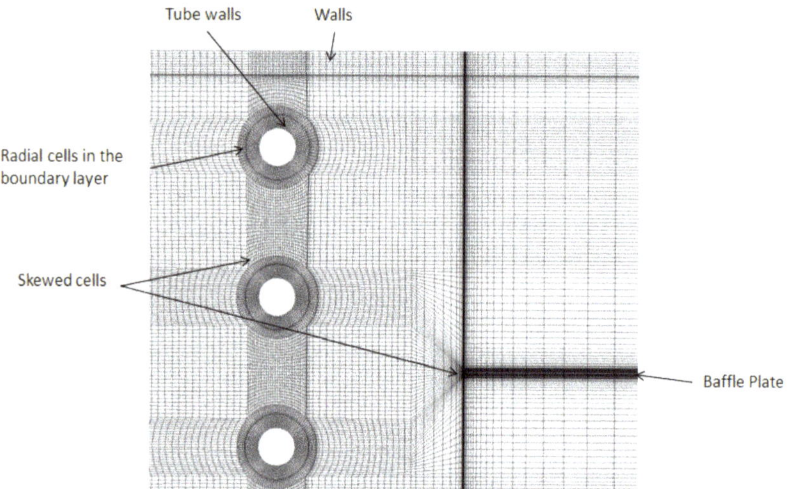

Figure 6. Detailed view of the computational mesh used for the $Ha = 1000$ case.

3. Computational Results

Transient magneto-convective simulations have been performed using the MHD solver integrated in the ANSYS-Fluent platform. This solver has been tested against the experimental results in a NaK loop [22] under pure MHD conditions. Moreover, a benchmarking exercise with another 4 MHD codes was successfully conducted under magneto-convective conditions [23].

Results are presented for the $Ha = 1000$ case and for the $Ha = 2000$ case after 200 s and 100 s, respectively. The gravity field has been scaled according with (10). In both cases, the 3D solutions obtained present a quite good symmetry along the toroidal direction. This is expected not only because of the toroidal symmetry of the geometry but also because the magnetic field tends to align the convective vortexes along its direction. This kind of behavior is called quasi-two-dimensional (Q2D) turbulence [24]. For simplicity, results are presented only in the central radial-poloidal plane. Small deviation from these results can be found in other radial-poloidal planes but they are not significant. Weak temperature toroidal dependence has been found as well in hydrodynamics works in the regions where the tubes are toroidally oriented [21,25].

Figure 7 depicts a comparison between the temperature distribution of the purely conducting model (or the initial condition) and the results for $Ha = 1000$ and $Ha = 2000$ cases in the frontal part of the domain. Convective heat-transfer distorts the conductive temperature map. However, the effect is quite moderate which points to a heat transfer scenario dominated by conduction.

In any case, the flow motion boosts heat transfer near the cooling tubes decreasing peak and average temperatures. High temperature regions are slightly displaced towards the upper part of the domain due to the buoyancy force. There are relatively small but appreciable differences between both magneto convective cases. This indicates that even when keeping the overall ratio between buoyancy and Lorentz forces, local effects play a role on heat transfer. For example, higher velocity jets are developed next to the conducting walls at higher Ha numbers.

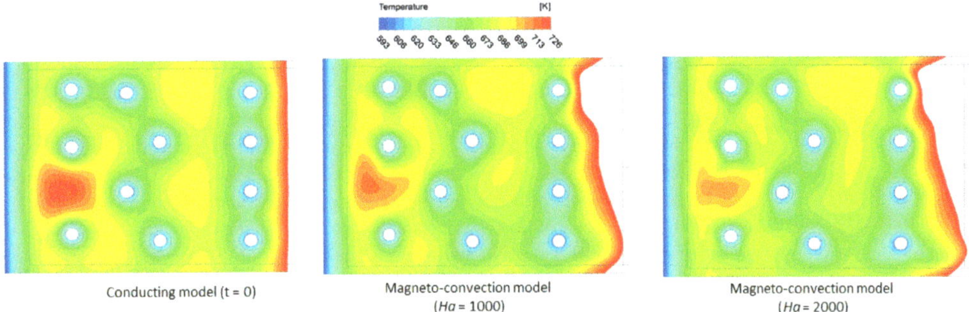

Conducting model (t = 0)   Magneto-convection model (Ha = 1000)   Magneto-convection model (Ha = 2000)

Figure 7. Comparison between the temperature distribution of the conducting and the magneto-convection model close to the FW.

The computed velocity vector field is exposed in Figure 8. In the first and second row of tubes, medium sized vortexes, (size comparable with the tube diameter) appear at both sides of each tube. Similar structures also appear next to the FW as a result of being the only cooled wall of the system. The vector field is qualitatively rather similar in both cases. Quantitatively, the velocity scale is of the order of $\sim 10^{-3}$ m/s ($Re \sim 10^2$) in both cases as well. However, peak velocities are higher (factor 2) in the $Ha = 2000$ case as a result of the electrical interaction between the fluid and conducting tube walls.

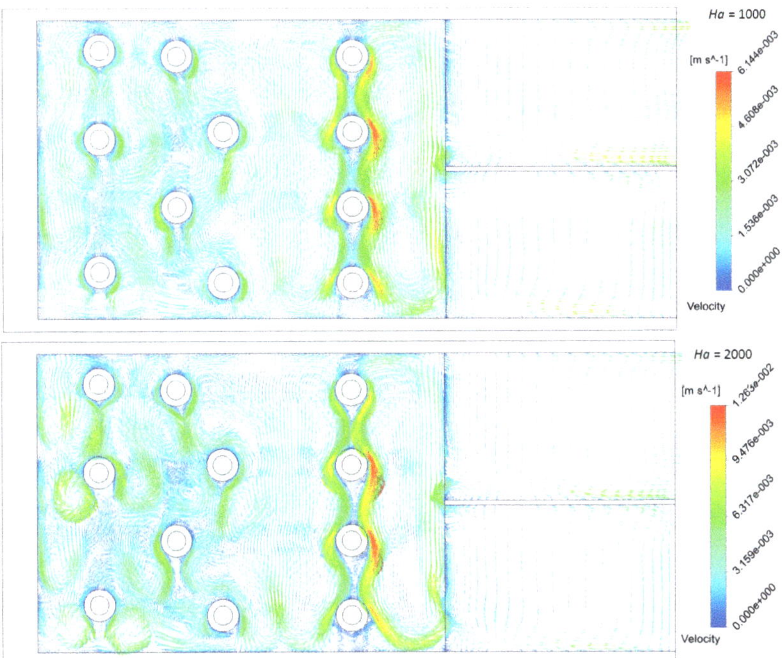

Figure 8. Velocity vector field in the central radial-poloidal plane for the $Ha = 1000$ case (**top**) and $Ha = 2000$ case (**bottom**).

Results also show a weak connection between the pressure driven flow in the radial channels and the rest of the domain. Indeed, a big recirculation region is developed in between the third row of tubes and the radial channels blocking the flow. As a result, the majority of the flow that comes from the inlet channel goes to the outlet channel through a narrow region close to the tip of the baffle plate. This effect was observed in pure MHD flows in simpler geometries that consider no tubes [26]. Recirculations regions are also observed in both the inlet and outlet radial channels. To confirm their apparation, the recirculation in the radial channels needs to be further investigated. In the present analyses, the proximity of the PbLi inlet and outlet faces might be introducing unrealistic effects. Other studies of the WCLL [17] predict recirculations in the radial channels as well. However, those seem to be less pronounced. The dissimilarities might be caused by differences in the heat source or the geometrical approximations employed in both works. Recirculations in the radial channels have been also computed for the HCLL TBM [16]. Nevertheless, the latter work considers a natural magneto-convection flow influenced by the HCLL horizontal cooling plates. These conditions are significantly different than in the WCLL radial channels.

Figure 9 depicts the electric potential distribution in both cases. Electric potential differences arises at both sides of the tubes. Comparing the potential contours with the vector velocity field, it is observed that local maxima and minima of potential are located in the center of the vortical structures. The vortexes extend along the whole toroidal direction of the computational domain. The alignment of the vortexes with the magnetic field direction is a characteristic of Q2D flows. In this flows, the contours of the electric potential coincide approximately with the stream lines of the flow. Indeed, assuming that the flow can be described in 2D, the electric potential is proportional to the stream function (ψ):

$$\nabla_{2D}^2 \phi = \nabla_{2D}(\vec{u} \times \vec{B_0}) = B_0(\partial_x u_y - \partial_y u_x) = -B_0 \nabla_{2D}^2 \psi \tag{11}$$

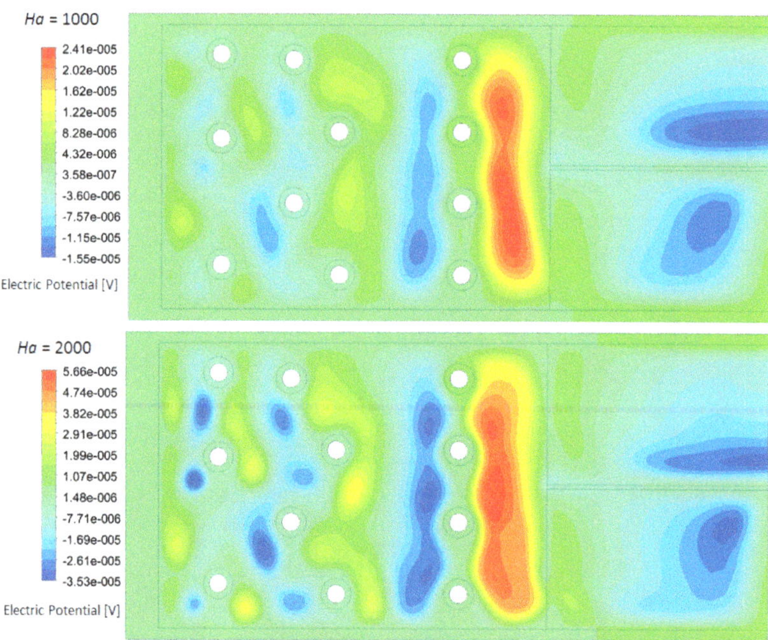

Figure 9. Electric potential in the central radial-poloidal plane for the $Ha = 1000$ case (**top**) and $Ha = 2000$ case (**bottom**).

3D potential iso-surfaces are exposed in Figure 10. The iso-surfaces are a good representation of the Q2D flow whose vortical structures are elongated along the magnetic field direction. This behavior is allowed and enhanced by symmetry of the geometry along this direction. This kind of elongated potential isosurfaces and their relation with the velocity stream-lines have been also obtained for the former EU-HCLL TBM conditions where the cooling of the PbLi was made by some radial-toroidal cooling plates [16].

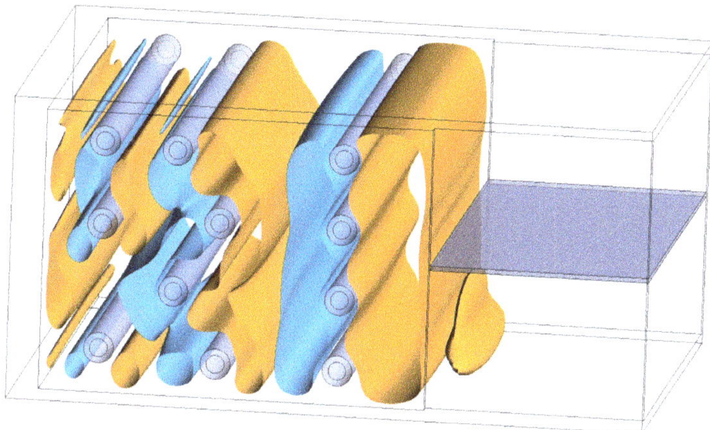

Figure 10. Isosurfaces of the electric potential ($\pm 10^{-5}$ V).

The sizes of the potential isosurfaces are dependent on the ratio $\frac{Gr}{Ha^2}$. Increasing the magnetic field while keeping Gr constant is expected to reduce the size of the vortexes [11].

4. Estimation of the Nusselt Number

Estimations of the Nusselt number (Nu) are of great interest for design purposes. Knowing Nu or the heat transfer coefficient h that appears in its definition allows performing system level heat transfer analyses. Nu represents the ratio between the convective and conductive processes in the fluid-solid interface:

$$Nu := \frac{h \cdot L}{\kappa} \tag{12}$$

The heat transfer coefficient can be estimated from the computational results by evaluating the average heat flux at the surface (q_s [W/m^2]) and the superficial temperature (T_s) of each tube (and the FW) of the system:

$$h = \frac{q_s}{T_s - T_\infty} \tag{13}$$

The temperature T_∞ is defined as a temperature sufficiently far away from the interface. Before computing, it is always necessary to specify the way in which the characteristic length (L) and the temperature far away (T_∞) have been defined. In this case, T_∞ has been defined as the average temperature of a cylindrical surface (a plane in the case of the FW) at a distance $2L$ from the tube of interest. The tube surface and the cylindrical surface form a hollow cylinder (a prism in the case of the FW). The characteristic length is defined as the ratio between the volume and the area of external surfaces ($L = V/S$).

The distance between the tube and cylindrical surface ($2L$) has been picked using the purely conducting model. Indeed, it is defined as the minimum length needed to obtain $Nu = 1$ with the conductive model results. This way, the definition is consistent with the extreme case of purely conductive heat transfer scenario. As a consequence of this definition, the characteristic length is slightly different for each tube but it is around 10 mm in every case which is roughly half the distance between tubes.

Table 2 depicts the heat transfer coefficients and Nu computed from the simulations results. The tubes of the different rows are labeled from the top to the bottom of the WCLL cell. This way the tube 1 of the row 1 is the tube located in the top left corner of Figure 4 and tube 4 of row 3 the one located in the bottom right corner.

Table 2. Average heat transfer coefficient and *Nu* in the fluid-solid interfaces.

		$Ha = 1000$		$Ha = 2000$	
		h (W/m^2 K)	*Nu*	h (W/m^2 K)	*Nu*
FW		4121	1.05	4344	1.10
	tube 1	4802	1.02	5221	1.11
1st row of tubes	tube 2	4612	1.03	5078	1.14
	tube 3	4643	1.05	5213	1.18
	tube 1	4332	1.06	4778	1.18
2nd row of tubes	tube 2	3983	1.04	4418	1.15
	tube 3	4273	1.08	4568	1.16
	tube 4	3931	1.02	4322	1.12
	tube 1	4271	1.23	4735	1.37
3rd row of tubes	tube 2	3847	1.11	4083	1.18
	tube 3	3845	1.11	4063	1.18
	tube 4	3569	1.03	3869	1.12

In agreement with the qualitative conclusions deduced from the obtained temperature maps (Figure 7), the quantitative analyses confirm that heat transfer is mostly ruled by conduction in the studied region. Indeed, *Nu* is very close to unity for all interfaces. There are small differences between the tubes and it is clear that the third row of tubes is slightly more affected by convection than the others. The third row of tubes is closer to the radial channels and therefore it is more influenced by the PbLi flow coming from them. In the other tubes, the flow is essentially moving by natural or free convection and unaffected by the flow of the channels.

It can also be deduced that *Nu* is around 5–10% higher when increasing *Ha* (and *Gr*). This small difference is related with the increase of the velocity jets in the vicinity of the conducting walls with *Ha*. In the real blanket conditions (*Ha* $\sim$ 8800 and the real gravity field) the heat transfer by convection might play a more important role but it is not expected to be comparable with the conducting mechanism.

5. Estimation of the Grashof Number

The value of the Grashof number is dependent on the characteristic length (*b*) and temperature differences (ΔT) picked in the definition (8). In the case of study this definition is not trivial since there are multiple heat sinks (each tube and the cooled FW) and the heat source is extended along the whole domain in a non-homogeneous way.

A conservative approach is considering the maximum temperature difference inside the PbLi (approximately 180 K) and the total radial length of the frontal cavity as characteristic length. With this strategy it is obtained that $Gr_r = 1.59 \times 10^9$ in the $Ha = 2000$ case and $Gr = 2.93 \times 10^{10}$ in WCLL conditions. If the system were ruled by pure natural convection heat transfer, the Reynolds number would be $Re = \frac{Gr}{Ha^2} = 4 \times 10^2$ which is of the order of the average *Re* obtained in the simulations.

Alternatively, a local definition can be used based on the temperature differences between some previously defined toroidal-poloidal planes located at different radial positions. This definition is motivated by tubes disposition in rows and by the radial dependence of the heat source.

The planes are defined in the middle of each tube row and in between them. This means that distances between two adjacent planes are approximately the size of the Q2D vortexes obtained in the simulations. Multiple values of *Gr* have been derived considering the differences between the planes average temperatures (ΔT) and the distance (b) between them. The Gr computed this way are presented in Figure 11 for the $Ha = 2000$ case.

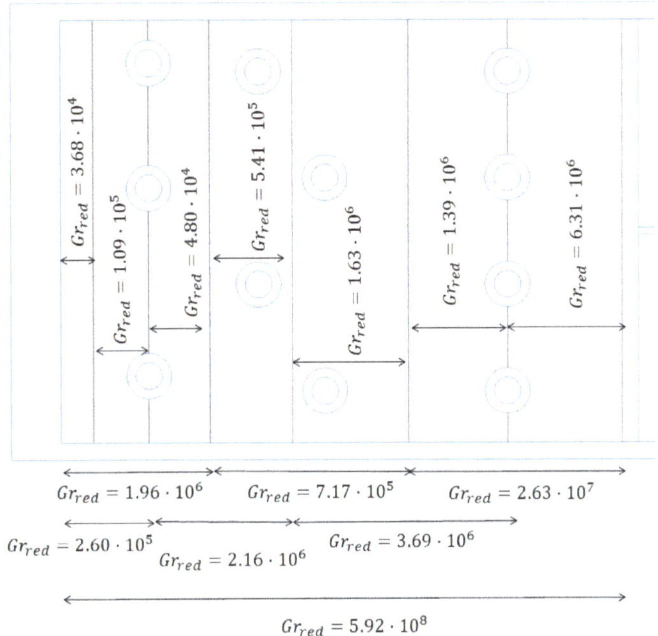

Figure 11. Reduced Grashof number for different toroidal-poloidal planes.

The values presented are calculated for a reduced gravity field (10). For obtaining the results in the real WCLL conditions, Gr has to be multiplied by the square of the ratio between target Ha (8800) and the reduced Ha (2000), in other words 4.4^2. After the re-scaling, the values obtained in the real conditions vary between 10^5 and 10^{10}, depending on the planes considered. Most likely, the most relevant Gr are the ones computed using two adjacent planes since this is the size of the convective vortexes associated to the problem. This implies a maximum value of 1.22×10^8 in between the third row of tubes and the radial channels and a minimum value of 7.13×10^5 in between the FW plane and the first plane. It is worth noting that the Gr obtained in between two rows of tubes is of the order of 10^7 which is one order of magnitude below the initial estimations.

6. Conclusions

This work presents a magneto-convective simulation of the EU-WCLL blanket central outboard elementary cell. The analyses are focused on the frontal region which is close to the FW and subjected to a very high volumetric heat source and to a very high magnetic field. Geometrically, this region is characterized by a group of cooling tubes that crosses the fluid domain mostly in the perpendicular direction to the flow (parallel to the magnetic field). Simulations have been implemented using the MHD solver of ANSYS-Fluent.

According to the results, the dynamics of this region is driven mostly by natural convection. Little influence of the colder radial PbLi flow was found in the frontal region. Indeed, most of the pressure driven flow goes from the inlet to the outlet channel very close to the tip of the baffle plate interacting weakly with the rest of the domain. Medium size vortexes (dimension of the order of the tube diameter) appear in the frontal part of the domain. Having vortexes isolated in the frontal region can have important implications for tritium transport since high concentrations can arise in these regions.

Heat transfer is mostly driven by conduction in the system. Indeed, temperature profiles are relatively similar to the purely conductive solution used as an initial condition. Hot spots are a bit distorted and displaced towards the top part of the domain but not in a strong way. This effect is a consequence of the magnetic field that suppresses the heat transfer via advection. The suppression of the advective heat transfer mechanism by the magnetic field was an expected result in agreement with previous experiences. Despite the conductive nature of the heat transfer, the dense network of water tubes are able to keep the PbLi and more importantly the structural steel below critical temperatures (550 °C).

Nu numbers and heat transfer coefficients have been computed for each cooling tube and for the FW. Values very close to unity have been obtained for Nu in every case. Gr numbers have been also estimated using the simulations outcomes. A global Gr of $\sim 2.93 \times 10^{10}$ has been obtained from the maximum temperature difference obtained in the WCLL cell. Local estimations of Gr in between rows of tubes ($\sim$size of the vortexes) provides a value two or three orders of magnitude smaller: $Gr \sim 10^7, 10^8$.

Future developments should be focused on reaching the actual values of the magnetic field and gravity field in the blanket. Moreover, the influence of the tube curvature should be evaluated since they will break the toroidal symmetry and possibly the Q2D structures. In fact, it is expected that curvature of the third row of tubes will break the toroidal extension of the bigger vortexes next to the baffle plate. Moreover, some of this tubes penetrate in the PbLi radial channels which will mitigate the hot spot located next to the end of the baffle plate. The effects could significantly affect the recirculation found in the radial channels. This circumstance has to be further investigated.

To include the tube curvature, unstructured tetrahedral elements will be most likely unavoidable. An investigation of the numerical stability of magneto-convective problems with this kind of meshes should be performed. Finally, different orientations of the gravity and the magnetic field should be analyzed. This would allow studying other scenarios in WCLL cells different from the central one. These have variable orientations with respect to the horizontal plane. Besides, the poloidal component of the magnetic field might play also a significant role on magneto-convective results.

Author Contributions: Conceptualization, F.R.U. and I.F.-B.; methodology, F.R.U.; software, F.R.U.; formal analysis, F.R.U.; writing—original draft preparation, F.R.U.; writing—review and editing, I.F.-B. and D.R.; visualization, F.R.U. and I.F.-B.; supervision, I.F.-B. and D.R.; project administration, D.R.; funding acquisition, D.R. All authors have read and agreed to the published version of the manuscript.

Funding: This work has been carried out within the framework of the EUROfusion Consortium and has received funding from the Euratom research and training programme 2014–2018 and 2019–2020 under grant agreement No 633053. The views and opinions expressed herein do not necessarily reflect those of the European Commission.

Institutional Review Board Statement: Not applicable.

Informed Consent Statement: Not applicable.

Data Availability Statement: Not applicable.

Acknowledgments: This work was partially supported by the computing facilities of Extremadura Research Centre for Advanced Technologies (CETA-CIEMAT), funded by the European Regional Development Fund (ERDF). CETA-CIEMAT belongs to CIEMAT and the Government of Spain. The authors thank Alessandro Del Nevo and the rest of the WCLL design team for providing the CAD files, heat source functions and boundary values needed for this work.

Conflicts of Interest: The authors declare no conflict of interest.

Abbreviations

The following abbreviations are used in this manuscript:

CFD	Computational Fluid Dynamics
DCLL	Dual Coolant Lithium Lead
DEMO	Demonstration Reactor
FW	First Wall
HCLL	Helium Cooled Lithium Lead
MHD	MagnetoHydroDynamics
Q2D	Quasi-two-dimensional
TBM	Test Blanket Module
WCLL	Water Cooled Lithium Lead

References

1. Boccaccini, L.; Aiello, G.; Aubert, J.; Bachmann, C.; Barrett, T.; Del Nevo, A.; Demange, D.; Forest, L.; Hernandez, F.; Norajitra, P.; et al. Objectives and status of EUROfusion DEMO blanket studies. *Fusion Eng. Des.* **2016**, *109–111*, 1199–1206. [CrossRef]
2. Federici, G.; Boccaccini, L.; Cismondi, F.; Gasparotto, M.; Poitevin, Y.; Ricapito, I. An overview of the EU breeding blanket design strategy as an integral part of the DEMO design effort. *Fusion Eng. Des.* **2019**, *141*, 30–42. [CrossRef]
3. Del Nevo, A.; Arena, P.; Caruso, G.; Chiovaro, P.; Di Maio, P.; Eboli, M.; Edemetti, F.; Forgione, N.; Forte, R.; Froio, A.; et al. Recent progress in developing a feasible and integrated conceptual design of the WCLL BB in EUROfusion project. *Fusion Eng. Des.* **2019**, *146*, 1805–1809. [CrossRef]
4. Moro, F.; Arena, P.; Catanzaro, I.; Colangeli, A.; Del Nevo, A.; Flammini, D.; Fonnesu, N.; Forte, R.; Imbriani, V.; Mariano, G.; et al. Nuclear performances of the water-cooled lithium lead DEMO reactor: Neutronic analysis on a fully heterogeneous model. *Fusion Eng. Des.* **2021**, *168*, 112514. [CrossRef]
5. Smolentsev, S.; Morley, N.B.; Abdou, M. Magnetohydrodynamic and Thermal Issues of the SiCf/SiC Flow Channel Insert. *Fusion Sci. Technol.* **2006**, *50*, 107–119. [CrossRef]
6. Urgorri, F.R.; Smolentsev, S.; Fernández-Berceruelo, I.; Rapisarda, D.; Palermo, I.; Ibarra, A. Magnetohydrodynamic and thermal analysis of PbLi flows in poloidal channels with flow channel insert for the EU-DCLL blanket. *Nucl. Fusion* **2018**, *58*, 106001. [CrossRef]
7. Bühler, L.; Mistrangelo, C. MHD flow and heat transfer in model geometries for WCLL blankets. *Fusion Eng. Des.* **2017**, *124*, 919–923. [CrossRef]
8. Mistrangelo, C.; Bühler, L. Numerical Study of Fundamental Magnetoconvection Phenomena in Electrically Conducting Ducts. *IEEE Trans. Plasma Sci.* **2012**, *40*, 584–589. [CrossRef]
9. Müller, U.; Bühler, L. *Magnetohydrodynamics in Channels and Containers*; Springer: Berlin/Heidelberg, Germany, 2001.
10. Zhang, X.; Zikanov, O. Mixed convection in a horizontal duct with bottom heating and strong transverse magnetic field. *J. Fluid Mech.* **2014**, *757*, 33–56. [CrossRef]
11. Smolentsev, S.; Vetcha, N.; Abdou, M. Effect of a magnetic field on stability and transitions in liquid breeder flows in a blanket. *Fusion Eng. Des.* **2013**, *88*, 607–610. [CrossRef]
12. Vetcha, N.; Smolentsev, S.; Abdou, M.; Moreau, R. Study of instabilities and quasi-two-dimensional turbulence in volumetrically heated magnetohydrodynamic flows in a vertical rectangular duct. *Phys. Fluids* **2013**, *25*, 024102. [CrossRef]
13. Vetcha, N.; Smolentsev, S.; Abdou, M. Theoretical Study of Mixed Convection in Poloidal Flows of DCLL Blanket. *Fusion Sci. Technol.* **2009**, *56*, 851–855. [CrossRef]
14. Vetcha, N.; Smolentsev, S.; Abdou, M. Stability Analysis for Buoyancy-Opposed Flows in Poloidal Ducts of the DCLL Blanket. *Fusion Sci. Technol.* **2011**, *60*, 518–522. [CrossRef]
15. Zikanov, O.; Listratov, Y.; Sviridov, V. Natural convection in horizontal pipe flow with a strong transverse magnetic field. *J. Fluid Mech.* **2013**, *720*, 486–516. [CrossRef]
16. Mistrangelo, C.; Bühler, L.; Aiello, G. Buoyant-MHD Flows in HCLL Blankets Caused by Spatially Varying Thermal Loads. *IEEE Trans. Plasma Sci.* **2014**, *42*, 1407–1412. [CrossRef]
17. Yan, Y.; Ying, A.; Abdou, M. Numerical study of magneto-convection flows in a complex prototypical liquid-metal fusion blanket geometry. *Fusion Eng. Des.* **2020**, *159*, 111688. [CrossRef]
18. Tassone, A.; Caruso, G.; Giannetti, F.; Del Nevo, A. MHD mixed convection flow in the WCLL: Heat transfer analysis and cooling system optimization. *Fusion Eng. Des.* **2019**, *146*, 809–813. [CrossRef]
19. Martelli, D.; Venturini, A.; Utili, M. Literature review of lead-lithium thermophysical properties. *Fusion Eng. Des.* **2019**, *138*, 183–195. [CrossRef]
20. Guillemot, F. *Material Property Handbook Pilot Project on EUROFER97 (MTA EK, KIT)*; Technical Report EFDA-D-2MRP77; EUROfusion: Garching, Germany, 2016.
21. Edemetti, F.; Martelli, E.; Tassone, A.; Caruso, G.; Nevo, A.D. DEMO WCLL breeding zone cooling system design: Analysis and discussion. *Fusion Eng. Des.* **2019**, *146*, 2632–2638. [CrossRef]

22. Satyamurthy, P.; Swain, P.; Tiwari, V.; Kirillov, I.; Obukhov, D.; Pertsev, D. Experiments and numerical MHD analysis of LLCB TBM Test-section with NaK at 1T magnetic field. *Fusion Eng. Des.* **2015**, *91*, 44–51. [CrossRef]
23. Smolentsev, S.; Rhodes, T.; Yan, Y.; Tassone, A.; Mistrangelo, C.; Bühler, L.; Urgorri, F.R. Code-to-Code Comparison for a PbLi Mixed-Convection MHD Flow. *Fusion Sci. Technol.* **2020**, *76*, 653–669. [CrossRef]
24. Sommeria, J.; Moreau, R. Why, how, and when, MHD turbulence becomes two-dimensional. *J. Fluid Mech.* **1982**, *118*, 507–518. [CrossRef]
25. Edemetti, F.; Martelli, E.; Del Nevo, A.; Giannetti, F.; Arena, P.; Forte, R.; Di Maio, P.A.; Caruso, G. On the impact of the heat transfer modelling approach on the prediction of EU-DEMO WCLL breeding blanket thermal performances. *Fusion Eng. Des.* **2020**, *161*, 112051. [CrossRef]
26. Tassone, A.; Caruso, G.; Del Nevo, A.; Di Piazza, I. CFD simulation of the magnetohydrodynamic flow inside the WCLL breeding blanket module. *Fusion Eng. Des.* **2017**, *124*, 705–709. [CrossRef]

MDPI

Article

Verification and Validation of COMSOL Magnetohydrodynamic Models for Liquid Metal Breeding Blankets Technologies

Ciro Alberghi [1,*], **Luigi Candido** [1], **Raffaella Testoni** [1], **Marco Utili** [2] **and Massimo Zucchetti** [1,3]

[1] ESSENTIAL Group, Politecnico di Torino—Corso Duca degli Abruzzi, 24, 10129 Torino, Italy; luigi.candido@polito.it (L.C.); raffaella.testoni@polito.it (R.T.); massimo.zucchetti@polito.it or zucchett@mit.edu (M.Z.)

[2] ENEA FSN-PROIN, C.R. Brasimone—Località Brasimone, 40043 Camugnano, Italy; marco.utili@enea.it

[3] PSFC, MIT, Massachusetts Institute of Technology, Cambridge, MA 02139, USA

* Correspondence: ciro.alberghi@polito.it

Abstract: Liquid metal breeding blankets are extensively studied in nuclear fusion. In the main proposed systems, the Water Cooled Lithium Lead (WCLL) and the Dual Coolant Lithium Lead (DCLL), the liquid metal flows under an intense transverse magnetic field, for which a magnetohydrodynamic (MHD) effect is produced. The result is the alteration of all the flow features and the increase in the pressure drops. Although the latter issue can be evaluated with system models, 3D MHD codes are of extreme importance both in the design phase and for safety analyses. To test the reliability of COMSOL Multiphysics for the development of MHD models, a method for verification and validation of magnetohydrodynamic codes is followed. The benchmark problems solved regard steady state, fully developed flows in rectangular ducts, non-isothermal flows, flow in a spatially varying transverse magnetic field and two different unsteady turbulent problems, quasi-two-dimensional MHD turbulent flow and 3D turbulent MHD flow entering a magnetic obstacle. The computed results show good agreement with the reference solutions for all the addressed problems, suggesting that COMSOL can be used as software to study liquid metal MHD problems under the flow regimes typical of fusion power reactors.

Keywords: liquid metal blanket; MHD benchmarking; COMSOL multiphysics; magneto-convection; turbulent MHD; large eddy simulations

Citation: Alberghi, C.; Candido, L.; Testoni, R.; Utili, M.; Zucchetti, M. Verification and Validation of COMSOL Magnetohydrodynamic Models for Liquid Metal Breeding Blankets Technologies. *Energies* **2021**, *14*, 5413. https://doi.org/10.3390/en14175413

Academic Editor: Anouar Belahcen

Received: 21 July 2021
Accepted: 26 August 2021
Published: 31 August 2021

Publisher's Note: MDPI stays neutral with regard to jurisdictional claims in published maps and institutional affiliations.

1. Introduction

In the nuclear fusion framework, the breeding blanket is a key system devoted to power extraction, shielding and tritium production. Among the different designs of the blanket, in the Water-Cooled Lithium-Lead (WCLL) breeding blanket of DEMO [1] and in the WCLL test blanket module of ITER, the liquid, electrically conducting LiPb is adopted as working fluid to address the above-mentioned functions. The intense magnetic field used in fusion reactors to confine the burning plasma has a strong influence on the flow behavior, producing a magnetohydrodynamic (MHD) effect. In addition, serious temperature gradients are present in the breeding blanket, giving rise to buoyancy forces. The presence of the external magnetic field produces an additional MHD pressure drop, and although rougher MHD studies can predict the Δp [2], other phenomena, such as turbulence and buoyancy-driven convection, have a drastic impact on blanket performance and require a deeper analysis. In addition, tritium transport mechanisms [3–5] are influenced by the magnetic field; therefore, a detailed solution of magnetohydrodynamics is necessary, and it is obtainable with 3D multiphysics models for the breeding blanket. Smolentsev et al. [6] proposed activity for verification and validation of MHD codes for fusion applications, consisting of a series of benchmark problems whose results are known from experimental data or trusted analytical and numerical solutions. In particular, the five problems cover a wide range of magnetohydrodynamic flows which are of interest for fusion applications:

1. 2D fully developed laminar steady flow;
2. 3D laminar, steady developing flow in a nonuniform magnetic field;
3. Buoyancy-driven flow in a square cavity;
4. Quasi-two-dimensional (Q2D) turbulent flow;
5. 3D turbulent flow.

In this work, a verification and validation procedure of the developed 3D codes is performed, taking as reference [6]. Verification and validation activities of MHD codes were conducted by different authors for 1D codes [7], 3D codes under the COMSOL environment [8] and other codes [9–12]. Here, in order to further extend the analysis performed by [8], two benchmark cases involving unsteady flows are solved. The Q2D turbulence case proposed by Burr [13] was solved by adopting a modified version of the $k - \varepsilon$ model, derived by [14], while the fully 3D turbulent problem was tackled using the Large Eddy Simulation (LES) Residual Based Variational Multiscale (RBVM) method [15–17]. For the magnetoconvection case, the problems selected are the ones proposed by Di Piazza and Buhler [18], which are of particular interest for liquid metal breeding blanket technologies. In effect, in the first problem, the non-isothermal condition is due to differentially heated boundaries, while in the second case, temperature gradients are produced by internal heat generation. Both conditions are typical in breeding blanket systems, where strong temperature gradients and intense volumetric heat generation are present.

2. Governing Equations

The flow of liquid metal in breeding blanket conditions is characterized by a small magnetic Reynolds number $R_m = \mu \sigma L U$ (−). Here, μ (Pa s) and σ (S m^{-1}) are, respectively, the dynamic viscosity and the electrical conductivity of the fluid, L (m) and U (m s^{-1}) are the characteristic length and velocity of the flow. The magnetic Reynolds number represents the ratio between induction and diffusion of the magnetic field, so, in the low-R_m approximation, the magnetic field transport can be considered purely diffusive [19]. For an incompressible fluid under an imposed time-independent magnetic field with $R_m << 1$ the MHD equations, mass conservation Equation (1), momentum conservation Equation (2), current conservation Equation (3) and Ohm's law Equation (4) are presented.

$$\nabla \cdot \vec{u} = 0 \tag{1}$$

$$\rho \frac{\partial \vec{u}}{\partial t} + \rho \left(\vec{u} \cdot \nabla \right) \vec{u} = -\nabla p + \mu \nabla^2 \vec{u} + \vec{J} \times \vec{B} \tag{2}$$

$$\nabla \cdot \vec{J} = 0 \tag{3}$$

$$\vec{J} = \sigma \left(-\nabla \phi + \vec{u} \times \vec{B} \right) \tag{4}$$

where $\vec{u}$ (m s^{-1}) is the velocity vector, p (Pa) is the pressure, ρ (kg m^{-3}) is the density of the fluid, μ (Pa s) is the dynamic viscosity, $\vec{J}$ (A m^{-2}) is the current density vector, $\vec{B}$ (T) is the magnetic flux density, σ (S m^{-1}) is the electrical conductivity and ϕ (V m^{-1}) is the electrostatic potential.

For non-isothermal problems, the buoyancy force contribution must be added to momentum conservation Equation (2), which, under the Boussinesq hypothesis, becomes:

$$\rho_0 \frac{\partial \vec{u}}{\partial t} + \rho_0 \left(\vec{u} \cdot \nabla \right) \vec{u} = -\nabla p + \mu \nabla^2 \vec{u} + \vec{J} \times \vec{B} + (\rho_0 + \Delta\rho) \vec{g} \tag{5}$$

where ρ_0 (kg m^{-3}) is the reference density, $\Delta\rho = \rho - \rho_0$ is the density variation with respect to the reference density ρ_0 and $\vec{g}$ (m s^{-2}) is the gravity vector. $\Delta\rho$ can be further expressed as $-\rho_0\beta(T - T_0)$, where β (T^{-1}) is the thermal expansion coefficient and T_0 (K) is the reference temperature. The general heat transfer equation must be solved simultaneously:

$$\rho c_p\left(\frac{\partial T}{\partial t} + \vec{u}\cdot\nabla T\right) + \nabla\cdot(-\lambda\nabla T) = \dot{Q} \tag{6}$$

where T (K) is the temperature, c_p (J kg^{-1} K^{-1}) is the specific heat at constant pressure, λ (W m^{-1} K^{-1}) is the thermal conductivity and $\dot{Q}$ (W m^{-3}) is the volumetric heat generation rate.

The MHD flow is characterized, in addition to the magnetic Reynolds number, by additional nondimensional numbers. Some of the most important is the interaction parameter:

$$N = \frac{\sigma L B^2}{\rho U} \tag{7}$$

that expresses the ratio of the Lorentz force to inertia force and the Hartmann number:

$$Ha = LB(\sigma/\mu)^{1/2} \tag{8}$$

whose square represents the ratio of the Lorentz force to viscous forces. The Hartmann number and the interaction parameter are related to the Reynolds number, $Re = \rho LU = Ha^2/N$. For the thermal problems, it is interesting to recall the Grashof number that expresses the square of the ratio between buoyant and viscous forces in the fluid:

$$Gr = \frac{L^3\rho^2\beta\Delta T g}{\mu^2} \tag{9}$$

where ΔT is a characteristic temperature difference, depending on the problem considered.

3. Verification and Validation of COMSOL Code

The procedure proposed by [6] was followed in order to verify the applicability of COMSOL models, and in the next sections, the benchmark cases are presented, as well as the results of the computations. The first problem is the 2D fully developed MHD flow in a rectangular duct analytically addressed by Shercliff [20] and Hunt [21] (Section 3.1). The second is an experimental case, proposed by Picologlou et al. [22–24], involving the study of flows under a fringing magnetic field that investigates the transition from magnetohydrodynamics to ordinary hydrodynamics, presented in Section 3.2. Then, two non-isothermal flow problems are considered, solved numerically by Di Piazza and Bühler [18] (Section 3.3). Lastly, two experimental turbulent flow cases [13,25] are resolved, presented in Sections 3.4 and 3.5, respectively.

The equations introduced in Section 2 were implemented in COMSOL, exploiting the single-phase flow, heat transfer and electric current modules.

3.1. Two-Dimensional Fully Developed Laminar Steady MHD Flow

The laminar, fully developed, incompressible flow of a conducting fluid driven by a pressure gradient along a rectangular duct under an imposed transverse magnetic field is considered. Shercliff [20] and Hunt [21] solved this problem analytically, using different boundary conditions. Particularly, for Shercliff's case, the four walls of the duct are nonconducting, while for Hunt's case, the two walls perpendicular to the magnetic field, called Hartmann walls, are conducting, while the walls parallel to $\vec{B}$, called side walls, are electrically insulated. The wall conductance ratio:

$$c_w = \frac{\sigma_w t_w}{\sigma L} \tag{10}$$

expresses the ratio between the electrical conductivity σ_w (S m^{-1}) and thickness t_w (m) of the walls and the electrical conductivity of the fluid and the characteristic length of the flow. For Shercliff's case $c_w = 0$ for all the four walls, for Hunt's case $c_w = 0$ for the side walls, while is non-null for the Hartmann walls. As suggested by [6], the wall conductance ratio considered is $c_w = 0.01$. Four values of the Hartmann number were selected: $Ha = 500, 5000, 10,000, 15,000$.

The problem was solved in dimensionless form, using a hexahedral, structured mesh, analog to the one proposed by Sahu et al. [8], shown in Figure 1. To minimize the computational cost, considering that the solution is symmetric with respect to x and y axis, just a quarter of the domain is considered, applying proper boundary conditions. Elements are generated in x and y direction with a geometric distribution, maximizing the number of cells in the side and Hartmann layers. The mesh, selected following a grid convergence study [5], consists of 50 elements in the x-direction and 75 in the y-direction for $Ha = 500$, 64×100 for $Ha = 5000$ and 78×125 for $Ha = 10,000$ and $15,000$; eight cells were always ensured in the Hartmann and side layers. Although the problem is invariant in z-direction, one element is generated in the direction of the flow to consider the vector products involving the magnetic field. The velocity boundary conditions adopted in the study are non-slip at the duct walls and periodic flow conditions at the inlet and outlet of the duct. The electrical boundary conditions are electrical insulation for the side walls and thin wall condition $\vec{J} \cdot \hat{n} = c_w \nabla^2 \phi$, with $\hat{n}$ unit vector perpendicular to the wall, which represents conservation of electric charge in the plane of the wall.

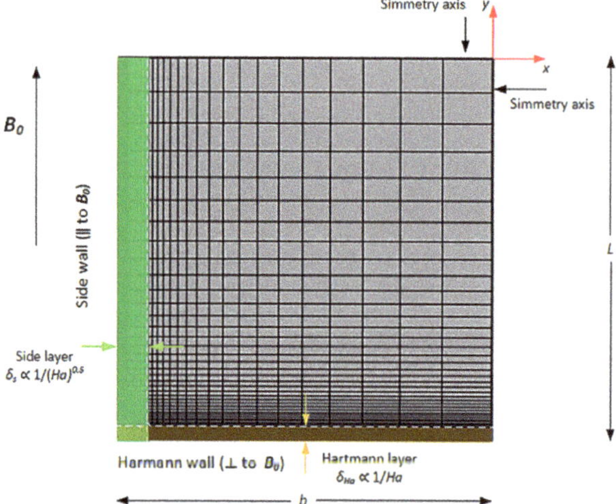

Figure 1. Example of mesh adopted for Shercliff's and Hunt's cases, corresponding to $Ha = 5000$. The elements are distributed, increasing the cell size moving inward to the duct.

Solutions obtained are compared to those reported by Smolentsev, choosing the dimensionless flow rate $\widetilde{Q}$ as comparison parameter, defined as

$$\widetilde{Q} = \frac{4}{b^2} \cdot \frac{\mu u}{(-\partial p / \partial x)} \tag{11}$$

where u (m s^{-1}) is the x-component of the velocity vector, b (m) is half the Hartmann wall length and $\partial p / \partial x$ (Pa m^{-1}) is the imposed pressure gradient in the direction of the flow. The relative error between COMSOL results and the analytical solutions by Shercliff and Hunt is evaluated as:

$$\varepsilon_{rel} = \left| 1 - \frac{\widetilde{Q}}{\widetilde{Q}_{an}} \right| \tag{12}$$

where $\widetilde{Q}\,(-)$ is the COMSOL solution and $\widetilde{Q}_{an}\,(-)$ is the analytical solution. The calculations are reported in Table 1. The presented table shows good agreement between analytical and numerical values.

Table 1. Comparison between the results of Shercliff's case and Hunt's case for 2D fully developed laminar steady flow. Results in terms of $\widetilde{Q}\,(-)$.

Shercliff Case			
$Ha\,(-)$	Analytical $\widetilde{Q}\,(-)$	COMSOL $\widetilde{Q}\,(-)$	$\varepsilon_{rel}\,(\%)$
500	7.680×10^{-3}	7.690×10^{-3}	0.130
5000	7.902×10^{-4}	7.906×10^{-4}	0.0456
10,000	3.965×10^{-4}	3.946×10^{-4}	0.478
15,000	2.648×10^{-4}	2.660×10^{-4}	0.453
Hunt Case			
$Ha\,(-)$	Analytical $\widetilde{Q}\,(-)$	COMSOL $\widetilde{Q}\,(-)$	$\varepsilon_{rel}\,(\%)$
500	1.405×10^{-3}	1.406×10^{-3}	0.0356
5000	1.907×10^{-5}	1.904×10^{-5}	0.184
10,000	5.169×10^{-6}	5.163×10^{-6}	0.118
15,000	2.425×10^{-6}	2.410×10^{-6}	0.635

3.2. Three-Dimensional Laminar, Steady Developing MHD Flow in a Nonuniform Magnetic Field

In the second benchmark case, a conducting fluid flows in two different ducts, with rectangular and circular cross sections, in the presence of a nonuniform magnetic field at the exit from a magnet. This case was experimentally investigated at the Argonne National Laboratory on ALEX (Argonne's Liquid metal EXperiment) facility [22–24]. The system employed eutectic NaK as a working fluid in a room temperature closed loop.

In this problem, the magnetic field changes in the direction of the flow x, $\vec{B} = B(x)\hat{y}$, with $\hat{y}$ unit vector in the y-direction, and this requires, considering the previously analyzed 2D case, the additional discretization of the domain in the x-direction. The velocity boundary conditions adopted in the study are non-slip at the duct walls and imposed average velocity at the inlet. The electrical boundary condition is a thin wall condition on the walls.

3.2.1. Rectangular Duct

The symmetry of the problem is exploited, and only a quarter of the duct cross section is considered. The mesh is constituted by a symmetric distribution of elements in the direction of the flow, maximizing the number of cells in the central region, where the magnetic field is changing the most. In y and z directions, the mesh is analog to the one proposed in Section 3.1. The total number of elements is 2.79×10^5. The equations are solved in dimensionless form.

The parameters adopted for the study are $Ha = 2900$, $N = 540$ and $c_w = 0.07$. The quantity selected for the comparison with the experimental results is the dimensionless axial pressure difference, that is, the pressure difference developed in the axis of the duct, scaled by $\sigma U B_0^2$. The results are presented in Figure 2, where the magnetic field profile scaled by B_0 and the axial pressure difference obtained by Picologlou et al. [22] are shown in the present work. Good agreement between the curves can be appreciated. The biggest discrepancy appears in $-5 < x/L < 0$, where COMSOL tends to overestimate the pressure difference. This behavior is also found in work by Sahu [8] and from the HIMAG Code calculations [26,27]. The difference between the two solutions is calculated using the integral of the curves with the following relation, called integral error index:

$$\varepsilon_{int} = \left| 1 - \frac{\int_{x_{min}}^{x_{max}} \Delta p(x)\,dx}{\int_{x_{min}}^{x_{max}} \Delta p_A(x)\,dx} \right| \tag{13}$$

where Δp_A (−) and Δp (−) are, respectively, the ALEX experiment and the present work nondimensional axial pressure difference. The integrals are computed numerically, using the trapezoidal rule, and the resulting error is 1.10%.

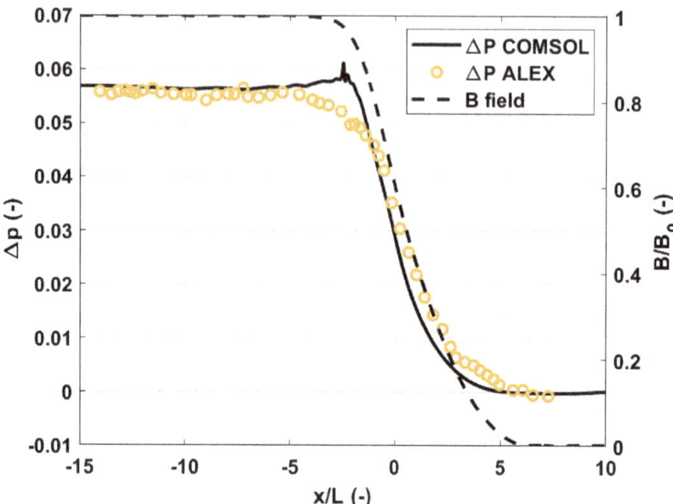

Figure 2. Comparison of the COMSOL code results against ALEX experiment at Argonne National Laboratory [22], rectangular duct.

3.2.2. Circular Duct

In x-direction the mesh adopted is equivalent to the previous case, whereas, in the y and z plane, 25 boundary layers are considered, generated from the first layer of thickness 10^{-6} m, with a growth rate of 1.3. The total number of elements is 3.03×10^5.

The parameters adopted for the study are $Ha = 6600$, $N = 10,700$ and $c_w = 0.027$. In Figure 3, the results are presented. The curves are matching very well, and the error is 0.913%.

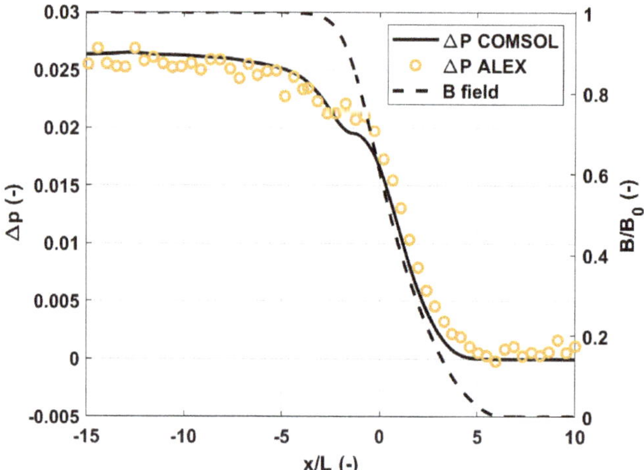

Figure 3. Comparison of the COMSOL code results against ALEX experiment at Argonne National Laboratory [22], circular duct.

3.3. Magneto-Convection

The following benchmark cases were developed with the aim to include also representative cases for the liquid metals breeding blankets, such as the WCLL, being characterized by non-isothermal conditions and internal volumetric heating.

The flow of an electrically conducting fluid in a long vertical channel of the rectangular cross section was considered [16]. The flow is promoted by buoyancy forces arising from non-isothermal conditions; hence, we refer to this as magnetoconvection. With reference to Figure 4, the imposed magnetic field is $\vec{B} = B_0 \hat{y}$ and the gravitational acceleration $\vec{g} = -g\hat{x}$, with $\hat{x}$ unit vector in x-direction, is aligned with the channel axis. Within this frame, two cases are considered: a differentially heated duct and a uniformly heated duct, solved by Di Piazza and Bühler [18] with the CFX commercial code (currently Ansys CFX). For both the problems, the Hartmann number is $Ha = 100$.

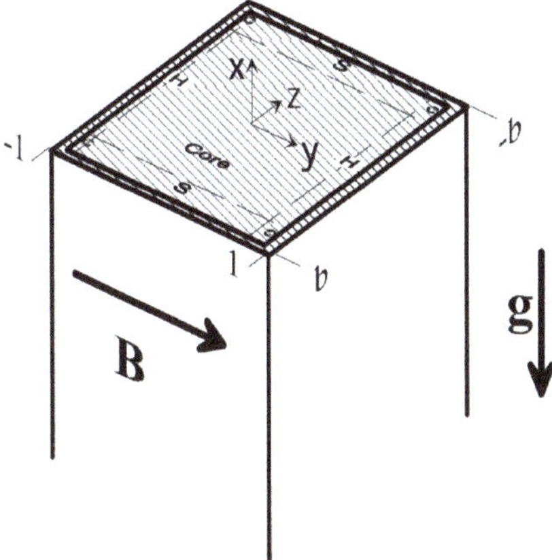

Figure 4. Geometry sketch for the two benchmark cases related to magneto-convective motion.

The problem is 2D, and the COMSOL solution is obtained for the nondimensional problem, expressed in depth in [28]. For the cases considered, $Gr \ll Ha^4$, from which results that the inertial term of the Navier–Stokes equation became negligible. The mesh adopted is analog to the one shown in Figure 1, where only one element is generated in the direction of $\vec{g}$. The total number of elements for the mesh selected is 5120, with 64 elements in the x-direction and 80 in the y-direction; eight cells were always ensured in the Hartmann and side layers. The velocity boundary conditions adopted in the study are non-slip at the duct walls and period flow conditions at the inlet–outlet. The electrical boundary condition is a thin wall condition on the walls. The temperature boundary conditions are defined, for the two different cases, in the next sub-sections.

3.3.1. Differentially Heated Duct

The two boundaries placed in the side walls ($z = -b$ and $z = +b$) are kept at different temperatures, while the Hartmann walls are thermally insulated, and there is no internal heat generation.

A sensitivity analysis, with wall conductance ratio c_w as a changing parameter, was carried out, and the nondimensional velocity profile at $y = 0$ as a function of z for half duct is shown in Figure 5 for $Ha = 100$.

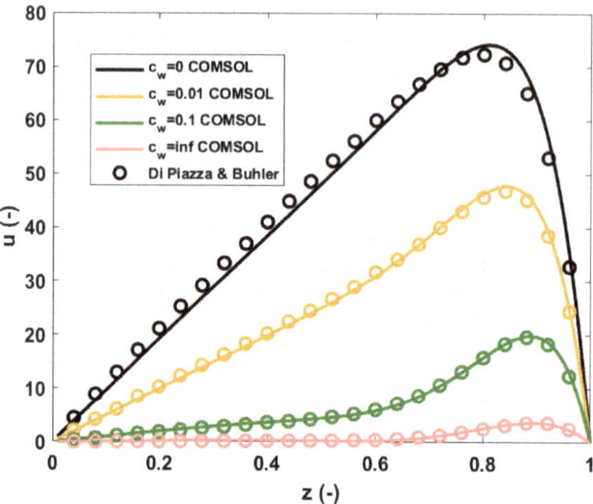

Figure 5. Numerical solutions of differentially heated duct case from [18] and from the COMSOL model.

It is interesting to notice that for the lower values of c_w, the damping effect of magnetohydrodynamics is less evident, while in the core region, the solution is still dominated by buoyancy and Lorentz forces and exhibits a linear behavior, with a slope of $\sim Ha$ for the perfectly insulating walls case [29]. In the lower conductivity cases, jets are not present. This is due to the fact that for low values of c_w the side layer becomes better conducting than the side walls, and high current jets are now present in the layers parallel to the side walls. They are also parallel to $\vec{B}$, so they do not interact with the magnetic field; therefore, the electromagnetic forces in the side layers become negligible, and the dominant effect is due to viscous dissipation.

For these cases, a comparison was made with respect to the numerical solutions [18] obtained with the CFX code. The integral of the curves is selected as a comparison parameter, and the relative difference is calculated with the integral error index:

$$\varepsilon_{int} = \left| 1 - \frac{\int_{z_{min}}^{z_{max}} u(0,z)\mathrm{d}z}{\int_{z_{min}}^{z_{max}} u_B(0,z)\mathrm{d}z} \right| \tag{14}$$

where $u_B(0,z)$ $(-)$ is [18] x-direction dimensionless velocity profile and $u(0,z)$ $(-)$ is the current work profile, both taken at $y = 0$. The error comparison is reported in Table 2. As can be appreciated, for all the cases, the maximum difference is lower than 2%.

Table 2. Comparison between COMSOL code and Di Piazza and Buhler solutions for $Ha = 100$. Results in terms of integral error index.

	Differentially Heated Duct	Uniformly Heated Duct
c_w $(-)$	ε_{rel} (%)	ε_{rel} (%)
0	0.957	1.50
0.01	0.326	4.78
0.1	1.77	0.585
∞	1.36	0.770

3.3.2. Uniformly Heated Duct

For the internally heated duct case, a volumetric heat generation $\dot{Q}$ is present, and the boundary at the side walls is kept at a fixed and equal temperature, while the Hartmann walls are thermally insulated.

The nondimensional velocity profiles at $y = 0$ and as a function of z-coordinate, a result of an analog sensitivity analysis to the one presented for the differentially heated duct case, are shown in Figure 6 for $Ha = 100$. For high wall conductivity ratios, the additional forces damp the velocity profile in the core region, and velocity jets are present in the side layers. For small values of c_w, jets are no more present, and the solution at the side layers is dominated by viscous effects.

The error comparison is reported in Table 2, where the maximum difference is related to the analysis with $c_w = 0.01$, that, nevertheless, presents a value below 5%. It should be recalled that in both differentially heated duct and uniformly heated duct cases, the comparison was addressed on the numerical solutions of the codes.

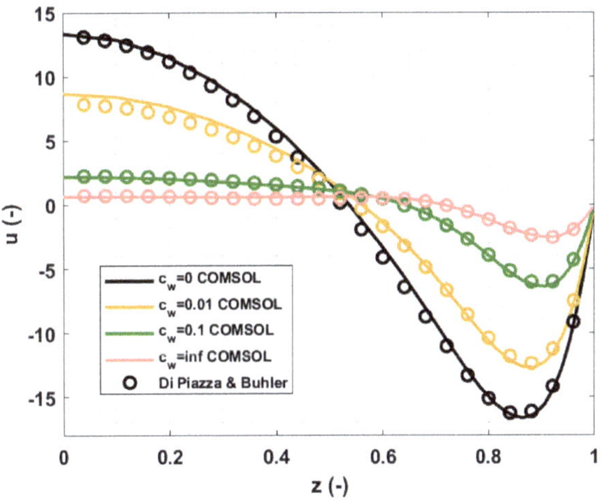

Figure 6. Numerical solutions of uniformly heated duct case from [18] and from the COMSOL model.

3.4. Quasi-Two-Dimensional MHD Turbulent Flow

This case regards a quasi-two-dimensional MHD turbulent flow as proposed in [6]. Burr et al. [13] developed an experimental setup consisting of a rectangular stainless steel channel of side length 0.04 m and wall thickness 6 mm where the eutectic sodium–potassium alloy is circulated under the presence of a magnetic field. NaK, with density 865 kg m^{-3} and kinetic viscosity 9.5×10^{-7} m^2 s^{-1}, flows in the x-direction, and $\vec{B}$ is oriented in z and can be varied from 0.25 T to 2.5 T. The electric conductivity of the wall is 1.39×10^6 S m^{-1}, whereas the one of the NaK is 2.8×10^6 S m^{-1}, from which it results in wall conductance ratios of the side and Hartmann walls of $c_{w,S} = 0.0714$ and $c_{w,H} = 0.0119$, respectively. The Hartmann numbers investigated are 600, 1200, 2400 and 4800 for Reynolds numbers between 3.3×10^3 and 1.0×10^5.

The problem is solved numerically using a RANS $k - \varepsilon$ turbulence model that includes two transport equations for the turbulent kinetic energy k (m^2 s^{-2}) and for the dissipation of turbulent kinetic energy ε (m^2 s^{-3}). For $k - \varepsilon$, Equation (2) becomes:

$$\rho \frac{\partial \vec{u}}{\partial t} + \rho \left(\vec{u} \cdot \nabla \right) \vec{u} = -\nabla p + (\mu + \mu_T) \nabla^2 \vec{u} + \vec{J} \times \vec{B}_0 \qquad (15)$$

where μ_T [Pa s] is the eddy viscosity. The two closure equations of the model are:

$$\left(\vec{u}\cdot\nabla\right)k = \nabla\cdot\left[\left(\mu + \frac{\mu_t}{\sigma_k}\right)\nabla k\right] + P_k - \rho\varepsilon + S_k^L \tag{16}$$

$$\rho\left(\vec{u}\cdot\nabla\right)\varepsilon = \nabla\cdot\left[\left(\mu + \frac{\mu_t}{\sigma_\varepsilon}\right)\nabla\varepsilon\right] + C_{\varepsilon 1}\frac{\varepsilon}{k}P_k - C_{\varepsilon 2}\rho\frac{\varepsilon^2}{k} + S_\varepsilon^L \tag{17}$$

Here, P_k (W m^{-3}) is a source term, σ_k ($-$), σ_ε ($-$), $C_{\varepsilon 1}$ ($-$) and $C_{\varepsilon 2}$ ($-$) are turbulent model parameters. S_k^L (W m^{-3}) and S_ε^L (W m^{-3}) are source terms that include the damping of the turbulent kinetic energy and the dissipation of the turbulent kinetic energy due to the Lorentz force and were modeled by different authors [30–32]. The relations selected are the ones proposed by Meng et al. [14], expressed by the following equations:

$$S_k^L = -\sigma B^2 k e^{-C_1^M\sqrt{\frac{\sigma}{\rho}B^2\frac{\nu}{k}}} \tag{18}$$

$$S_\varepsilon^L = -\sigma B^2 \varepsilon e^{-C_1^M\sqrt{\frac{\sigma}{\rho}B^2\frac{\nu}{k}}} \tag{19}$$

where C_1^M ($-$) is a constant with value 30, and ν (m^2 s^{-1}) is the kinematic viscosity. In these relations, $\sqrt{\sigma/\rho B^2\nu/k}$ is the characteristic turbulence damping time, and $\exp\left(-C_1^M\sqrt{\sigma/\rho B^2\nu/k}\right)$ is the decay rate of the turbulent kinetic energy.

The comparison parameters between Burr and the current study are the mean velocity profiles and turbulent kinetic energy of two-dimensional turbulence k_{2D} (m^2 s^{-2}), defined as

$$k_{2D} = \frac{1}{2}\left(\overline{u'^2} + \overline{w'^2}\right) \tag{20}$$

where u' (m s^{-1}) and w' (m s^{-1}) are the fluctuating term of the velocity for the x-th and z-th components.

The mesh refinement was carried out until an appropriate wall lift-off in viscous units δ_w^+ was obtained. In particular, $\delta_w^+ = 11.06$ for every case analyzed, that is the lower limit for COMSOL $k - \varepsilon$ turbulence model [33]. This value corresponds to the dimensionless wall distance y^+, where the viscous sublayer meets the logarithmic layer. The total number of elements is 1.28×10^6. The velocity boundary conditions adopted in the study are non-slip at the duct walls and imposed average velocity at the inlet. The electrical boundary condition is a thin wall condition on the walls.

The mean velocity in x-direction u, calculated for $Ha = 4800$ and various Reynolds numbers, is now compared with Burr results. In Figure 7, the COMSOL solutions and Burr experimental results are displayed.

The main characteristics of the flow are well expressed, and the influence of the Reynolds number on the flow is evident. This is a characteristic of turbulent MHD flows, while the velocity distribution of laminar MHD flows is governed only by Ha. Turbulence smoothens out velocity peaks in the side walls that are reduced for increasing Reynolds numbers, and the width of the side layer increases with Re due to turbulent transfer of momentum.

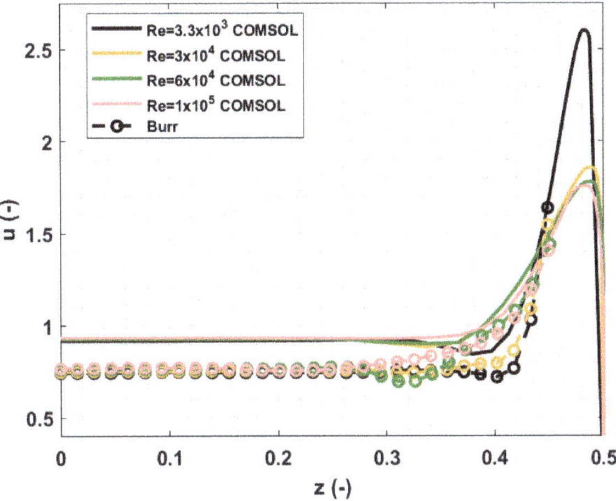

Figure 7. Comparison of the numerical results against Burr experiment [13]. Mean velocities at the midplane $y = 0$ for $Ha = 4800$.

In Table 3, the local relative errors between COMSOL and the experimental results, calculated with the following equation, are presented.

$$\varepsilon_{rel} = \left|1 - u/u_{exp}\right| \tag{21}$$

Here, u $(-)$ is the x-direction velocity magnitude calculated with COMSOL code and u_{exp} $(-)$ by the experiment, both evaluated at $z = 0.45$, that is, the closest point to the side layer, obtained in the experiment. As it can be observed, values agree very well, with a maximum error of about 3.52%. Comparing the velocity at $z = 0$, there is a relative difference of about 12% between the numerical and experimental values. The code overestimated the bulk velocity, and the same behavior is reported in [14], which uses the same strategy to model the dissipation of the turbulent kinetic energy due to Lorentz forces.

Table 3. Local error for the quasi-two-dimensional turbulent flow for $Ha = 4800$ and different Re. Results in terms of u $(-)$.

Re $(-)$	Experimental u $(-)$	COMSOL u $(-)$	ε_{rel} (%)
$3.3 \cdot 10^3$	1.638	1.644	0.368%
$3 \cdot 10^4$	1.547	1.492	3.52%
$6 \cdot 10^4$	1.442	1.482	3.31%
$1 \cdot 10^5$	1.405	1.447	2.98%

In Figure 8, the comparison between the distributions of the turbulent kinetic energy of two-dimensional turbulence k_{2D} reported by Burr [13] and calculated with COMSOL are presented for $Re = 1.0 \times 10^5$ and Hartmann numbers between 600 and 4800. The values are captured along the z-axis at the midplane $y = 0$. The increase in the turbulent kinetic energy as Ha increases can be appreciated, proving that turbulence is promoted by the magnetic field. In both [14] and COMSOL results, turbulence is restrained to the side layers, decreasing fast moving towards the core region, where, in the experimental results, the flow, although weakly, remains turbulent. The anisotropicity of the turbulent flow is particularly evident for the high Hartmann number cases, where $k_{2D} \approx k_{3D}$.

As shown by the results, the code can tackle quasi-two-dimensional MHD flow problems, giving reliable results, particularly in the side layer region. Further improvements are needed to better compute the bulk turbulence that is underestimated by the code.

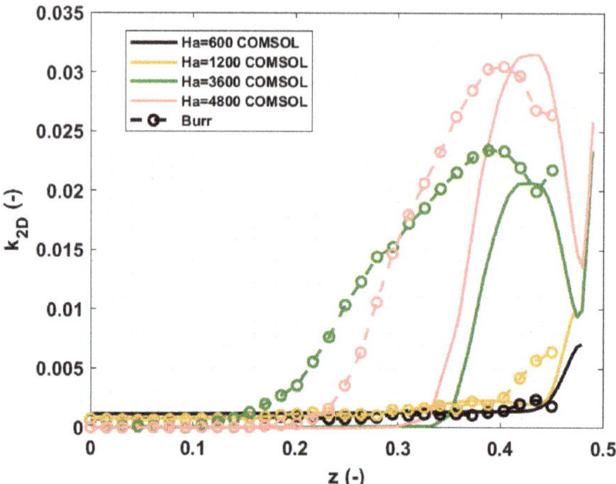

Figure 8. Comparison of the numerical results against Burr experiment [13]. Turbulent kinetic energy of two-dimensional turbulence k_{2D} at the midplane $y = 0$ for $Re = 1.0 \times 10^5$.

3.5. Three-Dimensional Turbulent MHD Flow

The benchmark problem on 3D turbulent MHD flow addressed is the one proposed by Smolentsev et al. [6]. The eutectic GaInSn, with density 6360 kg m^{-3}, the electrical conductivity of 3.46×10^{-6} S^2 m^{-1} and kinetic viscosity of 3.4×10^{-7} m^2 s^{-1}, flows with a maximum flow rate of 2×10^{-3} m^{-3} s^{-1} in plexiglass (insulating) rectangular channel of length 0.5 m and 100 mm × 20 mm cross section, and starting from pure hydrodynamics conditions, is subjected to a nonuniform magnetic field generated by a magnetic obstacle. This is an experimental problem addressed by Andreev et al. [25]. The flow direction is x-oriented, and the magnetic field is in the z-direction. Starting from a zero value, the magnetic field monotonically increases until it reaches the maximum value of $B_0 = 0.504$ T at the center of the duct, corresponding to $Ha = 400$, then it decreases to zero. The magnetic field is slightly nonuniform also in the y-direction, but this feature is neglected in the COMSOL model. Further information on the B profile can be found in [25]. The Reynolds number selected is $Re = 4000$; therefore, the interaction parameter is equal to $N = 40$.

The problem was solved using the Large Eddy Simulation (LES) model [34,35], as suggested by Smolentsev et al. [6]. In particular, the Residual Based Multiscale Variational (RBMV) method [16,34,35] was implemented. In this model, the velocity and pressure fields are decomposed into resolved and unresolved scales:

$$\vec{U} = \vec{u} + \vec{u}'$$ (22)

$$P = p + p'$$ (23)

where $\vec{u}$ (m s^{-1}) and $\vec{u}'$ (m s^{-1}) are the resolved scale and the unresolved scale velocities, respectively, and p (Pa) and p' (Pa) are the resolved scale and the unresolved scale pressures. In the RBMV method, the unresolved velocity and pressure scales are modeled in terms of the equation residuals for the resolved scales. Further information on RBMV LES modeling can be found [33].

To ensure adequate space discretization, the resolution of wall layers was checked by $\frac{u_\tau h_w}{\nu} < 1$, where the left term is the dimensionless wall distance evaluated at the first mesh cell next to the wall. Here, u_τ is the friction velocity, h_w is the thickness of the first mesh cell next to the wall, ν is the kinematic viscosity. Time discretization was checked using the relation $C = \frac{U\Delta t}{h_U} < 0.5$, where C is the Courant number, with U flow velocity magnitude, Δt time step and h_U mesh size in the streamline direction. The total number of elements of the selected mesh is 1.8×10^6. The velocity boundary conditions adopted in the study are non-slip at the duct walls and imposed average velocity at the inlet. The electrical boundary condition is a thin wall condition on the walls.

The selected comparison parameter is the mean velocity profile and the mean electric potential, evaluated at different distances along the channel. In particular, the selected locations correspond to the main flow regions, as described by Andreev. In Figure 9, a comparison between COMSOL and Andreev's velocity profile at $\frac{x}{H} = -5.3$, with H channel height in z-direction, is presented.

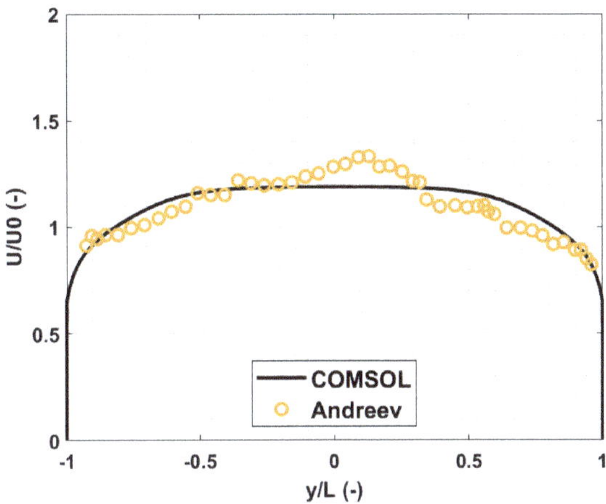

Figure 9. Comparison of the numerical results against Andreev experiment [25]. Velocity profile at $z/H = 0$ and $x/H = -5/3$ (turbulence suppression region).

The profile is referred to as the first region indicated by the author, characterized by the increasing magnetic field that influences the flow and damps its perturbations, called "turbulence suppression region". As evident, the agreement between the curves is good, and the integral error index is 0.947%. In Figure 10, the electric potential profile, placed at $\frac{x}{H} = 0$, is shown.

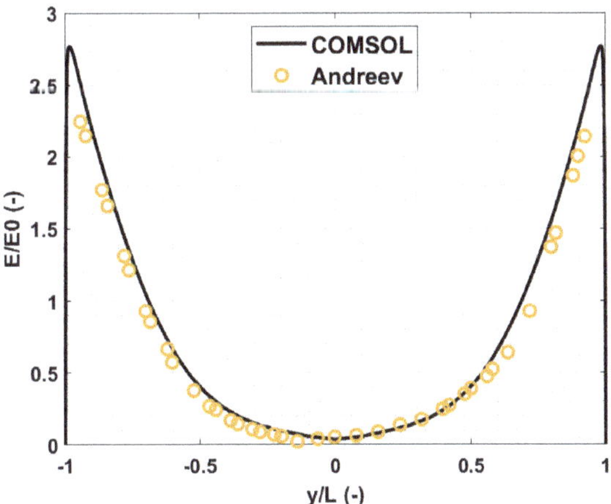

Figure 10. Comparison of the numerical results against Andreev experiment [25]. Electric potential profile at $z/H = 0$ and $x/H = 0$ (vortical region).

This region, around the center of the duct where B is maximum, is called "vortical region". The magnetic field suppresses the fluctuations in the direction parallel to the magnetic field, and the flow becomes quasi-two-dimensional. The integral error index is 4.21%. In Figures 11 and 12, the results for the last region, named "wall jet region", are reported.

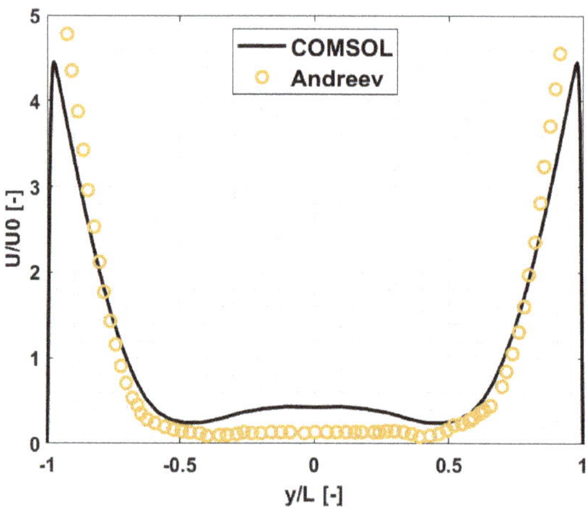

Figure 11. Comparison of the numerical results against Andreev experiment [25]. Velocity profile at $z/H = 0$ and $x/H = 3$ (wall jet region).

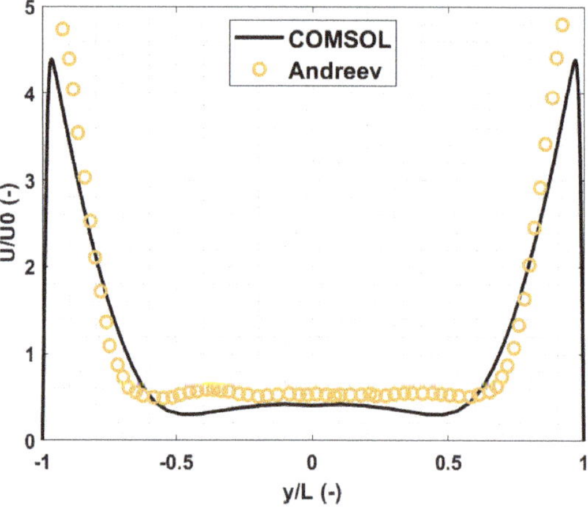

Figure 12. Comparison of the numerical results against Andreev experiment [25]. Velocity profile at $z/H = 0$ and $x/H = 6$ (wall jet region).

This region is located on the remaining part of the channel, where the magnetic field decreases. As observable, the velocity in the middle plane increases greatly, going from $x/H = 3$ to $x/H = 6$ thanks to the drop of the intensity of the magnetic field, and the quasi-two-dimensional profile stretches to the core of the flow. The agreement between the velocity fields is quite good, and the relative errors in percentage are 6.72% at $x/H = 3$ and 7.81% at $x/H = 6$.

As presented, the code is capable of representing the characteristic regions of the experimental problem, and the errors obtained are, for every case, well below 10%. The results are summarized in Table 4, and they provide confidence in the capabilities of COMSOL to simulate fully 3D turbulent flows.

Table 4. Comparison between COMSOL code and [25] experiment. Results in terms of integral error index.

x/H (−)	−5.3	0	3	6
ε_{rel} (%)	0.947%	4.21%	6.72%	7.81%

4. Discussion

In this paper, a verification and validation procedure was followed, as proposed by Smolentsev et al. [6], and different liquid metal MHD problems were solved, with the aim of verifying the developed models using the commercial software COMSOL Multiphysics. For the magnetoconvection case, the considered benchmarks were the ones tackled by Di Piazza and Bühler [18], presenting the typical conditions that are expected in liquid metal breeding blankets.

The compared parameters showed great agreement for laminar flow problems, both isothermal and non-isothermal. As far as the turbulent cases are concerned, Q2D and 3D, a modified version of the RANS $k - \varepsilon$ and the LES RBMV models were adopted, respectively, and the deriving results, in terms of relative errors and the capability of describing the flow features, are very promising.

Author Contributions: Conceptualization, C.A., L.C. and R.T.; methodology, C.A.; software, C.A.; validation, C.A.; formal analysis, C.A.; investigation, C.A.; data curation, C.A.; writing—original draft preparation, C.A.; writing—review and editing, C.A., L.C. and R.T.; supervision, R.T., M.U. and M.Z. All authors have read and agreed to the published version of the manuscript.

Funding: This work has been carried out within the framework of the EUROfusion Consortium and has received funding from the Euratom research and training programme 2014–2018 and 2019–2020 under grant agreement No 633053. The views and opinions expressed herein do not necessarily reflect those of the European Commission.

Institutional Review Board Statement: Not applicable.

Informed Consent Statement: Not applicable.

Data Availability Statement: The study didn't report and data.

Conflicts of Interest: The authors declare no conflict of interest.

References

1. Del Nevo, A.; Arena, P.; Caruso, G.; Chiovaro, P.; Di Maio, P.; Eboli, M.; Edemetti, F.; Forgione, N.; Forte, R.; Froio, A.; et al. Recent progress in developing a feasible and integrated conceptual design of the WCLL BB in EUROfusion project. *Fusion Eng. Des.* **2019**, *146*, 1805–1809. [CrossRef]
2. Miyazaki, K.; Inoue, S.; Yamaoka, N. MHD Pressure Drop of Liquid Metal Flow in Circular and Rectangular Ducts under Transverse Magnetic Field. In *Liquid Metal Magnetohydrodynamics*, 2nd ed.; Springer: Dordrecht, The Netherlands, 1989; Volume 10, pp. 29–36. [CrossRef]
3. Candido, L.; Testoni, R.; Utili, M.; Zucchetti, M. Tritium transport model at breeder unit level for WCLL breeding blanket. *Fusion Eng. Des.* **2019**, *146*, 1207–1210. [CrossRef]
4. Alberghi, C.; Candido, L.; Testoni, R.; Utili, M.; Zucchetti, M. Magneto-convective effect on tritium transport at breeder unit level for the WCLL breeding blanket of DEMO. *Fusion Eng. Des.* **2019**, *146*, 2319–2322. [CrossRef]
5. Candido, L.; Alberghi, C.; Moro, F.; Noce, S.; Testoni, R.; Utili, M.; Zucchetti, M. A novel approach to the study of magnetohydro-dynamic effect on tritium transport in WCLL breeding blanket of DEMO. *Fusion Eng. Des.* **2021**, *167*, 112234. [CrossRef]
6. Smolentsev, S.; Badia, S.; Bhattacharyay, R.; Bühler, L.; Chen, L.; Huang, Q.; Jin, H.-G.; Krasnov, D.; Lee, D.-W.; Valls, E.M.D.L.; et al. An approach to verification and validation of MHD codes for fusion applications. *Fusion Eng. Des.* **2015**, *100*, 65–72. [CrossRef]
7. Kim, S.; Kim, M.; Lee, D.; Choi, C. Code validation and development for MHD analysis of liquid metal flow in Korean TBM. *Fusion Eng. Des.* **2012**, *87*, 951–955. [CrossRef]

8.   Sahu, S.; Bhattacharyay, R. Validation of COMSOL code for analyzing liquid metal magnetohydrodynamic flow. *Fusion Eng. Des.* **2018**, *127*, 151–159. [CrossRef]

9.   Yan, Y. Validation of COMSOL Multiphysics for magneto-hydro-dynamics (MHD) flows in fusion applications. In Proceedings of the COMSOL Conference, Boston, MA, USA, 5 October 2017.

10.   Feng, J.; Chen, H.; He, Q.; Ye, M. Further validation of liquid metal MHD code for unstructured grid based on OpenFOAM. *Fusion Eng. Des.* **2015**, *100*, 260–264. [CrossRef]

11.   He, Q.; Chen, H.; Feng, J. Acceleration of the OpenFOAM-based MHD solver using graphics processing units. *Fusion Eng. Des.* **2015**, *101*, 88–93. [CrossRef]

12.   Gajbhiye, P.; Throvagunta, P.; Eswaran, V. Validation and verification of a robust 3-D MHD code. *Fusion Eng. Des.* **2018**, *128*, 7–22. [CrossRef]

13.   Burr, U.; Barleon, L.; Muller, U.; Tsinober, A. Turbulent transport of momentum and heat in magnetohydrodynamic rectangular duct flow with strong sidewall jets. *J. Fluid Mech.* **2000**, *406*, 247–279. [CrossRef]

14.   Meng, Z.; Zhang, S.; Jia, J.; Chen, Z.; Ni, M. A K-Epsilon RANS turbulence model for incompressible MHD flow at high Hartmann number in fusion liquid metal blankets. *Int. J. Energy Res.* **2018**, *42*, 314–320. [CrossRef]

15.   Bazilevs, Y.; Calo, V.; Cottrell, J.; Hughes, T.J.R.; Reali, A.; Scovazzi, G. Variational multiscale residual-based turbulence modeling for large eddy simulation of incompressible flows. *Comp. Meth. Appl. Mech. Eng.* **2007**, *197*, 173–201. [CrossRef]

16.   Liu, J. Residual-Based Variational Multiscale Models for the Large Eddy Simulation of Compressible and Incompressible Turbulent Flows. Ph.D. Thesis, Rensselaer Polytechnic Institute, Troy, NY, USA, 2012.

17.   Volker, J.; Songul, K. A finite element variational multiscale method for the Navier-Stokes equations. *SIAM J. Sci. Comp.* **2005**, *26*, 1485–1503. [CrossRef]

18.   Di Piazza, I.; Bühler, L. *Numerical Simulations of Buoyant Magnetohydrodynamic Flows Using the CFX Code*; Furschungszentrum Karlsruhe GmbH: Karlsruhe, Germany, 1999.

19.   Davidson, P.A. *An Introduction to Magnetohydrodynamics*, 1st ed.; Cambridge University Press: Cambridge, UK, 2001. [CrossRef]

20.   Shercliff, J.A. Steady motion of conducting fluids in pipes under transverse magnetic fields. *Math. Proc. Camb. Philos. Soc.* **1953**, *49*, 136–144. [CrossRef]

21.   Hunt, J.C.R. Magnetohydrodynamic flow in rectangular ducts. *J. Fluid Mech.* **1965**, *21*, 577–590. [CrossRef]

22.   Picologlou, B.F.; Reed, C.B. Experimental investigation of 3-D MHD flows at high Harmann numbers and interaction parameters. In *Liquid Metal Magnetohydrodynamics*, 2nd ed.; Springer: Dordrecht, The Netherlands, 1989; Volume 10, pp. 71–77. [CrossRef]

23.   Reed, C.B.; Picologlou, B.F.; Dauzvardis, P.V. Experimental facility for studying MHD effects in liquid metal. *Fusion Tech.* **1985**, *8*, 257–263. [CrossRef]

24.   Reed, C.B.; Picologlou, B.F.; Hua, T.Q.; Walker, J.S. ALEX results: A comparison of measurements from a round and a rectangular duct with 3-D code predictions. In Proceedings of the Symposium on Fusion Engineering, Monterey, CA, USA, 12–16 October 1987.

25.   Andreev, O.; Kolesnikov, Y.; Thess, A. Experimental study of liquid metal channel flow under the influence of a nonuniform magnatic field. *Phys. Fluids* **2006**, *18*, 065108. [CrossRef]

26.   Ni, M.-J.; Munipalli, R.; Huang, P.; Morley, N.B.; Abdou, M.A. A current density conservative scheme for incompressible MHD flows at a low magnetic Reynolds number. *Part II: On an arbitrary collocated mesh. J. Comp. Phys.* **2007**, *227*, 205–228. [CrossRef]

27.   Munipalli, R.; Shankar, V.; Ni, M.-J.; Morley, N. Development of a 3-D incompressible free surface MHD computational environment for arbitrary geometries: HIMAG. In *DOE Phase-II SBIR Final Report*; DOE: Washington, DC, USA, 2003.

28.   Buhler, L. Laminar buoyant magnetohydrodynamic flow in vertical rectangular ducts. *Phys. Fluids* **1998**, *10*, 223–236. [CrossRef]

29.   Bojarevics, V. Buoyancy driven flow and its stability in a horizontal rectangular channel with an arbitrary oriented transversal magnetic field. *Magnetohydrodynamics* **1996**, *31*, 245–253.

30.   Ni, H.C.; Gradner, R.A. Numerical analysis of turbulent pipe flow in a transfer magnetic field. *Int. J. Heat Mass Transf.* **1997**, *40*, 1839–1851. [CrossRef]

31.   Kenjeres, S.; Hanjalic, K. On the implementation of effects of Lorentz force in turbulence closure models. *J. Heat Fluid Flow* **2000**, *21*, 329–337. [CrossRef]

32.   Smolentsev, S.; Abdou, M.; Morley, N.; Ying, A.; Kunugi, T. Application of the "K-eps" model to open channel flows in a magnetic field. *Int. J. Eng. Sci.* **2002**, *40*, 693–711. [CrossRef]

33.   COMSOL Multiphysics User Guide. Available online: https://doc.comsol.com/5.5/doc/com.comsol.help.comsol/COMSOL_ReferenceManual.pdf (accessed on 21 July 2021).

34.   Chapman, D.R. Computational Aerodynamics Development and Outlook. *AIAA J.* **1979**, *17*, 1293–1313. [CrossRef]

35.   Choi, H.; Moin, P.A. Grid-point requirements for large eddy simulation: Chapman's estimates. *Phys. Fluids* **2012**, *24*, 011702. [CrossRef]

Article

Development of a RELAP5/MOD3.3 Module for MHD Pressure Drop Analysis in Liquid Metals Loops: Verification and Validation

Lorenzo Melchiorri *, Vincenzo Narcisi, Fabio Giannetti, Gianfranco Caruso and Alessandro Tassone

Dipartimento di Ingegneria Astronautica, Elettrica ed Energetica, Sapienza University of Rome, 00184 Rome, Italy; vincenzo.narcisi@uniroma1.it (V.N.); fabio.giannetti@uniroma1.it (F.G.); gianfranco.caruso@uniroma1.it (G.C.); alessandro.tassone@uniroma1.it (A.T.)
* Correspondence: lorenzo.melchiorri@uniroma1.it; Tel.: +39-06-4991-8639

Abstract: Magnetohydrodynamic (MHD) phenomena, due to the interaction between a magnetic field and a moving electro-conductive fluid, are crucial for the design of magnetic-confinement fusion reactors and, specifically, for the design of the breeding blanket concepts that adopt liquid metals (LMs) as working fluids. Computational tools are employed to lead fusion-relevant physical analysis, but a dedicated MHD code able to simulate all the phenomena involved in a blanket is still not available and there is a dearth of systems code featuring MHD modelling capabilities. In this paper, models to predict both 2D and 3D MHD pressure drop, derived by experimental and numerical works, have been implemented in the thermal-hydraulic system code RELAP5/MOD3.3 (RELAP5). The verification and validation procedure of the MHD module involves the comparison of the results obtained by the code with those of direct numerical simulation tools and data obtained by experimental works. As relevant examples, RELAP5 is used to recreate the results obtained by the analysis of two test blanket modules: Lithium Lead Ceramic Breeder and Helium-Cooled Lithium Lead. The novel MHD subroutines are proven reliable in the prediction of the pressure drop for both simple and complex geometries related to LM circuits at high magnetic field intensity (error range ±10%).

Keywords: magnetohydrodynamics (MHD); MHD pressure drop; system codes; RELAP5; breeding blanket; liquid metal technology

Citation: Melchiorri, L.; Narcisi, V.; Giannetti, F.; Caruso, G.; Tassone, A. Development of a RELAP5/MOD3.3 Module for MHD Pressure Drop Analysis in Liquid Metals Loops: Verification and Validation. *Energies* **2021**, *14*, 5538. https://doi.org/10.3390/en14175538

Academic Editor: Antonio D'angola

Received: 26 July 2021
Accepted: 30 August 2021
Published: 4 September 2021

Publisher's Note: MDPI stays neutral with regard to jurisdictional claims in published maps and institutional affiliations.

1. Introduction

The Breeding Blanket (BB) is a crucial component in the project of magnetic-confinement nuclear fusion reactors. The BB carries out three main functions: it conveys the heat produced by the plasma to the primary cooling system of the reactor, provides radiation shielding for personnel and components and produces enough tritium to ensure the fuel self-sufficiency of the reactor. Several blanket design solutions have been proposed where liquid metals (LMs), such as Lithium (Li) or Lithium–Lead alloys (Li-Pb), are used as working fluids [1]. One of the critical issues in the project of a LM blanket is represented by magnetohydrodynamic (MHD) phenomena that occur in the piping network of the blanket [2]. MHD effects are due to the interaction between the flowing liquid metal, that is an electro-conductive material, with the high magnetic field employed to confine the plasma in the reactor chamber. Electrical currents are induced in the conductive fluid volume and, in turn, Lorentz forces are generated, altering the flow behaviour compared with the ordinary hydrodynamic (OHD) case. For instance, MHD phenomena modify the velocity distribution and mass transport inside the ducts, enhance pressure losses, affect heat transfer mechanisms, etc. Estimating the impact of all those effects on the component performance is essential for an efficient project of a liquid metal breeding blanket and, therefore, the development of numerical tools able to predict them is extremely desirable [3]. An extensive overview of MHD phenomena that affect the BB design is available in Ref. [4].

In this work, the analysis is focused on the additional pressure drops introduced by the electromagnetic drag. Generally, it is possible to assume that MHD pressure losses are composed by the sum of two terms [5]:

$$\Delta P_{MHD} = \Delta P_{2D} + \Delta P_{3D} \, [Pa] \tag{1}$$

In Equation (1), the first right-hand term (ΔP_{2D}) quantifies the pressure losses due to Lorentz forces opposite to flow direction, induced by electrical currents that are confined in a plane perpendicular to fluid velocity. Two-dimensional MHD pressure drop is the only term driving the head loss in a fully developed flow configuration, i.e., in straight conduits. The second right-hand term (ΔP_{3D}) stands for the contribution of not "cross-sectional" currents (three-dimensional currents) that, interacting with the magnetic field, produce Lorentz forces in other direction than the stream one. Three-dimensional currents arise whenever axial electric potentials are induced within the fluid body either due to the channel complex geometry (cross-section variation, change of stream direction, etc.), or non-uniform electromagnetic boundary conditions, may that be a discontinuity of wall conductivity or a strong magnetic field gradient [6]. Formally, the total head loss (ΔP_{TOT}) in a closed system is composed by the sum of the hydrodynamic pressure drop, caused by distributed and concentrated friction losses, and the overall MHD loss due to the electromagnetic drag:

$$\Delta P_{TOT} = \Delta P_{MHD} + \Delta P_{OHD} \approx \Delta P_{MHD} \, [Pa] \tag{2}$$

Considering the typical magnitude of the magnetic field encountered in fusion reactors ($\approx$4–9 T), it can be demonstrated that the pressure drop due to MHD phenomena are dominating the other contribution, so that $\Delta P_{TOT} \approx \Delta P_{MHD}$. For instance, the MHD pressure loss in the Li-Pb loop of the European Demonstrator Reactor Water-Cooled Lithium Lead (EU DEMO WCLL) is estimated at $\approx$1.5 and $\approx$2.5 MPa for the outboard and inboard segment, whereas the OHD loss does not exceed 0.1 MPa [7].

The flow path of the liquid metal in the piping system of a blanket comprises a multitude of complex elements, such as manifolds and junctions, alongside with straight ducts. Hence, in the pressure drop assessment in a BB, both contributions of 2D and 3D MHD pressure loss must be taken into account. Generally, blanket-scale analyses for MHD phenomena are performed adopting a semi-analytical approach, in which empirical and semi-empirical correlations are supplemented with data extrapolated by direct numerical simulations [5,7,8]. This methodology, although effective to a certain degree, happens to be extremely time-consuming, not flexible in its scope, and with serious limitations in the achievable spatial and temporal resolution. Several computational fluid dynamic (CFD) tools have dedicated MHD modules, such as ANSYS-CFX, ANSYS-FLUENT and Open-FOAM but a comprehensive and mature code able to simulate all the MHD phenomena in the blanket is not yet available [9]. Unfortunately, reactor-scale analysis is not possible with CFD codes and, in any case, will be prohibitive in terms of calculation time and computational effort. Conversely, best-estimate systems thermal-hydraulic (BE SYS-TH) codes could enable the efficient and quick simulation of the blanket piping network level but, currently, they have limited or non-existent MHD capability.

This kind of codes are crucial for the assessment of thermal-hydraulics phenomena occurring within complex nuclear systems. BE codes are mainly developed for the analysis of incidental or accidental transients but are also employed for the characterization of operational transients or steady-states configurations [10]. Generally, to compute the underlying physics in such challenging scenarios, SYS-TH tools use 0-D or 1-D models derived by the analysis of numerous experimental campaigns. They are often used for safety demonstration analyses and BE SYS-TH codes are considered the reference numerical tools for the licensing of fission nuclear power plants [11]. A typical example of SYS-TH code is RELAP5 that, despite being initially developed for applications in light water reactors, has been extended in recent years to employ other fluids (i.e., liquid metals, molten salts, etc.). Rudimentary MHD correlations are implemented in RELAP5-3D [12] and MARS-FR [13] but are limited to straight rectangular/circular ducts and, thus, do not support any treatment for junctions, manifolds, etc. Basic MHD modelling capabilities have been implemented in MELCOR 1.8.5 and 1.8.6, which have been modified for the study of fusion-relevant systems [14–16]. Moreover, pursuing the quest for the realization of a MHD system-level tool, a novel code called MHD-SYS has been developed that features models for the simulation of multiple electrical-coupled ducts and heat transfer for basic layouts, whereas coupling with CFD codes is employed to supply the system code with reliable input data for the behaviour of the flow in complex geometrical elements [17]. The coupled approach for component-level analysis, although somewhat effective, nullifies the main benefit of a stand-alone BE code, hampering its agile functioning. More recently, basic MHD features have been implemented in GETTHEM, a tool developed specifically for tokamak fusion reactors, to model pressure drops in the WCLL Li-Pb loop [18,19].

The aim of this work is to present the first phase of the development of a comprehensive and robust numerical tool able to handle all the fundamental MHD effects occurring in a LM breeding blanket, ranging from pressure loss to mass transport, in order to support fusion reactor design. It is important to underline that at this early stage, attention will be focused on MHD features that influence the nominal operation of the reactor and for which the state of the knowledge is deemed sufficient to support the implementation of a SYS-TH module. MHD impact on relevant accidental transients for the BB (for instance, how an anticipated accidental pressure transient is affected by the tokamak magnetic field) is still under discussion and, as such, it will be taken into account once more data become available.

The prediction of the MHD pressure drop is one of the main concerns for liquid metal BB design and has been given priority in the code implementation. In the following, we discuss the models that, derived by experimental and numerical works, have been included in the system code RELAP5/MOD3.3. A verification and validation (V&V) procedure is reported to assess the confidence of our numerical method against the benchmark of high-quality numerical simulations and experimental data, as suggested by Ref. [20].

2. MHD Formulation

The incompressible electrically conducting fluid flow, under the influence of a magnetic field, is properly described by the combination of Navier–Stokes and Maxwell equations. For liquid metal flows in fusion reactor loops, the induction-less or low magnetic Reynolds number approximation is allowed, so that the self-induced magnetic field effect is neglected, and the velocity/magnetic field coupling is simplified reducing the latter to a boundary condition of the problem [21]. For the scope of this study, it is useful to focus our attention on the coupled momentum balance equation in its nondimensional form:

$$\frac{1}{N}\left[\frac{\partial \mathbf{v}}{\partial t} + (\mathbf{v} \cdot \nabla)\mathbf{v}\right] = -\nabla p + \frac{1}{Ha^2}\nabla^2 \mathbf{v} + \mathbf{j} \times \mathbf{B} \tag{3}$$

In Equation (3), vectors $\mathbf{v}$, $\mathbf{j}$ and $\mathbf{B}$ stand for velocity, current density and magnetic field, respectively, and p is the pressure. They are made nondimensional by scaling with the mean velocity v_0, $j_0 = \sigma v_0 B_0$ and the external magnetic field magnitude B_0; the nondimensional pressure p is scaled through the value $p_0 = \sigma v_0 a B_0^2$.

Parameter a stands for the typical length scale for the specific case of study (i.e., half-length of a duct along magnetic field direction), whereas σ is the electrical conductivity of the fluid. The flow is governed by two fundamental parameters: the Hartmann number (Ha) and the Stuart number, or interaction parameter (N).

$$Ha = aB_0\sqrt{\frac{\sigma}{\mu}} \tag{4}$$

$$N = \frac{\sigma a B_0^2}{\rho v_0} \tag{5}$$

The square of Ha sets the ratio of electromagnetic to viscous forces, whereas N gives the ratio of electromagnetic to inertial forces. Symbols μ and ρ denote dynamic viscosity and density of the fluid, respectively. Those parameters may be combined to return the Reynolds number (Re) in its classical formulation:

$$Re = \frac{Ha^2}{N} = \frac{\rho v_0 a}{\mu} \tag{6}$$

In LM loops for blanket applications, Ha and N reach very high values ($\approx 10^4$, [5]), so that MHD flows can be approximated as viscous-less and inertia-less in most cases. Therefore, in the core region of fluid domain, the motion is governed by the balance between Lorentz and pressure force, whereas inertial and viscous effects are confined in thin fluid layers of thickness $O(Ha^{-1})$ and $O(Ha^{-1/2})$ close to solid walls perpendicular or parallel to the magnetic field direction [21].

Another fundamental nondimensional parameter must be defined to represent the tendency of the induced currents to close either through the fluid or the bounding walls. The pipe system that hosts the liquid metal, generally, is itself made of electro-conductive material. Thus, electric currents arising in the fluid tend to close their path through the electrically conducting duct walls. Moreover, if different channels have common walls, they may interact by exchanging electric currents across those walls. This effect is known as Madarame effect or electromagnetic coupling [22]. The wall conductivity is usually expressed through the wall conductance ratio, c:

$$c = \frac{\sigma_w t_w}{\sigma a} \tag{7}$$

where σ_w and t_w are the electrical conductivity of the duct walls and their thickness.

Magnetohydrodynamic pressure drops in liquid metal circuits depend on all parameters discussed above. As already pointed in Section 1, 2D MHD pressure drop can be considered as the analogue of hydrodynamic friction loss (also referred as distributed head loss). It is the "electromagnetic drag" that dissipates fluid kinetic energy via Joule effect. This phenomenon has been studied in-depth hence its characterization can rely on an exhaustive amount of data. Therefore, extensively validated correlations are available in the literature for fully developed magneto-hydraulic loss of load [6,23,24]. Nevertheless, it is crucial to underline that the current version of RELAP5 can model only ducts which are assumed to be immersed in a dielectric medium, i.e., air. This assumption is not usually verified in a reactor where, unless the liquid metal is insulated from the structural material through some technical means, the conduits are in electrical contact with each other. Theoretical understanding of electromagnetic coupling is still limited; it is not possible, at this stage, to develop a suitable system code model to account for its effect (i.e., altered pressure loss and flow distribution). However, the above-mentioned assumption, although portraying a simplified picture, is still useful to predict pressure loss figures that are representative of the behaviour of a more realistic system, as it is demonstrated in Section 4.

Three-dimensional pressure losses, instead, can be treated as the MHD analogue of hydrodynamic localized losses. They are caused by electrical current closing their paths in stream-wise direction, that arise whenever the flow meets complex geometries (such as bends, cross-section variations, obstacles), different walls electrical conductivity and non-uniform magnetic field. All those are common features in the liquid metal piping network of a fusion reactor. However, data for 3D MHD pressure losses are relatively scarce compared with these available for fully developed MHD flow. Evaluating those kinds of losses is much more challenging since they strongly depend on the flow geometry and governing parameters. For these reasons, only configurations that have been studied and characterized the most are taken into account: expanding/contracting pipes, bending conduits and ducts with discontinuities in the conductance ratio (Flow Channel Inserts). In other words, the current RELAP5/MOD3.3 implementation for MHD 3D models has been devised as a readily extendable framework, which is amenable to be improved as soon as the knowledge progresses in this field. We anticipate revising the code implementation in the next years to increase its level of detail and accuracy.

3. RELAP5/MOD3.3 Overview and Models

RELAP5/MOD3.3 (referred as RELAP5 in the following) is a 1D thermo-hydraulic code, developed in Fortran language, that is based on a non-homogeneous and non-equilibrium model for the two-phase (liquid-vapour) system that is solved by a partially implicit numerical scheme to permit fast calculation of system transients. In particular, for the basic hydrodynamic model, which is solved numerically using a semi-implicit finite-difference technique, the two-fluid momentum equations are formulated in terms of volume and time-averaged parameters of the flow. Phenomena that are described by non-axial gradients, such as friction losses or heat transfer, are modelled in terms of fluids bulk properties employing empirically-derived transfer coefficient correlations [25].

RELAP5 has been developed for BE transient simulations of light water reactor coolant systems during postulated accidents. The code models the coupled behaviour of the reactor coolant system and the core for loss-of-coolant accidents and operational transients. A generic modelling approach is used that permits simulating a variety of thermal-hydraulic systems [25].

As such, RELAP5 has been extensively validated for a wide range of fission reactor applications and it is regarded as the "gold standard" for those activities. For those reasons, it is considered the best candidate to be improved and modified to become an essential instrument for fusion reactor design. Magnetohydrodynamic features are not implemented in the original version of the code since liquid metal MHD is not relevant to the design of fission reactors.

All the modifications for including magnetohydrodynamic effects on pressure drop are performed starting from an enhanced version of the RELAP5 system code [26–28]. This version was developed at the Department of Astronautical, Electrical and Energy Engineering (DIAEE) of "Sapienza" University of Rome in collaboration with ENEA, following a similar approach of other previous versions developed by ENEA and University of Pisa for liquid metals, in which the modelling capability for fusion reactors and their primary cooling systems is enhanced [29]. In this version of the code, updated physical properties for Li-Pb and sodium-potassium (NaK) liquid metals are implemented. These metals are used in this paper for the validation and verification of the MHD model. This choice is made because Li-Pb happens to be the most promising alloy (as coolant and/or breeder) for fusion reactors [30,31]; NaK, on the other hand, is extensively used as model fluid in experimental campaigns for the analysis of magnetohydrodynamic phenomena [32–34]. Other breeder fluids of interest, like the FLiBe molten salt, could be implemented in the future. Thermo-physical properties for those materials are inserted in the code following the approach employed in [35] for lead (Pb) and lead-bismuth (LBE) properties. For lithium-lead properties, the work performed by Martelli et al. is taken as reference [36], whereas the one developed by Foust is employed for sodium-potassium [37].

Moreover, electrical conductivity calculation for these and other relevant structural materials (Eurofer97, stainless steel, alumina, etc.) is currently performed by RELAP5 through a new specific subroutine. However, solid wall electrical conductivity is automatically calculated for Eurofer97 only, from the correlation outlined by Mergia and Boukos in [38]. For the analysis presented in this paper, whenever the input wall material needs to be different from the latter, the source code is changed. In a further release, this procedure is foreseen to be automated, alongside with extending the database of ducts wall materials to other commonly encountered in fusion blanket design: vanadium alloys, Inconel steels, alumina, silicon carbide composite, etc. In Table 1, the electrical conductivities for relevant materials in the framework of this activity are reported.

Table 1. Electrical conductivity for different materials. (*) Temperature T in K. (**) Temperature T in °C.

Material	Electrical Conductivity σ [S/m]
Li-Pb [36]	$\sigma = \dfrac{1}{(1.0333 \times 10^{-6} - 6.75 \times 10^{-11} \times T + 4.18 \times 10^{-13} \times T^2)}$ (*)
NaK [37]	$\sigma = \dfrac{10^8}{(37.66 + 2.307 \times 10^{-2} \times T + 7.187 \times 10^{-5} \cdot T^2)}$ (**)
Eurofer97 [38]	$\sigma = \dfrac{10^8}{(8.536 + 0.1484 \times T - 2.84 \times 10^{-5} \times T^2)}$ (*)
Stainless-Steel	$\sigma = 1.26 \times 10^6$

The program takes in account magnetohydrodynamic effects by computing two different factors: one related to friction losses and cross-sectional currents (2D friction factor), and the other one considered as a local loss factor (3D coefficient) caused by the induction of axial currents [2]. Each factor is calculated by different subroutines that are called whenever a liquid metal is defined in input. The implementation is designed to sum MHD pressure loss factors with the hydrodynamic ones, either for distributed (FWF_{MHD}) or local (K_{MHD}) pressure drops, as is shown in Equation (8), which represents the basic differential one-dimensional single-phase (liquid) formulation adopted for momentum conservation, where α_l stands for liquid volume fraction and the term BF takes in account body forces effects. As reasonable, the ordinary behaviour is recovered when the magnetic field is set to zero.

$$\alpha_l \rho \frac{\partial v}{\partial t} + \frac{1}{2}\alpha_l \rho \frac{\partial v^2}{\partial x} = -\alpha_l \frac{\partial p}{\partial x} + \alpha_l \rho BF - (\rho FWF + FWF_{MHD})\alpha_l v - \frac{1}{2}\alpha_l \rho \frac{\partial v^2}{\partial x}(K + K_{MHD}) \quad (8)$$

Exclusively liquid-phase equations are affected by MHD phenomena, no attempt has been made to develop a multi-phase MHD model. This choice is justified since only sub-cooled liquid metal flows are considered relevant for near-term fusion applications and helium bubble formation in the breeder, albeit important locally is not expected to change its overall behaviour. At last, it is worth underlining that, during the whole implementation campaign, compatibility tests (involving steam/liquid water) have been carried out and it has been verified that the new routines do not affect the original capabilities of the code, whenever they are not called.

3.1. 2D Magnetohydrodynamic Pressure Drop Factor

MHD pressure drops due to cross-sectional currents are modelled in the code adding a coefficient (FWF_{MHD}) to the hydrodynamic friction factor computed by RELAP5 to estimate the hydraulic head loss:

$$\frac{\partial p}{\partial x} = FWF_{MHD} \cdot v \,[Pa/m] \quad (9)$$

MHD pressure drop factor for 2D effects is implemented considering the magnetic field as uniform, transverse to the flow direction and aligned with at least a pair of duct walls (see Figure 1). Furthermore, viscous forces are always supposed negligible, namely $Ha >> 1$. Under those assumptions, MHD pressure loss in a straight channel can be evaluated according to Equation (9) where v is the average fluid velocity and x is the axial length direction of the component.

The FWF_{MHD} factor is a function of channel shape, geometry parameters and fluid/walls material. For square/rectangular cross-section conduits, it is derived from engineering correlations reported by Kirillov et al. [6] and it is expressed as:

$$FWF_{MHD} = \frac{\sigma_f B^2}{1 + c^{-1} + \frac{\delta}{2}(c_1^{-1} + c_2^{-1})}\,[kg/m^3 s] \quad (10)$$

In Equation (10), c is the wall conductance ratio of the Hartmann walls (walls perpendicular to B, see Figure 1b), c_1 and c_2 are the wall conductance ratio for the side walls (walls parallel to B); all these quantities are calculated with Equation (7). The symbol $\delta = a/3b$ stands for the channel aspect ratio, where b is the channel half-width perpendicular to the magnetic field. This correlation has been developed for ducts that may have different walls thickness or, equivalently, non-uniform wall conductance ratio ($c \neq c_1 \neq c_2$), that is a common feature in liquid metal circuits. It is valid only for channels that have a uniform Hartmann wall conductivity (c). For duct geometries that cannot be treated with this approach, the wall with the highest conductive ratio should be conservatively assumed to be representative of the whole channel.

If the duct walls have a uniform thickness, Equation (10) can be simplified [6,23,24]:

$$FWF_{MHD} = \frac{\sigma_f B^2}{1 + c^{-1}(1 + \delta)} \, [kg/m^3 s] \tag{11}$$

Similarly, a conduit with circular cross section and uniform wall conductivity is treated with another simplified expression of Equation (10) [6,23,24]:

$$FWF_{MHD} = \frac{\sigma_f B^2}{1 + c^{-1}} \, [kg/m^3 s] \tag{12}$$

It is important to keep in mind that predictions from Equations (10)–(12) show slight deviation from experimental data (i.e., $\pm 15\%$), according to the overview provided by Tassone et al. in [5]. In Table 2, all parameters needed for FWF_{MHD} computation are collected. Electrical conductivity, in $[S/m]$, for both fluid (σ_f) and walls material (σ_w) are computed by the code.

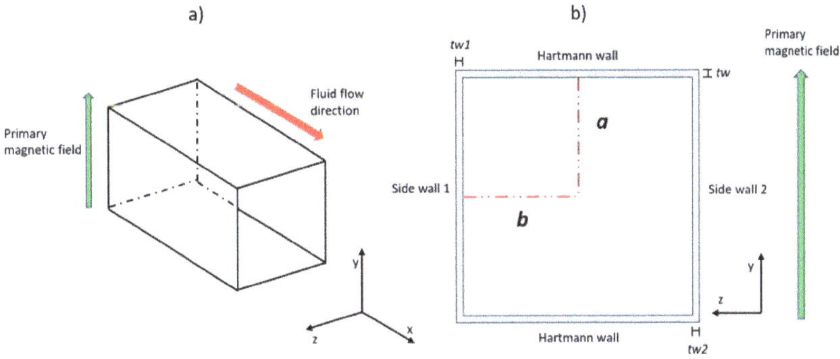

Figure 1. (**a**) Three-dimensional view of a generic hydraulic element (single volume). The green vector is the magnetic field applied and is considered perpendicular with respect to the fluid stream direction (red arrow). (**b**) Cross-sectional view of a hydraulic element. Walls perpendicular to the magnetic field are called Hartmann walls, while those aligned to the magnetic field are referred as Side walls. Characteristic lengths of the component a and b are underlined, and thickness of the walls (t_w) as well.

Table 2. Parameters used to compute Magnetohydrodynamic 2D additional pressure drop factor.

Parameter	Description	Units
Primary magnetic field B (defined in input by the user).	Magnetic field, always considered perpendicular to the stream direction.	[T]
Characteristic length a (defined in input by the user).	Component half-length along magnetic field direction.	[m]
Characteristic length b (defined in input by the user).	Component half-length perpendicular to magnetic field direction.	[m]

Table 2. *Cont.*

Parameter	Description	Units
Wall thickness t_w (defined in input by the user).	Thickness of component walls perpendicular to magnetic field direction	[m]
Wall thickness t_{w1} and t_{w2} (defined in input by the user).	Thickness of component walls parallel to magnetic field direction.	[m]
Electrical conductance ratio for Hartmann walls, c.	$c = \frac{\sigma_w t_w}{\sigma_f a}$	Nondimensional
Electrical conductance ratio for Side wall, c_1.	$c = \frac{\sigma_w t_{w1}}{\sigma_f a}$	Nondimensional
Electrical conductance ratio for Side wall, c_2.	$c = \frac{\sigma_w t_{w2}}{\sigma_f a}$	Nondimensional
Aspect ratio of the component divided by 3, Δ	$\delta = a/3b$	Nondimensional
Fluid and wall electrical conductivity, σ_f and σ_w	Electrical conductivity of fluid and wall, depending on input temperature.	[S/m]

3.2. 3D Magnetohydrodynamic Pressure Drop Factor

Magnetohydrodynamic local pressure drop factor (3D) is computed by the code when it detects between two hydraulic elements a difference of: cross section, stream direction or electrical conductivity. The local factor K_{MHD} represents the 3D effects caused by the induction of axial currents and it is nondimensional. Three-dimensional pressure drops ΔP_{3D} are calculated according to Equation (13) [4,5]:

$$\Delta P_{3D} = \frac{1}{2} \cdot \rho \cdot v^2 \cdot K_{MHD}[Pa] \tag{13}$$

where ρ is the fluid density, dependent by the temperature, and v is the average fluid velocity. Furthermore, K_{MHD} is the sum of three separate K factors, each one representative of a single condition that causes the appearance of 3D MHD pressure losses.

$$K_{MHD} = K_{EXP/CONTR} + K_{BEND} + K_{FCI} \tag{14}$$

In Equation (14), $K_{EXP/CONTR}$ is the factor correlated to cross section variation, K_{BEND} is the one referred to change of stream direction, and K_{FCI} is the coefficient that takes into account pressure drop when there is a large difference in electrical conductance between two components.

3.2.1. Bends

Correlations for K_{BEND}, to evaluate MHD pressure drop in bends (Figure 2a), are derived mainly referring to Kirillov et al. [6] for circular pipes ($K_{BEND-circ||}$) and Reimann experimental study [39] for square conduits ($K_{BEND-sq||}$).

$$K_{BEND-circ||} = 0.125 \cdot \frac{\Psi_{BEND}}{90} \cdot N \tag{15}$$

$$K_{BEND-sq||} = \frac{1.063 \cdot c}{4/3 + c} \cdot \frac{\Psi_{BEND}}{90} \cdot N \tag{16}$$

Equation (15) is conceived for 90° bend pressure drop in insulated circular channels, therefore is considered suitable also for electro-conductive ones, being conservative. Conversely, Equation (16) is derived for 90° bends in square/rectangular ducts. However, assuming an inertia-less flow, they both can be used for any curvature angle (Ψ_{BEND}, expressed in degrees) via the correction factor $\Psi_{BEND}/90$. This means that, for example, a 180° curve can be treated as the composition of two following 90° bends. Previous studies have analysed the case of a bend in the flow path occurring in a plane parallel or perpendicular to the magnetic field direction. It is known that pressure loss associated with a bend

240

perpendicular to the magnetic field direction ($\perp B$) is always smaller (if not negligible) compared with the same case featuring a parallel magnetic field ($\|B$) [6]. The underlying physics of such behaviour is quite complex, however, a qualitative explanation is offered in the following. In a MHD flow, the magnetic field tends to dampen the angular momentum components that are not parallel to the field itself. Therefore, liquid sub-structures (e.g., vortexes) that own an angular momentum which is preferentially perpendicular to B are rapidly dissipated, at the expense of the overall energy of the fluid. Only the angular momentum parallel to the field is preserved [40]. When the fluid is forced to bend in a direction $\perp B$ the angular momentum is mostly aligned with the field, thus, a relatively low amount of energy will be lost to rearrange the liquid sub-structures. Conversely, when the flow meets a $\|B$ alterations along its path the angular momentum of the fluid will be preferentially perpendicular to the magnetic field and consequently a substantial loss of energy will occur to establish again fully developed conditions.

$$K_{BEND-circ\perp} = \frac{1}{3} \cdot K_{BEND-circ\|} \tag{17}$$

$$K_{BEND-sq\perp} = \frac{1}{3} \cdot K_{BEND-sq\|} \tag{18}$$

Correlations (15) and (16), originally developed for this latter case, are adapted to perpendicular bends by adopting a scaling factor 1/3 [5]. Such corrective factor has been arbitrarily chosen to take in account the minor impact that $\perp B$ bends have on the overall loss of load if compared with $\|B$ bends. This formulation has been proven reasonably reliable for modelling complex piping systems involving bending conduits, as discussed in the following (i.e., Secetion 4). When more data will be available in the literature, which is currently lacking comparative analysis for those configurations, it is planned to reassess the scale in the interest of a better fidelity to physical behaviour. $K_{BEND-circ\perp}$ and $K_{BEND-sq\perp}$ are shown in Equations (17) and (18), respectively.

Therefore, RELAP5 is currently able to treat $\perp B$ and $\|B$ bends for any curvature ratio and angle.

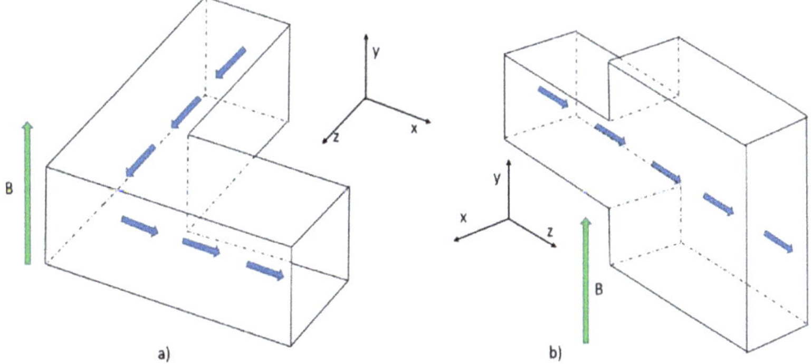

Figure 2. Geometry configurations examples that can be modelled with the new features implemented in the code. With the green vector is pointed the external magnetic field, B. (**a**) Bending square pipe geometry, with bending plane perpendicular to the magnetic field direction. (**b**) Square-to-rectangular duct expansion along B direction.

3.2.2. Sudden Cross-Section Variations

The model for the calculation of MHD local pressure loss coefficient ($K_{EXP/CONTR}$) caused by the variation of the cross section between two volumes is developed on the basis of the experimental work led by Bühler and Horanyi described in [32].

They considered the case of a rectangular duct undergoing a sudden cross-section expansion parallel to the magnetic field direction (see Figure 2b) and developed an empirical correlation to predict the 3D pressure drop, reported in a simplified form below:

$$\Delta p_{3D} = \Delta p_{3D\infty} + \Delta p_{3D,viscous}(Ha^{-1/2}) + \Delta p_{3D,inertial}(Re^{1/3}) \tag{19}$$

In essence, the pressure drop introduced by expansion geometry is the sum of three contributions: $\Delta p_{3D\infty}$ is a term derived by an asymptotic analysis for $Ha \to \infty$ and $N \to \infty$, whereas terms $\Delta p_{3D,viscous}$ and $\Delta p_{3D,inertial}$ assess the influence of viscous effects and inertial phenomena. Equation (19) is valid for $c = 0.028$ and expansion ratio $Z_c = 4$ (formally, the ratio between the characteristic length of the downstream duct and upstream one), which were used during the experimental campaign. Nevertheless, starting from Equation (19), the model has been extended to cover a wider range of expansion ratios and contraction ratios. The model for $K_{EXP/CONTR}$ relies on the assumption that expansions and contractions can be considered as reversible configurations (an expansion, if the flow through it is reversed, became a contraction or vice versa), hence they yield the same pressure drop when the expansion ratio and the contraction ratio are inverse. This is strictly true only for a flow where inertial effects are negligible $N \to \infty$. In other words, $K_{EXP/CONTR}$ is supposed symmetrical with respect to the cross-section variation ratio. Furthermore, an upper limit value is imposed for the local MHD factor. As matter of fact, according to the numerical analysis led by Feng et al. [41], the additional pressure drop caused by expansion ratio higher than 8 is independent by Z_c. Hence, a maximum $K_{EXP/CONTR} = 2 \times N$, as suggested by Smolentsev et al. in [4], is assumed. At last, for simulating cross-section variations occurring perpendicular to B ($\perp B$), a conservative scaling factor is employed, following the same approach discussed for bends (see Section 3.2.1).

The $K_{EXP/CONTR}$ model is currently able to treat $0 < Z_c < \infty$ for B parallel and perpendicular to the cross-section variation. The influence of wall conductivity and cross-section shape on the 3D loss is not currently implemented and this issue is planned to be addressed in the future, considering as starting point the numerical analysis performed by Rhodes et al. in [42], dealing with an electrically insulated sudden expansion.

3.2.3. Flow Channel Inserts

For discontinuity in wall electrical conductivity, the correlations implemented for K_{FCI} are extracted from experimental and numerical analyses, in several works regarding Flow Channel Inserts (FCIs) configuration [34,43,44]. An FCI is used to reduce MHD pressure loss in a conduit by electrically decoupling the fluid and the duct wall. A typical FCI is composed by a thin layer of dielectric material (alumina) enclosed between steel sheets and loosely inserted within the duct to be insulated (Figure 3). Manufacturing technology is currently limited to the production of insert length ≈ 0.5 m only for simple geometries; therefore, it is impossible to achieve a perfectly insulated piping system, especially accounting for the complex geometrical elements present in a typical fusion reactor blanket [45]. Hence, insulation discontinuities are always present when FCI are employed and they cause additional MHD pressure drops due to the sudden transition from a weakly to a well-conducting wall section. In RELAP5, two models of discontinuities are implemented. An inlet/outlet discontinuity class is encountered in the transition between a pipe section with an FCI and a second one which is lacking it, or vice versa. Conversely, a gap discontinuity is encountered when two neighbouring ducts, both insulated by FCIs, are separated by a smaller area where the liquid metal can enter in contact with the "naked" pipe surface.

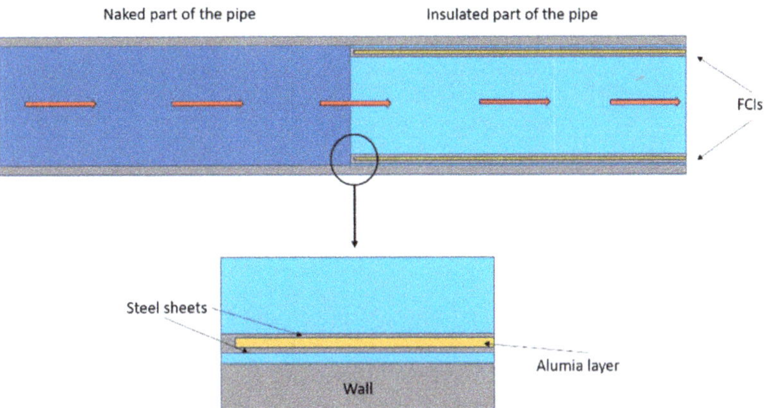

Figure 3. Sketch illustrating a Flow channel insert example configuration in a pipe (entrance in insulated region). Red arrows define metal liquid path. The FCI "sandwich" structure is highlighted in the bottom.

4. Verification and Validation (V&V)

IAEA guidelines suggest that computer-based tools or calculation methods that could be possibly used in safety analysis must undergo Verification and Validation (V&V) process to assess their reliability and efficiency [20].

The V&V procedure of the RELAP5 MHD module involves the comparison of the results deduced by the code with those of direct numerical simulation tools and data obtained by experimental works. Notable examples of V&V for MHD numerical solvers are found in References [46–50]. At first, simple test cases are employed to verify the response of the program in terms of magnetohydrodynamic pressure drops, either for 2D and 3D phenomena, thus testing the reproduction of a single MHD phenomenon separately. In itself, this is not considered satisfactory to assess the reliability of the code, thus more complex systems are analysed, i.e., two Test Blanket Modules (TBMs) concepts, to demonstrate the capability of the various subroutines of the MHD module to operate together and produce physical results. For this purpose, the numerical analysis performed by Swain et al. in [51] on the Lithium Lead Ceramic Breeder (LLCB) TBM at fusion-relevant magnetic field intensity is repeated with RELAP5 and an excellent quantitative agreement is found on the total pressure drop and local pressure profile.After that, the experimental data produced by the campaign carried out by KIT researchers on a Helium Cooled Lithium Lead (HCLL) TBM mock-up have been considered to further demonstrate the performance of the code against a more realistic benchmark [33]. The module geometry is simulated with the code and, again, an excellent agreement is found in terms of pressure drop prediction. It should be stressed that the HCLL TBM mock-up can be considered by all means a state-of-the-art scaled-down integrated effect test (IET) facility since it represented all relevant isothermal MHD phenomena but for a skewed (toroidal-poloidal) magnetic field that, however, is unlikely to be achieved before the start of the experimental TBM phase in ITER.

4.1. Single Effect Validation

4.1.1. Flow Channel Insert Test Case

The experimental study performed by Bühler et al. in [43] has been employed as test case, to verify the correct behaviour of the code when it comes to treat FCI configurations and fully developed flow pressure drops in circular cross-section conduits.

The reference geometry is a straight pipe (Figures 3 and 4) which is divided in a bare region and in an FCI region (darker in Figure 4). The liquid metal streams from the naked part to the one where the insulation layer is located. Therefore, the LM meets a discontinuity in the wall electrical conductivity that induces an electric potential gradient along the flow direction. For this reason, currents with a three-dimensional path arise in the fluid domain, extra Lorentz forces occur and cause additional loss of load compared with the fully developed flow. This 3D drop must be evaluated properly since, considering it occurs at the exit of FCIs or at gaps between insulation layers, it could significantly reduce the efficacy of flow channel inserts.

The test case is representative of the breeder ingress in the insulated part of the Li-Pb loop since, to minimize complexity and cost, only a small section of it is going to be fitted with inserts.

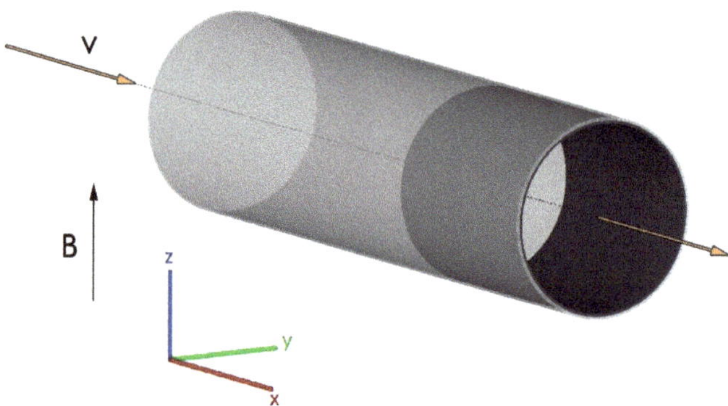

Figure 4. 3D sketch of the experimental section. The darker portion of the pipe is where insulation is applied (FCI). Magnetic field vector B and velocity v are depicted.

The molten metal involved in the experiment is a sodium-potassium alloy (NaK, mass fraction: Na 22% K 78%), whereas the horizontal pipe is made of stainless-steel with uniform wall thickness. The flow is treated as isothermal; therefore, materials properties are taken at the reference temperature of 293.15 K (Table 3).

Table 3. Material properties at 293.15 K.

Material	Electrical Conductivity [S/m]	Density [kg/m³]	Dynamic Viscosity [Pa · s]
Na22K78	2.62×10^6	874.15	8.5×10^{-4}
Stainless Steel	1.26×10^6	-	-

The magnetic field applied is uniform and unidirectional within the test section domain. Its intensity may be varied, as well as NaK mass flow rate. Thus, the experiment is carried out in a range of nondimensional numbers relevant for fusion reactor applications, that is $10^3 < Ha < 5 \times 10^3$ and $10^4 < Re < 4.8 \times 10^4$; hence, in the MHD analysis viscous and inertial effects are considered negligible in most of the fluid volume. As crucial results, the experimental campaign observed that the pressure drop (Δp_{3D}), introduced by the LM entering in an insulated region, corresponds to the pressure loss that would occur in the same insulated pipe about four characteristic lengths (a) long.

$$\Delta p_{3D} \approx 4 \cdot (k_{pFCI}\sigma v B^2 a) \tag{20}$$

In Equation (20), the internal radius of the bare duct is chosen as characteristic length. Factor $k_p = c/(1+c)$ stands for the nondimensional fully developed flow pressure gradient, defined by Miyazaki et al. in [23]. Moreover, it is worth underlining that Equation (20) seems to be valid in the whole range of the experiment, namely for each Hartmann number and Reynolds number investigated [43]. In Figure 5, the meshing scheme employed for calculations is shown. Inlet conditions of the liquid metal are imposed by a time-dependent volume (TMDPVOL #10) and a time-dependent junction (TMDPJUN #15). The TMDPVOL sets NaK inlet temperature and the TMDPJUN imposes the liquid metal mass flow rate. The liquid outlet pressure is fixed by an additional time-dependent volume (TMDPVOL #30). A pipe component (PIPE #20) is used to model the conduit, that is divided in 15 single volumes, thus, linked by 14 single junctions. In the last 10 volumes, the isolation layer is applied; this is accomplished by modelling thinner stainless-steel walls, in order to obtain the same conductance ratio that is reported by Bühler, $c_{FCI} = 0.00476$ [43]. In Table 4, the main parameters are reported.

Table 4. Model geometry parameters.

Parameter	Bare Pipe	Insulated Pipe
Internal radius [mm]	48.6	47
Axial length [mm]	200	486
t_w [mm]	7.32	0.46
c	0.0727	0.00476

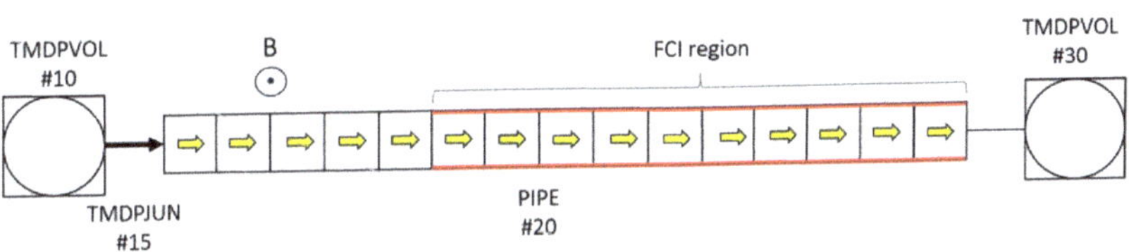

Figure 5. Sketch of the model, outlining RELAP5 discretization. The region where flow channel insert is applied is underlined in red. With yellow arrows is delineated the fluid path.

The trend of the scaled pressure versus the nondimensional axial length (x) is shown in Figure 6. The pressure is scaled with reference value $p_0 = \sigma f v a B^2$ Pa, whereas all geometrical parameters are scaled through characteristic length a.

Is important to underline that the general behaviour of the nondimensional pressure loss found by Bühler is very similar for every Ha and Re values; for this reason, results for other governing parameters are not reported. In the whole pipe, it can be observed the very good agreement that RELAP5 has with both theoretically predicted pressure loss and experimental results. The nondimensional pressure gradient computed by the code completely matches the one reported by Miyazaki for circular conduits (black line in Figure 6).

In Table 5, are gathered numerical results for several calculations performed. The code shows an excellent accuracy (relative errors $< +0.25\%$) in evaluating fully developed flow loss of load and the absolute value of three-dimensional pressure drop is detected with good precision (discrepancy $< +10\%$). The slight over-valuation regarding localized losses can be explained as follows: beside a discontinuity in wall electrical conductance, the flow also faces a moderate cross-section contraction due to the presence of insulation structure. Hence total 3D head loss, expressed by Equation (20), comprehends the contribution of a sudden cross-section contraction that, however, is challenging to extrapolate. Therefore, considering that the RELAP5 FCI model is based on experimental outcomes that usually include these contraction/expansion contributions, the discrepancy of Δp_{3D} with the experimental data in the present test case is because of the cross-section variation being taken into account twice by RELAP5: with the activation of the flow channel insert model and the cross-section restriction model.

Figure 6. Total nondimensional pressure drop versus scaled axial length. Vertical dashed line states the beginning of the insulated part of the conduit. The deviation, between fully developed flow theory and numerical/experimental outcomes, that exists after $x = 0$ is due to three-dimensional pressure drop.

Table 5. Relative errors of numerical results compared with theory/experimental outcomes.

Nondimensional Parameters	Pressure Gradient Error in Bare Pipe [%]	Pressure Gradient Error in Insulated Pipe [%]	3D Pressure Loss Error [%]
$Ha = 2000$, $Re = 10{,}186$	0.0076	0.06	9.4
$Ha = 2000$, $Re = 20{,}155$	0.008	0.021	9.43
$Ha = 3000$, $Re = 10{,}147$	0.25	0.17	9.7

4.1.2. Sudden Cross-Section Variations

The experimental analysis carried out by Buhler and Horanyi in [32] is used as a benchmark for verifying RELAP5 model for 3D MHD pressure drop due to a sudden change of the duct cross-section. Furthermore, it is also useful to demonstrate the capability of the code to predict 2D pressure loss occurring in square/rectangular conduits. The experiment characterizes a liquid metal (NaK) flow within a rectangular duct that undergoes a sudden cross-section expansion, after which the conduit cross-section is a square (Figure 7). The duct expands only along the magnetic field direction; therefore, the cross-section variation is purely parallel. The expansion ratio is defined as Z = 4 and the conductance ratio of the expanded channel is $c = 0.028$ (see Table 6 for geometrical details). When the fluid approaches the variation, it undergoes a velocity redistribution to rearrange its structure that induces an axial potential difference. Consequently, stream-wise electrical currents arise and, interacting with the strong magnetic field, they generate extra Lorentz forces that are responsible for the additional pressure drop (Δp_{3D}). The test section is made of stainless-steel with uniform wall thickness, both upstream and downstream of the expansion. The flow is horizontal and isothermal, so that thermo-physical properties of the materials are evaluated at the constant temperature of 293.15 K (see Table 3). The magnetic field, considered uniform and unidirectional, is applied along the expansion direction. The experiment is conducted in a range of nondimensional numbers $10^3 < Ha < 5 \times 10^3$ and $30 < N < 1.3 \times 10^5$ [32].

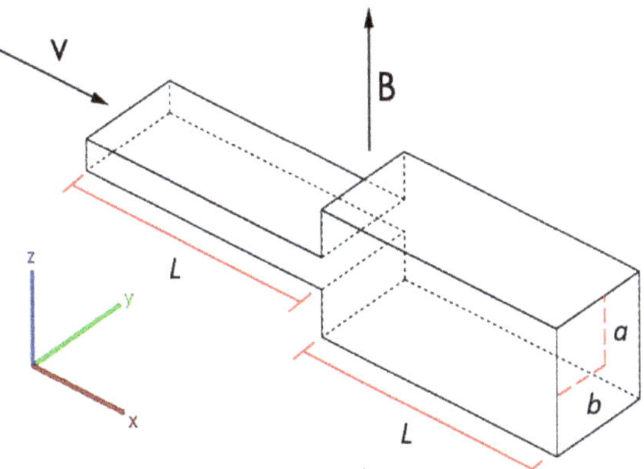

Figure 7. Geometry sketch of the experimental section.

In Figure 8, is shown the nodalization scheme of the model. Fluid inlet conditions are established by a time-dependent volume component (TMDPVOL #10) and a time-dependent junction (TMDPJUN #15). The TMDPVOL provides metal inlet temperature and the TMDPJUN sets the mass flow rate. The liquid outlet pressure is fixed by another time-dependent volume (TMDPVOL #30). A pipe component (PIPE #20) simulates the duct, that is divided in 12 single volumes linked by 11 single junctions. In Table 6, the main parameters of the model are reported.

In Figure 9, is shown a representative trend of the scaled pressure versus the nondimensional axial length. The pressure is scaled with $p_0 = \sigma f v a B^2$ Pa and all geometrical parameters through the characteristic length $a = 0.047$ m. In the region where the flow is fully developed, according to Bühler, the nondimensional pressure gradient is $k_p = 0.37479$ for the rectangular pipe, whereas $k_p = 0.02057$ in the square one, for the governing parameters $Ha = 4000$ and $N = 29{,}604$.

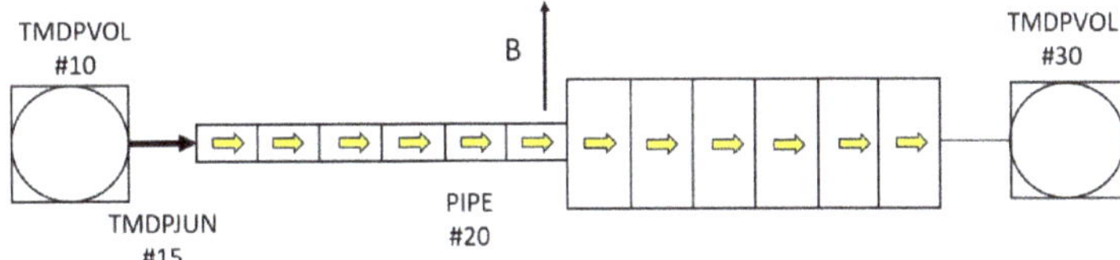

Figure 8. Sketch of the RELAP5 nodalization. Yellow arrows mark the fluid path. Magnetic field (B) is shown with a black vector.

Table 6. Model geometry parameters.

Parameter	Rectangular Pipe	Square Pipe
a [mm]	11.75	47
b [mm]	47	47
L [mm]	300	300
t_w [mm]	3	3
c	0.112	0.028

As reported in Table 7, RELAP5 detects with great accuracy the pressure gradient due to MHD 2D phenomena in square/rectangular conduits, with a relative error always lower than +0.5%. Furthermore, also the total 3D loss is well predicted by the code, that provides accurate results if compared with the ones of Equation (19).

Figure 9. Total nondimensional pressure drop versus scaled axial length for $Ha = 4000$ and $N = 29{,}608$. Vertical dashed line points the sudden expansion.

Table 7. 2D and 3D pressure losses errors, for different Ha and N.

Parameters	Numerical k_p Rectangular Pipe	Pressure Grad. Error RECT. Pipe [%]	Numerical k_p Square Pipe	Pressure Grad. Error Sq. Pipe [%]	Experiment ΔP_{3D} [kPa]	Numerical ΔP_{3D} [kPa]	Relative Error [%]
$Ha = 2000, N = 7455$	0.375	0.056	0.02053	0.1	1.46	1.462	0.13
$Ha = 3000, N = 3160$	0.37512	0.09	0.02057	0.29	4.9	4.9	0.54
$Ha = 4000, N = 29{,}608$	0.3754	0.17	0.0206	0.46	3.9	3.87	−0.12

4.2. Multiple Effect Validation

4.2.1. Lithium Lead Ceramic Breeder TBM

Computational fluid-dynamic analysis of MHD flow in a full-scale variant of the proposed Lithium–Lead cooled Ceramic Breeder Test Blanket Module proposed by India for the ITER experimental campaign, has been performed, at fusion-relevant Hartmann number ($Ha \approx 1.8 \times 10^4$), by Swain et al. in [51].

Numerical MHD studies at high magnetic field for such complex configurations are rare and they provide precious results for the system codes verification phase. In [51], an exhaustive thermal-hydraulic analysis is performed; for the aim of this work, however, the attention is focused on the pressure drop outcomes.

In Figure 10a, a two-dimensional sketch that describes the TBM geometry and flow path is represented. The LM involved is Li-Pb and enters the module through an inlet header channel (segment A-B). After that, the fluid is driven upward through five vertical (poloidal) channels (segment B-C). The poloidal channels are numbered from 1 to 5 moving away from the return channel, with Channel 1 being the closest to the latter. Consequently, the fluid is collected in top header region and then it goes down in the poloidal return channel (segment D-E). At last, the lithium-lead gathers in the outlet header (E-F) before leaving the TBM. All Li-Pb ducts have rectangular cross-section. The LM is the coolant used to extract thermal power from the breeding zone, which is constituted by boxes filled with the solid breeder (lithium-titanate) and enclosed by thin Reduced-Activation Ferritic-Martensitic steel (RAFMS) walls. For the purpose of MHD calculations, the breeder modules can be assumed as perfectly insulating and only the bounding walls must be considered to estimate the equivalent wall conductivity. Geometrical parameters are consistent with those reported in [51], and are collected in Table 8. Thermo-physical properties of RAFMS and Li-Pb are evaluated at the reference temperature of 623 K and the flow is assumed to be purely isothermal. Magnetic field B is set at 4 T and assumed to be unidirectional and aligned with the toroidal axis. Mass flow rate imposed at the inlet is equal to 12 kg/s.

The nodalization scheme, developed on the geometry presented in Figure 10a, is reported in Figure 10b. It is composed of 118 control volumes and 121 hydraulic junctions. The sliced modelling approach is applied, assuming the same length for the vertically oriented volumes at the same level. Inlet conditions of the liquid metal are imposed by a time-dependent volume (TMDPVOL # 10) and a time-dependent junction (TMDPJUN # 15). The TMDPVOL sets Li-Pb inlet temperature and the TMDPJUN imposes the liquid metal mass flow rate. The Li-Pb outlet pressure is fixed by an additional time-dependent volume (TMDPVOL # 130). All the sections composing the TBM have been modelled with equivalent pipe components. The modelling approach is to keep actual inventory and flow area for each section. Hydraulic k-loss coefficients for abrupt area changes, bends and tees have been evaluated using the correlations presented in Idel'chik handbook [52] and are used to obtain a reliable figure for the ordinary hydrodynamics. The coloured markers in Figure 10b (refer to the online version of the paper) identify the acquisition points of the pressure within the model. The same colours are associated with the characteristic hydraulic elements reported in Figure 10a: red (inlet header), blue (poloidal channel), green (top header), purple (return channel), orange (outlet header). The differential pressure for the i-th control volume is evaluated as follows: $dp_i = p_i - p_F + \rho g h_{i-F}$, where p_i is the pressure acquired in the i-th control volume, p_F is the pressure acquired in the F position, ρ is the Li-Pb density, g is the gravitational acceleration and h_{i-F} is the elevation difference between the i-th control volume and the F level.

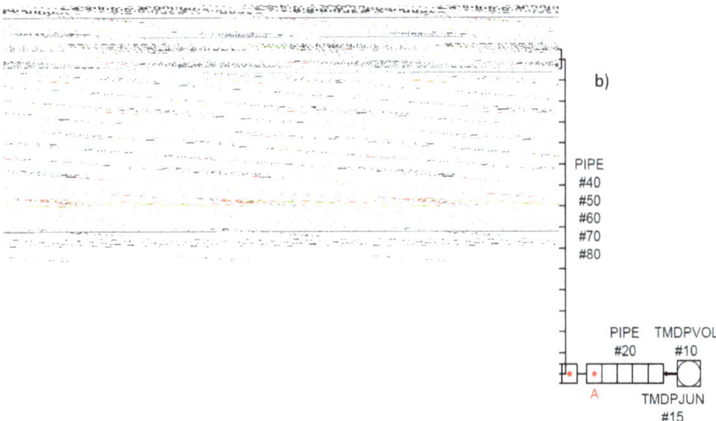

Figure 10. (**a**) 2D scheme of the LLCB TBM. Shaded area represents the breeding modules. (**b**) Nodalization of the geometry.

Table 8. Main geometrical parameters of LLCB flow path [51].

Component	a[m]	b[m]	Axial Length[m]	$t_w(Ha)$[m]	$t_w(Side1)$[m]	$t_w(Side2)$[m]
Inlet header	0.214	0.025	0.373	0.026	0.004	0.004
Poloidal channels	0.214	0.012	1.47	0.026	0.004	0.004
Top header	0.214	0.025	0.391	0.026	0.0025	0.002
Return channel	0.214	0.038	1.545	0.026	0.004	0.004
Outlet header	0.214	0.025	0.373	0.026	0.004	0.004

Before the main numerical campaign, a purely hydrodynamic simulation, imposing $B = 0$ T, is performed in [51] to quantify the weight of magnetohydrodynamic effects over inertial and viscous ones. The same simulation is repeated with RELAP5 to assess the efficacy of the geometrical and nodalization model chosen. Excellent results, in terms of mass flow rate distribution in vertical channels, are obtained with the code; the discrepancy from the numerical analysis is lower than 3.7% in each channel. In Table 9, the comparison is summarized. As a reference, the pressure loss in the hydrodynamic LLCB TBM is estimated at 0.432 kPa, approximately 1.2% of the pressure loss estimated by the MHD model.

Table 9. Mass flow rate outcomes for ordinary hydro-dynamic case ($B = 0T$), comparison between CFD analysis and RELAP5. (*) Inlet total mass flow rate as boundary condition.

Channel Number	Flow Rate [kg/s] Swain [51]	Flow Rate [kg/s] RELAP5	Error [%]
Channel-1	4.0068	4.0043	-0.0624
Channel-2	3.0568	3.0167	-1.3118
Channel-3	2.2790	2.3378	2.5801
Channel-4	1.6100	1.5632	-2.9068
Channel-5	1.0400	1.0780	3.6538
Return channel	12.0 (*)	12.0 (*)	0

Once the input geometry is proven efficient, the calculation with $B = 4T$ is carried out. Reference governing parameters are Ha = 17,845 and N = 4697; the characteristic length is the toroidal half-length of the whole module, namely a = 0.214 m. In Figure 11, it is reported the dimensional pressure drop versus axial length, occurring in the fluid path that is underlined with coloured segments in Figure 10a. It can be noticed that RELAP5 accurately predicts the total pressure drop, with a relative error that is $\epsilon = -0.97\%$. It is worth reminding that magnetic field is always directed perpendicular to the blanket module. Therefore, this calculation demonstrates that the 3D MHD models implemented for bends and cross-section variations occurring perpendicular to B are reliable to simulate even a complex geometry like the one considered in this study.

Nevertheless, it is important to point out a significant shortcoming of our code. As mentioned, RELAP5 is not currently capable of recreating electromagnetic coupling effects, like the ones that exist in the LLCB due to the sharing of an electrically conductive wall between Channel-1 and the return channel. The lack of this feature is made quite noticeable by the huge difference that is found between the RELAP5 and CFD predictions of the mass flow rate distribution across the TBM vertical channels, as it is shown in Table 10.

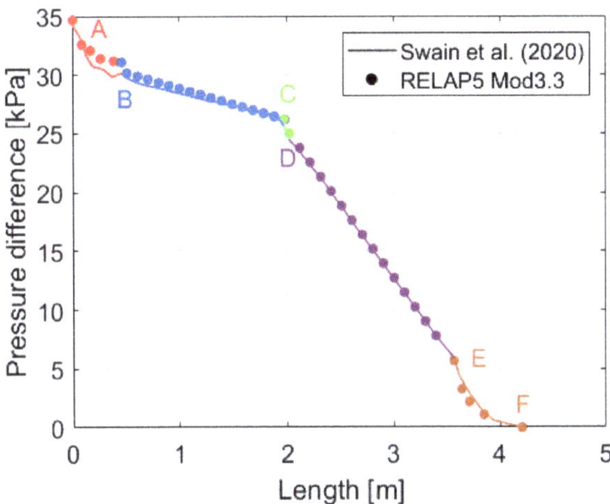

Figure 11. Pressure drop trend along the specified fluid path, for $Ha = 17{,}845$ and $N = 4697$.

Table 10. Mass flow rate outcomes for MHD ($B = 4T$), comparison between CFD analysis and RELAP5.

Channel Number	Flow Rate [kg/s] Swain [51]	Flow Rate [kg/s] RELAP5	Error [%]
Channel-1	7.2600	2.8323	−60.9876
Channel-2	1.4300	2.5061	75.25175
Channel-3	1.0100	2.3328	130.9703
Channel-4	1.1500	2.0886	81.61739
Channel-5	1.1500	2.2402	94.8
Return channel	12.0	12.0	0

According to Swain's CFD calculation, the presence of electromagnetic coupling phenomena accentuates the imbalance of mass flow rate in poloidal channels, that in the OHD case is due to inertial forces only. As a matter of fact, the numerical pressure gradient of Channel-1 is significantly decreased by local electrical coupling effects due to its adjacency with the counter-flowing return channel that, assuming uniform flow distribution, carries as much as five times its flow rate. A qualitative explanation, cfr. Figure 12, can be given as follow. Due to the electrical contact between the two channels, currents generated in one can leak and close through the other. Since $j \propto u_0$, the currents induced in the return channel are stronger in magnitude than those in Channel-1, due to the larger flow rate there. When these currents close through Channel-1, they promote the fluid movement rather than opposing it, as they would have in their original conduit, since $j = u \times B$. Velocity increases in Channel-1 until the resistive Lorentz force generated by currents induced in is enough to compensate the leakage currents.

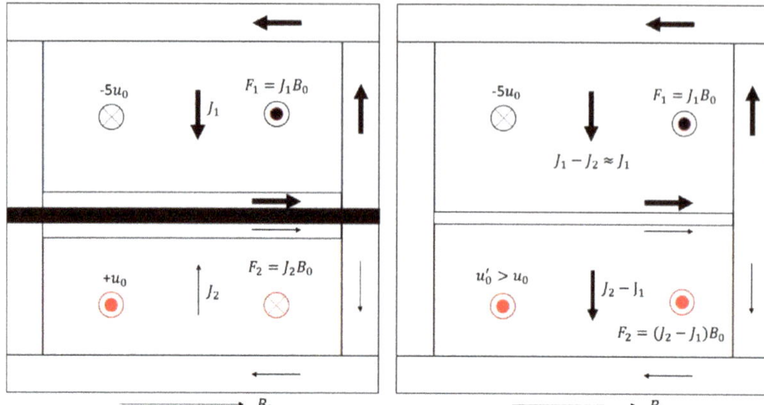

Figure 12. (Left) If the return (**top**) and Channel-1 (**bottom**) are not in electrical contact, i.e., separated by a thin layer of perfect electrical insulation (black band), no leakage currents can be exchanged and, in general, $j_1 > j_2$ due to the different flow rate. Lorentz force self-induced in the two channels opposes the flow movement. (**Right**) if the channels are in electrical contact, currents from the return channel can close through Channel-1 and induce a net promoting Lorentz force that increases the flow rate in it. The increase in average velocity eventually makes $j_1 > j_2$ and the balance is restored with the reappearance of a retarding Lorentz force in Channel-1.

The macroscopic effect is that Channel-1 draws more flow rate than the other poloidal channels, from which the flow imbalance observed by Swain et al. in [51]. Furthermore, counter-flow coupling effects influence also the flow in the inlet and outlet header where a substantial flow rate difference exists, in particular, in the region close to Channel-1 and Channel-2. This behaviour further favours a net decrease of the resistance offered by the route through Channel-1 (and, to a lesser degree, Channel-2). It is worth noticing that, instead, electro-coupling does not radically affect poloidal channels 3–5, since they are divided by solid breeder modules.

Aforementioned phenomena cannot be modelled by RELAP5 that, therefore, is going to compute the flow distribution across the poloidal channels exclusively according to the 2D and 3D pressure losses occurring therein. Inertial effects are almost negligible at $N = O(10^3)$ and $Ha = O(10^4)$ and, consequently, the result is an almost uniform flow distribution. Differently put, the main source of mass flow rate disequilibrium in hydro-dynamic calculation (inertia), is completely overcome by magnetohydrodynamic effects that, anyhow, are significantly affected by electromagnetic coupling in this specific configuration. It can be concluded that in its current state RELAP5 MHD module is not suitable to estimate the flow rate distribution in a manifold featuring one or more channels in electrical contact with an adjacent-counter-flowing duct with a very large flow rate. Nevertheless, it should be highlighted that the feature that the code is currently lacking, seems to not affect the pressure drop calculation in this configuration significantly, as can be observed by the results reported in Table 11, which surprisingly match quite well the prediction from the CFD code. A possible explanation of this unintuitive behaviour is provided in the following.

The major part of the overall pressure drops of the TBM, more than 50%, takes place in the return channel as is shown in Table 11. This happens because in the return channel the velocity is larger than the one in the poloidal ascending conduits and the expansions/contractions contributions are more important. Conversely, the effect of electrical coupling within the descending channel is weak (negligible), as it is also highlighted by the results of Swain et al. (Table 4 in [51]), where the numerical pressure gradient is observed to be very close to the theoretical one (which, of course, neglects coupling). Consequently, RE-LAP5 is able to compute consistently the "leading" pressure drop source within the whole

TBM. Moreover, it is important to keep in mind that electrocoupling primary decreases distributed pressure losses (2D MHD), which are linearly dependent by the the fluid mean velocity. On the other hand concentrated drops (3D MHD), that are also linear with velocity, are not significantly affected by the coupling. The good agreement in terms of total pressure drop in the poloidal Channel-1 between RELAP5 and CFD computation (see Table 11), despite a substantial mass flow rate difference, could be due to an overall compensation of 2D and 3D MHD losses (referring to the relations expressed in Equations (1) and (2)). In [51], 2D MHD drops are reduced by coupling, whereas 3D MHD ones are increased due to the high mean velocity in the conduit; RELAP5 numerical model instead provides higher MHD distributed losses since the coupling is not taken in account but the local MHD drops are less influential considering the moderate average fluid velocity. However, it is crucial to underline that this discussion is strictly valid only for the specific geometrical configuration of the LLCB TBM, since electrocoupling phenomena deeply depend on multi-channel arrangement.

To summarize, the MHD module of RELAP5 has demonstrated the capacity to provide results which are consistent with the prediction from a direct numerical simulation code in terms of electromagnetic pressure loss. A lack of fidelity is highlighted in the estimate of mass flow distribution for a manifold feeding channels which are in electrical contact with adjacent counter-flowing ducts with large flow rates. This is caused by the absence of a dedicated electromagnetic coupling model.

Table 11. Pressure drop differences in counterflow electro-coupled conduits ($B = 4$ T).

Channel Number	Swain [51] Δp [kPa]	RELAP5 Δp [kPa]	Error [%]
Channel-1	4.5700	4.8770	6.717724
Return channel	18.5000	19.3270	4.47027

4.2.2. Helium Cooled Lithium Lead TBM Mock-Up

MHD flow in a scaled mock-up of a Helium Cooled Lithium Lead test blanket module for ITER has been experimentally investigated by Bühler et al. in [33] to gather experimental data to support the proposed design concept and as a benchmark for validation of numerical tools. As one of the few experimental campaigns involving a geometry representative of the complexity of a fusion reactor blanket design, this benchmark is considered the most suitable to demonstrate the validity of the numerical approach adopted by RELAP5. The HCLL mock-up is scaled down by a factor of 2 compared with the TBM to fit into the experimental facility MEKKA. Experimental data have been collected in a wide range of parameters: $5 \times 10^2 < Ha < 5 \times 10^3$ and $2 \times 10^2 < Re < 10^4$.

The fluid adopted for the experiment is NaK, whereas the structural material is stainless steel. Thermo-physical properties of involved materials are taken at the operating temperature of 293 K (see Table 3). The HCLL geometry is very complicated and the fluid undergoes numerous bends and sudden cross-section variations. A two-dimensional scheme of the flow path is depicted in Figure 13. The mock-up encompasses two parallel breeding units (BUs), each one being composed by an inflow unit (D-E) and an outflow unit (F-G). Liquid metal enters the TBM at position A, which corresponds to the beginning of the feeding pipe (with circular cross section). After, it is collected in the inlet manifold, that serves to distribute the flow rate in the two breeding units. The manifold is composed by a tall and narrow rectangular duct realized between the BU back plate and the mock-up external plate (Figure 14). The connection with the feeding pipe is realized in a larger cavity that undergoes a sudden contraction of about 50% in the magnetic field direction at the height of the first dashed line in Figure 13. The narrower conduit conveys the fluid to the second BU and it is gradually tapered to 50% of its original cross-section until the third dashed line in Figure 13. The geometry of the outlet manifold mirrors the previous one: gradually increasing cross-section, sudden expansion and connection with the draining pipe. It should be noted that helium manifolds, that will reduce the available cross-section for the LM in its manifold, were not included in the mock-up and they are, similarly, not included in RELAP5 [33]. The interface between manifold and BUs is ensured through an elongated narrow orifice in the BU back plate. Once inside the inflow unit, the fluid is subjected to another subdivision due to the presence of thin metal plates that partitions the volume in 6 parallel channels, simulating the presence of helium cooling plates. At the first wall (FW), the liquid meets an elongated and narrow toroidal-radial orifice that allows hydraulic connection with the outflow unit. The LM is then gathered in the outlet manifold and leaves the TBM through the draining pipe [33]. In Table 12, are gathered the dimensions of the mock-up.

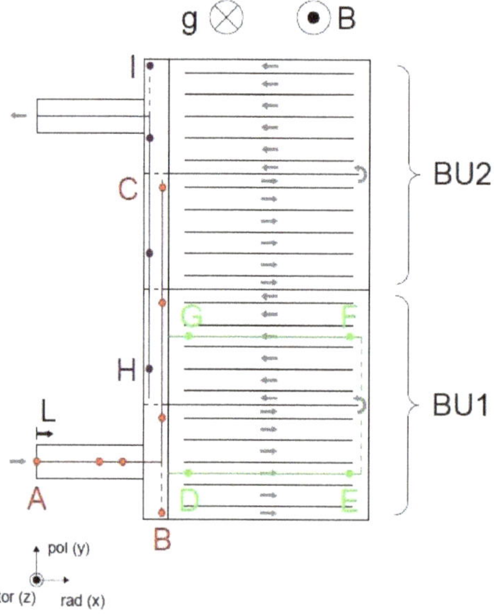

Figure 13. 2D view (radial-poloidal) of the TBM geometry. The flow path underlined with colored segments (A-I) is used in the following for pressure drop calculation.

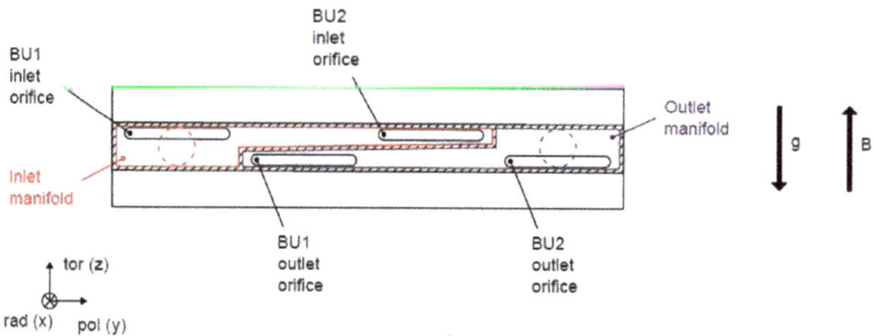

Figure 14. Toroidal-poloidal view of the mock-up. In red is outlined the feeding manifold, whereas in purple the collecting one. Cross-section shape of both inlet/outlet pipes and manifold/BU interface are reported.

Table 12. Main geometrical parameters. (*) Average toroidal length of the narrow manifold along poloidal direction. (**) Cooling plates dimensions are not reported [33].

Component	Toroidal Length [mm]	Poloidal Length [mm]	Radial Length [mm]	$t_w(Ha)$ [mm]	$t_w(Side1)$ [mm]	$t_w(Side2)$ [mm]
Feeding/Draining Pipe	14.75	14.75	181.5	3.88	3.88	3.88
Larger Manifold (In/out)	31	101.48	13	3	2.8	7.5
Narrow Manifold (in/out)	14 (*)	203	13	3	2.8	7.5
BU (**)	90	9.48	171	1.38	2.75	2.75

The nodalization scheme adopted is reported in Figure 15. Two time-dependent volumes define the inlet temperature and the outlet pressure of the NaK alloy. The liquid metal mass flow rate is imposed by the time-dependent junction, connecting the inlet TMDPVOL and the inlet conduit (PIPE #20 in Figure 15). Inlet and outlet manifolds are simulated with two equivalent pipes. Each one is composed of three control volumes (CVs), reproducing the relevant configuration of the component. Focusing on the inlet manifold, the first and the third CVs simulate the inlet chambers, adjacent to the first interface between manifold and breeding units. The second CV reproduces the connection between those. For the first and the third control volumes, the flow along the radial direction is selected. This approach allows to estimate the actual abrupt area change in the interface with the breeding zone (BZ). The two BUs are simulated with two pipe components. The six channels composing the BU are collapsed in an equivalent single pipe, characterized by total flow area, equivalent hydraulic diameter and geometrical parameters relevant for MHD analysis of a single representative channel, namely one of the four at the centre of the cell. Each pipe component consists of seven CVs, reproducing inlet and outlet holes, the radial channels and the inversion chamber. Coloured markers, in both Figures 13 and 15, mark the position of pressure acquisition data points within the mock-up and RELAP5 numerical model.

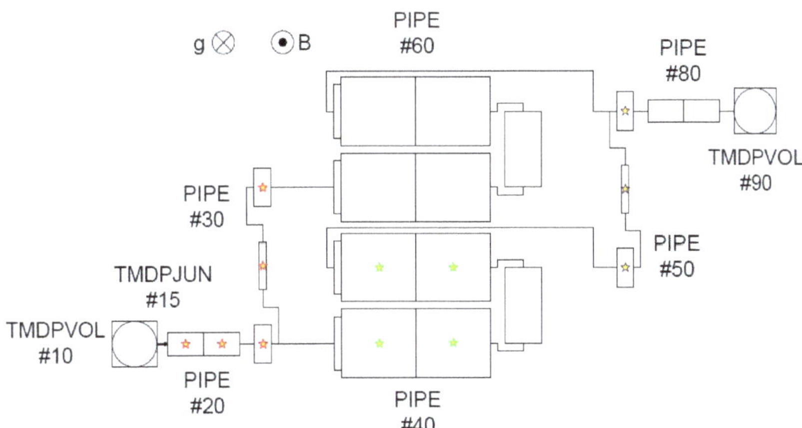

Figure 15. Meshing scheme of the geometry.

Figure 16 shows the comparison between experimental and RELAP5 data for nondimensional pressure drop along the flow path A-I (i.e., coloured lines in Figure 13) with respect to the pressure measured in point A. the characteristic length chosen is the toroidal half-length of BU, namely $a = 0.045$ m, whereas pressure is scaled with $p_0 = \sigma f v a B^2$ Pa. As for cases previously studied, the pressure scaling employed allows, for a single Hartman number investigated, to condense the results for various Reynolds numbers in a unique representation. Furthermore, it confirms that MHD flows at fusion conditions are not affected by relevant inertia phenomena. For this reason, only few results are reported in graphical form.

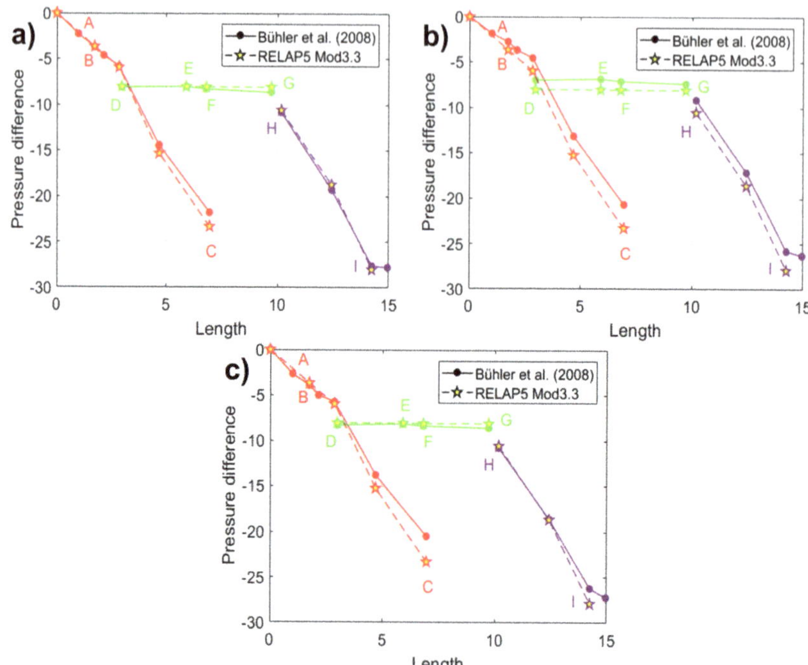

Figure 16. Nondimensional pressure drop versus scaled length. Comparison of results between R5 and reference experiment. (**a**) $Ha = 3000, Re = 3360$. (**b**) $Ha = 4000, Re = 1140$. (**c**) $Ha = 5000, Re = 2000$.

As can be observed, pressure drop mostly occurs in the circular inlet conduit (A–B) and in the inlet/outlet manifolds (B–C and H–I), since the velocities here are much higher than in BUs due to the smaller cross sections. Moreover, the walls of pipes and manifolds are thicker, hence their wall conductance ratios generally increased compared with the ones in breeder units ($c_{MAN} \approx 0.09, c_{BU} \approx 0.01$). Those features provide higher currents and, therefore, more intense Lorentz forces causing greater pressure head losses. Furthermore, additional contributions to pressure loss occur because the fluid is forced to pass through narrow orifices into and out of the breeder units (B–D and G–H). Here, three-dimensional MHD effects arise and induce an additional pressure drop. The contraction (and the following expansion) of the fluid into the narrow gap between inflow/outflow units (dashed line E–F) and the 180° bend at the first wall, have no substantial effect on the loss of load, since they occur in a plane perpendicular to B. It is noticed that the pressure loss in the breeding cells (D–E and F–G) where the fluid has a moderate velocity ($\approx$1 mm/s) is negligible, especially when compared with the one occurring in pipes/manifolds.

RELAP5 shows an excellent agreement with experimental results. In Table 13, as example, outcomes and relative errors for two the calculations performed at $Ha = 3000$ are reported. In general, is observed an excellent prediction of the total pressure loss, with a discrepancy that is $-1\% < \epsilon < +14\%$. This is a crucial result, since it clearly demonstrates that RELAP5 is suitable to predict pressure losses even in relatively complex geometries as the one proposed for the HCLL TBM.

As discussed for LLCB, the main drawback of the code in the current version is the lack of modelling for electromagnetic coupling, which is the main source of error of RELAP5 calculation. In [33], electrical coupling effects are demonstrated to be influential in the flow rate distribution, in particular among internal channels of BUs. Nevertheless, the pressure drop is significantly reduced in the breeding unit, thus the lack of a coupling model in the code does not essentially influence the prediction of pressure loss. It should also be noted that the code slightly over-predicts the pressure loss due to the cross-section variations; this is particularly evident in the connection between manifold and breeding unit (B–D and G–H). This is a consequence of the conservative assumptions adopted for the development of the 3D pressure loss model for this component; the estimate is likely to improve once more data will be available to fine tune it. As a final note, it is worth mentioning that the lower accuracy that the code shows for results at $Ha = 4000$ could be attributed to an anomaly related to experimental data acquisition, considering that RELAP5 performance is substantially the same for $Ha = 3000$ and $Ha = 5000$.

Table 13. Pressure drop results (nondimensional) for every position of flow path and total loss. Relative Errors are reported.

	$Ha = 3000, Re = 666$				$Ha = 3000, Re = 1172$		
Position	Δp [33]	Δp RELAP5	Relative Error	Position	Δp [33]	Δp RELAP5	Relative Error
B	−3.24	−3.661	12.99%	B	−3.73	−3.662	−1.82%
C	−21.84	−23.29	6.64%	C	−21.97	−23.29	6.01%
D	−8.09	−8.019	−0.88%	D	−8.16	−8.018	−1.74%
E	−7.82	−8.03	2.69%	E	−8.06	−8.03	−0.37%
F	−8.23	−8.061	−2.05%	F	−8.38	−8.06	−3.82%
G	−8.63	−8.072	−6.47%	G	−8.74	−8.072	−7.64%
H	−10.79	−10.55	−2.22%	H	−10.79	−10.56	−2.13%
I	−27.64	−27.89	0.90%	I	−28.26	−27.91	−1.24%

5. Conclusions and Further Works

A new module for the best-estimate SYS-TH code RELAP5/MOD3.3 has been developed at Sapienza, University of Rome, allowing the program to model MHD pressure drop in liquid metal circuits. Two-dimensional and three-dimensional MHD pressure losses, akin to distributed and concentrated friction losses in ordinary hydrodynamic flow, are modelled for a wide range of geometries. These features are of particular interest for fusion related activities and, in particular, for the design of liquid breeding blankets, where MHD pressure loss is the dominant component. In its current state, the code is deemed suitable for the simulation of liquid metal breeding blankets and circulating loops in normal operation and postulated transients that do not involve a rapid time variation of the reactor magnetic field. Application of the code to the prediction of non-isothermal transients for these systems must be carefully considered since MHD effects on heat transfer have not yet been implemented.

A Verification and Validation (V&V) campaign is performed, using several numerical and experimental studies as benchmark, to demonstrate the capacity of the code of reproducing physical results. In a first step, sub-routines are validated independently (single effect) to ensure their correct operation. Subsequently, RELAP5 is used to recreate the results obtained by direct numerical simulation tools and experimental campaigns in two Test Blanket Modules, i.e., the LLCB [51] and HCLL [33]. It is found that MHD subroutines of RELAP5/MOD3.3 are reliable in modelling electromagnetic effects that are crucial in the pressure drop assessment of fusion related LM circuits. Two-dimensional loss is well predicted by the code, with a discrepancy with respect to the numerical/experimental outcomes that is always around 1%. This certainly makes the code competitive, in terms of capability, with other numerical tools that can handle this level of analysis, such as MARS-FR [13] and MHD-SYS [17] (Section 1). Furthermore, RELAP5 three-dimensional MHD pressure drop module assures good agreement with TBM validation test cases results, with a discrepancy that is generally $\epsilon < 10\%$; this is mostly due to the conservative approach that has been employed to develop 3D pressure drop models, needed to overcome the physical uncertainties that still exist in the prediction of those phenomena. Albeit the code is lacking a dedicated subroutine to represent electro-coupling phenomena, the absence of such a model does not affect significantly the pressure drop analysis, but rather the capacity of the code in predicting mass flow rate distribution when multichannel configurations occur. It should be remarked that the excellent agreement with the HCLL TBM experimental data is a result that demonstrates the code scalability to reactor conditions since the HCLL mock-up can be considered by all means a IET facility. In conclusion, the code development discussed in this paper has significantly increased the capability of the modified RELAP5, that now can be considered a reliable tool for magnetohydrodynamic pressure drop evaluations in liquid metal circuits.

The current version of the code is still far from an ideal implementation since several important features, such as an electromagnetic coupling model, are still lacking and correlations for 3D MHD losses must be refined to extend their flexibility and range of validity. It should be stressed that these limitations reflect a lack of theoretical understanding for the former case and inconsistent experimental data for the latter and, thus, are likely to be improved as the knowledge progresses. Future activities will be focused on the expansion of bend and cross-section variation modelling for arbitrary wall conductivity. Preliminary work has already been done for the definition of an electromagnetic coupling model based on the analytical solution presented by Bluck et al. [53]. Implementation of other relevant configurations that imply localized loss of load, such as fringing magnetic field or MHD flow around obstacles, is currently being considered, as well as to extend the support to the modelling of perfectly insulated walls. To complete the picture of MHD phenomena in the breeding blanket of a fusion reactor, the next phase of development will be dedicated to implement a MHD heat transfer models. It is planned to derive, at first, a basic model for MHD forced convection heat exchange mechanism that, afterwards, could be extended to include buoyancy effects (mixed convection). Subsequently, a model for evaluating MHD influence on tritium transport is foreseen to be developed if the research about this topic will be intensified, as proposed by Seo et al. in [10]. At last, whenever MHD phenomena caused by a strong time-varying magnetic field will be better characterize in the literature, the models are foreseen to be updated so that a wider range of scenarios (operational or incidental) could be possibly simulated by the code.

Author Contributions: Conceptualization, A.T., L.M., F.G. and G.C.; methodology, A.T., L.M.; software, L.M. and V.N.; validation, V.N., L.M. and A.T.; formal analysis, V.N., L.M. and A.T.; investigation, L.M., A.T. and V.N.; writing—original draft preparation, L.M.; writing—review and editing, A.T., L.M., V.N. and G.C.; visualization, L.M. and V.N.; supervision, A.T.; project administration, A.T. and G.C.; funding acquisition, F.G., G.C. All authors have read and agreed to the published version of the manuscript.

Funding: This work has been carried out within the framework of the EUROfusion Consortium and has received funding from the Euratom research and training program 2014–2018 and 2019–2020 under grant agreement No 633053. The views and opinions expressed herein do not necessarily reflect those of the European Commission.

Institutional Review Board Statement: Not applicable.

Informed Consent Statement: Not applicable.

Data Availability Statement: Not applicable.

Acknowledgments: The authors wish to thank Joelle Vallory and Italo Ricapito, for their preliminary review of the paper and effective suggestions. A heartfelt thanks to Elisa Martino, for her precious help with several images reported in this paper.

Conflicts of Interest: The authors declare no conflict of interest.

Abbreviations

The following abbreviations are used in this manuscript:

BB	Breeding Blanket
BE	Best Estimate
BU	Breeding Unit
BZ	Breeding Zone
CFD	Computational Fluid-Dynamic
DIAEE	Dipartimento di Ingegneria Astronautica, Elettrica ed Energetica
ENEA	Agenzia Nazionale Energie Alternative
EU-DEMO	EUropean DEMOnstration Power Plant
FCI	Flow Channel Insert
FW	First Wall
HCLL	Helium-Cooled Lithium–Lead
ITER	International Thermonuclear Experimental Reactor
KIT	Karlsruher Institut für Technologie
LLCB	Lithium–Lead Ceramic-Breeder
LM	Liquid Metal
MEKKA	Magneto-hydrodynamische Experimente in Natrium Kalium Karlsruhe
MHD	MagnetoHydroDynamic
OHD	OrdinaryHydroDynamic
RAFMS	Reduced-activation Ferritic/Martensitic Steel
RELAP5	Reactor Excursion Leak Analysis Program
SYS-TH	SYStem Thermal-Hydraulic
TBM	Test Blanket Module
TMDPJUN	TiMe-DePendent JUNction
TMDPVOL	TiMe-DePendent VOLume
V&V	Verification and Validation
WCLL	Water-Cooled Lithium–Lead

References

1. EUROfusion. *European Research Roadmap to the Realisation of Fusion Energy*; EUROfusion Programme Management Unit: Garching/Munich, Germany, 2018.
2. Smolentsev, S. Physical background, computations and practical issues of the magnetohydrodynamic pressure drop in a fusion liquid metal blanket. *Fluids* **2021**, *6*, 110. [CrossRef]
3. Smolentsev, S.; Spagnuolo, G.A.; Serikov, A.; Rasmussen, J.J.; Nielsen, A.H.; Naulin, V.; Marian, J.; Coleman, M.; Malerba, L. On the role of integrated computer modelling in fusion technology. *Fusion Eng. Des.* **2020**, *157*, 111671. [CrossRef]
4. Smolentsev, S.; Moreau, R.; Bühler, L.; Mistrangelo, C. MHD thermofluid issues of liquid-metal blankets: Phenomena and advances. *Fusion Eng. Des.* **2010**, *85*, 1196–1205. [CrossRef]
5. Tassone, A.; Caruso, G.; Del Nevo, A. Influence of PbLi hydraulic path and integration layout on MHD pressure losses. *Fusion Eng. Des.* **2020**, *155*, 111517. [CrossRef]
6. Kirillov, I.R.; Reed, C.B.; Barleon, L.; Miyazaki, K. Present understanding of MHD and heat transfer phenomena for liquid metal blankets. *Fusion Eng. Des.* **1995**, *27*, 553–569. [CrossRef]
7. Tassone, A.; Siriano, S.; Caruso, G.; Utili, M.; Del Nevo, A. MHD pressure drop estimate for the WCLL in-magnet PbLi loop. *Fusion Eng. Des.* **2020**, *160*, 111830. [CrossRef]
8. Reimann, J.; Benamati, G.; Moreau, R. *Report of Working Group MHD for the Blanket Concept Selection Exercise (BSE)*; Forschungszentrum Karlsruhe GmbH: Karlsruhe, Germany, 1995.
9. Smolentsev, S.; Badia, S.; Bhattacharyay, R.; Bühler, L.; Chen, L.; Huang, Q.; Jin, H.G.; Krasnov, D.; Lee, D.W.; De Les Valls, E.M.; et al. An approach to verification and validation of MHD codes for fusion applications. *Fusion Eng. Des.* **2015**, *100*, 65–72. [CrossRef]
10. Seo, S.B.; Hernandez, R.; Neal, M.O.; Meehan, N.; Novais, F.S.; Rizk, M.; Maldonado, G.I.; Brown, N.R. A review of thermal hydraulics systems analysis for breeding blanket design and future needs for fusion engineering demonstration facility design and licensing Figure of Merit. *Fusion Eng. Des.* **2021**, *172*, 112769. [CrossRef]
11. Bestion, D. *The Structure of System Thermal-Hydraulic (SYS-TH) Code for Nuclear Energy Applications*; Elsevier Ltd.: Amsterdam, The Netherlands, 2017; pp. 639–727. [CrossRef]
12. Code Development Team. *RELAP5-3D Code Manual: Code Structure, System Models and Solution Methods.*; Idaho National Laboratory (INL): Idaho Falls, ID, USA, 2015; Volume I.

13. Kim, S.H.; Kim, M.H.; Lee, D.W.; Choi, C. Code validation and development for MHD analysis of liquid metal flow in Korean TBM. *Fusion Eng. Des.* **2012**, *87*, 951–955. [CrossRef]
14. Panayotov, D.; Grief, A.; Merrill, B.J.; Humrickhouse, P.; Trow, M.; Dillistone, M.; Murgatroyd, J.T.; Owen, S.; Poitevin, Y.; Peers, K.; et al. Methodology for accident analyses of fusion breeder blankets and its application to helium-cooled pebble bed blanket. *Fusion Eng. Des.* **2016**, *109–111*, 1574–1580. [CrossRef]
15. Grief, A.; Owen, S.; Murgatroyd, J.; Panayotov, D.; Merrill, B.; Humrickhouse, P.; Saunders, C. Qualification of MELCOR and RELAP5 models for EU HCLL TBS accident analyses. *Fusion Eng. Des.* **2017**, *124*, 1165–1170. [CrossRef]
16. Panayotov, D.; Grief, A.; Merrill, B.J.; Humrickhouse, P.W.; Murgatroyd, J.T.; Owen, S.; Saunders, C. Uncertainties identification and initial evaluation in the accident analyses of fusion breeder blankets. *Fusion Eng. Des.* **2018**, *136*, 993–999. [CrossRef]
17. Wolfendale, M.J.; Bluck, M.J. A coupled systems code-CFD MHD solver for fusion blanket design. *Fusion Eng. Des.* **2015**, *98-99*, 1902–1906. [CrossRef]
18. Froio, A.; Batti, A.; Nevo, A.D.; Savoldi, L.; Spagnuolo, G.A.; Zanino, R.; Energia, D.; Torino, P.; Kessel, C.; Staack, G.; et al. Implementation of a System-Level Magnetohydrodynamic Model in the GETTHEM Code for the Analysis of the EU DEMO WCLL Breeding Blanket. In Proceedings of the Technology Of Fusion Energy (TOFE) Conference, Charleston, SC, USA, 20–23 April 2020.
19. Batti, A. *System-Level Hydraulic Modeling of the PbLi Loop for the Breeding Blanket of a Tokamak*; Politecnico di Torino: Turin, Italy, 2020.
20. International Atomic Energy Agency (IAEA). *Safety Assessment for Facilities and Activities*; Technical Report; IAEA: Vienna, Austria, 2016.
21. Müller, U.; Bühler, L. *Magnetofluiddynamics in Channels and Containers*; Springer: Berlin, Germany, 2001. [CrossRef]
22. Madarame, H.; Taghavi, K.; Tillack, M.S. Influence of Leakage Currents on Mhd Pressure Drop. *Fusion Technol.* **1985**, *8*, 264–269. [CrossRef]
23. Miyazaki, K.; Kotake, S.; Yamaoka, N.; Inoue, S.; Fujiie, Y. MHD pressure drop of NaK flow in stainless steel pipe. *Nucl. Technol.* **1983**, *4*, 447–452. [CrossRef]
24. Miyazaki, K.; Inoue, S.; Yamaoka, N.; Horiba, T.; Yokomizo, K. Magneto-Hydro-Dynamic Pressure Drop of Lithium Flow in Rectangular Ducts. *Fusion Technol.* **1986**, *10*, 830–836. [CrossRef]
25. Nuclear Safety Analysis Division. *RELAP5/MOD3.3 Code Manual Volume I: Code Structure, System Models and Solution Methods*; Idaho National Laboratory: Idaho Falls, ID, USA, 2003; Volume I.
26. Giannetti, F.; D'Alessandro, T.; Ciurluini, C. *Development of a RELAP5 Mod3.3 Version for FUSION Applications*; Sapienza DIAEE internal report D1902_ENBR_T01; Sapienza University of Rome: Rome, Italy, 2019.
27. Narcisi, V.; Melchiorri, L.; Giannetti, F.; Caruso, G. Assessment of relap5-3d for application on in-pool passive power removal systems. In Proceedings of the 30th European Safety and Reliability Conference and the 15th Probabilistic Safety Assessment and Management Conference, Venice, Italy, 1–5 November 2020; Research Publishing: Singapore, 2020; pp. 1135–1142. [CrossRef]
28. Narcisi, V.; Melchiorri, L.; Giannetti, F. Improvements of RELAP5/Mod3.3 heat transfer capabilities for simulation of in-pool passive power removal systems. *Ann. Nucl. Energy* **2021**, *160*, 108436. [CrossRef]
29. Barone, G.; Martelli, D.; Forgione, N. Implementation of Lead-Lithium as working fluid in RELAP5/Mod3.3. *Fusion Eng. Des.* **2019**, *146*, 1308–1312. [CrossRef]
30. Tas, H.; Malang, S.; Reiter, F.; Sannier, J. Liquid breeder materials. *J. Nucl. Mater.* **1988**, *155–157*, 178–187. [CrossRef]
31. Malang, S.; Mattas, R. Comparison of lithium and the eutectic lead-lithium alloy, two candidate liquid metal breeder materials for self-cooled blankets. *Fusion Eng. Des.* **1995**, *27*, 399–406. [CrossRef]
32. Bühler, L.; Horanyi, S. Experimental investigations of MHD flows in a sudden expansion. *Wiss. Berichte FZKA* **2006**, *7245*, 1–65.
33. Bühler, L.; Brinkmann, H.J.; Horanyi, S.; Starke, K. *Magnetohydrodynamic Flow in a Mock-Up of a HCLL Blanket. Part II. Experiments*; Technical Report FZKA 7424; Forschungszentrum Karlsruhe GmbH: Karlsruhe, Germany, 2008.
34. Bühler, L.; Brinkmann, H.J.; Koehly, C. Experimental study of liquid metal magnetohydrodynamic flows near gaps between flow channel inserts. *Fusion Eng. Des.* **2019**, *146*, 1399–1402. [CrossRef]
35. Balestra, P.; Giannetti, F.; Caruso, G.; Alfonsi, A. New RELAP5-3D lead and LBE thermophysical properties implementation for safety analysis of Gen IV reactors. *Sci. Technol. Nucl. Install.* **2016**, *2016*. [CrossRef]
36. Martelli, D.; Venturini, A.; Utili, M. Literature review of lead-lithium thermophysical properties. *Fusion Eng. Des.* **2019**, *138*, 183–195. [CrossRef]
37. Foust, O. *Sodium-NaK Engineering Handbook*; Gordon and Breach: New York, NY, USA, 1972; Volume I.
38. Mergia, K.; Boukos, N. Structural, thermal, electrical and magnetic properties of Eurofer 97 steel. *J. Nucl. Mater.* **2008**, *373*, 1–8. [CrossRef]
39. Reimann, J.; Molokov, S.; Platnieks, I.; Platacis, E. Mhd-Flow in Multichannel U-Bends: Screening Experiments and Theoretical Analysis. *Fusion Technol.* **1993**, *1992*, 1454–1458. [CrossRef]
40. Davidson, P.A. Magnetic damping of jets and vortices. *J. Fluid Mech.* **1995**, *299*, 153–186. [CrossRef]
41. Feng, J.; He, Q.; Chen, H.; Ye, M. Numerical Investigation of magnetohydrodynamic flow through Sudden expansion pipes in Liquid Metal Blankets. *Fusion Eng. Des.* **2016**, *109–111*, 1360–1364. [CrossRef]
42. Rhodes, T.J.; Smolentsev, S.; Abdou, M. Magnetohydrodynamic pressure drop and flow balancing of liquid metal flow in a prototypic fusion blanket manifold. *Phys. Fluids* **2018**, *30*, 057101. [CrossRef]

43. Bühler, L.; Mistrangelo, C. Pressure drop and velocity changes in MHD pipe flows due to a local interruption of the insulation. *Fusion Eng. Des.* **2018**, *127*, 185–191. [CrossRef]
44. Bühler, L.; Mistrangelo, C.; Brinkmann, H.J. Experimental investigation of liquid metal MHD flow entering a flow channel insert. *Fusion Eng. Des.* **2020**, *154*, 111484. [CrossRef]
45. Koehly, C.; Neuberger, H.; Bühler, L. Fabrication of thin-walled fusion blanket components like flow channel inserts by selective laser melting. *Fusion Eng. Des.* **2019**, *143*, 171–179. [CrossRef]
46. Patel, A.; Pulugundla, G.; Smolentsev, S.; Abdou, M.; Bhattacharyay, R. Validation of numerical solvers for liquid metal flow in a complex geometry in the presence of a strong magnetic field. *Theor. Comput. Fluid Dyn.* **2018**, *32*, 165–178. [CrossRef]
47. Smolentsev, S.; Rhodes, T.; Yan, Y.; Tassone, A.; Mistrangelo, C.; Bühler, L.; Urgorri, F.R. Code-to-Code Comparison for a PbLi Mixed-Convection MHD Flow. *Fusion Sci. Technol.* **2020**, 653–669. [CrossRef]
48. Sahu, S.; Bhattacharyay, R. Validation of COMSOL code for analyzing liquid metal magnetohydrodynamic flow. *Fusion Eng. Des.* **2018**, *127*, 151–159. [CrossRef]
49. Feng, J.; Chen, H.; He, Q.; Ye, M. Further validation of liquid metal MHD code for unstructured grid based on OpenFOAM. *Fusion Eng. Des.* **2015**, *100*, 260–264. [CrossRef]
50. Gajbhiye, N.L.; Throvagunta, P.; Eswaran, V. Validation and verification of a robust 3-D MHD code. *Fusion Eng. Des.* **2018**, *128*, 7–22. [CrossRef]
51. Swain, P.K.; Koli, P.; Ghorui, S.; Mukherjee, P.; Deshpande, A.V. Thermofluid MHD studies in a model of Indian LLCB TBM at high magnetic field relevant to ITER. *Fusion Eng. Des.* **2020**, *150*. [CrossRef]
52. Idel'chik, I.E. *Handbook of Hydraulic Resistance-Coefficients of Local Resistance and of Friction*; Israel Program for Scientific Translations: Jerusalem, Israel, 1966.
53. Bluck, M.J.; Wolfendale, M.J. An analytical solution to electromagnetically coupled duct flow in MHD. *J. Fluid Mech.* **2015**, *771*, 595–623. [CrossRef]

Article

Overview of Thermal Hydraulic Optimization and Verification for the EU-DEMO HCPB BOP ICD Variant

Wolfgang Hering *, Evaldas Bubelis, Sara Perez-Martin and Maria-Victoria Bologa

Karlsruhe Institute of Technology (KIT), Institute for Neutron Physics and Reactor Technology (INR), Hermann-von-Helmholtz-Platz 1, 76344 Eggenstein-Leopoldshafen, Germany; evaldas.bubelis@kit.edu (E.B.); sara.martin@kit.edu (S.P.-M.); maria-victoria.bologa@kit.edu (M.-V.B.)
* Correspondence: wolfgang.hering@kit.edu; Tel.: +49-721-608-22556

Abstract: When progressing from the International Thermonuclear Experimental Reactor (ITER) to the Demonstration Fusion Reactor (DEMO), a system for transferring plasma heat exhaust to a power conversion system is necessary for the so-called Balance of Plant (BOP). During the preconceptual phase of the EU-DEMO project, different BOP concepts were investigated in order to identify the main requirements and feasible architectures to achieve that goal in the most efficient way. This paper comprises the investigations performed during the DEMO preconceptual design phase (p-CDP) and compares the different variants. The main aspect was focused on the helium-cooled pebble bed (HCPB) breeding blanket (BB) concept. After all assessments were performed, the indirect coupled design (ICD) was chosen as the reference configuration for the DEMO HCPB BOP for further development and optimization. The ICD provides decoupling using a molten salt storage loop, which accumulates thermal power during plasma pulses that are released during dwell periods. The work is supported by simulations using design codes EBSILON and MATLAB/SIMULINK, providing the basis for the next design phase.

Keywords: EU-DEMO; helium-cooled pebble bed; balance of plant; thermal storage; indirect coupled design; energy balance; power conversion system; simulation

Citation: Hering, W.; Bubelis, E.; Perez-Martin, S.; Bologa, M.-V. Overview of Thermal Hydraulic Optimization and Verification for the EU-DEMO HCPB BOP ICD Variant. *Energies* **2021**, *14*, 7894. https://doi.org/10.3390/en14237894

Academic Editors: Alessandro Del Nevo and Marica Eboli

Received: 29 October 2021
Accepted: 19 November 2021
Published: 25 November 2021

Publisher's Note: MDPI stays neutral with regard to jurisdictional claims in published maps and institutional affiliations.

1. Introduction

Following the Karlsruhe Institute of Technology (KIT) activities with respect to the helium-cooled pebble bed (HCPB) breeding blanket (BB) design, the contribution to the EUROfusion balance of plant (BOP) work package was initiated in 2012. The focus was on the development of the future DEMO plant to address the needs of the future. The BOP of DEMO was one of the new main topics to be investigated, since ITER is not designed to generate any electrical power. The DEMO power plant has to demonstrate electricity production for future electrical grids in a stable, predictable, and reliable manner [1]. Since EUROfusion uses many abbreviations, an explanative acronym list is added at the end of the paper.

DEMO as a TOKAMAK fusion reactor operates in pulsed mode with expected cycles of 2 h long plasma operation (pulse) followed by ~10 min dwell time, necessary to clean, refuel, and reload the central solenoid. Although the dwell time between pulses has been significantly reduced thanks to cutting-edge technologies for solenoid loading and vacuum pump capacity, the intermittent thermal power implies an intermittent electrical power output. Various options were explored along the years to handle the requirements and demands of the future DEMO powerplant, leading to direct coupled designs (DCDs) and indirect coupled designs (ICDs) [2]. Among the proposed solutions, that including an energy storage system (ESS) based on commercial systems already operating in concentrating solar power (CSP) plants was the most robust. However, other options were proposed, assuming smaller or even no energy storage in order to evaluate robustness, feasibility, reliability, and technological readiness.

The preconceptual design phase (p-CDP) considers not only the required characteristics of the BOP system, but also the framework in which the DEMO plant will be constructed and operated by the middle of the century. Taking into account the EU projection for the greenhouse gas emission reduction by 2050 [3], the renewable energy share will provide most of the electric energy. Bringing fusion powerplants (FPP) to the market will imply added value to power production, especially by balancing the enormous share of variable renewable energy sources (VRES) by load following operation and sector coupling. This means that FPPs will be welcomed if they support grid stability. Today, this concept is considered in advanced nuclear powerplants, where innovative reactor designs feature energy storage systems to handle the dynamics of the electrical grid [4].

The balance of plant work package in the EU-DEMO Project [1] investigates the transfer of plasma power from the breeding blanket to electricity to be delivered to plant systems, as well as to the grid via the plant electrical system (PES) [2]. This involves adequate cooling systems of different fluids and an efficient heat transfer across heat exchangers. Additionally, secondary heat sources from divertor (DIV) and vacuum vessel (VV) are used as additional feedwater heating to enhance the efficiency of the power conversion system (PCS) and avoiding investment toward additional cooling capacity of the component cooling water system. The system was designed on the basis of existing industrial technology.

In the selected design variant, an intermediate heat transfer and storage (IHTS) system including a thermal energy storage system decouples the helium-based primary heat transfer system (PHTS) from a highly efficient PCS designed by industry.

In this paper, the design solutions and the challenges are discussed for the helium-cooled pebble bed concept described in Section 2. In Section 3, the reference version is described. In Section 4, the focus is on the experimental facility, necessary to select BOP components and to test the interplay of the different thermal hydraulic cooling systems (i.e., PHTS, IHTS and PCS). In Section 5, a summary and conclusions complete the paper.

2. DEMO HCPB BOP Architecture

In the preconceptual design phase, several variants were investigated focusing on an efficient operation and easy integration into future energy systems in Europe and worldwide. This implies some requirements since the climate crisis stimulates steadily increasing shares of variable renewable energy sources and the replacement of fossil fuels in the heat market by electricity or solar heat, if available. This requires active control and an energy storage system. To select the best DEMO HCPB BOP option for the given boundary conditions, a total of four variants were investigated to different extents.

BOP interacts with most of the other DEMO systems as indicated in Figure 1. This indicates some constraints with respect to plant power demand during different operating phases, especially during the dwell time when only 1% of the full power is released to the PHTS.

The most critical interface is between the breeding blanket (inside vacuum vessel) and BOP, which is located geometrically at the water-cooled VV wall. From the BOP control point of view, the BB is a passive system, transferring plasma power to the heat transfer fluid (HTF). BB designers specify the tolerable temperature range and the power to be exhausted to the BB-PHTS. BOP, via the BB-PHTS, delivers a mass flow rate of helium to keep the temperature within the tolerable range. Another important requirement is the system pressure, which has to be controlled by BOP.

BOP design should provide operability of the powerplant during all possible DEMO plant states, i.e., plasma operation (pulse: 2 h) and dwell time (10 min). Maintenance (short- and long-term) was not considered in the p-CDP. In addition, safety functions were defined to manage failures and to reduce risks. The transitions between plasma operation states (pulse-to-dwell: P2D and dwell-to-pulse: D2P) are the most challenging issues for DEMO design. Thus, the BOP investigation was focused on (i) identifying the critical items of the whole BOP concept, (ii) proposing and assessing solutions to minimize the effects of the

P2D and D2P transitions, and (iii) concluding and recommending the best solution on the basis of validated system code results and industrial experience. The other main interface is the DEMO plant electrical system. It was defined to be located at the shaft between the turbine and electric generator, providing power to the plant internal demand and the grid.

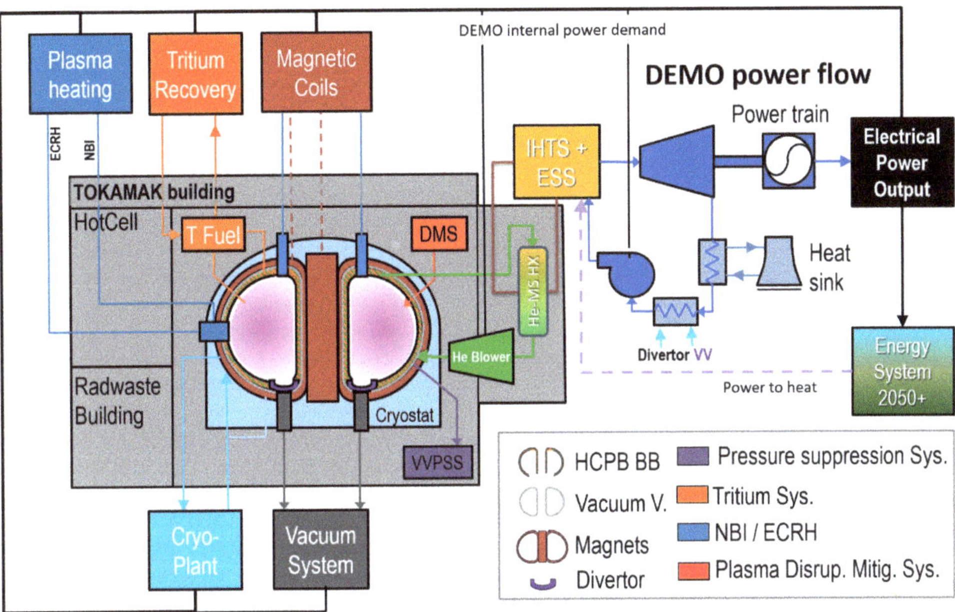

Figure 1. Overview and interfaces of DEMO systems (selected variant HCPB BOP ICD).

Variants Investigated: The two main requirements considered in the variant assessment (see Table 1) were (i) to avoid disconnection from the grid for each pulse/dwell transition, and (ii) to reduce the impact of frequent temperature transients to structures. Moreover, integration, performance, safety, and cost aspects were taken into account in the integral analysis performed for each variant.

Table 1. DEMO HCPB BOP variants investigated during pre-conceptual design phase [5].

Name	Explanation
DIRECT-AUXB (DCD-1)	Direct PHTS–PCS coupling but including a gas fired auxiliary boiler
DIRECT-Small ESS (DCD-s1)	Direct PHTS–PCS coupling but including a small concrete ESS heated by PHTS
DIRECT-Small ESS (DCD-s2)	Direct PHTS–PCS coupling but including a small ESS loaded by PHTS during pulse
INDIRECT (ICD)	Indirect PHTS–PCS coupling design with an IHTS and an industrially proven ESS

Generally, all possible solutions were investigated focusing on feasibility, applicability, and safety consequences. The three direct coupled variants: Direct-AUXB (DCD-1), Direct-Small ESS (DCD-s1), and Direct-Small ESS (DCD-s2), had several drawbacks with respect to the requirements mentioned above.

The first DCD case investigated was Direct-AUXB (DCD-1), where the avoidance of loss of synchronization during dwell was accomplished using a gas-fired boiler that provides steam flow to the turbine and maintains the power train in operation. The boiler size is directly proportional to the minimum steam flow rate needed by the turbine. Depending on the turbine concept considered, different lowest operation power levels can be achieved keeping the frequency constant. The key challenge in this DCD-1 case is, however, to manage the fast P2D and D2P transients, while operating the turbine in a safe way. On the other hand, the auxiliary boiler power has to reach the level of several hundred MW during dwell time, which implies an additional infrastructure (included in the auxiliary heater section (AHS)), similar to a small gas-fired power station requiring rather large gas pipelines. Another drawback of this case is the temperature and pressure transients achieved in the boiler during pulsed operation, which add additional difficulties to the boiler operation. All these facts together with the needs to assess the costs, dimensioning, and thermal power limitations were the reasons to keep this option as a secondary solution.

The second DCD case investigated was Direct-Small ESS (DCD-s1). This variant collects part of the BB thermal energy during the pulse and stores it in a solid-state ESS for use during the dwell time. This option reduces the boiler size and DEMO power output such that this case is more realistic. A main challenge is, however, that the high-temperature-concrete ESS has difficulties in providing the thermal energy stored in a relatively short time as required for the dwell time. Additionally, the control system and pipelines increase the complexity of the system. Another aspect considered applies to the safety function. The solid ESS working as an HX stores heat from PHTS helium on one side and provides heat to the PCS water/steam during dwell time. Thus, the PHTS safety function could not be assured due to the spatial needs of the ESS. Therefore, further investigations will be needed, and this case was classified as a possible back-up solution, as with the previous DCD-1 case.

The third DCD case investigated was Direct-Small ESS (DCD-s2). This variant developed by industry uses HITEC™ molten salt (approximately 400 m^3) and an electrical heater (41 MWe) ahead of the hot ESS tank. This allows maximizing the electrical power production during pulse while maintaining the electrical generator synchronized to the grid during dwell phases, whereby a future steam turbine is being designed to operate at a minimum load of 10%. All auxiliary components belonging to PCS are concentrated in the AHS section.

The HCPB BOP reference design variant was the indirect coupled design called HCBP-ICD, which uses an IHTS operating with HITEC™ to decouple the thermal power coming from the BB/FW from the PCS operation. The IHTS design was based on the current technology used in concentrating solar power (CSP) plants (150 MWe power and up to 1 GWh$_{th}$ storage energy). The overall analysis of this variant performed so far considered not only the feedback from the industry to improve the design focusing on the different primary heat transfer systems from BB, DIV, and VV, but also the PCS configuration to find reasonable answers to the challenges and requirements. Thus, this HCPB-ICD option was selected as the reference variant for the next step of DEMO development. Further details of this DEMO HCPB BOP variant are presented in Section 3.

The main characteristics and critical issues for each studied HCPB variant are compiled in Table 2, which includes the major topics of each heat transfer loop, namely, PHTS, IHTS-ESS, and PCS. The colors identify the current feasibility of the corresponding component, where a green color denotes the highest degree of present industrial readiness, as supported by industry. Considering all technological aspects, as well as the response of the various BOP variants to the DEMO requirements, the highest ranked design was the ICD variant. It provides not only the most reliable and stable electricity production without any supplementary source, but also inherently provides more safety barriers against tritium and ACP release.

Table 2. DEMO HCPB BOP variants investigated during pre-conceptual phase [5].

System	Subsystem/Component	DCD-l	DCD-s1	DCD-s2	ICD
PHTS	BB PHTS-IHTS/PCS	He–water SG	He–water SG	He–water SG	He–MS HX
	BB PHTS HX/SG pressure	High	High	High	Low ~ 6 bar
AHX/IHTS	IHTS/ESS fluid	No	No	MS	MS
	IHTS/ESS storage capacity	-	-	$2 \times 400\ \text{m}^3$	$2 \times 3000\ \text{m}^3$
	Other thermal storage	No	SS Concrete	No	No
	Auxiliary heating system	Gas (220 MWth)	Gas (93 MWth)	Electric (41.2 MWe)	No
	Gas-fired boiler supply	Large	Medium	No	No
	Space for aux. heating syst./IHTS	Large (aux.)	Large (aux. + conc.)	Medium (elec. + ESS)	Large (IHTS–ESS)
PCS	Turbine loads	Frequent ramps	Frequent ramps	Steep ramps	Steady
	System feasibility	TBI	No	TBI	Yes
	Tolerant to frequent transients	TBI	No	TBI	TBI
Summary	Critical components	Ext. gas boiler He–water HXs (FW) ST	SS concrete ESS He–water HXs (FW)	He SG MS SG He–water HXs (FW) ST	He–MS HX MS SG PCS HXs (FWH)
	Supplementary power needed	+Gas	+Gas	+Ele. Power	No
Safety	Safety barriers (T, ACP)	1	1	1	2
Red:	Critical issue (size, feasibility, …)		Blue:	producible but not on shelf	
Orange:	Near or at present feasible and producible		Green:	component from shelf	
	TBI—to be investigated further				

3. Reference Version: Indirect Coupled Design (HCPB-ICD)

To prove the feasibility of the selected reference variant, industrial expertise was included to verify the most appropriate BOP variant for the DEMO fusion reactor—HCPB-ICD. Schematically, this variant is presented in Figure 2.

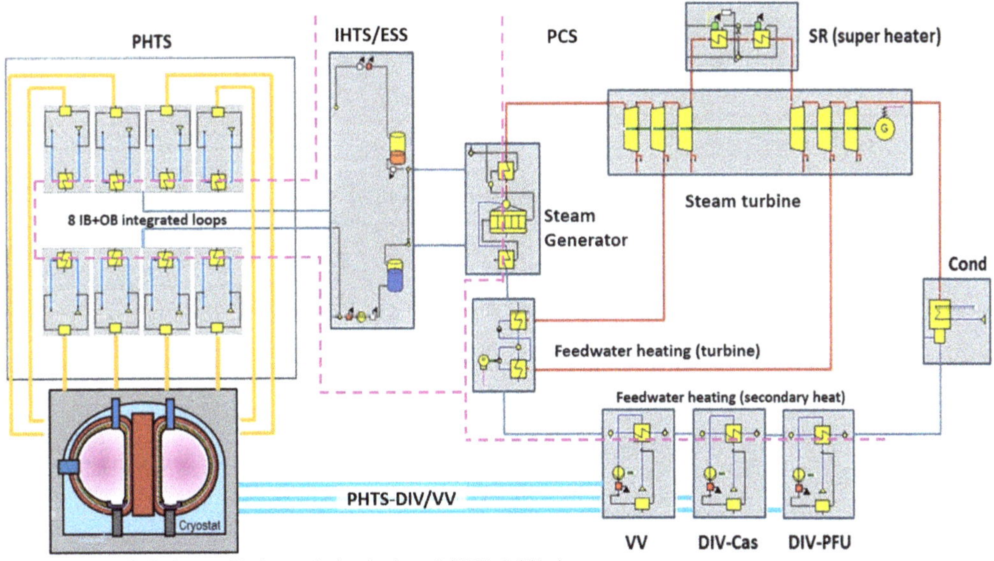

Figure 2. Overview of the DEMO HCPB ICD variant with PHTS, IHTS, and PCS indicating the different HTFs and the interfaces (dashed lines).

3.1. Results of Simulations

As the first step, simulations of the HCPB-ICD variant were performed using EBSILON software in order to come up with a feasible BOP configuration. Later, the fixed HCPB-ICD configuration was tested dynamically using a MATLAB/SIMULINK model.

3.1.1. EBSILON Simulation Results

The HCPB-ICD conceptual design was developed assisted by the industrial tool EBSILON as shown in Figure 3 for the pulse time operation. During pulse time, 90% of the generated power by the plasma in the BB is delivered to the PCS, while 10% is stored in the ESS. During dwell time, the ESS releases energy to the PCS, thus suppling 104% (~890 MWe) of the nominal power, and compensating for the missing power from the plasma (decay heat during dwell time is only 1% of the nominal power) [6,7]. The results of EBSILON were used for analyses with thermal hydraulic system codes such as APROS [8] or RELAP5-3D (PhD thesis in print).

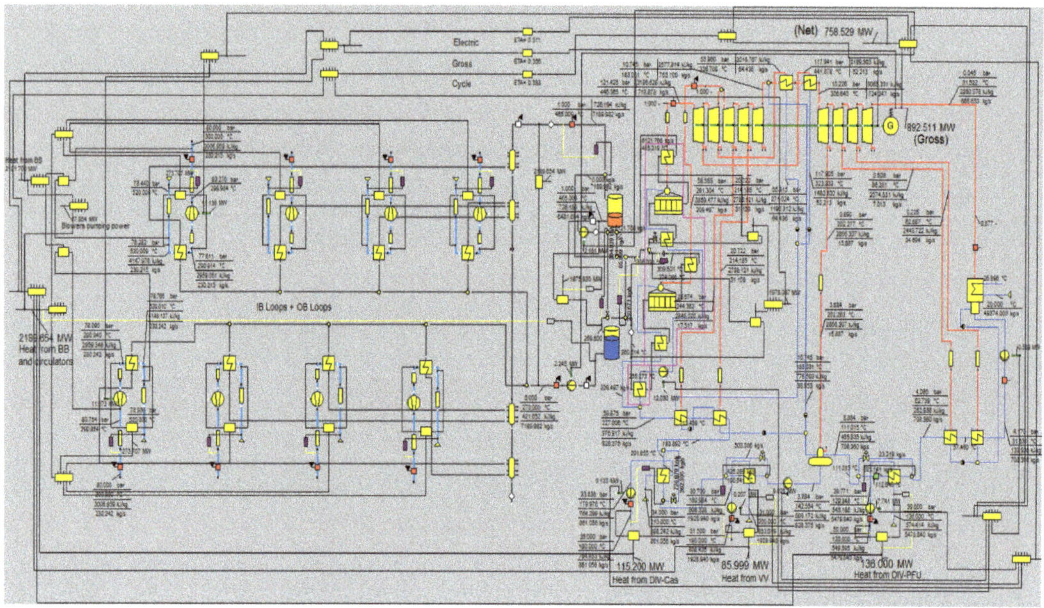

Figure 3. DEMO HCPB BOP ICD with optimized PCS and 90% power output during pulse and dwell.

3.1.2. MATLAB/SIMULINK Model Results

Starting from the EBSILON results, a more detailed model was established using MATLAB/SIMULINK [9]. The schema for the HCPB-ICD variant between PHTS and PCS via IHTS is presented in Figure 4. The simulation model of the IHTS was joined with the model of the modified PCS Rankine cycle. The coupling of IHTS with PCS takes place via the thermal ports for the heat transfer from the side of the preheater (PH), steam generator (SG), and steam heater (SH) of IHTS to that of the PH, SG, and SH of the PCS. Only heat transfer was simulated, taking into consideration the heat transfer from HITEC molten salt to water. The temperature source as the heat input was eliminated and was substituted with solid connection lines via the ports H. Due to modeling conditions, the standard water/steam property block was excluded, as only one solver block can be used. Hence, in this case the customized block of newly developed HITEC properties remained. The properties of water/steam were considered as constants due to constant heat capacity and mass flows from the side of IHTS SG.

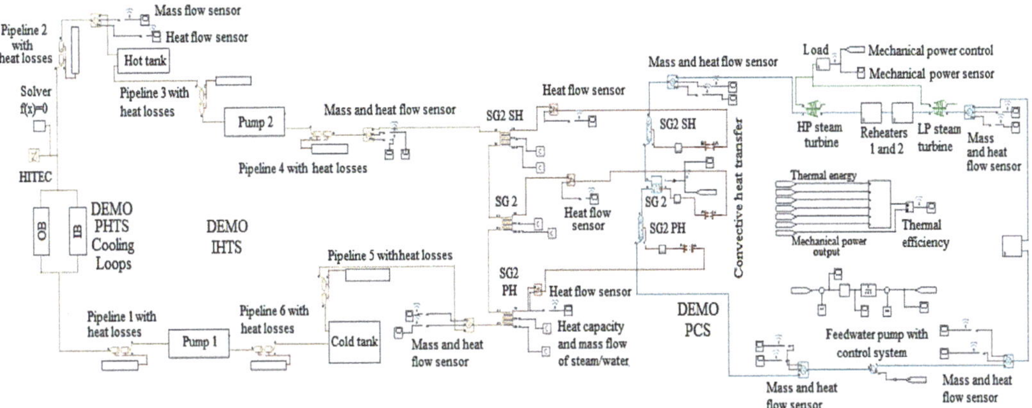

Figure 4. MATLAB/SIMULINK model of DEMO BOP ICD concept: the connection of DEMO PHTS with PCS via IHTS.

In a second step, the IHTS model was coupled to a DEMO PCS based on the Rankine cycle concept, using a specific MATLAB/SIMULINK coupling tool named thermal port. The obtained energy balance for the DEMO BOP energy transfer chain during the DEMO pulse operation revealed that the dynamic simulations enabled a good evaluation of the DEMO BOP operation parameters, in order to optimize the process in terms of power change rate. The results of the simulations confirmed the stability of DEMO BOP energy transfer chain during the pulse and dwell operation of the TOKAMAK reactor using the ICD concept. The simulations confirmed the fact that the application of the IHTS is an essential solution to ensure a stable operation of the steam turbines and electric generators [9]. To realize stable operation of the energy transfer chain, it is important to ensure the operation stability of the generator both from the "heat–mechanical energy side" of the DEMO constellation and from the "electrical energy side" of the grid [10].

Taking into account these promising results, in a next step, system codes such as RELAP5-3D or TRACE were used to provide overall simulations and support for the experimental verification using the new research infrastructure HELOKA-US, as described in in detail in Section 4.

3.2. Design and Layout

The reference version of HCPB-ICD (see Figure 5) uses an IHTS-ESS operating with molten salt (HITEC) to decouple regular plasma strokes from the PCS. The IHTS design is based on the current technology used in CSP plants (150 MWe power and up to 1 GWh$_{th}$ storage energy). As already mentioned previously, for such a concept, detailed plant functional design was performed by KIT using EBSILON software (see Figure 3 for pulse time operation) supported by industry with respect to the power conversion system.

3.2.1. Main Heat Train PHTS–IHTS/ESS–PCS

As shown in Figure 2, BOP has to supply cooling to the FW and BB using helium as the HTF and to the divertor and the vacuum vessel using water. For the BB PHTS, the heat sink is the IHTS, while the energy of the DIV PHTS and the VV PHTS is transferred via heat exchangers to the feedwater train of the PCS, thus enhancing the overall system efficiency.

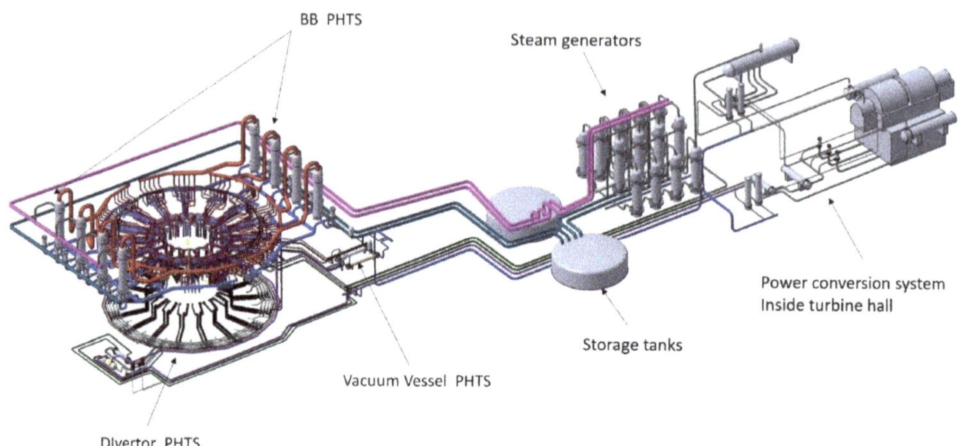

Figure 5. Schematic layout of DEMO HCBP ICD design including all PHTS systems, the IHTS with two tank ESS, and the PCS up to the AC generator.

The FW/BB-PHTS (helium) is segmented into eight loops (see Figure 5, left side) each providing heat transfer to two sectors (six outboard plus four inboard segments) via top/upper ports [6]. This configuration was necessary to keep the dimensions of the components small and to minimize challenges and risks of loss-of-coolant accidents, as well as blower failure, with subsequent damage of the first wall due to a lack of cooling. The drawback is a more complicated pipe routing from the TOKAMAK to the heat exchanger and blower compartments. The segmentation offers benefits with respect to safety, maintenance, and component size, thus restricting valves to fill and drain lines. The HTF velocity in the main coolant line is kept limited, thus leading to feasible pipe diameters (<1.3 m). For safety reasons, each loop incorporates two helium blowers. They can be aligned in parallel or in series and incorporate an internal bypass valve, such that one blower alone can provide sufficient cooling to the BB to ramp down correctly without unintended plasma breakdown. For the heat exchanger between helium and solar salt (HITEC), two possible solutions exist (spiral and once through) [7]. For a final selection, the dynamic behavior of both solutions has to be investigated.

As depicted in Figure 5, the IHTS gathers the thermal energy from the BB PHTS in the ESS during pulse, manages the inlet temperature to the BB-PHTS via the HX secondary-side inlet temperature, and then transfers the thermal energy to the steam generator/superheater as requested by the PCS. During dwell, the HITEC flow rate is adjusted to the need of the BB decay heat removal using a dedicated small pump. On the right side, as in the pulse time, it follows the requests of the PCS. To achieve such a decoupling function, two or three HITEC pumps are foreseen, operating independently of each other. For operation and safety reasons, each pump is foreseen as a twin-pump. For simplification, the ESS is realized as a classical two-tank solution. For the molten salt (HITEC) steam generator, a technical offer and a price indication from Company Siemens AG are available.

During p-CDP, the PCS (Figure 3, right side) was optimized on the basis of the available energy sources. All relevant usable energy (exergy) sources are included at optimal temperature and power level, even if additional components become necessary to avoid investment toward additional cooling capacity. Lastly, the detailed design proposed by an industrial partner (SIEMENS AG) provided substantial progress in efficiency due to the fine optimized turbine–feedwater train. The company Siemens AG supplied the design for both the turbogenerator and the steam generator, as the interaction of these two large components has a high impact on system performance and space and cost optimization.

The gross output of the SIEMENS SST5-6000 turbogenerator during dwell time is now even higher due to the reduced BB PHTS circulation power.

3.2.2. Secondary Heat Train PHTS–PCS

In the DEMO HCPB BOP, secondary heat sources are water-cooling systems for the divertor and the vacuum vessel. As for the DIV-CAS and DIV-PFU PHTS, BOP design follows the requirement of the DIV, thus defining system pressure, as well as inlet and max. tolerable outlet temperature for the divertor cassettes and plasma facing units. The heat is transferred via heat exchangers in the TOKAMAK building to the feedwater loop as indicated in Figure 5, thus contributing to the PCS power output.

VV PHTS provides water-cooling to the vacuum vessel at an inlet temperature of 190 °C by forced convection. The heat is transferred via heat exchanger in the TOKAMAK building to the feedwater loop, thus also contributing to the PCS power output.

3.3. Open Issues

The DEMO BOP project is on a solid path and is close to reaching the conceptual design phase. One of the aims of the preconceptual design phase was to assess the readiness level at different scales and to review the maturity of subsystems and components. In order to be able to make an adequate and useful statement here, it is necessary to consider not only the TRL for new technologies, but also the SRL for the BOP overall-/sub-systems, as well as the IRL for proven technologies, which must be integrated into the system (according to the operating conditions).

One result of the investigation was that, with regard to the IRL, the main interfaces between PHTS and IHTS (Main HX) and between IHTS and PCS (MSSG) are to be assessed as critical. Another result was that the IHTS, the He-compressors, and their operating behavior must be investigated in an experimental facility in order to achieve a higher IRL. The PCS was also rated as being noncritical.

Another aim of the preconceptual design phase was to identify design optimizations or alternatives for the identified critical components. For the important and critical interface component "Main-HX", four design variants were examined, compared, and evaluated. It was found that the "plate and shell design", which was not previously investigated, is not only compact but also proven in nuclear powerplants and, therefore, has a high reliability. It is recommended to further investigate this design alternative under DEMO conditions.

The other important and critical interface component "molten salt steam generator (MSSG)" is currently based on an SG design from the CSP industry, which was indeed not upscaled or adapted, but multiplied in a modular approach to match DEMO requirements. This enables various optimization possibilities, including a reduction in the number of MSSGs and, thus, the complexity and the space requirement of a modular arrangement. Another optimization aspect is to be found in the actual design, i.e., the compactness of the MSSG in terms of surface area per unit volume. In addition to the aforementioned performance optimizations, the costs of the MSSG system can be reduced to one-third of the price by changing the material from high-alloy stainless steel to common carbon steel. One prerequisite is to assure that no chloride impurities in the HITEC molten salt are present.

The molten salt storage and transfer system components of the current IHTS concept are based on today's CSP technology. In addition to the common two-tank-thermal storage system, the pros and cons of a single-tank (i.e., thermocline) and a modular approach with several smaller tanks (i.e., multi-tank) were indicated. The comparison showed that a thermocline setup has a comparatively low TRL and IRL, whereas a multi-tank setup does have several advantages compared to the currently planned common two-tank system. In the context of the multi-tank system, the required molten salt transfer system can also be optimized. Here, centrifugal pumps can be used at ground level or on a small buffer tank, which reduces operating costs and simplifies maintenance and operation.

Another aim of the preconceptual design phase was a market survey in order to identify potential suppliers for the He-compressors that are required for the FW/BB-PHTS loop. In general, several companies are available that can provide He-compressors for DEMO application. However, there is currently no focus on producing such large He-compressors due to currently missing application. Nevertheless, positive feedback from the market with several companies was received, while also obtaining budgetary price offers.

4. Experimental Verification

While the secondary heat train uses technology with an industrial readiness level, some primary heat components of the transfer train such as the helium HITEC primary heat exchanger, the HITEC steam generator, or the helium blowers have to be evaluated and optimized in the conceptual design phase. This includes scaling and selecting the type of the components to fulfill the DEMO requirements. Furthermore, the interplay of the different systems has to be investigated and optimized. Here, the expertise at KIT on helium system operation (HELOKA-HP [11], KATHELO [12], and HEMAT [13]), as well as the expertise in the high-temperature molten salt field (LIVE [14]), was favorably combined within the HELOKA-US project, as explained below.

For HELOKA-US a scaling factor of ~1000 for power (compared to DEMO HCPB BOP) was selected, bringing one FW/BB-PHTS loop down to the values listed in Table 3. The vertical scaling, which is only necessary for the HITEC section, was set to <10, limited by the available location in the HELOKA-HP building.

Table 3. Main data for HELOKA-US dimensioning.

	IHX Test Section	Molten Salt Loop	Cooling Loop (Water)
Thermal power (kW)	250–280		<300
Temperature (°C)	300–550	300–465	10–80
Pressure (bar)	80	<6	5–12
Mass flow rate (kg/s)	0.24	1.10	<3

4.1. Component Test

HELOKA-US Phase 1a (see Figure 6, left) comprises the construction and commissioning of the molten salt (MS) loop (Intermediate Heat Transfer System—IHTS) to test and optimize the molten salt side of the Helium-HITEC heat exchanger. The main issue in the MS loop is the heat transfer to the molten salt, which has to be checked and optimized for conditions corresponding to all DEMO states of operation. Thus, an electrical heating system allows simulating the frequent operational transitions between pulse and dwell phases. That approach has the advantage that several configurations can be tested and easily coupled to the heating device without operating of costly helium loop.

4.2. System Tests

After a scientific review of Phase 1 around 2024, HELOKA-US Phase 2 will be focused on a prototypical dynamic temperature-adaptive helium blower test. Here, a prototypic design for DEMO will be scaled down and connected to the helium supply and pressure control system of HELOKA on one side and to the molten salt loop on the other side. To simulate the IHTS, a thermal energy storage system (ESS) will be enclosed as shown in the schematics of Figure 7. The full pilot plant will then allow simulating fast transients and operational transitions to qualify the DEMO HCPB BOP design and optimization, as well as the system and safety codes necessary to foster BOP simulation and, later on, safety assessments.

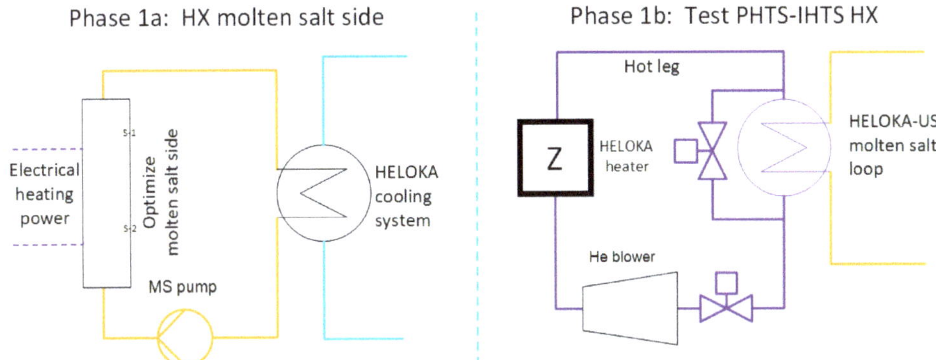

Figure 6. HELOKA-US Phase 1 subdivided into HX qualification and He HX plus blower tests.

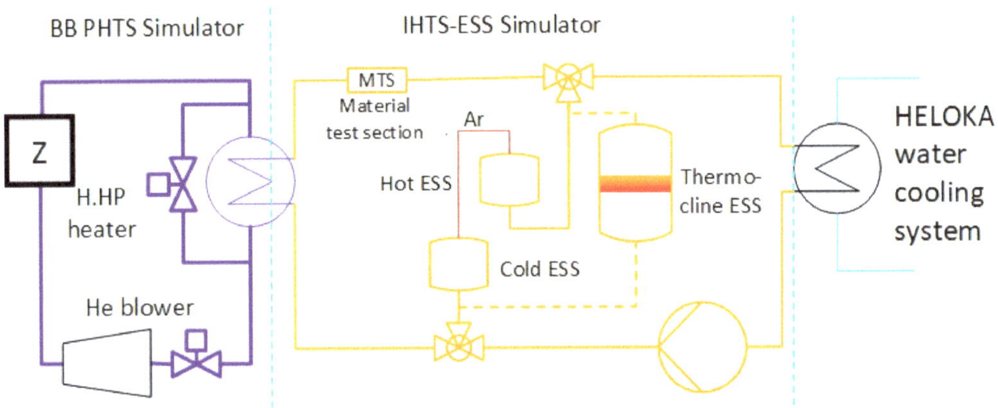

Figure 7. Configuration of HELOKA-US Phase 2 indicating the three systems: PHTS–IHTS/ESS–PCS (only cooling).

As a future extension of Phase 2, the connection to the HELOKA cooling system can be replaced by a power conversion system to demonstrate the whole heat and power train for DEMO BOP.

5. Summary and Conclusions

Within the pre-conceptual design phase, the DEMO balance of plant of the HCPB breeding blanket concept was developed and constantly adapted to the requirements of the different heat sources. In addition to the primary heat source from the BB, which is transferred to the IHTS with the ESS and to the PCS, the secondary heat sources are directed via HX to the feedwater train of the PCS. The selected variant ICD was considered as a reference design for the DEMO HCPB BOP concept.

Following the technological approach, the first step was to use simulating tools (EB-SILON and MATLAB/SIMULINK) to come up with a feasible HCPB-ICD configuration. Afterward, the best technological and available components were found, where feedback from our industrial partners played an important role. After the assessment and comparison of the different variants investigated, the various criteria used (TRL, IRL, and SRL) led to selecting the reference configuration as the indirect coupling design. Among the four investigated variants, the ICD concept provided the highest and most flexible electrical power output of DEMO, varying between 90% (pulse: ~760 MWe) and 104% (dwell: ~890 MWe). The higher value in the dwell time originates from the lower required

He circulator power, due to fusion power lacking. Therefore, the subsequent steps in the consolidation of this DEMO HCPB BOP design involve the experimental validation of those critical components not yet demonstrated to perform successfully under DEMO operation conditions. This will be achieved in the HELOKA-US project for the Helium blower and the intermediate heat exchanger, whereas Phase 2 will be devoted to the interplay of the different systems and components.

Author Contributions: Conceptualization, W.H., E.B. and S.P.-M.; EBSILON analysis and results, E.B.; MATLAB/SIMULINK analysis and results, M.-V.B.; Experimental verification, W.H. and S.P.-M.; writing—original draft preparation, W.H., E.B. and S.P.-M.; writing—review and editing, W.H., E.B. and S.P.-M.; supervision, W.H.; project administration, W.H. All authors have read and agreed to the published version of the manuscript.

Funding: This work was carried out within the framework of the EUROfusion Consortium and received funding from the Euratom research and training program 2014–2018 and 2019–2020 under grant agreement No 633053. The investment of HELOKA-US is sponsored by EUROfusion BOP (FP9 from 2021 to 2027, contract to be signed) and by the KIT funding from the Helmholtz Program FUSION.

Institutional Review Board Statement: Not applicable.

Informed Consent Statement: Not applicable.

Data Availability Statement: Data sharing is not applicable to this article.

Acknowledgments: This work was carried out within the framework of the EUROfusion Consortium and received funding from the Euratom research and training program 2014–2018 and 2019–2020 under grant agreement No 633053. The views and opinions expressed herein do not necessarily reflect those of the European Commission.

Conflicts of Interest: The authors declare no conflict of interest.

Abbreviations

ACP	Activated corrosion products
AHS	Auxiliary Heater System (to cope with dwell time power lacking)
AUXB	Auxiliary boiler
BB	Breeder blanket
BOP	Balance of Plant
CAS	Cassettes (part of divertor)
CDP	Conceptual design phase
CSP	Concentrating solar power
CWS	Coolant water system
DCD	Direct coupled design
DIV	Divertor
ESS	Energy storage system
FPP	Fusion powerplant
FW	First wall
HCPB	Helium cooled pebble bed (blanket)
HCPB ICD BOP	HCPB indirect coupling design balance of plant
HELOKA-US	Helium Loop Karlsruhe Upgrade Storage
HP	High pressure
HT	High temperature
HTF	Heat transfer fluid
HX	Heat exchanger
ICD	Indirect coupled design
IHTS	Intermediate heat transfer system (based on molten salt as HTF)
IHX	Intermediate heat exchanger

IRL	Integration readiness level
LP	Low pressure
MS	Molten salt (HITEC)
NPP	Nuclear powerplant
OTGS	Once through steam generator
p-CDP	Preconceptual design phase
PCS	Power conversion system
PES	Plant electrical system
PFU	Plasma facing units
PH	Preheater
PHTS	Primary heat transfer system
SG	Steam generator
SH	Steam heater
SR	Steam reheater
SRL	System readiness level
ST	Steam turbine
TRL	Technical readiness level
VV	Vacuum vessel
VVPSS	Vacuum vessel pressure suppression system

References

1. Frederici, G. Overview of the design approach and prioritization of R&D activities towards an EU DEMO. *Fusion Eng. Des.* **2016**, *109*, 1464–1474. Available online: https://www.sciencedirect.com/science/article/pii/S0920379615303835 (accessed on 11 October 2021).
2. Barucca, L.; Bubelis, E.; Ciattaglia, S.; D'Alessandro, A.; Del Nevo, A.; Giannetti, F.; Hering, W.; Lorusso, P.; Martelli, E.; Moscato, I.; et al. Pre-conceptual design of EU DEMO balance of plant systems: Objectives and challenges. *Fusion Eng. Des.* **2021**, *16*, 112504. Available online: https://www.sciencedirect.com/science/article/abs/pii/S0920379621002805 (accessed on 11 October 2021). [CrossRef]
3. 8th Action Plan of The European Green Deal. Available online: https://ec.europa.eu/environment/pdf/8EAP/2020/10/8EAP-draft.pdf (accessed on 11 October 2021).
4. International Workshop on Advanced Reactor Systems and Future Energy Market Needs, Paris, France, 12 April 2017. Available online: https://www.oecd-nea.org/jcms/pl_30808/workshop-arfem-19-terrestrial-energy (accessed on 11 October 2021).
5. Hering, W.; Bubelis, E. *BOP-DEF-2-CD1, BOP-2.1-T060-D001: HCPB PHTS & PCS Rationale for the Selection of the Variant, EUROfusion Pre-CD Phase*; KIT Internal Report: FUSION 517, INR 13/20; KIT: Karlsruhe, Germany, in preparation.
6. Hering, W.; Jin, X.Z.; Bubelis, E.; Perez-Martin, S.; Ghidersa, B.E. Operation of the Helium Cooled Demo Fusion Power Plant and Related Safety Aspects, IAEA-Tecdoc-13657. 2020. Available online: https://www.iaea.org/publications/13657/challenges-for-coolants-in-fast-neutron-spectrum-systems (accessed on 11 October 2021).
7. Bubelis, E.; Hering, W.; Perez-Martin, S. Industry supported improved design of DEMO BOP for HCPB BB concept with energy storage system. *Fusion Eng. Des.* **2019**, *146*, 2334–2337. [CrossRef]
8. Malinowski, L.; Lewandowska, M.; Bubelis, E.; Hering, W. Design and analysis of the secondary circuit of the DEMO fusion power plant for the HCPB BB option without the energy storage system and with the auxiliary boiler. *Fusion Eng. Des.* **2020**, *160*, 112003. [CrossRef]
9. Bologa, M.-V.; Stieglitz, R.; Hering, W.; Bubelis, E. Parameter Study and Dynamic Simulation of Current DEMO Intermediate Heat Transfer and Storage System Design via MATLAB/Simulink, Poster Session at 14th International Symposium on Fusion Nuclear Technology (ISFNT-14), Budapest, Hungary, 2019. Available online: https://publikationen.bibliothek.kit.edu/1000099197/45106288 (accessed on 11 October 2021).
10. Bologa, M.-V.; Bubelis, E.; Hering, W. Parameter Study and Dynamic Simulation of the DEMO Intermediate Heat Transfer and Storage System Design Using MATLAB/Simulink. *J. Fusion Eng. Des.* **2021**, *166*, 112291. [CrossRef]
11. Ghidersa, B.E.; Ionescu-Bujor, M.; Janeschitz, G. Helium Loop Karlsruhe (HELOKA): A valuable tool for testing and qualifying ITER components and their He cooling circuits. *Fusion Eng. Des.* **2006**, *81*, 1471–1476. [CrossRef]
12. Ghidersa, B.E.; Jin, X.; Rieth, M.; Ionescu-Bujor, M. KATHELO: A new high heat flux component testing facility. *Fusion Eng. Des.* **2013**, *88*, 854–857. [CrossRef]
13. KIT-HEMAT Facility. Available online: https://www.inr.kit.edu/english/630.php (accessed on 11 October 2021).
14. Gaus-Liu, X.; Bigot, B.; Journeau, C.; Payot, F.; Cron, T.; Clavier, R.; Peybernes, M.; Angeli, P.E.; Fluhrer, B. *Experiment and Numerical Simulations on SFR Core-catcher Safety Analysis after Relocation of Corium*; FR-22: IAEA-CN-291/336; IAEA: Vienna, Austria, in preparation.

Article

Model Development and Transient Analysis of the HCPB BB BOP DEMO Configuration Using the Apros System Code

Marton Szogradi * and Sixten Norrman

VTT Technical Research Centre of Finland Ltd., 02044 Espoo, Finland; sixten.norrman@vtt.fi
* Correspondence: marton.szogradi@vtt.fi

Abstract: Extensive modeling and analytical work has been carried out considering the Helium-Cooled Pebble Bed Breeding Blanket (HCPB BB) Balance Of Plant (BOP) configuration of the Demonstration Power Plant (DEMO) using the Apros system code, developed by VTT Technical Research Centre of Finland Ltd. and Fortum. The integral plant model of the HCPB BB plant has been improved with respect to the blanket and steam generator models. Based on HCPB-BL2017 v1 data, reported in 2019, the blanket has been remodeled by separate Apros process components, dedicated to average inboard and outboard segments, where the power deposition scheme of the breeding units took into account the output of high-fidelity neutronic analyses. A new helical coil steam generator model has been developed for primary–secondary system coupling using CAD data provided by EUROfusion partner University of Palermo. Transient analyses have been performed with Apros on the plant configuration that utilizes a molten salt technology-based small Energy Storage System (ESS).

Keywords: DEMO; HCPB BB; small ESS; transient; Apros

Citation: Szogradi, M.; Norrman, S. Model Development and Transient Analysis of the HCPB BB BOP DEMO Configuration Using the Apros System Code. *Energies* **2021**, *14*, 7214. https://doi.org/10.3390/en14217214

Academic Editors: Alessandro Del Nevo and Marica Eboli

Received: 26 August 2021
Accepted: 25 October 2021
Published: 2 November 2021

Publisher's Note: MDPI stays neutral with regard to jurisdictional claims in published maps and institutional affiliations.

1. Introduction

The Demonstration Power Plant (DEMO) Balance Of Plant (BOP) integral thermal-hydraulic models have been developed by VTT Technical Research Centre of Finland Ltd. using the Apros system code. Such activities have been facilitated by the BOP work package of the EUROfusion Consortium [1], elaborating on two blanket concepts, namely the Helium-Cooled Pebble Bed (HCPB) and the Water-Cooled Lithium–Lead (WCLL) Breeding Blanket (BB) configurations [2,3]. In order to map the potential advantages and shortcomings of these plants, several BOP layouts have been investigated focusing on different Primary Heat Transfer System (PHTS) and Power Conversion System (PCS) coupling techniques.

The subsequent variants can be divided into direct and indirect coupling schemes. In the case of direct coupling, the PHTS is directly connected to the PCS via steam generators, whereas in the indirect layout, the primary–secondary systems' interface is represented by an Intermediate Heat Transfer System (IHTS), equipped with an Energy Storage System (ESS) [4]. The direct coupling approach has two models, a pure direct version with an AUXiliary Boiler (AUXB) and another, incorporating a small ESS. The former variant served only as a benchmark case, with the ultimate goal of studying the pulsed operation regime of the power plant using a gas-fired boiler. In dwell, the boiler provided a modest steam supply, alleviating the effects of the pulsed operation of DEMO heat sources [5]. Referring back to the small ESS configuration, instead of an auxiliary boiler, a molten salt loop was implemented, connected to the PCS via a Molten Salt Steam Generator (MSSG).

Although the basic engineering principles of the DEMO plants have been appropriated from conventional nuclear power plant design, the operating regime of the tokamak (*toroidal chamber with magnetic coils*, Rus.: ТОроидальная КАмера с МАгнитными Катушками) is a fundamental departure. The plasma current in the reactor chamber is driven by the discharge of the central solenoid until the plasma current reaches its opposite peak current. This implies that the tokamak has to operate in pulsed mode where a pulse period (~7200 s) is followed by a dwell phase (~600 s) when only the decay heat is produced (~1–3% P_{nom}). Such a pulsed operation connotes challenges in the design with regard to electricity production in a commercially and technically viable way.

VTT has been contributing to the BOP work package with various analyses, concerning the outlined plant variants using the Apros system code [6]. The present article disseminates the results of the analytical work related to the HCPB BOP small ESS configuration. Section 2 introduces the general architecture of the PHTS (§2.1) and the PCS with an emphasis on the Helical Coil Steam Generator (HeSG) model. The corresponding control logics are given in §2.2, and the small ESS is introduced briefly in §2.3. Section 3 gives a depiction of the integral Apros model detailing the main subsystems in a similar manner as Section 2 with more focus on the mentioned key components (§3.1 to 3.4). The results of the transient analysis are laid out in Section 4, finally leading the reader to the conclusions and outlook of the work in Sections 5 and 6, respectively.

2. HCPB BOP Small ESS Configuration

The HCPB BOP operates with a helium-cooled primary system, composed of eight coolant loops with a pair of circulators in each loop. The coolant loops pass through the First Wall (FW) and Breeding Zone (BZ) compartments of the breeding blanket, i.e., unlike the WCLL BB, the helium-cooled system incorporates concurrent heat exhaust systems for FW and BZ domains. The PHTS layout implies that the eight helical coil steam generators serve as the primary–secondary system interface with helium as the primary and pressurized water as the secondary coolant. The design of the secondary power sources is shared with the WCLL configuration, and the Divertor Cassette (DIV-CAS), Divertor Plasma Facing Unit (DIV-PFU), and Vacuum Vessel (VV) loops are connected to the PCS via three Heat exchanger (HX) pairs. The helical coil SGs represent a parallel circuit, where eight HeSGs are placed on the longer (FAR) and shorter (NEAR) feedwater collectors, four on each collector. The Main Steam Line (MSL) mixes fresh steam from the SG headers before reaching the inlet of the HP stage of the Steam Turbine (ST). The CAD layout of the facility is depicted in Figure 1, highlighting the mentioned subsystems.

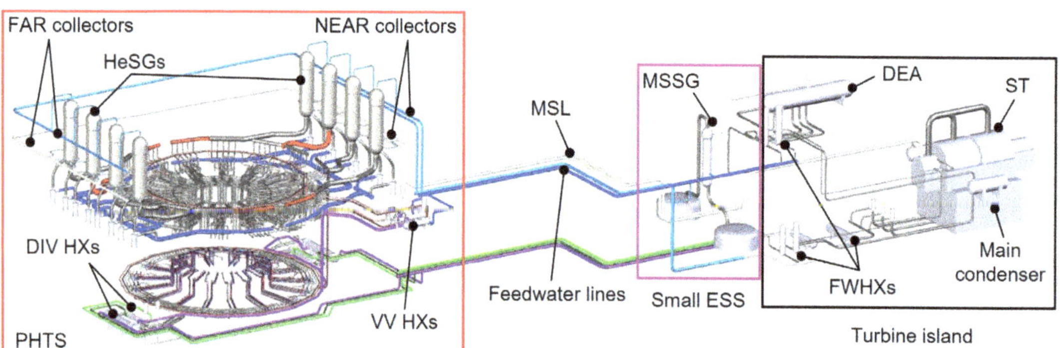

Figure 1. CAD layout of the HCPB small ESS DEMO.

The transient scenarios enveloped two consecutive pulse-dwell phases where the pulse (P_{fus}= 100%, P_{MSEH}= 100% (molten salt electrical heater), $P_{MSSG} \approx 5\%$) and dwell section (P_{fus}= 1%, $P_{MSEH} \approx 38\%$, P_{MSSG}= 100%) lasted 7200 s and 600 s, respectively. The plasma ramps represent asymptotic time-dependent power functions describing the reactor's thermal power variation between 1% and 100% under 100 s. The unloading of the turbine initiates 500 s ahead of the plasma power ramp-down, achieving an acceptable −10%/min power gradient on the turbine. As the reactor power winds down, radiative and volumetric heat loads drop in the blanket and secondary power sources. Balancing this power reduction, the small ESS discharge line is activated, loading up the molten salt SG. During the 10 min dwell phase, a 10% load has to be maintained on the turbine throughout the dwell, by a suitable re-alignment of the whole power conversion system.

2.1. Primary Heat Transfer System

The PHTS consists of four short and four long coolant loops giving eight loops for sixteen blanket sectors. The coolant medium is pressurized helium, circulated by a compressor pair in each loop. A closer depiction of a 1/16 BB sector can be seen in Figure 2, where the inlet and outlet manifolds are highlighted with respect to Inboard (Left, LIB; Right, RIB) and Outboard (Left, LOB; Center, COB; Right, ROB) fingers. The cold helium (marked with blue in Figure 2b) flows through the inlet piping, reaching the segments (2 IB + 3 OB/sector), where the gas enters the FW channels via the inlet manifolds before passing through the fuel-pin breeder channels (green in Figure 2b). At the end of these triple-wall tubes, the warm helium (red in Figure 2b) takes a U-turn, reaching the outlet chamber behind the BZ volume that encases the $Be_{12}Ti$ Neutron Multiplier Material (NMM). Penetrations provide a flow path for the hot gas into the outlet manifold from the outlet chamber. After passing through the Breeder Units (BUs), the gas is collected in the hot manifolds above the reactor, finally leading to the HeSGs via the hot legs. The helium ports are located on the top of the IB segments, while for OB segments, they were placed at $\sim 2/3$ of the segment height.

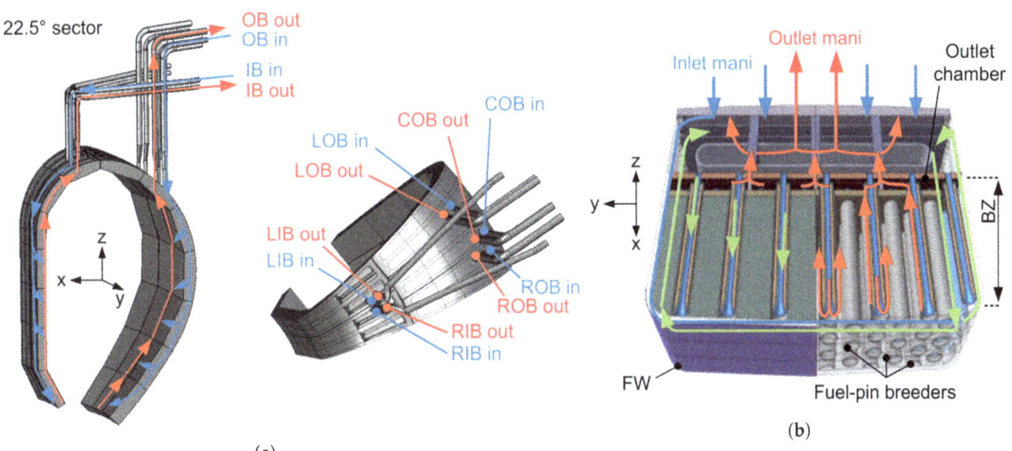

Figure 2. Helium flow paths in the segments (**a**) and in an OB BU (**b**).

The operation scheme of the circulators has been subject to revision; in the current arrangement, the compressors are running at full speed during the pulse, while the dwell phase control is dependent on the PHTS Thermal-Hydraulic (TH) conditions; more details on these logics are provided in §3.4. The general steady-state parameters of the PHTS are listed in Table 1 with regard to pulse and dwell phase conditions.

Table 1. PHTS steady state parameters.

Param. / Phase	P_{BB} [MW$_{th}$]	P_{DIV+VV} [MW$_{th}$]	$\dot{m}_{He}$ [kg/s]	$p_{CL,He}/p_{HL,He}$ [bar]	$T_{CL,He}/T_{HL,He}$ [°C]
Pulse	2101.7	337.2	1777.6	78.2/78.9	300/520
Dwell	21.0	3.4	$\sim$17.8	77.7/77.7	300/450

2.2. Power Conversion System

The PCS was developed on the basis of common Rankine cycles of commercial (nuclear) power plants, comprising a preheater line, boiler section, steam turbine, and condenser. The preheater line can be distributed into Low-Pressure (LP) and High-Pressure (HP) sections, connected by the Deaerator (DEA). Feedwater preheaters FW(1,2)HX and the DIV-PFU HX pair are located upstream from the DEA, while high-pressure preheaters FW(3,4)HX and the DIV-CAS and VV heat exchangers are downstream from the DEA. Since there is no dedicated pressurizer system in the PCS, such a role is fulfilled by the DEA. The boiler section houses two branches, the NEAR and FAR collectors, where the labeling refers to the difference in the distance from the FW4HX outlet junction (the FAR collector is 164 m longer than the NEAR). Each feedwater collector supplies a chain of four helical coil SGs with 228.2 °C water at 132.4 bar. The coolant is evaporated and superheated to ∼430–450 °C in the helices of the HeSGs. The molten salt SG uses the same helical coil layout as the HeSGs, with HITEC® as the primary coolant. The HITEC® salt is a eutectic mixture of water-soluble, inorganic salts of $NaNO_3$–$NaNO_2$–KNO_3 [7]. The MSSG collectors are located closer to the turbine island; hence, its steam collector joins the main steam line downstream from the FAR and NEAR branches. The MSL supplies the turbine, which has two high-pressure and four low-pressure stages with an intermediate Moisture Separator (MS) and two Reheaters (RHs). Bleed lines from the MSL and extraction lines from the turbine stages feed the HXs of the preheater line, maintaining steady conditions throughout the entire operation. Closing the Rankine cycle of the PCS, the exhausted steam is condensed in the main condenser; hereafter, the Condensate Pump (CEP) delivers the fresh coolant to the FW1HX inlet. The layout of the HCPB small ESS plant is depicted in Figure 3 with respect to the subsystems marked also in Figure 1. The general properties are provided in Table 2 in the pulse and dwell phase conditions.

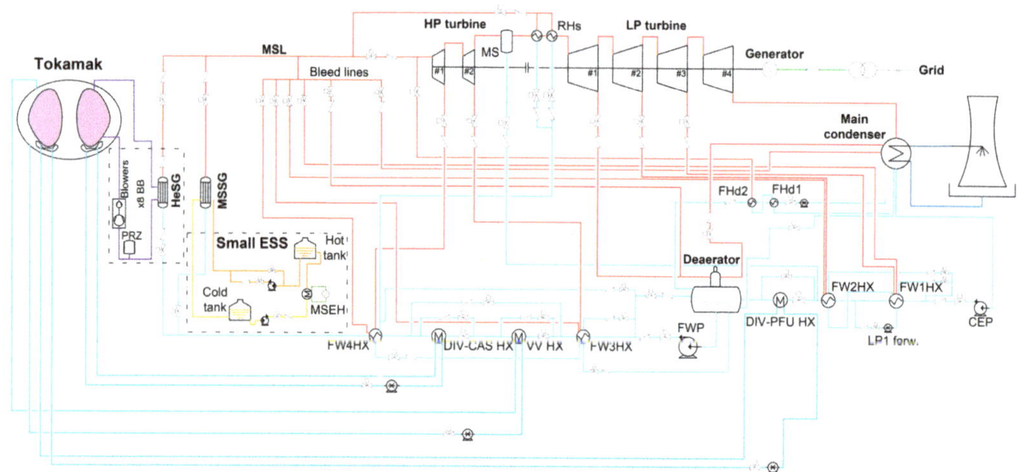

Figure 3. Power conversion system in the HCPB small ESS DEMO plant.

Table 2. HCPB PCS steady state parameters.

Param. Phase	P_{HeSGs} [MW$_{th}$]	$P_{gross,ST}$ [MW$_e$]	$\dot{m}_{fw}$ [kg/s]	p_{steam} [bar]	$T_{fw,max}/T_{steam}$ [°C]	p_{DEA} [bar]
Pulse	2101.7	1050.0	953.0	123.5	228/440	
Dwell *	∼21.0	∼10.5	214.5	110.0	228/400-500	3.5

** These values are static estimates obtained from steady state heat balances, transient trends are shown in §4.2.*

2.3. Small Energy Storage System

The small ESS design was based on large-scale industrial molten salt storage system technology utilizing HITEC® salt. The salt loop has no connection to the PHTS; hence, an electrical heater warms up the salt, and this component is connected to the plant's inner grid as other on-site consumers, e.g., coolant pumps. The pulse phase operation principle is as follows: The charging molten salt pump delivers the coolant of the cold tank (288 °C) to the electrical heater. After warming up the salt (to 500 °C), the flow is directed into the hot tank. The discharge pump supplies the MSSG with the hot salt; after passing through the SG, the cold HITEC® salt is collected in the cold tank. For the depiction of the system, the reader is advised to return to Figures 1 and 3. The total MS inventory yields 921 t, 884 t in total from the hot and cold tanks, and 37 t in the pipe system, which includes the salt masses of the recirculation line and the tube side of the MSEH.

3. Apros Model

The Advanced Process Simulation (Apros) system code has been developed by VTT Technical Research Centre of Finland Ltd., and Fortum since 1986. The code provides three-equation (homogeneous) and six-equation solutions for one-dimensional TH problems utilizing a staggered space discretization scheme. The state variables are calculated in the center of the mesh cells (nodes), the flow related variables are derived at the border of adjacent cells. Considering heat transfer modeling, a vast array of analytical and empirical correlations is available, in addition, a new formula can be implemented using Simantics Constraint Language (SCL) scripts. SCL is a functional programming language used for scripting purposes in Simantics-related products (https://www.simantics.org/, accessed on 4 June 2021). As a result of the EUROfusion-related work, nested User Components (UCs) were developed in order to provide a higher-fidelity solution with respect to blanket and helical coil SGs. In this case, the homogeneous model has been applied to the nodes of the small ESS and the PHTS, while the six-equation solution has been used in every other node, filled with water/steam.

The general structure of the integral model can be seen in Figure 3 featuring some additional elements compared to Figure 1; these extra components were requisites of satisfying plant control (e.g., ST steam dump line, make-up and let-down piping, and steam feed lines).

Providing a basis for further comparison, the cycle net efficiency (η_{cy}) has been derived for pulse and dwell phase, following the formulation given in [4] by:

$$\eta_{cy} = \frac{W_{gross} - W_{PCS,pump}}{P_{BB} + P_{DIV} + P_{VV} + P_{MSSG}} \tag{1}$$

where W_{gross} is the gross power on the ST shaft, $W_{PCS,pump}$ is the total pumping power of the PCS, and P_{BB}, P_{DIV}, P_{VV}, and P_{MSSG} are the heat inputs from the breeding blanket, divertor (PFU and CAS), and vacuum vessel HXs and MSSG, respectively. Furthermore, the overall plant net average efficiency (η_o) was also calculated as:

$$\eta_o = \frac{\int_0^{t_{cy}} (W_{gross} - W_{plant})\, dt}{\int_0^{t_{cy}} (P_{BB} + P_{DIV} + P_{VV} + P_{MSSG})\, dt} \tag{2}$$

where t_{cy} is the length of a full cycle (7800 s) and W_{plant} is the total power of on-site consumers (total pumping power and MSEH).

3.1. Primary Heat Transfer System

Following the evolution of the design, the 18-sector layout was updated based on 2017–2019 materials provided by EUROfusion partners [8]. The new BB model incorporates one IB and one OB model, describing average segments of a 22.5° sector. The arrangement of these nested Apros process components is shown in Figure 4, featuring also boundary conditions that represent another IB segment and two OB segments in terms of enthalpy and flow rates.

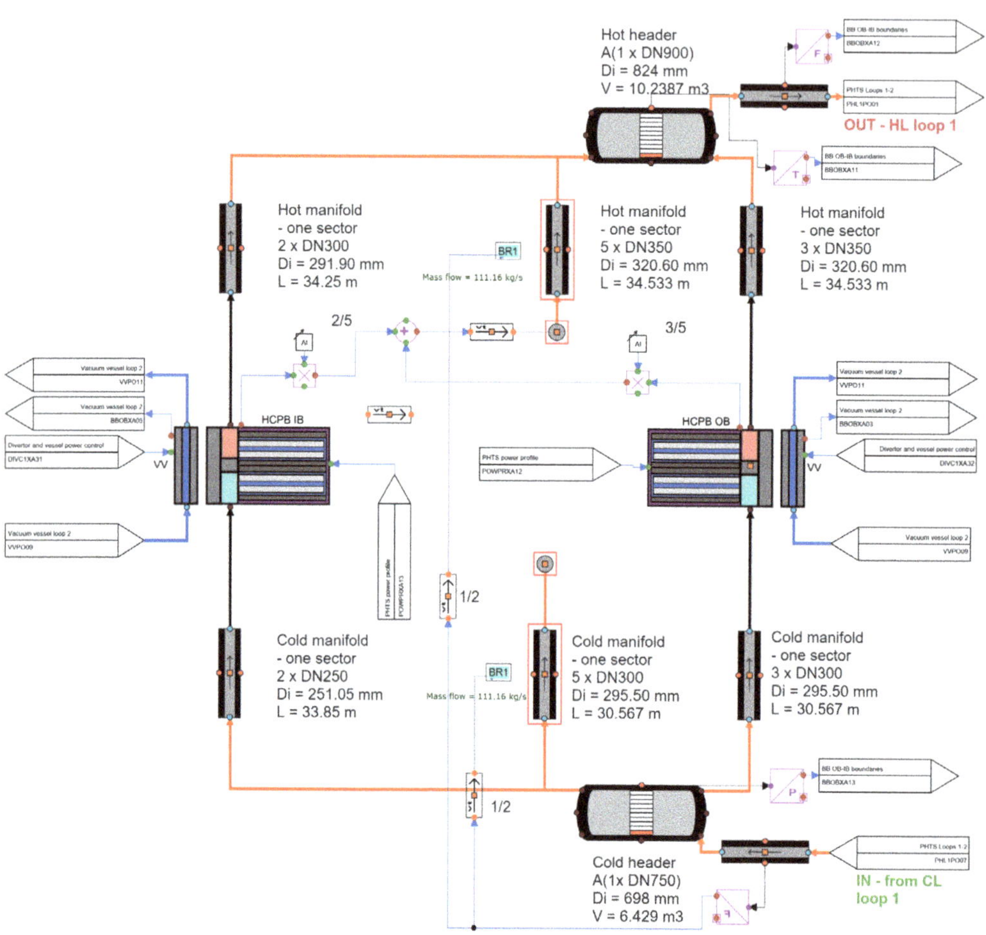

Figure 4. IB and OB segment models with collectors and TH boundary conditions in the Apros reactor model.

A homogeneous model has been applied in every TH node on the primary side; the thermophysical properties of helium were defined according to the report of the National Institute of Standards and Technology [9]. Each segment model contains a first wall, a BZ-Back Support Structure (BSS), and a vacuum vessel compartment. The FW and BZ Heat Structures (HSs) are coupled via heat conduction, while the BSS and VV sections are in radiative contact. The breeding zone domain is modeled by a lattice of solid heat structures, denoting the Advanced Ceramic Breeder (ACB) and neutron multiplier volumes. The thermophysical functions ($c_p \cdot \rho$, k) of the NMM material were derived by lumping together the properties of the multiplier material and structural steel elements, according to volume ratios that correspond to the IB and OB geometries. The coolant loops have been modeled explicitly considering four short and four long loops, and the relevant integral geometry parameters are listed in Table 3. As Section 3 pointed out, the TH nodes of the PHTS utilized the homogeneous model; the layout of Loop #1 is depicted in Figure 5.

Table 3. PHTS piping characteristics.

Section	Long Loop	Short Loop
Length of HL [m]	386.8	357.1
Length of CL [m]	363.6	353.7
Total length [m]	782.3	742.7
Volume of HL [m^3]	66.6	54.7
Volume of CL [m^3]	50.9	40.7
Total volume * [m^3]	129.8	107.7
Total piping length [m] (4 × long + 4 × short)	6100	
Total He volume [m^3] (4 × long + 4 × short)	950	

* *These values also include the circulators' piping.*

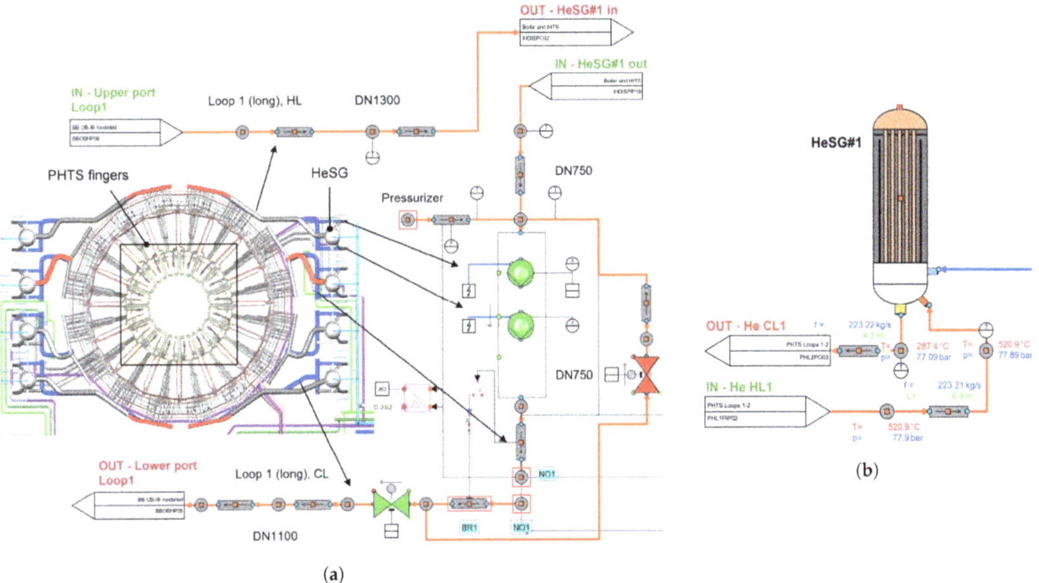

(a)

(b)

Figure 5. Loop #1 layout (**a**) and HeSG#1 in the Apros model (**b**).

3.2. Power Conversion System

The helical coil SG design was conceived of by colleagues at the University of Palermo; later, it was further developed in the cooperation of the designer and VTT. The ASME Boiler & Pressure Vessel Code standard's Class 1 relevant section was used for the sizing of the heat exchanger [10] since the DEMO is classified as a nuclear power plant. The current iteration of the SG incorporates 42 helical modules, each module composed of 18 helical tubes (helices), a central rod (fixing the helices' position), and an outer cylinder, providing a confined flow channel for the helium around the coils. The cross-section of the SG is depicted in Figure 6, where the primary coolant enters the lower plenum at the inlet ports. After passing through the annuli and the mixing chamber, the gas turns downward to flow through the modules around the helical tubes. After heat exchange, the cold gas arrives at the rest chamber and exits the SG frustum through its outlet port. Considering the secondary side, the feedwater is fed to the modules via 42 inlet tubes that deliver the coolant to the inlet spiders of the modules, which distribute the coolant among the helices. After evaporating and superheating, the secondary coolant is collected by the head spider at first, then passing through the upper tube sheet, the total steam flow exits the SG via the outlet port on the top of the upper head. As mentioned earlier, the primary side utilizes a homogeneous flow model, while the secondary side uses a six-equation heterogeneous solution. The Apros model describes one helical coil module explicitly in twelve helical process components. These process components accommodate the helium-filled volume around the tubes, the tubes themselves, the module wall, the central rod, and the shroud volume. The remaining 41/42 modules are represented by boundary conditions between the rest chamber and the upper tube sheet.

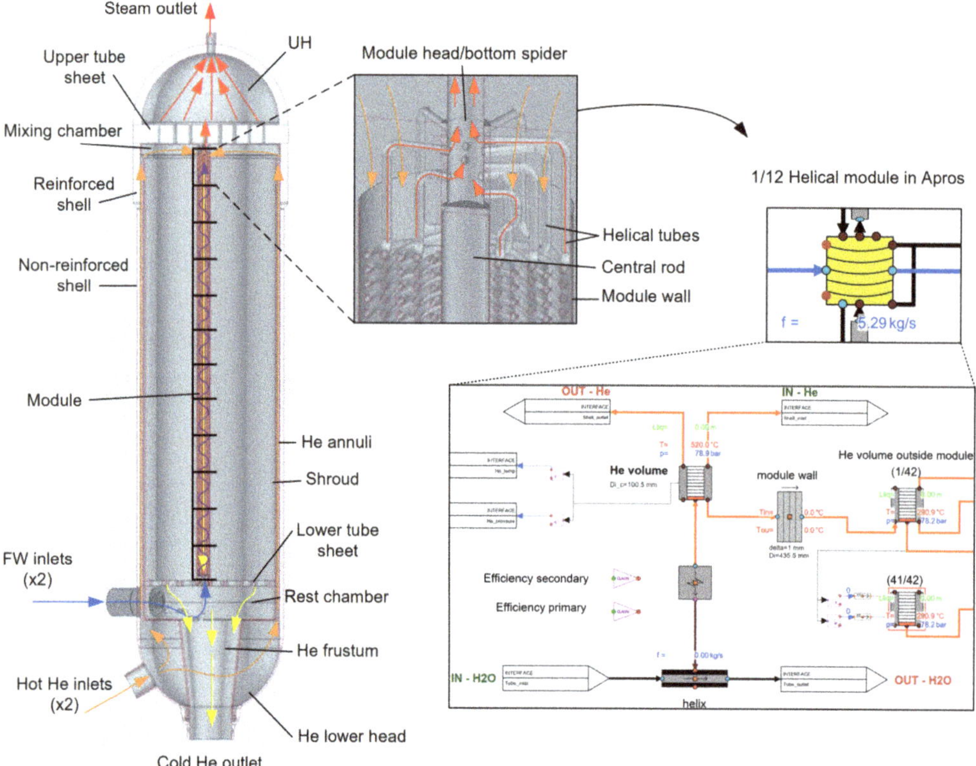

Figure 6. The layout of the helical coil steam generator.

Heat transfer between helium and the HX tubes' outer surface has been calculated using the Žukauskas correlation [11], while the treatment of secondary side heat transfer demanded a novel approach. The correlation package of Apros, relevant for helical geometries, has been utilized for such purposes [12]. The main parameters of the steam generators can be found in Table 4; note that the architecture of the MSSG is identical to the HeSG for the time being. Apart from the TH differences, the only departure from the helium–water SG is that the primary coolant medium was switched to HITEC®.

Table 4. SG nominal (pulse) parameters.

Param. SG	P_{SG} [MW$_{\text{th}}$]	m_{prim} [kg/s]	m_{fw} [kg/s]	A_{HT} [m^2]	$T_{fw,in}/T_{fw,out}$ [°C]
HeSG	263.3	229.0	118.2	3624.09	228/400-500
MSSG	14.0	42.3	7.4		

The remaining components of the PCS were modeled using basic Apros process components such as heat exchangers, various valves, pumps, tanks, and turbine elements.

3.3. Small Energy Storage System

The small ESS has been modeled explicitly featuring the elements described in §2.3. The recirculation line of the discharge pump enabled a soft control of the salt flow rate, where the motor speed is kept at a constant 100%. During the pulse phase, the Control Valve (CV) of the recirculation line is fully open; in dwell, the CV closes, directing the total salt flow to the primary inlet port of the MSSG. In Section 3, it was noted that the salt loop used the homogeneous solution for its TH nodes; nonetheless, due to the uncertainties of auxiliary systems (e.g., geometry, pressure control), the salt tanks have been modeled by boundary conditions where dynamic enthalpy calculations simulated the adequate transient behaviour. The pressure of the tanks was set to 8 bar, providing a reasonable pressure profile over the system. The general parameters of the small ESS model can be found in Table 5, where $m_{disch.}$ refers to the cold pump flow rate (cold tank $\rightarrow$ MSEH/hot tank) and $m_{ch.}$ denotes the hot pump flow rate (hot tank $\rightarrow$ recirc. line/MSSG/cold tank).

Table 5. Main steady state parameters of the small ESS in Apros.

Param. SG	P_{SG} [MW$_{th}$]	m_{prim} [kg/s]	m_{fw} [kg/s]	A_{HT} [m^2]	$T_{fw,in}/T_{fw,out}$ [°C]
Pulse	41.2	14.0	500.0/$\sim$287.7	42.3	124.5
Dwell	$\sim$15.7	$\sim$275.0		830.1	47.3

3.4. Logics

Considering power deposition, the reactor model used an asymptotic time-dependent curve for plasma ramps as the boundary condition, simultaneously in the blanket segments and secondary power sources as well.

The PHTS flow control's development is still in progress; the model introduced hereby describes the recent iteration of the logics, which fulfills the established requirements. The control scheme envisages Cold Leg (CL) flow rate modulation according to reactor power with the constraint of the $T_{in,comp}$ compressor inlet temperature, that is to keep $T_{in,comp}$ as close as reasonably possible to its nominal value. This has been achieved by integrating a recirculation line on each compressor station. As the compressors are running at 100% at all times, this recirculation line adjusts the flow rate that is being sent to the blanket or fed back to the compressor's inlet. Since the compressor power is relatively large ($\sim$11 MW$_e$/pair), an intercooler was also installed on the recirculation lines in the form of a boundary condition. This cooler decreases gas enthalpy in order to prevent the overheating of the machine and the recirculated gas.

On the secondary side, the turbine main control valve was following the reference heat balances' steady-state pressures (123.5$\leftrightarrow$119.0 bar in pulse$\leftrightarrow$dwell transitions), as the turbine unload commenced bleed lines took over the preheater line heat exchangers' pressurization from the extraction lines. Bypass lines of the FWHXs, DIV, and VV HXs guaranteed that the primary side average temperatures remained around the pulse phase values. The deaerator pressure was maintained at 3.5 bar throughout the entire operation.

4. Results

The Apros calculation envelopes two consecutive pulse–dwell cycles with a 1200 s pulse phase at the beginning. The pulse phase is 7200 s including two plasma ramps before and after the pulse period, and the dwell phase lasts 600 s; thus, one full cycle is 7800 s. The figures of §4.1 to 4.3 show this sequence of events with grey areas marking the dwell periods.

4.1. Primary System Behaviour

The breeding blanket, the lumped curve of DIV, and the VV loops (secondary) are compiled in Figure 7, featuring also pumping power demands and steam turbine gross power.

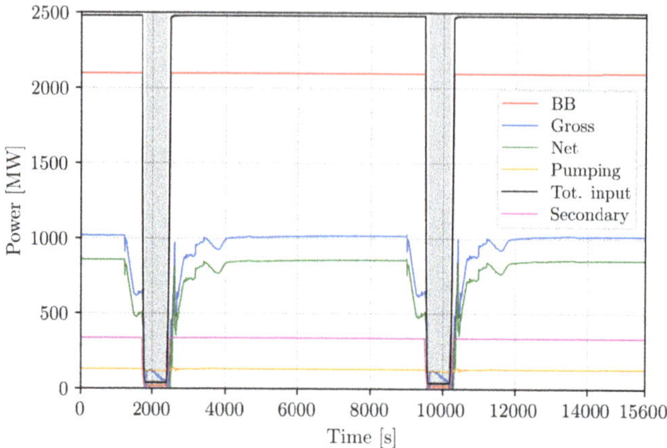

Figure 7. Power trends.

The Loop #1 helium temperatures are depicted in Figure 8, denoting measurements in the hot and cold legs, the collectors, and the compressor inlet. The upper limit for hot leg temperatures was set at 550 °C due to the Eurofer creep temperature, while another restriction was defined for the circulator inlet temperature. During transients and dwell, the inlet temperature has to be kept as close as possible at its pulse phase value (296 °C). The transient showed that gas temperatures could reach 532–533 °C in the HL, yielding a narrow safety margin considering the earlier mentioned limit (HL1 and hot collector trends overlap in Figure 8). The compressor inlet temperature reached 316 °C, giving a mild 20 °C increase compared to the pulse phase. Since even higher HL temperatures are permissible, the mentioned maximum compressor inlet temperature could be further lowered, should the compressor operation require so.

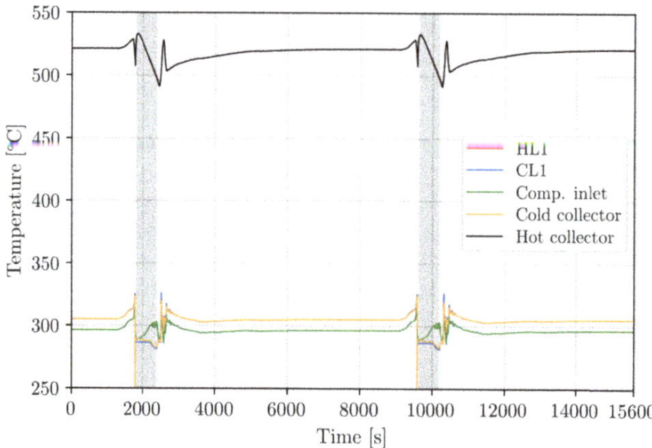

Figure 8. Helium temperatures in the PHTS.

The PHTS pressures showed that during transients, the CL and HL pressures decreased ~3 bar and ~1.6 bar, respectively, while no significant change could be observed at the circulator inlet. This indicated that the recirculation line did not induce perturbations apart from a momentary 0.6 bar depression when the recirculation valve opened at a nominal loop flow rate.

Gas flow rates in the coolant loops varied between 45 kg/s and 229 kg/s; the Loop #1 trends are shown in Figure 9. The ramp-down and ramp-up were rather asymmetric, governed by the thermal inertia of the blanket and steam generators. As for the ramp-up, the recirculation line throttled gas flow rates in the cold legs in order to ensure low circulator inlet temperatures, still maintaining acceptable HL temperatures. Cold and hot leg flow rates overlap in Figure 9, since in the pulse phase, the recirculation line is closed; in addition, the recirculation line is connected to the compressor suction; hence, the bypassed gas does not appear in the HL trends in dwell.

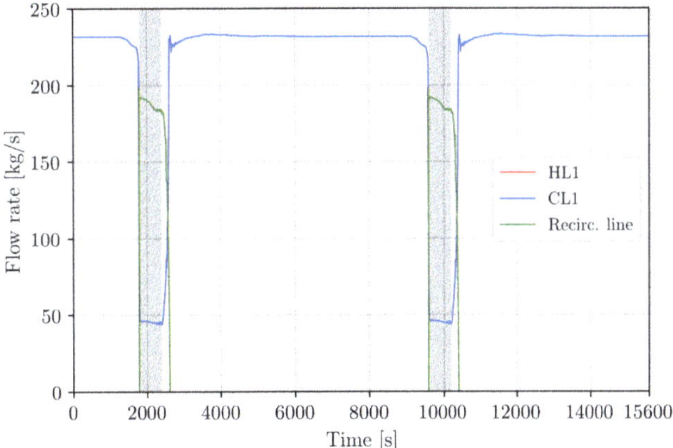

Figure 9. Helium flow rates in Loop #1.

The primary side temperatures of the DIV and VV heat exchangers are compiled in Figure 10, highlighting a notable change in the ΔT values. As the DIV and VV powers shift, from 100% to dwell time levels of $\approx$1%, the PCS (secondary) side flow rates of the corresponding HXs decrease as the bypass lines are activated. When HeSG chains restarted and flow rates rose in the feedwater collectors, the DIV-CAS and VV outlet temperatures decreased accordingly.

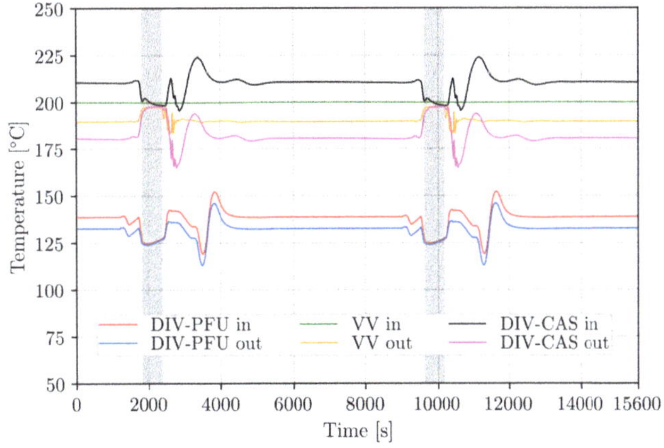

Figure 10. Divertor and vacuum vessel HX primary side temperatures.

4.2. Secondary System Behaviour

The MSL pressure was maintained at 119.0 bar during dwell, while the HP1 inlet pressure varied between 10 bar and 28 bar; the steam side pressures are given in Figure 11. The steam generators represent a substantial source of thermal inertia ($\sim$401 t dry mass/HeSG) aside from the BB heat capacity; eventually, a significant part of the secondary inventory was evaporated; consequently, collapsed water levels decreased in the helices by 95% by the end of dwell. The surplus steam was delivered to the main condenser via the Dump Line (DL) during turbine (un)loading, resulting in a smoother pressure regulation upstream from the turbine. The dump line was closed at the end of the ramp-down period; thus, the HP1 inlet pressure followed strictly the inlet steam flow rate.

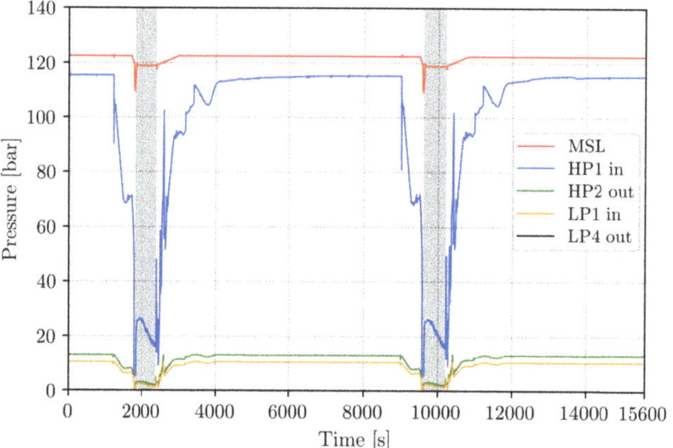

Figure 11. Steam pressures in the secondary system.

The transient trends showed steam qualities $\gg$99% in the collectors. As the fresh steam pressures and temperatures were sufficiently high in the MSL, the bleed lines could maintain hot conditions on the preheater line, as illustrated in Figure 12. During the unloading process, the maximum feedwater temperature $T_{FH4,out}$ fell $\sim$15 °C, although as flow rates stabilized on the preheater line, the $T_{FH4,out}$ recovered, and the dwell phase values varied between 230–233 °C.

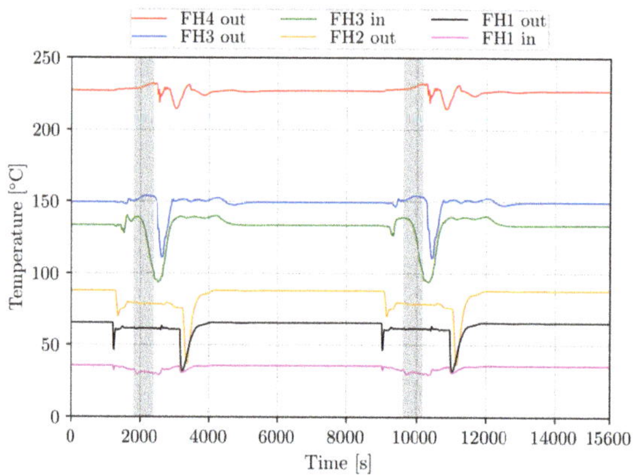

Figure 12. Feedwater temperatures in the PCS.

The deaerator collapsed water level varied between 3.0 m and 3.7 m corresponding to an $\sim$44 m^3 inventory displacement between plasma ramps. The pressure control was maintained at 3.5 bar during dwell with a $\pm$0.1 bar margin; thus, no significant disturbance was prompted on the preheater line; the DEA behaviour is depicted in Figure 13 with respect to the mentioned properties. Despite satisfying the PCS pressure control, the collapsed water level resembled notable variation during the transients, and such deviations could be mitigated by tuning the capacities of the DEA/condenser let-down lines and their corresponding logics.

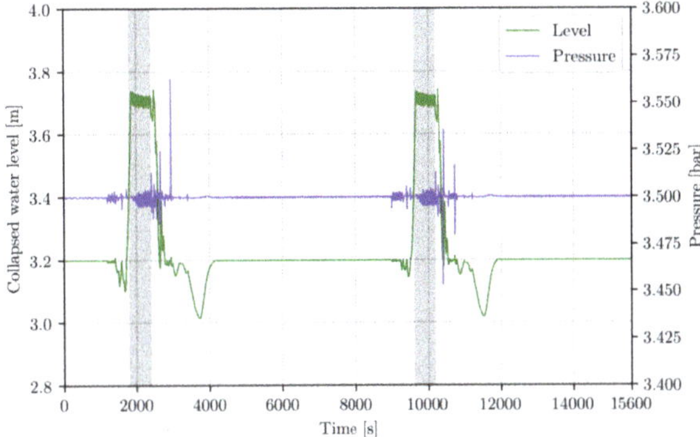

Figure 13. DEA pressure and collapsed water level trends.

The Apros cycle net efficiency for the PCS (η_{cy}) in pulse and dwell was calculated as 37.6% and 21.9%, while the overall plant net average efficiency (η_o) was 30.7% integrated over one cycle.

4.3. Small ESS Behaviour

The salt temperatures are depicted in Figure 14a, the electrical heater outlet temperature remained around 500 °C within a range of ± 4 °C, implying that the current control scheme was ample. Due to the small salt tank volume and MSSG warm-up, the cold tank temperature varied by ± 5 °C. In this layout, each tank has a volume of 418 m^3, which is much smaller compared to the 1362 m^3 tank volume in the small ESS of the WCLL plant. This disparity in salt inventories already indicates that the HCPB small ESS has indeed a smaller heat capacity compared to its WCLL counterpart. Figure 14b illustrates the molten salt levels with the top of the tanks, the level trends highlight a ± 2.7 m and ± 2.5 m variation for the hot and cold tank, respectively. The available free volumes above the salt highlight the challenge of future pressure control, albeit that the hereby described trends will assist developers. The results also showed that the minimum salt level at the end of the (dis)charge period yields ~ 0.66–1.04 m (18–28% fill level).

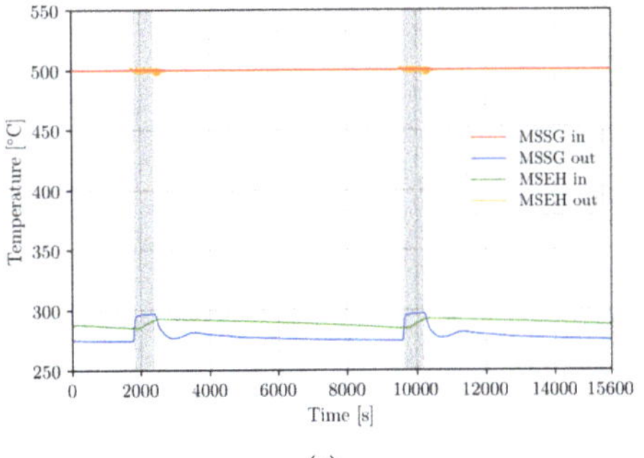

(a)

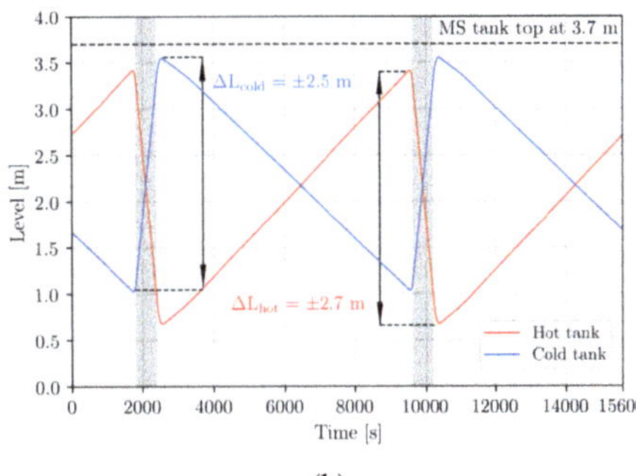

(b)

Figure 14. Temperatures in the small ESS (**a**) and salt levels (**b**).

5. Synopsis

Modeling efforts and transient analysis were presented considering Apros activities related to the HCPB small ESS configuration of the DEMO. A novel helical coil steam generator was introduced as the primary–secondary side interface with helium as the primary and water as the secondary coolant. The transient simulation, composed of two consecutive dwell–pulse periods, revealed that the HeSG design in its current state is a feasible option with proper gas circulator logics. It was emphasized that the various restraints, with respect to primary system operation, have a profound impact on optimization freedom regarding control logics, moreover on the PCS behaviour. The power conversion system showed noticeable, but manageable transients during plasma ramps and dwell (e.g., coolant inventory displacement), corresponding pressure and temperature variations can be considered acceptable keeping in mind the pressure decrease exceeding 100 bar at the inlet of the high-pressure turbine section.

6. Outlook

The circulator control scheme's optimization will be a high priority task in the upcoming conceptual design phase. Due to the high degree of uncertainty considering, e.g., manufacturer and component characteristics, the logics will have to be further tailored. Hence, secondary side control systems will also have to be monitored and adjusted, should the PCS performance suggest otherwise. It has to be noted that no heat losses have been taken into consideration thus far, due to the maturity of the design. Nonetheless, based on the substantial surface area and length of the primary and secondary system piping ($\sim$6.1 km and $\sim$3.5 km, respectively), when compared to commercial nuclear power plants, one has to pay close attention to the optimization of plant efficiency. One action in this direction was the addition of a pair of recuperative heat exchangers to the condenser–deaerator let-down line (see Figure 3), where the dumped steam heats the condensate, delivered to the DEA as a result of the condenser level control. Other areas can be also identified where energy could be recovered, for instance valve stations as the large steam dump valve where the integrated enthalpy loss amounts to $\approx$423 kWh over a cycle. Regarding the evolution of the small ESS, the implemented boundary conditions, responsible for dynamic enthalpy calculation, will be removed as the soon-to-be published Apros service pack will provide an upgraded iteration loop, dedicated to molten salt–inert gas systems.

Author Contributions: Conceptualization, formal analysis, data curation, writing, M.S.; supervision, S.N. All authors have read and agreed to the published version of the manuscript.

Funding: This work was carried out within the framework of the EUROfusion Consortium and received funding from the Euratom research and training program 2014–2018 and 2019–2020 under Grant Agreement No. 633053. The views and opinions expressed herein do not necessarily reflect those of the European Commission.

Abbreviations

The following abbreviations are used in this manuscript:

ACB	Advanced Ceramic Breeder	HL	Hot leg
Apros	Advanced Process Simulation	HP	High-pressure
AUXB	Auxiliary Boiler	HX	Heat Exchanger
BB	Breeding Blanket	IB	Inboard
BOP	Balance of Plant	IHTS	Intermediate Heat Transfer System
BSS	Back Support Structure	LIB	Left Inboard
BU	Breeding Unit	LOB	Left Outboard
BZ	Breeding Zone	LP	Low-pressure
CEP	Condensate Pump	MS	Moisture Separator
CL	Cold leg	MSEH	Molten Salt Electrical Heater

COB	Center Outboard		MSL	Main Steam Line
COND	Condenser		MSSG	Molten Salt Steam Generator
CV	Control Valve		NEAR	Shorter feedwater/steam collector
DC	Downcomer		NMM	Neutron Multiplier Material
DEA	Deaerator		OB	Outboard
DEMO	Demonstration Power Plant		PCS	Power Conversion System
DIV-CAS	Divertor Casette		PHTS	Primary Heat Transfer System
DIV-PFU	Divertor Plasma Facing Unit		RH	Reheater
DL	Dump line		RIB	Right Inboard
ESS	Energy Storage System		ROB	Right Outboard
FAR	Longer feedwater/steam collector		SCL	Simantics Constraint Language
FH	Feedwater Heater		SG	Steam Generator
FW	First Wall		ST	Steam Turbine
FWHX	Feedwater Heat Exchanger		UC	User Component
FWP	Feedwater Pump		VV	Vacuum Vessel
HCPB	Helium-Cooled Pebble Bed		WCLL	Water-Cooled Lithium-Lead
HeSG	Helical Coil Steam Generator			

Variables

The following variables are used in this manuscript:

η_{cy}	Cycle efficiency		$p_{HL,He}$	Helium pressure in HL
ρ	Density		P_{MSEH}	MSEH power
A_{HT}	Heat transfer area		P_{VV}	VV power
c_p	Isobaric heat capacity		t	Time
k	Heat conductivity		t_{cy}	Cycle length
$m_{ch.}$	Salt charge flow rate		$T_{CL,He}$	Helium temperature in CL
$m_{disch.}$	Salt discharge flow rate		$T_{cold,tank}$	Cold salt tank temperature
m_{fw}	Total feedwater flow		$T_{FH4,out}$	FH4 feedwater outlet temperature
m_{He}	Total primary He flow rate		$T_{fw,in}$	Feedwater inlet temperature
$p_{CL,He}$	Helium pressure in CL		$T_{fw,max}$	Max. feedwater temperature
p_{DEA}	DEA pressure		$T_{fw,out}$	Feedwater outlet temperature
$p_{HL,He}$	Helium pressure in HL		$T_{HL,He}$	Helium temperature in HL
p_{steam}	Fresh steam pressure		$T_{hot,tank}$	Hot salt tank temperature
P_{BB}	BB power		$T_{in,comp}$	Compressor inlet temperature
P_{DIV}	DIV power		T_{steam}	Fresh steam temperature
P_{fus}	Fusion power		W_{gross}	Gross power
$P_{gross,ST}$	ST gross power		$W_{PCS,pump}$	PCS pumping power
P_{HeSGs}	Total HeSG power		W_{plant}	Plant power

References

1. Turnyanskiy, M.; Neu, R.; Albanese, R.; Bachmann, C.; Brezinsek, S.; You, J.H. European roadmap to the realization of fusion energy: Mission for solution on heat-exhaust systems. *Fusion Eng. Des.* **2015**, *96–97*, 361–364. [CrossRef]
2. Hernández, F.; Pereslavtsev, P.; Kang, Q.; Norajitra, P.; Kiss, B.; Nádasi, G.; Bitz, O. A new HCPB breeding blanket for the EU DEMO: Evolution, rationale and preliminary performances. *Fusion Eng. Des.* **2017**, *124*, 882–886. [CrossRef]
3. Nevo, A.D.; Arena, P.; Caruso, G.; Chiovaro, P.; Di Maio, P.A.; Eboli, M.; Martelli, E. Recent progress in developing a feasible and integrated conceptual design of the WCLL BB in EUROfusion project. *Fusion Eng. Des.* **2019**, *146*, 1805–1809. [CrossRef]
4. Barucca, L.; Bubelis, E.; Ciattaglia, S.; D'Alessandro, A.; Del Nevo, A.; Giannetti, F.; Vallone, E. Pre-conceptual design of the EU DEMO balance of plant systems: Objectives and challenges. *Fusion Eng. Des.* **2021**, *169*, 112504. [CrossRef]
5. Szogradi, M.; Norrman, S.; Bubelis, E. Dynamic modeling of the helium-cooled DEMO fusion power plant with an auxiliary boiler in Apros. *Fusion Eng. Des.* **2020**, *160*, 11970. [CrossRef]
6. Official Apros Website. Available online: https://www.apros.fi/ (accessed on 3 June 2021).
7. Sohal, M.S.; Ebner, M.A.; Sabharwall, P.; Sharpe, P.B. *Engineering Database of Liquid Salt Thermophysical and Thermochemical Properties*; INL/EXT-10-18297 Rev. 1; Idaho National Laboratory: Idaho Falls, ID, USA, 2013; p. 24.
8. Hernández, F.A.; Pereslavtsev, P.; Zhou, G.; Neuberger, H.; Rey, J.; Kang, Q.; Dongiovanni, D. An enhanced, near-term HCPB design as driver blanket for the EU DEMO. *Fusion Eng. Des.* **2019**, *146*, 1186–1191. [CrossRef]
9. Arp, V.D.; McCarty, R.D.; Friend, D.G.B. *Thermophysical Properties of Helium-4 from 0.8 to 1500 K with Pressures to 2000 MPa*; NIST Technical Note 1334 (Revised); National Institute of Standards and Technology: Boulder, CO, USA, 1998; pp. 80303–83328.

10. American Society of Mechanical Engineers. *ASME Boiler & Pressure Vessel Code, Section III—Rules for Construction of Nuclear Facility Components, Division 1—Subsection NB for Class 1 Components*; American Society of Mechanical Engineers: New York, NY, USA, 2019.
11. Kakaç, S.; Liu, H.; Pramuanjaroenkij, A.B. *Heat Exchangers: Selection, Rating and Thermal Design*, 3rd ed.; CRC Press, Taylor & Francis Group: Boca Ration, FL, USA, 2012; pp. 33487–33742.
12. Leskinen, J.; Ylätalo, J.; Pettini, R. Extending thermal-hydraulic modeling capabilities of Apros into coiled geometries. *Nucl. Eng. Des.* **2020**, *357*, 110429. [CrossRef]

Article

Model Development and Transient Analysis of the WCLL BB BOP DEMO Configuration Using the Apros System Code

Marton Szogradi * and Sixten Norrman

VTT Technical Research Centre of Finland Ltd., 02044 Espoo, Finland; sixten.norrman@vtt.fi
* Correspondence: marton.szogradi@vtt.fi

Abstract: Extensive modelling and analytical work has been carried out considering the water-cooled lithium–lead breeding blanket (WCLL BB) balance of plant (BOP) configuration of the demonstration power plant (DEMO) using the Apros system code, developed by VTT Technical Research Centre of Finland Ltd. and Fortum. Contributing to the BOP work package of the EUROfusion Consortium, the integral plant model for dynamic analyses of the WCLL BB configuration has been updated with special attention to primary system components. Following trends of relevant neutronics modelling, a new BB model has been implemented in 2020 with the aim to obtain higher resolution output data and a more realistic thermalhydraulic feedback from the primary system. Once-through steam generator user components have been built based on CAD models conceived by BOP partners. Transient analyses have been performed providing a better picture regarding the behaviour of main components, e.g., the BB and the OTSGs, whilst highlighting possible ways to optimise the control scheme of the plant.

Keywords: DEMO; WCLL BB; small ESS; transient; Apros

Citation: Szogradi, M.; Norrman, S. Model Development and Transient Analysis of the WCLL BB BOP DEMO Configuration Using the Apros System Code. *Energies* **2021**, *14*, 5593. https://doi.org/10.3390/en14185593

Academic Editors: Alessandro Del Nevo and Marica Eboli

Received: 26 July 2021
Accepted: 27 August 2021
Published: 7 September 2021

1. Introduction

The balance of plant (BOP) work package of the EUROfusion Consortium is responsible for the development of cooling systems comprising the primary heat transfer system (PHTS), the intermediate heat transfer system (IHTS), and the power conversion system (PCS) concerning their technical description and control logics. Equipment and layouts of systems rely on designs used in conventional power plants albeit a stark departure can be highlighted with respect to the fusion power plant's normal operation. The initial break-down of hydrogen into plasma, afterwards the plasma current in the tokamak (Acronym for *toroidal chamber with magnetic coils*, Rus.: ТОроидальная КАмера с МАгнитными Катушками) are driven by the discharge of the central solenoid (located in the bore of the torus), until the plasma current reaches its opposite peak current. This means that the tokamak has to operate in pulsed mode, where a pulse period is followed by a dwell time when only decay heat is produced ($\sim$1–3% P_{nom}). The pulsed operation brought forth challenges in the design for producing electricity in a commercially and technically viable way.

Activities of the work package have been organised around two candidate blankets during the pre-conceptual design phase of DEMO, namely the helium-cooled pebble bed (HCPB) and the water-cooled lithium–lead (WCLL) breeding blanket (BB) configurations [1,2]. The various blanket concepts offer different benefits and challenges compared to one another, thus several plant configurations have been developed in order to explore the various aspects of primary-secondary system coupling.

Depending on the PHTS and PCS interface one can separate direct and indirect coupling schemes, in case of the direct layout the PHTS is directly connected to the PCS via steam generators (SGs) whereas the indirect adjective implies that PHTS-PCS coupling is realised by an IHTS [3]. Such an intermediate system utilises molten salt energy storage technology in a similar fashion as solar power plants and other large scale applications of the chemical industry.

The direct coupling branch has two further variants, one operating an auxiliary boiler (AUXB) and another adopting a small energy storage system (ESS). A former variant represented a benchmark case, with the purpose to verify the feasibility of the pulsed operation regime of the DEMO plant using a gas-fired boiler during dwell periods [4]. AUXB studies confirmed that the envisaged operation scheme can be implemented, although results underlined the necessity to develop further, e.g., SG and feedwater inventory management. In the small ESS configuration, the auxiliary boiler was replaced by a molten salt steam generator (MSSG) as the interface between the salt loop and the PCS.

VTT has been taken part in the development of the mentioned plant variants using the Apros system code [5], present article will report the results of the modelling and analytical work related to the WCLL BB small ESS configuration. Section 2 will introduce the architecture of the PHTS (§2.1), the PCS with an emphasis on the true OTSGs and corresponding control logics (§2.2) setting out the general small ESS design as well in §2.3. Section 3 will give a depiction of the integral Apros model detailing the main subsystems in a similar manner as Section 2 with more focus on the mentioned key components (§3.1–3.4). The results of the transient analysis are outlined in Section 4, finally, leading the reader to the conclusions and outlook of the work within Sections 5 and 6, respectively.

2. WCLL Small ESS BOP Configuration

The WCLL small ESS BOP features a pressurised water-cooled primary system with two separate loops representing the first wall (FW) and the breeding zone (BZ) cooling circuits. Both loops are connected to the PCS via two OTSGs giving four steam generators in total. Secondary power sources, namely the divertor casette (DIV-CAS), divertor plasma facing unit (DIV-PFU), and the vacuum vessel (VV) coolant loops have been connected to the PCS via three pairs of heat exchangers, where the DIV-PFU units are located on the low-pressure (LP), the DIV-CAS and VV units are placed on the high-pressure (HP) feedwater preheater line. The fresh steam is fed via the main steam line (MSL) to the steam turbine (ST) that consists of two high- and four low-pressure stages. Ultimately, the exhaust steam is condensed in the main condenser, closing the imperfect Rankine cycle of the power conversion system. The small ESS and the MSSG with their auxiliary systems are installed in between the tokamak and turbine islands. The CAD layout of the facility is depicted in Figure 1 highlighting the enlisted subsystems.

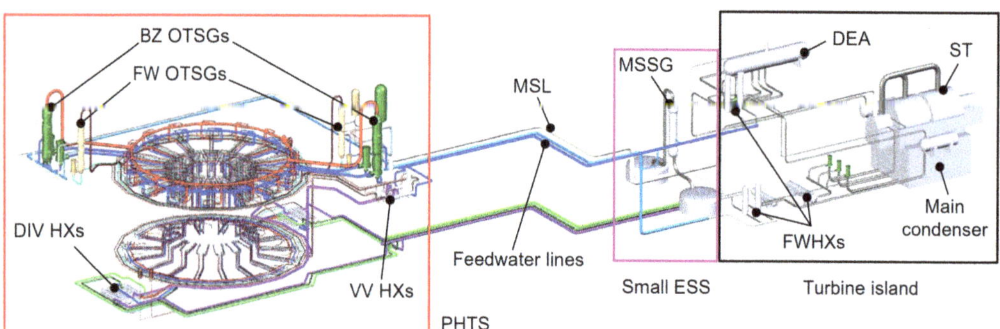

Figure 1. CAD layout of the WCLL small ESS DEMO.

Transient scenarios enveloped two consecutive pulse-dwell phases where the pulse (P_{fus} = 100%, P_{MSEH} = 100% (molten salt electrical heater), $P_{MSSG} \approx$ 5%) and dwell periods (P_{fus} = 1%, $P_{MSEH} \approx$ 40%, P_{MSSG} = 100%) lasted 7200 and 600 s, respectively. The plasma ramps represent asymptotic time-dependent power functions describing the reactor's thermal power variation between 1–100% under 100 s.

Prior to the plasma ramp-down the unloading of the turbine is initiated 500 s before the reactor rundown in order to achieve a permissible power gradient on the machine around $-10\%/min$. As dwell commences the power load rapidly decreases in the blanket and secondary power sources. In order to compensate the diminishing enthalpy difference between primary and secondary systems, the small ESS discharge line is activated loading the molten salt SG. The synchronisation of the OTSG and turbine unloading with MSSG control is a crucial aspect of successful pulse-dwell transition, where numerous safety requirements have to be fulfilled simultaneously, e.g., pressure restrictions in feedwater heat exchanger (FWHX) shells and the deaerator (DEA), hot leg (HL) temperature limitations in the PHTS ($T_{w,max} \ll T_{sat,HL}$). On the top of these requirements, various goals pose challenges as well, for instance a 10% load has to be maintained on the turbine throughout dwell by a suitable re-alignment of the whole PCS.

2.1. Primary Heat Transfer System

The primary system is responsible for the cooling of the BB, divertor, and VV components of the reactor. The coolant enters the FW and BZ inlet collectors at $\sim$295 °C, 155 bar, hereafter the flow is distributed among 16 sectors via the cold fingers that connect the segments' inlet ports and the collectors. As the 1/16 flow reaches the cold manifolds, the coolant is redistributed further among two inboard (IB) and three outboard (OB) segments. The cold water enters the segment at the top (IB) or $\sim$2/3 height (OB) of the segment, then the coolant passes downward via the FW and BZ spinal cords. In the lower section of each segment the flow path takes a U-turn, filling the FW and BZ inlet manifolds. As the coolant rises in the segments the inlet manifold flow rates gradually decrease in subsequent breeding units (BUs) whilst the outlet manifold flow rates increase correspondingly. In the head of the IB segments, the total segment inlet flow appears at the FW and BZ outlet ports, for OB segments the coolant takes another U-turn in the upper mixing volume, reverting finally to the water port at $\sim$2/3 height. The outlet flow enters the outlet manifolds at $\sim$328 °C eventually mixing the IB and OB loops' coolant inventory in the hot collectors. The hot leg piping connects these collectors to the steam generators' inlet junction on the top of the SGs at 152 bar. In the BZ loop each OTSG has a pair of pumps, while on the FW side one SG has one pump. Considering both loops, the pumps are running at a constant 100 % speed during the entire operation. Relevant thermalhydraulic properties are tabulated in Table 1 with respect to pulse and dwell phase values.

Table 1. Main parameters of the WCLL PHTS.

Param. Phase	P_{BB} [MW$_{th}$]	P_{DIV+VV} [MW$_{th}$]	$\Sigma m_{BZ}/\Sigma m_{FW}$ [kg/s] *	p_{HL} [bar]
Pulse	2019.8	337.0	$2 \times 3830.0/2 \times 1124.0$	155.0
Dwell	20.2	3.37		

* *The values refer to loop-wise flow rates hence the multiplication by a factor of two.*

Each 22.5° sector accommodates two, nearly identical left and right IB segments (left-LIB, right-RIB) and three, more distinct outboard segments (left-LOB, center-COB, right-ROB). In addition to upfront differences in waterport arrangements and structures, the breeding units also show noticeable variation in terms of material inventories, i.e., geometry, power loads, and TH conditions. The current 2020 design is depicted in Figure 2 below, where the left side was dedicated to the IB design while the right side to the COB segment, featuring the layout of the equatorial BU (region #4). Volumes marked with yellow contain PbLi eutectic, the blue, purple and red manifolds refer to water junctions with increasing temperature. All manifolds are enclosed by the back support structure (BSS), made of Eurofer alloy.

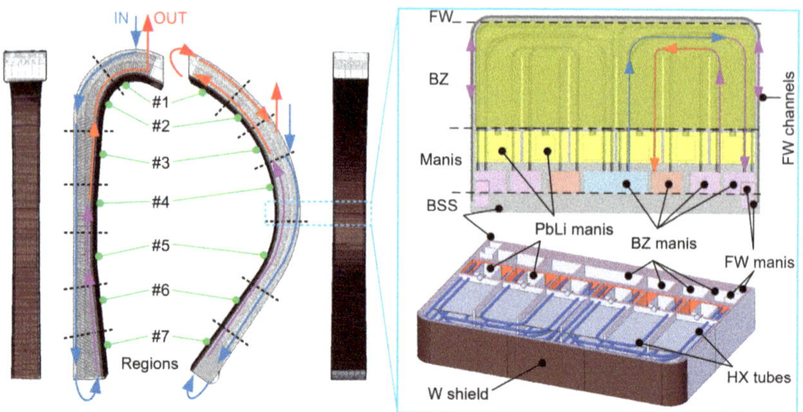

Figure 2. The CAD layout of the IB (**left**) and the COB segments (**right**) with flow paths.

Depending on the location, each segment contains 94–106 BUs stacked over each other between the lower mixing volume and the top of the segment. The layout of the equatorial COB unit is shown in Figure 2 featuring three arrays of coolant tubes (Σ22), the FW coolant channels (Σ20), the BZ filled with PbLi$_{17}$ (90 at.% ^{6}Li) and the coolant manifolds with the Eurofer skeleton. The PbLi coolant (yellow in Figure 2) enters the BZ via oval perforations in the manifold wall, the heated alloy exits the BZ via the outlet manifolds. Considering BZ loops, the cold water enters the inlet manifold (blue in Figure 2) before passing through the Eurofer heat exchanger tubes which lay across the PbLi manifolds and the BZ, leading the hot coolant to the outlet manifolds (red in Figure 2).

2.2. Power Conversion System

As mentioned before the PCS design was strongly influenced by commercial nuclear power plant architecture, where a two-stage feedwater preheater line supplies the boiler section with 228 °C water at 75 bar (see layout in Figure 3). The first stage incorporates the two LP heat exchangers (fed by the LP ST stages) and the twin DIV-PFU HXs. The coolant is collected in the DEA at ~140 °C with a tank fill level at 45%.

Apart from a major role in coolant inventory management a proper DEA pressure control is also capable of preventing undesired transients in the secondary system. The feedwater pump (FWP) delivers the coolant to the SG collectors via the two HP heat exchangers (fed by the HP ST stages and the reheaters) and the DIV-CAS and VV HX pairs. Bypass lines were also installed on the HP preheater section in order to achieve better temperature and flow rate control during dwell. The feedwater collectors divide the total flow into three branches: (1) MSSG line, (2) FAR, and (3) NEAR collectors. Two latter lines are connected to the OTSG chains, the designation of the branches is a reminder of the 168 m difference in length. The main steam line gathers the fresh steam (67–70 bar, 300–330 °C) from the SG collectors supplying the turbine and the bleed lines. During ST (un)loading the bleed lines pressurise the feedwater preheaters' shells in order to maintain the nominal feedwater temperature at the end of the preheater line throughout the entire operation.

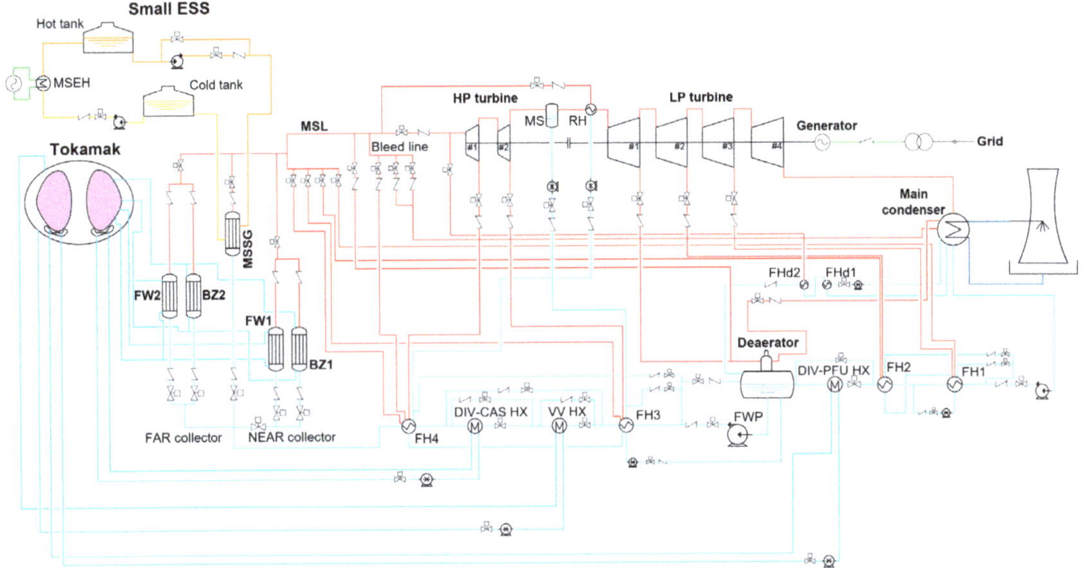

Figure 3. Power conversion system of the WCLL small ESS DEMO plant.

2.3. Small Energy Storage System

As mentioned earlier, the small ESS design was conceived based on the principles of large scale industrial molten salt storage system technology, the current configuration's coolant is HITEC®, an eutectic mixture of water-soluble, inorganic salts composed of $NaNO_3$–$NaNO_2$–KNO_3 [6]. Since the small ESS has no connection to the PHTS a molten salt electrical heater increases the enthalpy of the salt, this heater receives electricity from the plant as an onsite consumer, alike coolant pumps. The pulse phase operation principle is as follows: the charging pump delivers the coolant of the cold tank (282 °C) to the electrical heater, after the heating of the salt (to 330 °C) the coolant arrives at the hot tank. The discharge pump supplies the MSSG with fresh hot salt, after passing through the SG the salt is collected in the cold tank. For the depiction of the system the reader is advised to return to Figures 1 and 3.

3. Apros Model

The advanced process simulation (Apros) system code has been developed by VTT Technical Research Centre of Finland Ltd. and Fortum since 1986. The code provides three- and six-equation solutions for one-dimensional thermalhydraulic problems utilising a staggered space discretisation scheme. State variables are calculated in the center of the mesh cells (nodes), flow related variables are derived at the border of adjacent cells. Considering heat transfer modelling, a vast array of analytical and empirical correlations is available, in addition new formula can be implemented using Simantics Constraint Language (SCL) scripts. As a result of the EUROfusion-related work, nested user components were developed on top of the basic process component library of Apros, providing higher-fidelity solution with respect to the WCLL blanket and OTSGs. The homogeneous model has been applied considering nodes of the small ESS while the six-equation solution has been used in every other node filled with water or steam.

The general structure of the integral model can be seen in Figure 3 featuring some complementary elements compared to Figure 1, such elements were necessities to obtain satisfying plant control (e.g., steam dump line, make-up and let-down piping and steam bleed lines).

In order to provide basis for further comparison with other plant configurations the cycle net efficiency (η_{cy}) has been derived for pulse and dwell phase, as given in [3] by:

$$\eta_{cy} = \frac{W_{gross} - W_{PCS,pump}}{P_{BB} + P_{DIV} + P_{VV} + P_{MSSG}} \tag{1}$$

where W_{gross} is the gross power on the ST shaft, $W_{PCS,pump}$ is the total pumping power of the PCS, P_{BB}, P_{DIV}, P_{VV} and P_{MSSG} are heat inputs from the *BB*, *DIV* (PFU and CAS), *VV*, HXs, and *MSSG*, respectively. Furthermore the overall plant net average efficiency (η_o) was also calculated as:

$$\eta_o = \frac{\int_0^{t_{cy}} (W_{gross} - W_{plant})\, dt}{\int_0^{t_{cy}} (P_{BB} + P_{DIV} + P_{VV} + P_{MSSG})\, dt} \tag{2}$$

where t_{cy} is the length of a full cycle (7800 s), W_{plant} is the total power of onsite consumers (total pumping power and MSEH).

3.1. Primary Heat Transfer System

The blanket model was rebuilt in 2020 following the spatial discretisation principles of relevant plasma physics and neutronics models, also taking into account given CAD specifications. The complexity of the structure had to be tackled on a system code level whilst preserving the novelty of certain aspects, e.g., power deposition schemes, coolant, and dry mass inventories. In order to fulfil such requirements nested process components were created, with respect to the aforementioned factors, representing an interface between boundary conditions (BCs) and the thermalhydraulic model of the PHTS. The underlying algorithm of this BB model can be characterised as a non-linear onion scheme where the outermost shell is the interface between the PHTS/BC modules and the segments. Unlike in classical onion structures, process components on the lowest level receive information from the top (e.g., power) bypassing all the shells in between, thus the output information is directly affected by the input. As §2.1 pointed out, in total 5 segments compose one 22.5° sector, thus the reactor model contains 5 segment components (2nd shell). Each segment accommodates 7 BU components, corresponding mixing volumes, water ports, and interfaces between the BUs (3rd shell).

The poloidal power distribution has been defined for 7 regions, separately for IB and OB segments (uniform for OBs), thus the breeding units of these regions have been collapsed into 7 representative BUs, each corresponding to the average BU of its region. Inside the BU model three further user components have been placed: (1) FW (2) BZ, and the (3) VV components with their dependent interfaces (4th shell). These modules contain the heat structures (HSs), piping and power boundary conditions. The general structure can be seen in Figure 4 opening up the information flow and the shells of the reactor. Such an approach made it possible to customise FW, BZ, and VV compartments, independently from one another, in terms of nodalisation and power deposition. The radiative power distribution was derived by EUROfusion partners using the ASTRA transport code [7] developed by IPP for core radiative flux, the SOLPS code [8] for the scrape-off-layer radiative flux and the CHERAB code [9] developed by Culham Centre for Fusion Energy (CCFE). Nuclear heating deposition schemes were obtained with Monte Carlo N-particle analyses [10,11] concerning W, PbLi, water, Eurofer, and SS316L (VV) domains.

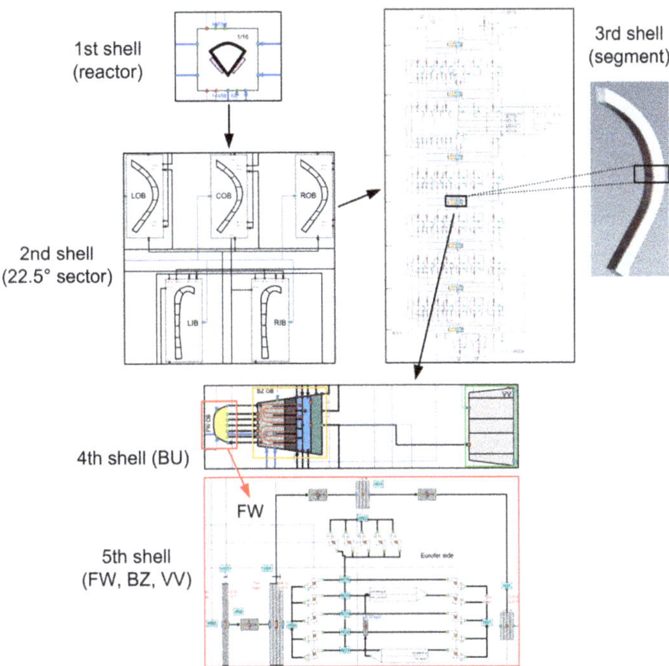

Figure 4. Structure of the BB sector user component.

3.2. Power Conversion System

The secondary system was built using basic Apros process components, such as heat exchangers, various valves, pumps, tanks, and turbine elements. Concerning the once-through steam generators, two new user components were dedicated to the coupling of the FW and BZ loops to the PCS. The BZ OTSG design was set up by the University of Napoli, as a member of the BOP work package, utilising a modified Babcock and Wilcox layout as a template while incorporating Westinghouse-type tubesheets. The SG is modified in the sense that no recirculation takes place between the downcomer annuli and riser volume, unlike in the generic layout. The development of the component is still in progress exploring alternatives as an attempt to find the most suitable configuration for the needs of the primary system, nonetheless present paper will discuss the behaviour of the first prototype. Taking into account foreseeable limitations in terms of geometry and TH parameters the BZ OTSG was scaled down by VTT to derive the FW OTSG. Since DEMO de jure represents a nuclear power plant, the ASME boiler and pressure vessel code standard's class 1 relevant section was applied during the sizing of both components [12].

The cross section of the BZ OTSG can be seen in Figure 5 also depicting the Apros component, note that FW and BZ OTSG components share their topology. The feedwater enters the riser volume via the downcomer annuli, the coolant is heated to its boiling point, evaporated, and superheated while rising through the 15 broached tube plates. The fresh steam leaves the SG via the outlet manifold connecting the OTSGs to the steam collectors (NEAR and FAR). General properties of the SGs are given in Table 2 with respect to nominal values, dwell phase powers, and flow rates are reduced to 1%.

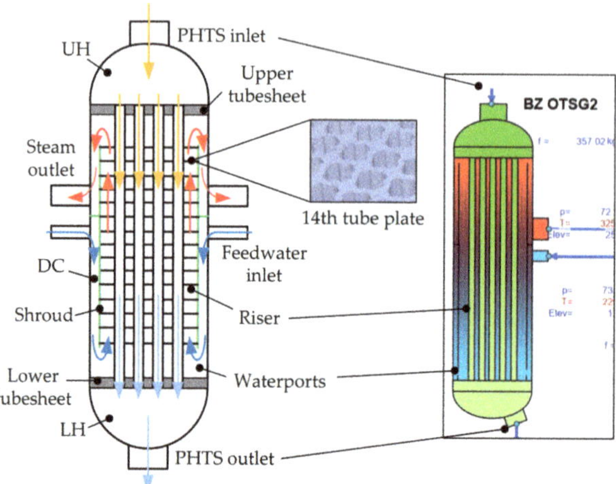

Figure 5. OTSG layout with Westinghouse-type broached plates.

Table 2. Characteristics of the OTSG UCs in pulse phase.

Param. UC	P_{SG} [MW$_{th}$]	p_{out} [bar]	T_{out} [°C]	A_{HT} [m^2]	m_{sec} [kg/s]
BZ OTSG	742			4904	3830
FW OTSG	220	70.0	320–330	1388	1137

3.3. Small Energy Storage System

The small ESS has been modelled explicitly featuring the elements described in §2.3. Earlier it was found that a recirculation line, installed on the discharge line, could enhance control quality significantly, thus an additional pipeline was connected to the discharge and suction points of the discharge salt pump. During pulse phase the control valve (CV) of this line is in fully open position, as for dwell the recirculation line closes, i.e., the full flow is directed to the MSSG. This control scheme was proven to be very reliable in terms of TH parameters throughout the pulse-dwell cycle. Referring back to Section 3, the molten salt loop used a homogeneous solution for its TH nodes filled with salt. Nonetheless due to uncertainties of auxiliary systems (pressure control), the salt tanks have been modelled by boundary conditions where dynamic enthalpy calculations have ensured that the tanks' proper transient behaviour is accounted for. The pressure of these tanks was set to 8 bar in order to provide a sufficiently high back pressure for salt pumps. General parameters of the small ESS model can be found in Table 3 where m$_{disch.}$ refers to the cold pump flow rate (cold tank→MSEH/hot tank) and m$_{ch.}$ denotes the hot pump flow rate (hot tank→recirc. line/MSSG/cold tank).

Table 3. Main parameters of the small ESS in Apros.

Phase Param.	Pulse	Dwell
P_{MSEH} [MW$_e$]	~41.2	~15.7
P_{MSSG} [MW$_{th}$]	14.21	270.11
$T_{hot,tank}$ [°C]	330.0	330.0
$T_{cold,tank}$ [°C]	282.4	282.4
$m_{disch.}$ [kg/s]	189.9	3590.0
$m_{ch.}$ [kg/s]	522.1	210.9

3.4. Logics

The control scheme of the plant revolves around the needs of the blanket, i.e., the main goal of the PHTS and PCS logics is to ensure the cooling of the BB at all times. In the current arrangement the PHTS pumps are running at full speed also in dwell according to design specification. The design objective, to maintain a constant average temperature on the SGs primary side ($T_{SG,p}$), has been simplified in the simulation with the requirement to preserve a ~17 °C subcooling in hot legs until a comprehensive regulation scheme is realised. Whilst providing a robust heat sink, a secondary objective of PCS logics is to keep the PCS in a low-load regime during dwell, ready for the subsequent plasma ramp-up. Due to the large, periodic displacement of the coolant inventory the DEA pressure is controlled, in this model, by a bleed line (maintaining 3.5 bar), FWHX shell pressures are also controlled from the bleed line since turbine extractions close down at the beginning of turbine unloading and open up only after reaching nominal conditions on the MSL.

4. Results

The transient analysis entails two consecutive pulse-dwell cycles with a 1200 s pulse phase at the beginning. The complete pulse phase is 7200 s including two plasma ramps before and after the flat-top period (100 s + 7000 s + 100 s), the dwell phase lasts 600 s thus one full cycle is 7800 s. Figures in Sections 4.1–4.3 follow this sequence of events where grey rectangles mark the two dwell periods. The reported results were recorded after driving several cycles and adjusting parameters, e.g., molten salt tank levels, valve driving times, and controller characteristics.

4.1. Primary System Behaviour

Power trends are depicted in Figure 6 featuring the blanket, secondary power sources and the molten salt SG. As the reactor power decreases from 100 to 1% the small ESS discharge line activates increasing the salt flow rate on the primary side of the MSSG, thus bringing its power up to 271 MW. As an improvement the MSSG loading time was shortened by 10 s (to 90 s) in order to achieve a smoother feed-transition on the MSL. Due to PHTS thermal inertia and OTSG behaviour, the MSSG unloading process had to be extended in order to supply the turbine for some extra time after dwell. As a solution, the discharge line CV followed a 200 s long asymptotic flow rate curve.

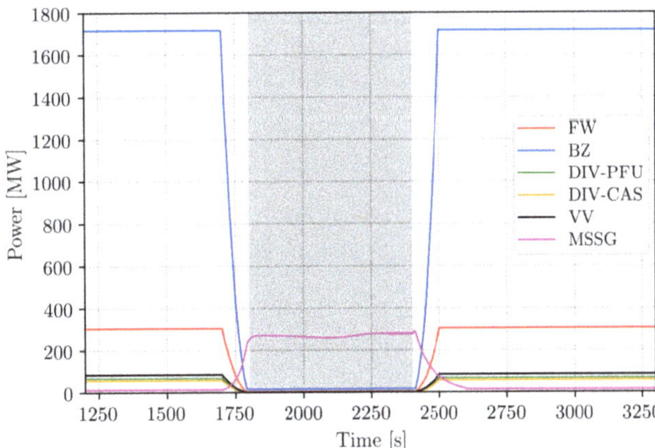

Figure 6. Power trends.

Regarding BB coolant loops and the primary side of DIV-VV loops, pressure deviations were in the range of 1 to 2 bar, illustrated in Figure 7, thus no significant perturbation was observed while BZ and FW loop trends practically overlapped. Therefore, it was concluded that the continuous pump operation and the pressurisers of the mentioned systems effectively damped perturbations coming from the reactor chamber and the PCS.

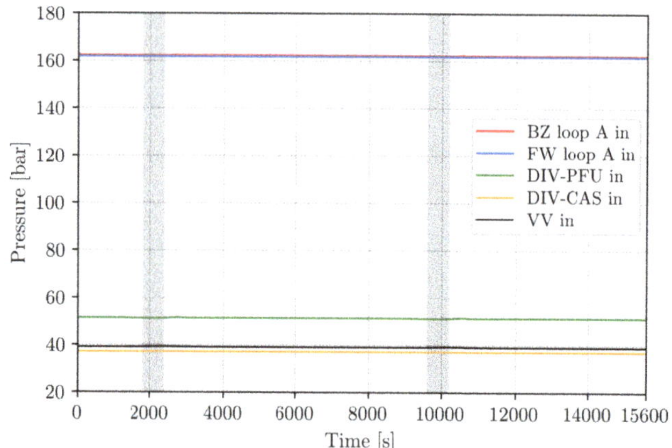

Figure 7. BB, DIV, and VV coolant loops' primary pressure trends.

FW and BZ coolant temperatures are given in Figure 8 with respect to hot and cold collector temperatures, i.e., SG inlet and outlet temperatures of the OTSGs where three distinct sections can be defined according to governing control logics. During turbine unloading temperatures varied by 1–2 °C, however, as plasma ramp-down commenced, HL temperatures (OTSG inlet) started to drop significantly. As the secondary coolant inventory of the SGs was reduced substantially, this cool-down trend constituted section #1 (1700–2000 s). Shortly after the beginning of dwell SG feedwater inventories stabilised as flow rates on the PCS-side feedwater collectors reached their dwell phase set-points, i.e., heat transfer rates saturated as primary side ΔT values nearly halved in both FW and BZ loops.

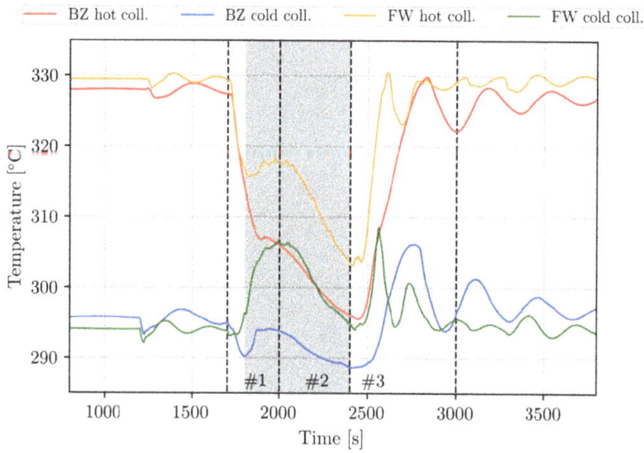

Figure 8. OTSG primary side inlet and outlet coolant temperatures.

In section #2 (2000–2400 s), hot leg temperatures continued to fall as OTSG water levels gradually recovered, nonetheless the primary FW OTSG ΔT did not change noticeably compared to #1 due to tranquil TH conditions in the SGs. In section #3 (2400–3000 s), the

OTSG feedwater flow rates were briefly decreased in order to increase primary ΔT prior to SG reload, thus the SG feedwater logic does not overcool the primary system while flooding the heat exchangers. After the short decrease, the feedwater flow rates were once again increased following the average primary loop temperatures until reaching nominal conditions around the end of the ST reload. The hereby outlined protocol highlighted that PHTS primary control was mainly realised by OTSG feedwater control, more light will be shed on the reason of such practice and its further consequences in §4.2 where the OTSG operation will be discussed in detail.

4.2. Secondary System Behaviour

Using Equation (1) the cycle net efficiency for the PCS (η_{cy}) in pulse and dwell was calculated as 30.3% and 25.6%, respectively, while Equation (2) yielded an overall plant net average efficiency (η_o) of 26.7%.

To some degree, pulse phase control schemes have relied on the output of preliminary (steady state) heat balance analyses, e.g., with respect to MSEH power, FWHX shell pressures, etc., while in other cases logics had to be tailored to meet the requirements of both phases. The OTSG feedwater control valves directly controlled PHTS coolant temperatures as described earlier, during pulse the average primary coolant temperature of the given OTSG was the control variable for the SG feedwater control valves. If negative discrepancy appeared ($\overline{T}_{SG,p} < 311.5\ °C$) the valves decreased flow rates, otherwise vice versa.

This method was a simple and robust solution to mitigate smaller anticipated perturbations during pulse, yet the challenges posed by plasma ramps had to be tackled using a more thorough approach that takes into account the following three main interdependent factors, affecting general system stability: (1) PHTS hot leg temperature has to remain well below saturation ($T_{HL} \ll T_{sat,HL}$), (2) the collapsed water level in OTSG risers has to be maintained above the upper rim of the waterports to avoid serious downcomer voiding ($L_{coll} > L_{wp}$—see in Figure 5), and (3) the steam quality in the MSL has to be kept as high as possible in order to prevent erosive damage on the turbine blades ($q_{MSL} > 99\%$). One can see that the enlisted variables can oppose one another, for instance a higher HL coolant temperature would induce flow rate increase on the SG secondary side, albeit process could lower steam qualities or even flood the SG under far-out operating conditions. Considering another scenario, one can assume that low HL temperatures would force the SG control valves to close, ultimately decreasing collapsed water levels in the riser. This action could lower the water level below the waterports' upper rim leading to rapid downcomer voiding further propagating and amplifying an initially minor perturbation (It has to be noted though that the process and automation design of the BOP model does not yet account for possible safety feature design aspects, thus the hereby considered factors represent a preliminary and conservative take on future safety criteria.).

In order to avoid unintentional transients and optimisation pitfalls of such a complex system, a workbench solution was conceived offering a helping hand to developers of the WCLL plant and control systems during the upcoming conceptual development phase of DEMO. The solution is based on a steam generator feedwater CV logic, following a flow rate function, hence decoupling primary and secondary side response from the OTSGs during plasma ramps and dwell. The normalised flow rate function is depicted in Figure 9 below with respect to FW and BZ OTSG control loops.

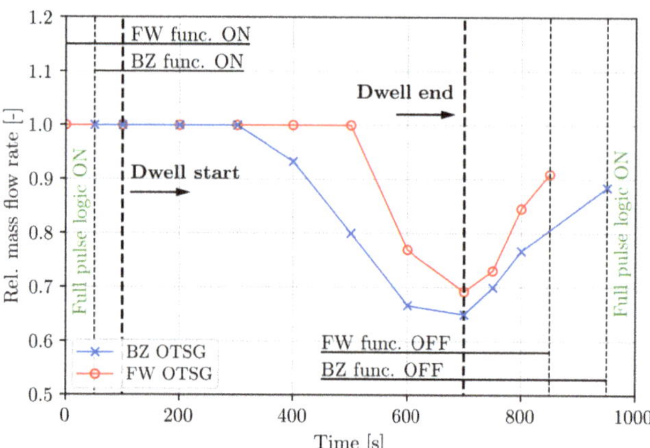

Figure 9. OTSG secondary side flow rate function.

The motivation behind the functions can be understood best by studying OTSG collapsed water levels, Figure 10 depicts OTSG-averaged riser and DC water levels. At 0 s the reactor rundown initiates and the dwell phase FW OTSG logic takes over, however pulse phase flow rates are maintained for an additional 50 s on the BZ branches in order to compensate PHTS inertia. At 50 s in the plasma ramp the dwell phase logic seizes control on both FW and BZ control branches, setting a constant flow rate on every OTSG feedwater CV. At 400 s after the start of dwell the flow rates are decreased on each SG reducing the cooling of the PHTS. Such action was necessary to decrease the difference between the after-dwell and pulse phase value of the average primary coolant temperature (311.5 °C).

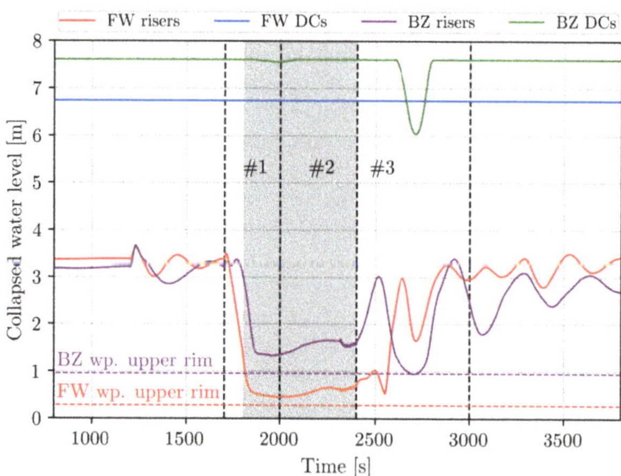

Figure 10. Averaged collapsed water levels in riser and DC volumes.

At the end of dwell feedwater flow rates are increased as plasma power rises between 700 and 800 s, ensuring also that collapsed water levels remain above the waterports' upper rim. This post-dwell protocol is enabled for 150 s in case of FW OTSGs and 250 s for BZ OTSGs, where the difference is due to the different inertia of the two loops. By maintaining acceptable water levels in the risers the primary coolant temperature inevitable decreases by the end of dwell to some extent. Lowering flow rates, then gradually increasing them at a lower gradient leaves time for the blanket to warm up the hot legs and approach pulse phase primary side coolant ΔT. At 850 s the FW OTSG dwell logic is disabled, the FW OTSG feedwater control valves are driven only by the pulse phase logic afterwards. Correspondingly at 950 s (250 s after dwell end) the BZ OTSG dwell logic hands over the control to the pulse phase logic. The closer the primary side ΔT and hot leg temperature to their nominal values the smoother the transition into the period where pulse phase logics have full control over OTSG CVs.

Steam flow rates are depicted in Figure 11 with respect to HP, LP stages, and the steam dump line (DL). At the beginning of the unloading process (600 s before dwell) extraction lines close along the shaft, thus the steam flow rate at the 4th LP stage (LP4) outlet converges to the value taken at HP1 inlet. As dwell is reached ($\Delta p_{MSL} = -4.5$ bar) the steam dump-line remains open to divert the excess flow to the main condenser while the thermal inertia of the BB provides some surplus steam in addition to the molten salt SG. In Figure 11, the LP2 flow rate is identical to LP1, this is due to the fact that earlier the LP1 extraction was used as primary feed for deaerator pressure control. Preliminary calculations showed that the periodic turbine operation posed challenges on the mentioned control loop, thus the extraction line was closed and the DEA bleed line was activated for pulse, consequently simplifying and improving control.

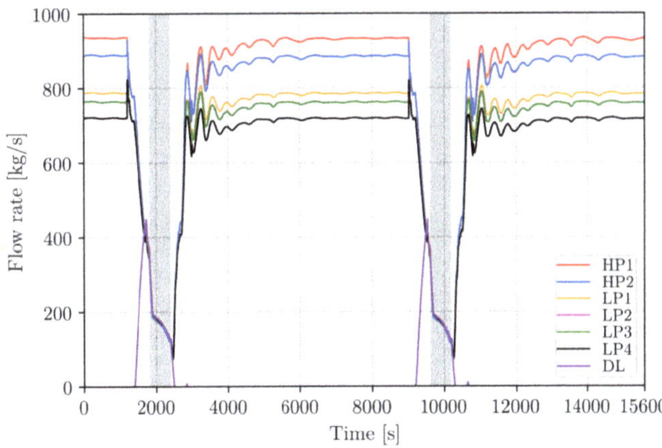

Figure 11. Steam flow rates on the ST and dump line.

Considering the main condenser the observed pressure transient was in close agreement with heat balance analysis where shell pressure varied between 27.2 to 50.0 mbar. In addition to incorporating the flow of the dump line, the condenser served also as a pressure buffer of the DEA. Originally the DEA was pressurised by an LP1 extraction however as turbine operation had been tested, such arrangement was not deemed to be feasible during transients thus the DEA feed was moved to the MSL, where no significant pressure transient was postulated. Since DEA inventory shows substantial change (± 39 t between pulse-dwell) a stable steam source (MSL) and a reliable low-pressure point (main condenser) are essential for a firm DEA pressure control. DEA collapsed water level and pressure trends are shown in Figure 12. It can be seen that the current system handled the DEA pressure with a margin of ± 0.08 bar despite the 0.7 m level variation.

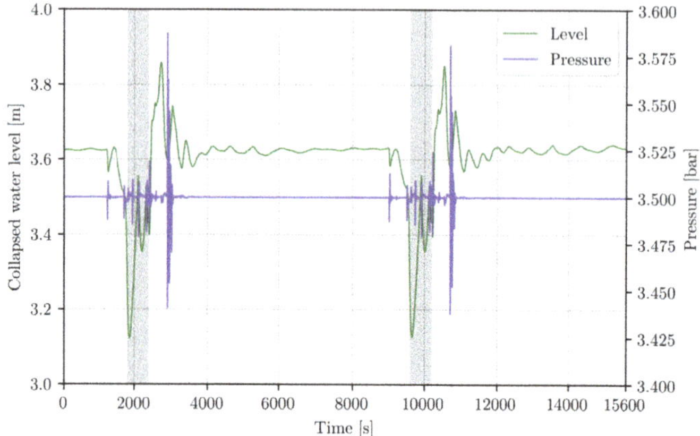

Figure 12. Deaerator collapsed water level and pressure.

Preheater line temperatures were adjusted via preheater HX shell pressures, considering FH1HX the pulse phase pressure was maintained in dwell, as for FH2HX the pressure was increased by 50%. Moving onto the high-pressure stage larger relative pressure corrections had to be made due to the order of HXs in the preheater chain. Since secondary power sources ramped down, their bypass lines opened up meaning that the pulse phase ΔT had to be preserved only by FH(3,4)HXs, hence the FH3HX shell pressure was increased by 45%, while maintaining pulse phase pressure in FH4HX. The feedwater temperature trends are depicted in Figure 13 indicating a $\pm 5\,°C$ variation in FH4HX outlet temperature.

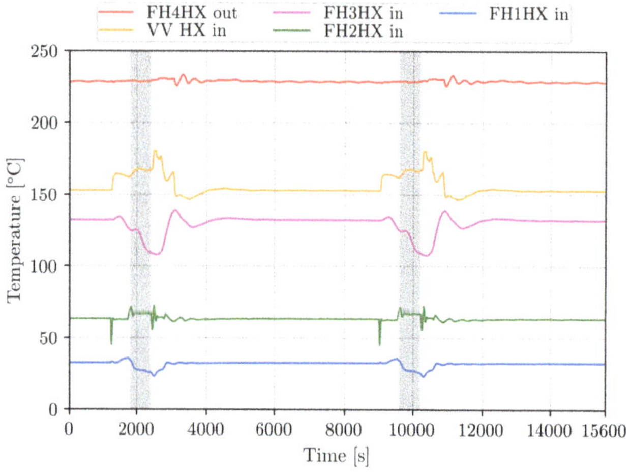

Figure 13. Feedwater temperatures along the preheater line.

4.3. Small ESS Behaviour

Considering the periodic, high amplitude variation in salt flow rates, as shown in Figure 14a, concerns might be raised regarding mechanical stresses of components located on the discharge line. Comparing flow rate gradients (F_g [kg/s^2] = [N/m]) to weight forces (F_w [N/m]) on the unit length it was found that the induced stresses were negligible. Corresponding tank levels are depicted in Figure 14b, where a $\sim$5.5 m ($\sim$2300 t) displacement can be seen between charge and discharge periods. Such a trend implies that the current 1362 m^3 salt inventory was mostly depleted leaving a residual salt reserve of

~440 t per tank per period. Note that in this configuration the cover gas volume was not taken into consideration, such parameters will be defined later on during the conceptual development phase.

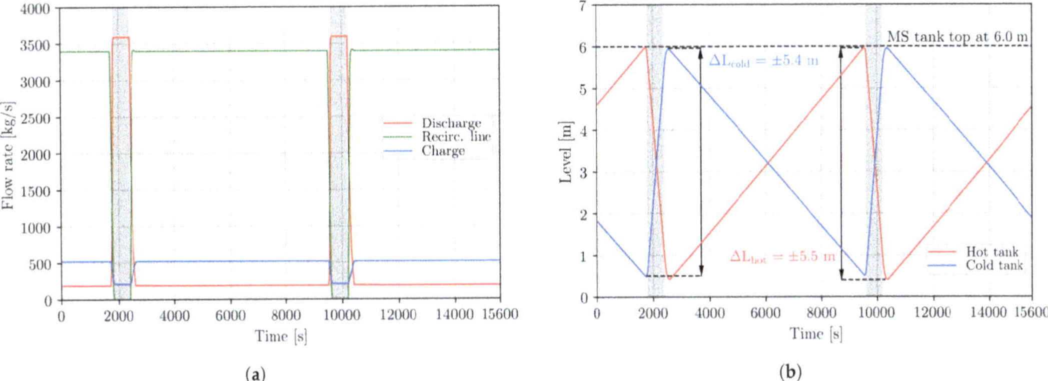

Figure 14. Flow rates in the small ESS (**a**) and molten salt tank levels (**b**).

5. Synopsis

Development and analytical work, carried out for the DEMO WCLL small ESS integral Apros thermalhydraulic model has been discussed. A new blanket concept was introduced utilising a layered structure where the segments' spatial discretisation and power deposition schemes follow the trends of neutronics models. Once-through steam generators have been implemented in Apros using a similar, nested architecture as the breeding blanket model. The transient simulations outlined the complexity of the challenges facing the control logics, where engineering solutions have to be derived respecting all safety criteria. In order to provide a reference case for future development a simplified OTSG control scheme was obtained highlighting the narrow operating margins of both PHTS and PCS.

6. Outlook

Carrying on with the development of the BB model the power deposition schemes have to be refined. In the current setup each material compartment has a homogeneous volumetric power distribution that is adjusted only in the toroidal direction, neglecting radial gradients. As the BU design matures, with respect to certain components, e.g., FW, BSS or in a broader context on segment level, the blanket model shall follow the specifications in any case. The OTSG design is still under development where alternative options are being considered, e.g., allowing internal circulation between riser and DC volumes. Model performance will be benchmarked during 2021–2022 and additional finite element method analyses will support the system code results. Keeping in mind the dimensions of the PHTS and PCS, where the total piping length yields ~6.9 km and ~2.1 km for primary and secondary systems, respectively, the issues of heat and pressure losses will have to be tackled as well. Aside from mitigating such losses, energy can still be recovered in various aspects, e.g., from dumped steam via recuperative heat exchangers (FHd(1,2)) on the condenser-deaerator let-down line, where the dumped steam heats the cold condensate before it enters the DEA (see Figure 3). This small system recycles only a portion of the total enthalpy loss on the steam dump valves, ≈168 kWh/cycle is still being ejected from the PCS via the main condenser. In recent Apros 6.11 release a new iteration scheme, dedicated to molten salt-inert gas systems will be available, thus the small ESS model will be refined accordingly, removing obsolete boundary conditions.

Author Contributions: Data curation, M.S.; Validation, S.N.; Writing—review & editing, M.S. All authors have read and agreed to the published version of the manuscript.

Funding: This research was funded by the Euratom research and training programme 2014–2018 and 2019–2020 under grant agreement No 633053.

Acknowledgments: This work has been carried out within the framework of the EUROfusion Consortium and has received funding from the Euratom research and training programme 2014–2018 and 2019–2020 under grant agreement No 633053. The views and opinions expressed herein do not necessarily reflect those of the European Commission.

Conflicts of Interest: The authors declare no conflict of interest.

Abbreviations

The following abbreviations are used in this manuscript:

Apros	Advanced Process Simulation	IB	Inboard
AUXB	Auxiliary Boiler	IHTS	Intermediate Heat Transfer System
BB	Breeding Blanket	LIB	Left Inboard
BC	Boundary Condition	LOB	Left Outboard
BOP	Balance Of Plant	LP	Low-pressure
BSS	Back Support Structure	MS	Moisture Separator
BU	Breeding Unit	MSEH	Molten Salt Electrical Heater
BZ	Breeding Zone	MSL	Main Steam Line
CL	Cold leg	MSSG	Molten Salt Steam Generator
COB	Central Outboard	OB	Outboard
CV	Control Valve	OTSG	Once-Through Steam Generator
DC	Downcomer	PCS	Power Conversion System
DEA	Deaerator	PHTS	Primary Heat Transfer System
DEMO	Demonstration Power Plant	RH	Reheater
DIV-CAS	Divertor Casette	RIB	Right Inboard
DIV-PFU	Divertor Plasma Facing Unit	ROB	Right Outboard
DL	Dump Line	SCL	Simantics Constraint Language
ESS	Energy Storage System	SG	Steam Generator
FH	Feedheater	ST	Steam Turbine
FW	First Wall	TH	Thermalhydraulic
FWHX	Feedwater Heat Exchanger	UC	User Component
HCPB	Helium-Cooled Pebble Bed	VV	Vacuum Vessel
HL	Hot leg	WCLL	Water-Cooled Lithium-Lead
HP	High-pressure		

Variables

Δp_{MSL}	MSL pressure loss	P_{MSEH}	MSEH power
η_{cy}	Cycle efficiency	P_{MSSG}	MSSG power
η_o	Overall efficiency	P_{nom}	Nominal power
A_{HT}	Heat transfer area	P_{SG}	SG power
F_g/F_w	Gravitational/weight force	q_{MSL}	Steam quality in MSL
L_{coll}	Collector water level	t	Time
L_{wp}	Waterport elevation	t_{cy}	Cycle length
m_{BZ}	BZ loop primary flow rate	$T_{cold,tank}$	Cold salt tank temperature
$m_{disch.}/m_{ch.}$	Discharge/charge salt flow rate	T_{HL}	Hot leg temperature
m_{FW}	FW loop primary flow rate	$T_{hot,tank}$	Hot salt tank temperature
m_{sec}	SG steam flow rate	T_{out}	SG steam temperature
p_{HL}	Hot leg pressure	$T_{sat,HL}$	Saturation temp. in HL
p_{out}	SG steam pressure	$T_{SG,p}$	Primary side SG temperature
P_{BB}	BB power	$T_{w,max}$	Max. water temperature in PHTS
P_{DIV}	DIV power	W_{gross}	Gross power
P_{VV}	VV power	$W_{PCS,pump}$	Pumping power of the PCS
P_{fus}	Fusion power	W_{plant}	Plant power

References

1. Hernández, F.; Pereslavtsev, P.; Kang, Q.; Norajitra, P.; Kiss, B.; Nádasi, G.; Bitz, O. A new HCPB breeding blanket for the EU DEMO: Evolution, rationale and preliminary performances. *Fusion Eng. Des.* **2017**, *124*, 882–886. [CrossRef]
2. Nevo, A.D.; Arena, P.; Caruso, G.; Chiovaro, P.; Maio, P.; Eboli, M.; Edemetti, F.; Forgione, N.; Forte, R.; Froio, A.; et al. Recent progress in developing a feasible and integrated conceptual design of the WCLL BB in EUROfusion project. *Fusion Eng. Des.* **2019**, *146*, 1805–1809. [CrossRef]
3. Barucca, L.; Bubelis, E.; Ciattaglia, S.; D'Alessandro, A.; Nevo, A.D.; Giannetti, F.; Hering, W.; Lorusso, P.; Martelli, E.; Moscato, I.; et al. Pre-conceptual design of the EU DEMO balance of plant systems: Objectives and challenges. *Fusion Eng. Des.* **2021**, *169*, 112504. [CrossRef]
4. Szogradi, M.; Norrman, S.; Bubelis, E. Dynamic modelling of the helium-cooled DEMO fusion power plant with an auxiliary boiler in Apros. *Fusion Eng. Des.* **2020**, *160*, 111970. [CrossRef]
5. Official Apros Website. Available online: https://www.apros.fi/ (accessed on 25 May 2021).
6. Sohal, M.S.; Ebner, M.A.; Sabharwall, P.; Sharpe, P.B. *Engineering Database of Liquid Salt Thermophysical and Thermochemical Properties*; INL/EXT-10-18297 Rev. 1; Idaho National Laboratory: Idaho Falls, ID, USA, 2013; p. 24.
7. Janky, F.; Fable, E.; Treutterer, W.; Zohm, H. Simulation of burn control for DEMO using ASTRA coupled with Simulink. *Fusion Eng. Des.* **2017**, *123*, 555–558. [CrossRef]
8. Wouter, X.B.D.; Richard, P.A.; David, C.; Sergey, V.; Sven, W. Presentation of the New SOLPS-ITER Code Package for Tokamak Plasma Edge Modelling. *Plasma Fusion Res.* **2016**, *11*, 1403102.
9. Moscheni, M.; Carr, M.; Dulla, S.; Maviglia, F.; Meakins, A.; Nallo, G.F.; Subba, F.; Zanino, R. Radiative heat load distribution on the EU-DEMO first wall due to mitigated disruptions. *Nucl. Mater. Energy* **2020**, *25*, 100824. [CrossRef]
10. Moro, F.; Colangeli, A.; Nevo, A.D.; Flammini, D.; Mariano, G.; Martelli, E.; Mozzillo, R.; Noce, S.; Villari, R. Nuclear analysis of the Water cooled lithium lead DEMO reactor. *Fusion Eng. Des.* **2020**, *160*, 111833. [CrossRef]
11. Moro, F.; Arena, P.; Catanzaro, I.; Colangeli, A.; Nevo, A.D.; Flammini, D.; Fonnesu, N.; Forte, R.; Imbriani, V.; Mariano, G.; et al. Nuclear performances of the water-cooled lithium lead DEMO reactor: Neutronic analysis on a fully heterogeneous model. *Fusion Eng. Des.* **2021**, *168*, 112514. [CrossRef]
12. American Society of Mechanical Engineers. *ASME Boiler & Pressure Vessel Code, Section III-Rules for Construction of Nuclear Facility Components, Division 1-Subsection NB for Class 1 Components*; ASME: New York, NY, USA, 2019.

Article

Loss of Liquid Lithium Coolant in an Accident in a DONES Test Cell Facility

Danilo Nicola Dongiovanni [1],* and Matteo D'Onorio [2]

[1] ENEA, CR. Frascati, FSN-TEN Department, via Enrico Fermi, 45, 00044 Frascati, Italy
[2] Department of Astronautical Electrical and Energy Engineering (DIAEE), Sapienza University of Rome, C.so Vittorio Emanuele II 244, 00186 Rome, Italy; matteo.donorio@uniroma1.it
* Correspondence: danilo.dongiovanni@enea.it

Abstract: A Demo-Oriented early NEutron Source (DONES) facility for material irradiation with nuclear is currently being designed. DONES aims to produce neutrons with fusion-relevant spectrum and fluence by means of D–Li stripping reactions occurring between a deuteron beam impacting a stable liquid lithium flowing film implementing the target. Given the hazard constituted by the liquid lithium inventory and the potential risk of reactions with water, air, and concrete eventually resulting in fire events, the Target Test Cell (TTC) is filled with helium and the reinforced concrete walls forming the bio-shield are covered with steel liners. A loss of Li in TTC, due to a large break in the Quench Tank, is postulated, and consequences are deterministically studied. With the TTC liner being water-cooled, the impact of the liner temperature rise following a leakage event is evaluated. Two separate MELCOR code models have been defined for the liquid lithium loop and water-cooled loop and are numerically coupled. The amount of leaked inventory dependent on the implemented safety logic and impact on TTC containment is evaluated. The water pressurization pattern within the liner cooling loop is studied to highlight possible risks of lithium–water/concrete reactions.

Keywords: fusion; DONES; liquid lithium; LOCA; Melcor; numeric coupling

Citation: Dongiovanni, D.N.; D'Onorio, M. Loss of Liquid Lithium Coolant in an Accident in a DONES Test Cell Facility. *Energies* **2021**, *14*, 6569. https://doi.org/10.3390/en14206569

Academic Editors: Dan Gabriel Cacuci and Marica Eboli

Received: 30 July 2021
Accepted: 4 October 2021
Published: 12 October 2021

Publisher's Note: MDPI stays neutral with regard to jurisdictional claims in published maps and institutional affiliations.

1. Introduction

Special materials able to withstand outstanding neutronic loads will be required for the exploitation of nuclear fusion energy. However, there is limited experience relating to materials exposed to such load conditions. In particular, in-vessel components will be subject to thermal and neutron loads with energy up to around 14.1 MeV and fluence leading to 5–15 displacements per atom (dpa) [1,2] in the irradiated components (first wall, divertor) able to alter their structural and functional behavior.

The IFMIF-DONES (International Fusion Material Irradiation Facility-DEMO Oriented NEutron Source) facility for material irradiation is currently being designed within the EUROfusion programme [3] to fill this knowledge gap in view of a DEMOnstration fusion power plant expected for the mid-2040s [4]. DONES achieves the production of neutrons with fusion–relevant spectrum and fluence through D–Li stripping reactions occurring between a 125 mA and 40 MeV deuteron beam impacting a stable liquid lithium flowing film implementing the target.

A safety analyses campaign [5] is ongoing within the DONES project to assess the DONES system ability to handle any abnormal event ensuring safe conditions for workers and the population. To reach the goal, a selection of the analyses has been done using the FMECA plan and FMEAs studies.

As part of the conceptual design, a selection of accident events has been studied deterministically for lithium systems and or the test systems interface either by MELCOR code [6] or RELAP code [7].

The exploitation of liquid lithium within the facility constitutes a hazard with potential risk of reactions with air, water, and concrete, eventually resulting in fire events [8].

311

Therefore, great attention has been paid to the DONES design to avoid any direct contact between reactants (air, water, concrete) such as the use of an inert atmosphere (helium, argon) or double-stage heat removal systems lithium–oil/oil–water. Safety mitigation provisions are also in place to prevent the possible reaction of lithium and the reinforced concrete TC walls forming the bio shield. In particular, all TC is covered by a steel liner layer to prevent liquid lithium–concrete reactions in the case of a loss of liquid lithium in an accident. Despite such provisions, deterministically assessing the consequences of possible accidents is of paramount importance in the early phases of the design to identify safety concerns. Assessments have been performed in the past for other test facilities exploiting liquid lithium [9,10]. Therefore, a preliminary analysis on the consequences of postulating fire events in the DONES lithium system has also been performed [11].

In the present analysis, a loss of Li in TTC (LTTC1 Postulated Initiating Event) due to large break in the Quench Tank is postulated and consequences deterministically studied with MELCOR [12–14]. In particular, the amount of leaked inventory and its impact on TTC containment and the atmosphere is evaluated. The TTC liner is cooled by a cooling water loop, and the effect of the temperature rise due to leaked lithium inventory on the liner floor and underlying cooling lines is evaluated. To overcome the MELCOR limitation for two cooling fluids contemporary usage, two different MELCOR models have been defined for the liquid lithium loop and water-cooled loop and are numerically coupled by means of external data exchange for interfacing heat structures. Possible water pressurization within the cooling loop has been studied.

2. Materials and Methods

2.1. System Description

The DONES facility is composed of three main systems (Figure 1) and several auxiliary systems [3]. The accelerator system is in charge of accelerating and focusing deuteron ions into a beam with a 125 mA current and 40 MeV energy. The beam flies within a vacuum conduit entering the Test Cell within a beam duct until a vacuum chamber (Figure 2) where it impacts the lithium film to have D-Li striping reactions and produces fusion-spectra neutrons [15]. A backplate (BP) is placed between the lithium film and the High Flux Testing Module [16]. The lithium system [17] main lithium loop provides liquid lithium flowing film for striping reactions. The lithium main loop is composed of the Target Assembly (TA) [18] located in the TC, which in turn is composed of pipework leading the lithium to the vacuum chamber area where the thin liquid lithium film is created by means of a flow straighter and reducing nozzle (ST-NZ). Below the target area where the beam impacts the lithium film, lithium is collected in a quench tank (QT) before exiting the TC towards the lithium system room. In order to obtain liquid lithium film, specific velocity/mass flow rate and thermodynamic conditions need to be maintained in the loop. This task is in charge of the part of the lithium main loop located in the lithium room. In particular, lithium flow is provided by an Electromagnetic Pump (EMP) and is measured by an EM flow meter. A Dump Tank (DT) collecting lithium at shutdown is also located therein. The Lithium System (LS) includes three main systems; a Heat Removal System (HRS) is deployed in two rooms housing the main Li loop. Beam heat deposited on the lithium film is exhausted within a primary Heat Exchanger (HX) exploiting a secondary oil loop to prevent lithium–water reactions and is located in the lithium main room. In turn, heat transferred to the oil secondary loop is transferred to a tertiary water loop located in a separate room, considered out of scope. The TC Atmosphere [19] is He at a pressure of 20 kPa, with a free volume of 47.1 m^3. The Lithium System (LS) [17] is Ar buffered during normal operation and is slightly depressurized (-140 Pa) with respect to standard atmospheric pressure.

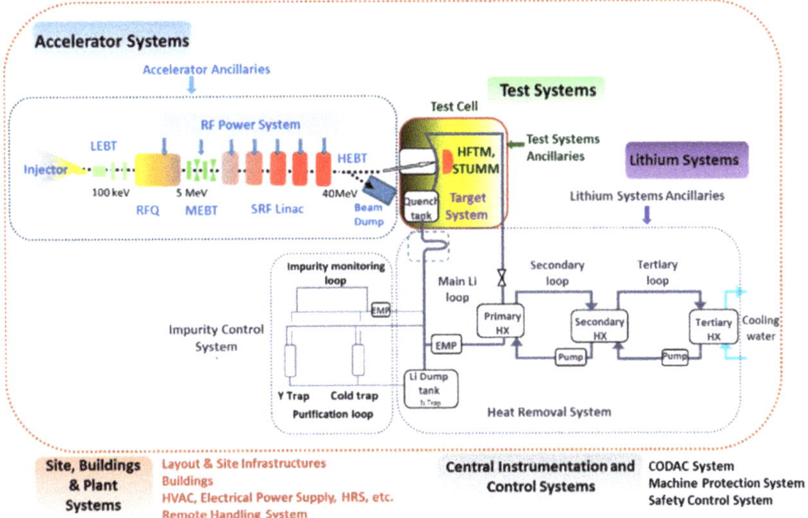

Figure 1. IFMIF-DONES plant system configuration.

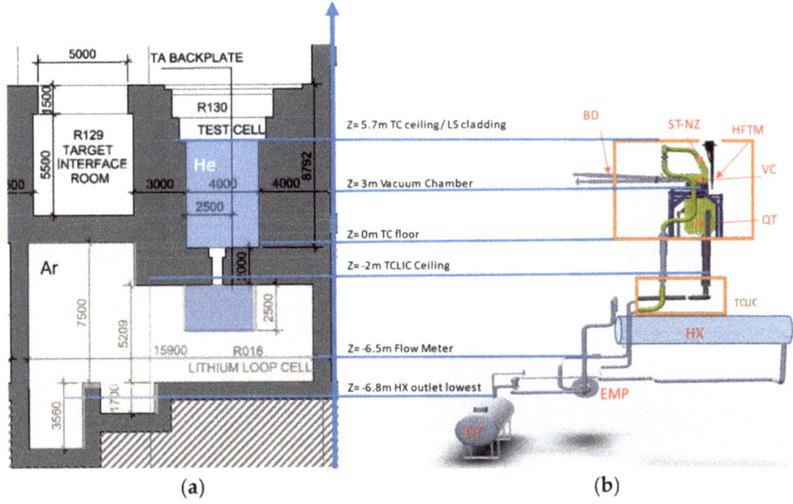

Figure 2. (**a**) Building layout and rooms in scope (light blue). (**b**) Related Melcor model elevations for considered systems.

2.2. Accident Specification for the LTTC1 Event

Accident Analysis Specifications for the accident have been developed to define the analysis scope, methodology, and expected goal. The PIE event considered is LTTC1 Loss of Li in TTC due to a large break in the Quench Tank. The transient sequence initiated by the rupture in the lithium loop running inside the TTC induces the release of liquid metal into the TTC. The TTC atmosphere is filled with inert (Helium) gas and kept at "sub-atmospheric/or slight" pressure; then, no Li–air reaction occurs. Nevertheless, the rupture of the QT induces the release of a large amount of hot lithium in the TTC floor. A liner is foreseen to prevent contact between Li and the concrete of the TTC structures to mitigate the risks of reactions between Li and the concrete elements (e.g., water humidity and air). Under the liner, incorporated in the concrete structure, a water-cooling circuit is foreseen to prevent heat-up of the TTC during the facility operations. The release of the large Li mass over the liner will start the heat exchanging between the hot lithium and the cold TTC structures. Over-heating and over-pressurization of the water cooling could occur, impairing the cooling loop integrity. The objective of the analysis is to study over-pressurization of the TTC volume because of the heating of the He gas contained inside the TTC, the over-heating of the TTC structures (liner and concrete), and the over-heating and over-pressurization of the TTC water cooling system.

Note that no Loss of Off-Site power is assumed to co-occur in the selected initiating event. As safety mitigation system assumptions, the EMP stop and BEAM shutdown have been assumed. The beam power switches off and EMP fully stops after 0.1 s and 5 s from the event detection, respectively. The accelerator beam duct is isolated from the target after beam shutdown. Note that in compliance with diversity criterion for safety important classified detection systems, several accident condition detection systems are currently being considered in design (e.g., QT level, pressure in selected volumes, etc.) and are expected to have different detection times depending on the accident.

No valves for the lithium main loop isolation are foreseen, so in order to investigate the possibility of reducing the lithium leaked inventory, the exploitation of a dump tank (normally used to store lithium during loop maintenance or off-line operation) to be opened as a consequence mitigation solution is still under discussion.

2.3. Simulation Models and Approaches

MELCOR is a code originally developed for the safety analysis of light water fission reactors. Over the past twenty years, a series of modifications have been implemented to account for nuclear fusion–relevant phenomena [12,13,20,21]. This new MELCOR version validated against available fusion experimental data or benchmarked against validated codes [22–25] has become a reference code for safety analysis in nuclear fusion studies. It has also been adopted for accident analyses exploited for licensing purposes in fusion facilities, the most representative example being ITER [13,26].

More recently, the MELCOR Fusion adapted version 1.8.6—2017 computer code [12] has been exploited for the analysis of integrated HTS containment as part of preliminary safety assessments in pre-conceptual designs of European DEMO concepts [27–30] as well as in Korean DEMO [31]. Being able to model phenomena relevant for fusion safety applications such as radionuclide transport (e.g., tritiated water) in the presence of a variety of cooling fluids (helium, water, liquid metals) as well as pressure conditions (e.g., vacuum), the MELCOR code has been considered for deterministic safety analyses aimed at the licensing of the DONES plant [5,6], showing good agreement with RELAP5 results [7,32,33].

Considering thermal–hydraulic analyses performed within the DONES/IFMIF facilities context, most of the studies have focused on CFD modeling of liquid lithium target with a focus on correctly modeling phenomena such as liquid metal flow, free surface flow under vacuum [34–37] or specific effects resulting from beam/target interactions [38]. This is justified by the many requirements for facility operation relying on the specific conditions that the flowing liquid lithium film will stably achieve. The present application assesses the impact on the containment of a liquid lithium large spillage event. Therefore, despite that liquid metal flow and free surface flow under vacuum are relevant phenomena to reproduce a steady-state regime before the postulated initiating event, the focus is on the heat transfer phenomena occurring at the lithium/liner (Figure 3) and liner/cooling pipe (Figure 4) interfaces.

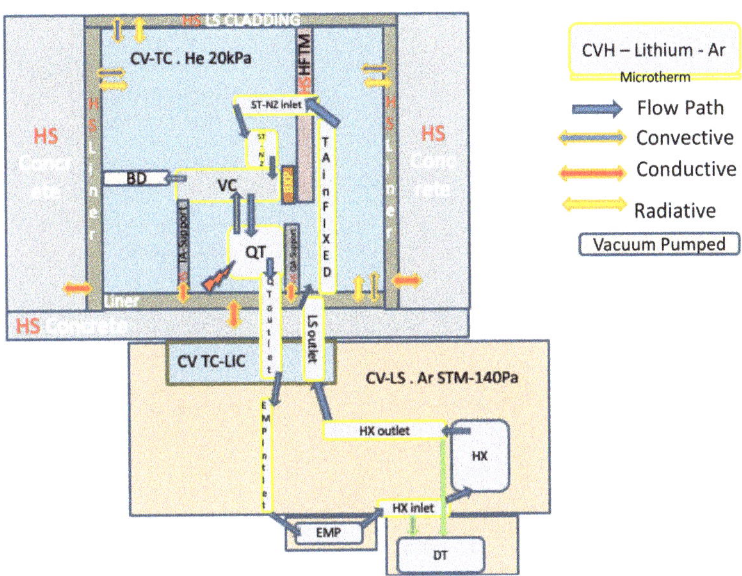

Figure 3. TC-LS Melcor model nodalization.

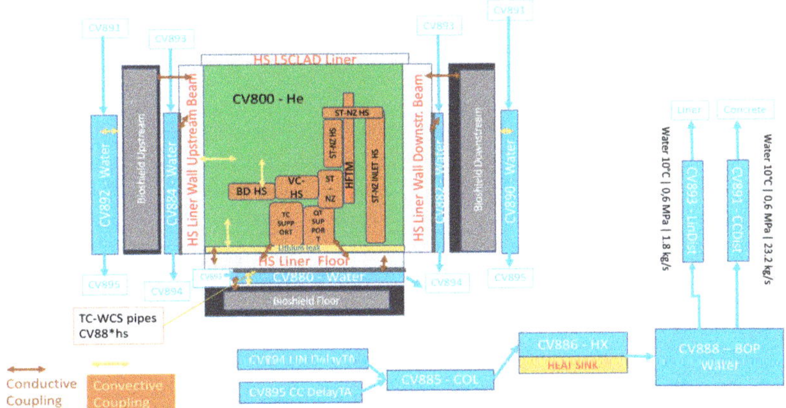

Figure 4. TC-WCS Melcor nodalization.

The MELCOR code has been reputed to be suitable for the present application modeling purpose. In fact, the MELCOR fusion version when lithium is selected as a working fluid includes in the model the thermodynamic and transport properties of lithium other than lithium/air reactions possibly occurring after spills events [10]. In particular, the Equation of State (EOS) was modified to integrate the liquid lithium properties from [39], extrapolation below the triple point, and film formation on cold structures. Refer to [40] for further details on numerical models. These modifications were validated by using Hanford Engineering Development Laboratory (HEDL) experimental data [41,42] as a benchmark.

An additional feature available in MELCOR fusion exploited in the present analysis is the user-defined function FUN1 to model direct heat transfer (radiation and conduction) between heat structures. Heat power exchanged across heat structures is equal to

$$Q = S \cdot \left(f \cdot \frac{k_s}{dx} \right) (T_1 - T_2) + (1 - f) \cdot \varepsilon \cdot \sigma \cdot \left(T_1^4 - T_2^4 \right) \qquad [W]$$

with S surface of heat exchange across structures, T_1 and T_2 are the considered heat structure surface temperatures, f is the fraction of such surfaces in conductive contact, ε is emissivity, the Stefan-Boltzmann constant is σ, dx is the distance across a solid, and k_s is the thermal conductivity. In the present analysis, the DBA version of the MELCOR code has been used.

Unfortunately, the current MELCOR version does not allow for contemporary usage of two cooling fluids such as lithium (used in the LS main loop) and water (used in TC -WCS). Therefore, to have a preliminary study providing future reference for the upcoming Melcor-TMAP integrated version able to treat multiple fluids [42], the exploited approach was:

1. Implement a MELCOR input deck to estimate the leaked lithium inventory, modeling TC, TA, and LS systems and using lithium as a working fluid.
2. Implement a separate MELCOR model including the TC and TC-WCS systems to simulate TC wall cooling and using water as a working fluid.
3. Implement a numerical coupling of the two models running in parallel and exchanging respective boundary conditions relative to the TC floor liner as an interfacing heat structure of interest.

Details about the MELCOR model nodalization and coupling approach are provided in the following subsections.

2.3.1. Nodalization of TC and LS with the Lithium Working Fluid

With reference to Figure 2, the first implemented model considered TC and TCLIC rooms with a Helium atmosphere and LS main loop room filled with an Argon atmosphere. As outlined in (Figure 3), the adopted nodalization includes a TC room, TA volumes, LS main loop volumes, beam duct (0.8 m^3), vacuum chamber (0.36 m^3), and some service volumes such as the environment or vacuum supply volume.

The main lithium loop has a total inventory (Table 1) of 7.4 m^3 of lithium in the liquid phase at 523.15 K, set as the working fluid for the MELCOR code. The loop initially contains an argon atmosphere; then, the vacuum is pumped at start-up and EMP is started. After start-up, the loop reaches a reference operative parameter, i.e.,: 49.8 kg/s of lithium mass flow rate, 1.43 m liquid level within QT. Then, the beam is operated and injects 5.E6 W power into the lithium film, increasing the liquid lithium temperature from 523.15 K at the HX outlet to 546.15 K within QT. Note that for comparability's sake and whenever possible, the same CV names and safety logic were used as in [6]. The break was assumed to occur at the bottom of the QT in correspondence with the QT outlet flange. The break area was assumed to be the outlet pipe section (0.0314 m^2), assumed to open at t = 1000 s after a stationary operation period. Figure 3 also shows the considered heat structure general scheme and heat transfer coupling phenomena. Each CV within the main lithium loop was provided with heat structures representing stainless steel piping or component walls. Pipes were covered with 0.04 m of MICROTHERM$^{\circledR}$ pipe-insulating material (shown in Figure 3 with a yellow contour) to reduce lithium loop heat losses towards the room atmosphere in the TC and lithium area. A set of additional structures representing TC structures were also defined. In particular, BP and HFTM structures (made of Eurofer) and TA and TC support structures (made of Stainless Steel) and assumed to be in thermal conductivity with the liner floor. LSP Cladding acting as a TC ceiling and HTFM module were helium-cooled. Since the helium cooling loop is not considered within scope, those HSs were modeled as boundary condition structures with a fixed temperature. Test cell heat structure details are reported in Table 2. Convective and radiative heat exchange between heat structures and the surrounding atmosphere was considered. A conductive coupling was considered across selected structures (as shown in Figure 3) and modeled using MELCOR FUN1. Two heat structures modeled the TC lateral wall, one upstream target plane with respect to the beam source and one downstream. This choice was made to account for different nuclear heating and decay heat conditions [43,44] in the two zones.

Table 1. Lithium inventory. * DT volume is empty during operation.

Lithium Inventory	Description	Length [m]	Vol [m^3]
Piping primary loop	Pipe LS—TLIC	14.7	0.27
	Pipe LS—EMP outlet	13.1	0.24
	Pipe LS—HX outlet	17.7	0.33
	Pipe LS outlet—TLIC	17.6	0.33
	Pipe LS—TLIC-LSRoom	8.4	0.16
Piping target assembly	ST-NZ Inlet Fixed pipe	5.8	0.11
	ST-NZ Inlet Removable pipe	2.2	0.04
Equipment	ST-NZ		0.06
	QT		1.20
	EMP		0.52
	HX		4.10
	DT		9 *
Total			7.4

Table 2. Heat structures within the Test Cell.

Description	Material	Equivalent Thickness [m]	Surface [m^2]	Initial Temperature [K]	Nuclear Heating [W]	Decay Heat [W]
Liner Floor (HS80001)	Stainless Steel	0.008	8.4	283.15	373.0	8.0
Bioshield Floor	Concrete	2.0	8.4	313.0	3701.0	74.0
Liner Wall downstream beam	Stainless Steel	0.008	33.1	283.15	4308.0	86.0
Liner Wall upstream beam	Stainless Steel	0.008	45.8	283.15	2532.0	50.0
Bioshield wall downstream beam	Concrete	1.0	33.1	293.15	56.0	1123.0
Bioshield wall upstream beam	Concrete	1.0	45.8	322.15	27.0	546.0
LSP Cladding	Stainless Steel	0.008	11.2	333.0		
HFTM	Eurofer	0.11	1.15	349.0		
Backplate	Eurofer	0.03		523.	1085.0	22.0
Qt Support	Stainless Steel	0.01	3.51	353.0	400.0	32.0
Tc Support	Stainless Steel	0.01	14.482	353.0	1448.0	116.0

2.3.2. Nodalization of TC Walls Water Cooling Loop

A second MELCOR model was implemented for the TC-WCS loop using water as a working fluid. Figure 4 shows the adopted nodalization for the TC/TC-WCS model, heat structures, and heat transfer phenomena. Table 3 reports control volume data. Liner and bio shield walls are cooled by parallel branches. In particular, a main distributor manifold splits the nominal 25 kg/s water mass flow into 23.2 kg/s flow for the bio shield concrete walls cooling branch and the remaining 1.8 kg/s for the liner wall cooling branch. Such branches converge into a unique collector after two separate delay tank volumes allowing for activated water decay before entering HX. An HX removes about 109 kW power, of which 90 kW is from the bio shield and the remaining part from the liner. Water at the HX outlet is at 283.15 K and 0.6 MPa. A pump compensates for an estimated pressure drop of 4500 hPa along the loop. Cooling pipes volumes were provided with a 2-mm-stainless steel wall representing pipe walls with which the liner and concrete walls have conductive exchange.

Table 3. Control Volumes in the TC-WCS loop.

Description	CV	Pool P [Pa]	Pool Temperature [K]	Volume [m^3]
Floor Liner cooling branch	880	3.50×10^5	3.50×10^5	4.5×10^{-3}
Downstream liner cooling branch	882	5.80×10^5	5.80×10^5	2.89×10^{-2}
Upstream liner cooling branch	884	4.00×10^5	4.00×10^5	3.85×10^{-2}
Main Collector	885	1.50×10^5	1.50×10^5	1.06
Heat Exchanger	886	1.50×10^5	1.50×10^5	7.50
Pressurizer	887	1.50×10^5	1.50×10^5	1.40
Balance volume	888	6.00×10^5	6.00×10^5	6.00
Downstream Bioshield cooling branch	890	5.80×10^5	5.80×10^5	1.23×10^{-1}
Bioshield branch distributor	891			1.00
Upstream Bioshield cooling branch	892	3.50×10^5	3.50×10^5	1.23×10^{-1}
Liner branch distributor	893	6.00×10^5	6.00×10^5	1.00
Liner Cooling branch delay tank	894	1.50×10^5	1.50×10^5	1.00
Concrete Cooling branch delay tank	895	1.50×10^5	1.50×10^5	1.00

2.3.3. Numerical Coupling of the Two Models

Several working fluids are available in the MELCOR 1.8.6 code; however, for a given problem, only one of these fluids can be used (together with non-condensable gases). Thus, obvious challenges have been encountered during the modeling phase of the DONES facility since both lithium and water coolants could not be used in a single run.

A possible solution is represented by the external coupling through a Python script of two separate MELCOR input decks, one working with lithium and the other one with water. The two models run concurrently, and information is exchanged between them. Figure 5 represents the approach implemented to couple the TC-LS and TC-WCS Melcor models.

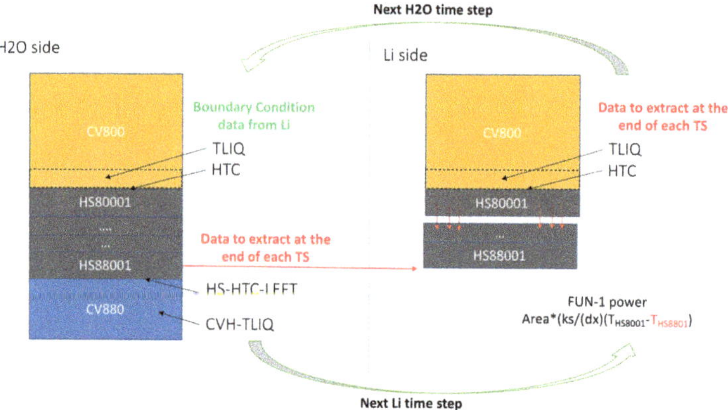

Figure 5. Scheme of numerical coupling between TC-LS and TC-WCS Melcor models.

In the Python script, information regarding the heat exchange between the two fluids is collected. In each MELCOR input, a heat structure and two control volumes are present. One of the control volumes contains the working fluid, while the other is a time-dependent control volume that simulates the other fluid involved in the heat transfer. In this control volume, as previously described, the properties are specified as a function of time.

During each MELCOR run, the update on the exchanged heat flux is passed from one input to the other, reporting the new heat transfer coefficients and the new temperatures in the heat structure and in the time-dependent control volume, respectively. Indeed, the temperature of the time-dependent volume will be the temperature assumed by the fluid of the other input at the previous time step. The exchange of information between the MELCOR inputs is obtained by writing in external files (External Data File or EDF) the variables of interest, i.e., temperatures and heat transfer coefficients in this case, after they have been saved in the Python script after a single run. Thus, the heat transfer coefficients are given as a type of boundary condition applied at the left or right boundary surface of the heat structure. In this case, a convective boundary condition is applied with the heat transfer coefficients specified by the Control Function in which the value is reported through the EDF. Instead, the temperatures are directly introduced into the time-dependent volumes directly through the EDF. To clarify this with an example referring to Figure 5 (left side), $CV800_{H2Oinput}$ is the time-dependent volume representing the CV containing lithium in the H2O input and the HS80001 heat structure simulating the liner, and the power transferred to HS80001 is equal to $Q = HTC_{LiInput} (TLIQ_{LiInput} - T_{HS80001}) A_{HS80001}$.

Note that the current design of liner wall cooling considers cooling coils welded to the liner surface. Therefore, cooling power is extracted by means of conduction between the coil pipe surface and liner surface. This was implemented in MELCOR (Figure 5) using a FUN1 function assuming a thermal conductivity $k_s = 10$ W/mK (order of SS316, accounting for thermal contact resistance) and a distance dx over which the thermal gradient was calculated to be 8 mm.

The MELCOR execution maximum time step parameter "dt" was set to 0.1 s as a trade-off between the computational time and amount of power exchanged within each time step across heat structures by means of FUN1 functions. In fact, higher values of time steps have been observed to lead to numerical instabilities. The coupling dt_{coup} for data extraction was then set accordingly.

Note that a 60,000 s stationary regime simulation period was considered to reach a steady-state for the TC-WCS loop before the PIE event in the TC-LS simulation. Numerical coupling starts at 60,000 s for the TC-WCS loop, 1000 s before PIE occurrence set at 61,000 s.

3. Results and Discussion

Model validation in the two cases was at first performed on the basis of selected parameters. For the main lithium loop, the adopted parameters were (i) Li temperature at target inlet 522.68 K (523 K nominal), (ii) Li mass flow rate 49.5 kg/s (49.8 kg/s nominal), (iii) Li level in QT 1.39 m (1.43 m nominal at steady state), and iv) Li temperature in QT at 546 K (546.7 K nominal). Concerning the TC-WCS cooling loop, it should be noted that its design stage is pre-conceptual and not detailed. Thus, the model validation during the stationary regime was performed on available dimensioning parameter expected values: (i) mass flow rate at distributors: 1.8 kg/s and 22.5 kg/s for liners and concrete distributors, respectively (in good agreement with nominal values reported in Section 2.3.2); (ii) liner temperatures (HS-80001, HS-80003, HS80004) within ±1% from nominal value of 283.15 K.

After model validation, a first assessment of the LTTC1 event in terms of released inventory and consequences on the TC atmosphere was performed. After a steady-state period with the loop fully active and beam operating, the PIE occurred at time t = 1000 s. A sensitivity analysis was performed on PIE detection sensors and the mitigation device. Note that for the considered PIE, detections upstream QT might result in late detection, since perturbation of the main LS loop parameters is firstly propagated to downstream QT. A selection of the main accident detection sensors considered to initiate the beam stop procedure is reported in Table 4.

Table 4. Accident detection.

Sensor	Triggering Condition	Time of Occurrence Since PIE [s]
Target Vacuum Chamber Gas Pressure	$TC_{pressure} \geq 1$ kPa (approx. 10 × nominal pressure)	0.22
Pressure in TC	Increase 10% with respect to nominal	2
Temperature in TC liner	Liner Floor Temperature > 393 K (100 K over nominal)	13.0
Quench Tank Li-Level	$QT_{level} \leq 0.715$ m (50% nominal value)	21.9

Given the goal of this simulation (i.e., to investigate the liner temperature rise and underlying cooling loop pressurization), the results from the case with a slower detection were conservatively considered.

Figure 6 shows the mass flow rate transient at PIE for selected flow paths judged relevant (break flow path, straightener inlet, and EMP outlet). Note that after EMP stop triggered by accident detection, the mass flow rate in the loop gradually decreases from the steady-state value (49.5 kg/s) until an inversion occurs due to relative elevations of EMP and the target (refer to Figure 2). This limits the released lithium mass inventory to about 729 kg (Figure 7), slightly more than the steady-state inventory within the QT (562 kg).

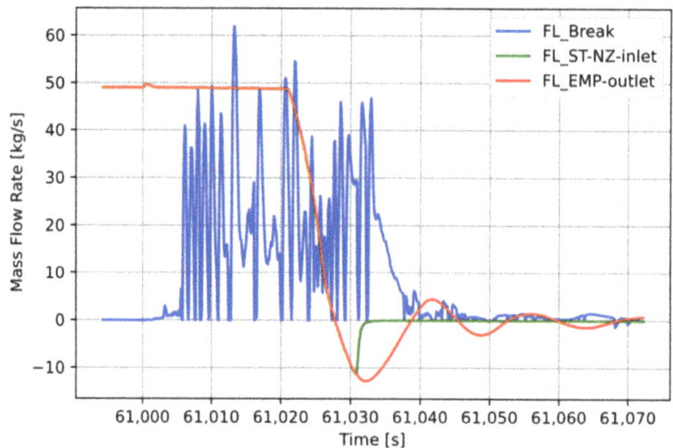

Figure 6. Mass flow rate for selected flow paths for break (FL801), removable ST-NZ inlet (FL852), and EMP (FL350).

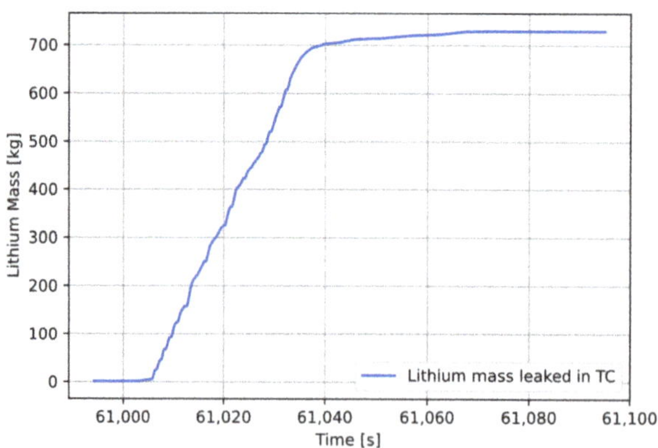

Figure 7. Inventory in QT(CVH850) and released in TC (CVH800).

The pressure wave form of the TC containment after PIE is shown in Figure 8. The pressure peak reaches 33 kPa within 8 s, then starts a decreasing pattern within 100 s from PIE.

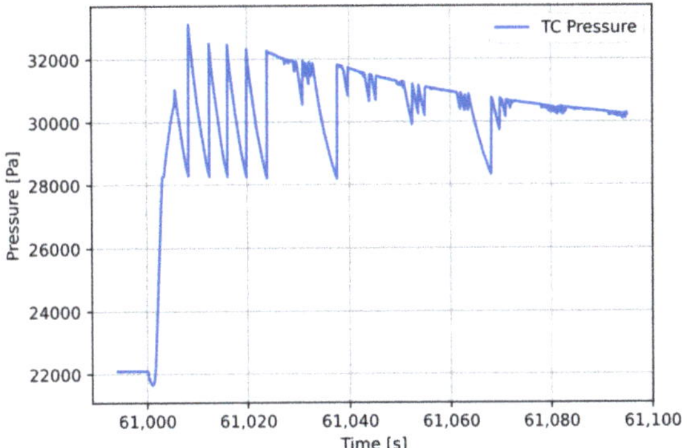

Figure 8. Pressurization of TC after PIE.

Figure 9 focuses on the temperatures on the liner floor and related cooling systems. Leaked lithium is released at a temperature of 546 K and then quickly cools down once in contact with the liner floor. In turn, the liner floor quickly passes from the 287 K temperature before PIE, to 463 K within 50 s, peaking at 472 K within 220 s. As a consequence, the underlying cooling pipe HS and water coolant reach 460 K.

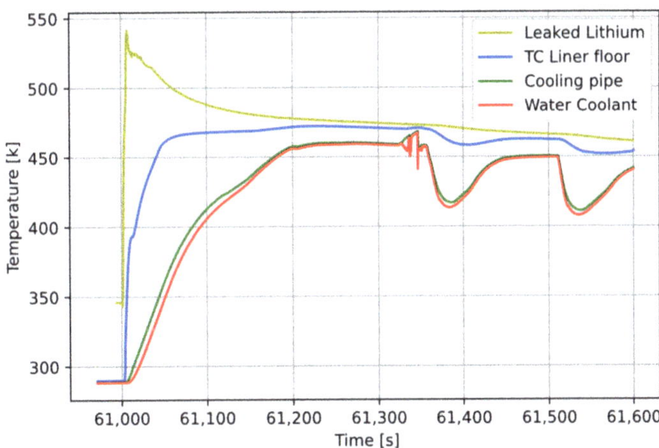

Figure 9. Liner floor, underlying water pipes, and water coolant temperature after PIE.

Figure 10 shows that water within the liner floor cooling loop vaporizes after about 350 s, causing a temporary loss of coolant mass from the CV880 volume, as shown in Figure 11. The mentioned vaporization results in the pressurization pattern shown in Figure 12, though gradually absorbed by the loop within 500 s from PIE.

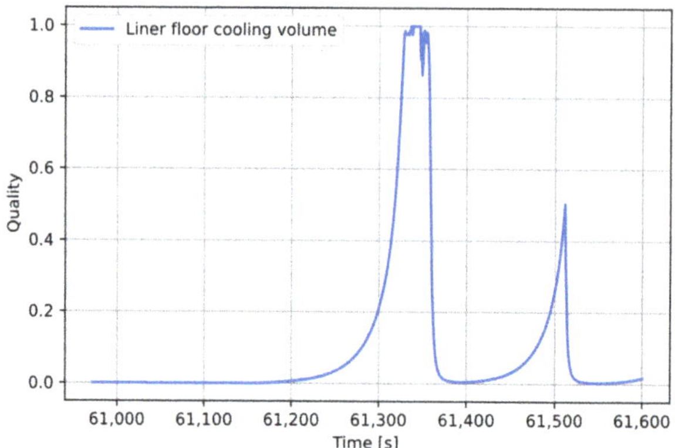

Figure 10. Water steam quality within liner floor cooling loop water pipes.

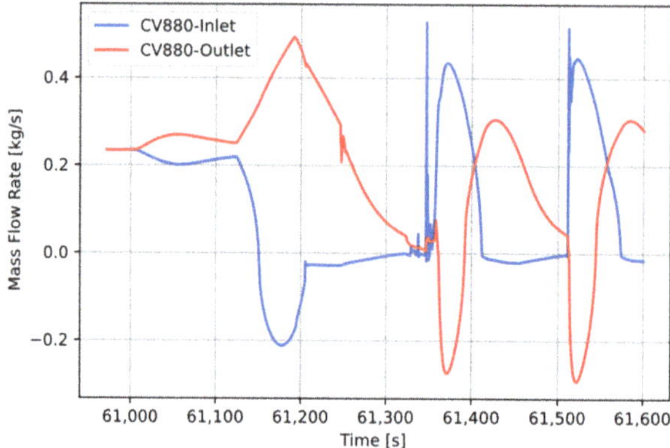

Figure 11. Flow mass rate at the liner floor cooling volume inlet/outlet.

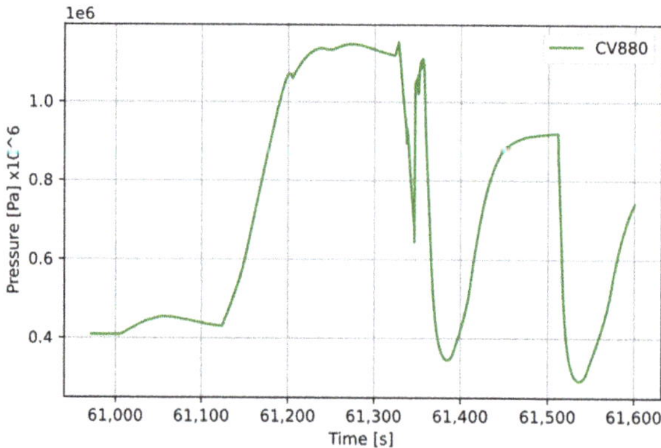

Figure 12. Water coolant pressure in the liner cooling loop water pipes after PIE.

4. Conclusions

An analysis of the impact of a liquid lithium spillage event on room containment has been proposed. In particular, the purpose of this study was to investigate consequences on the water-cooled steel liner floor of the containment room with a possible breach of the liner barrier enabling lithium to contact the underlying concrete or water itself in the case of cooling pipe rupture. Considering the impact of presented transients in terms of safety consequences, the TC pressurization does not appear to be a concern for the TC containment. In fact, despite the occurrence of water vaporization transient in the cooling loop under the floor liner, the reached pressure peak at about 1.14 MPa appears withstandable within the cooling pipe design pressure. Therefore, the break of cooling pipes and consequent release of water appear unlikely. Given the high temperature difference between released liquid lithium inventory (546 K) and liner temperature (287 K), the temperature rise has a steep pattern reaching more than 3 K/s temperature rise velocity. Future work shall investigate liner structural resistance to such thermal shock events [45–47], possibly leading to liner failure and lithium contacting the concrete. However, even in the case of such a concrete–lithium contact event, lithium at such an interface point would likely be in solid phase or below the minimal observed ignition point (453.15 K) [8], making a consequent fire event unlikely. Future work will also investigate model sensitivity [48] to specific parameters impacting the liner model as the interface point (e.g., thermal conductivity at the liner/cooling pipe contact point, cooling pipe exchange surface).

Author Contributions: Conceptualization, D.N.D.; methodology, D.N.D. and M.D.; software, D.N.D. and M.D.; validation, D.N.D. and M.D.; writing—original draft preparation, D.N.D.; writing—review and editing, D.N.D. and M.D. All authors have read and agreed to the published version of the manuscript.

Funding: This work has been carried out within the framework of the EUROfusion Consortium and has received funding from the Euratom research and training programme 2014–2018 and 2019–2020 under grant agreement No 633053. The views and opinions expressed herein do not necessarily reflect those of the European Commission.

Institutional Review Board Statement: Not applicable.

Informed Consent Statement: Not applicable.

Conflicts of Interest: The authors declare no conflict of interest. The funders had no role in the design of the study; in the collection, analyses, or interpretation of data; in the writing of the manuscript, or in the decision to publish the results.

Nomenclature

BP	Back Plate
CV	Control Volume
EMP	Electromagnetic Pump
HEBT	High Energy Beam Transport.
HFTM	High Flux Test Module
HX	Heat Exchanger
LEBT	Low Energy Beam Transport
LS	Lithium System
LSP	Lower Shielding Plug
MEBT	Medium Energy Beam Transport
QT	Quench Tank
RFQ	Radiofrequency Quadrupoles
SRF	Superconducting Radiofrequency
ST-NZ	Straightener Nozzle
STUMM	Start-up Monitoring Module
TA	Target Assembly
TC-WCS	Test Cell Water Cooling System
TTC	Target Test Cell
VC	Vacuum Chamber

References

1. Linke, J.; Du, J.; Loewenhoff, T.; Pintsuk, G.; Spilker, B.; Steudel, I.; Wirtz, M. Challenges for plasma-facing components in nuclear fusion. *Matter Radiat. Extrem.* **2019**, *4*, 056201. [CrossRef]
2. Federici, G.; Biel, W.; Gilbert, M.R.; Kemp, R.; Taylor, N.; Wenninger, R. European DEMO design strategy and consequences for materials. *EURATOM Nucl. Fusion* **2017**, *57*, 9. [CrossRef]
3. Ibarra, A.; Arbeiter, F.; Bernardi, D.; Cappelli, M.; Garcia, A.; Heidinger, R.; Krolas, W.; Fischer, U.; Martin-Fuertes, F.; Micciché, G.; et al. The IFMIF-DONES project: Preliminary engineering design. *Nucl. Fusion* **2018**, *58*, 105002. [CrossRef]
4. Federici, G.; Bachmann, C.; Barucca, L.; Biel, W.; Boccaccini, L.; Brown, R.; Bustreo, C.; Ciattaglia, S.; Cismondi, F.; Coleman, M.; et al. DEMO design activity in Europe: Progress and updates. *Fusion Eng. Des.* **2018**, *136*, 729–741. [CrossRef]
5. Martín-Fuertes, F.; García, M.E.; Fernández, C.; D'Ovidio, G.; Pinna, T.; Porfiri, M.T.; Fischer, U.; Ogando, F.; Mota, F. Integration of Safety in IFMIF-DONES Design. *Safety* **2019**, *5*, 74. [CrossRef]
6. D'Ovidio, G.; Martín-Fuertes, F. Accident analysis with MELCOR-fusion code for DONES lithium loop and accelerator. *Fusion Eng. Des.* **2019**, *146*, 473–477. [CrossRef]
7. Arena, P.; Di Maio, P.A.; Nitti, F.S. Safety analysis of the dones primary heat removal system. *Fusion Eng. Des.* **2020**, *161*, 112002. [CrossRef]
8. Piet, S.; Jeppson, D.; Muhlestein, L.; Kazimi, M.; Corradini, M. Liquid metal chemical reaction safety in fusion facilities. *Fusion Eng. Des.* **1987**, *5*, 273–298. [CrossRef]
9. Barnett, S.; Kazimi, M.J. *Consequences of a Lithium Spill in-Side the Containment and Vacuum Torus of a Fusion Reactor*; PFC/RR-87-9; MIT Plasma Fusion Center: Cambridge, MA, USA, 1987.
10. Merrill, B.J. A lithium-air reaction model for the melcor code for analyzing lithium fires in fusion reactors. *Fusion Eng. Des.* **2001**, *54*, 485–493. [CrossRef]
11. Dongiovanni, D.N.; Porfiri, M.T. Exploratory fire analysis in DONES lithium system. *Fusion Eng. Des.* **2020**, *156*, 111680. [CrossRef]
12. Merrill, B.J.; Humrickhouse, P.; Moore, R.L. A recent version of MELCOR for fusion safety applications. *Fusion Eng. Des.* **2010**, *85*, 1479–1483. [CrossRef]
13. Merrill, B.J.; Humrickhouse, P.; Shimada, M. Recent development and application of a new safety analysis code for fusion reactors. *Fusion Eng. Des.* **2016**, *109–111*, 970–974. [CrossRef]
14. Gauntt, R.O.; Cash, J.E.; Cole, R.K.; Erickson, C.M.; Humphries, L.L.; Rodriguez, S.B.; Young, M.F. *MELCOR Computer Code Manuals Vol. 1: Primer and Users*; Guide Version 1.8.6, NUREG/CR-6119, Volume 1, Rev. 3; Sandia National Laboratory: Princeton, NJ, USA, 2005.
15. Arbeiter, F.; Diegele, E.; Fischer, U.; Garcia, A.; Ibarra, A.; Molla, J.; Mota, F.; Möslang, A.; Qiu, Y.; Serrano, M.; et al. Planned material irradiation capabilities of IFMIF-DONES. *Nucl. Mater. Energy* **2018**, *16*, 245–248. [CrossRef]
16. Tian, K.; Ahedo, B.; Arbeiter, F.; Barrera, G.; Ciupinski, L.; Dézsi, T.; Horne, J.; Kovács, D.; Molla, J.; Mota, F.; et al. Overview of the current status of IFMIF-DONES test cell biological shielding design. *Fusion Eng. Des.* **2018**, *136*, 628–632. [CrossRef]
17. Nitti, F.; Ibarra, A.; Ida, M.; Favuzza, P.; Furukawa, T.; Groeschel, F.; Heidinger, R.; Kanemura, T.; Knaster, J.; Kondo, H.; et al. The design status of the liquid lithium target facility of IFMIF at the end of the engineering design activities. *Fusion Eng. Des.* **2015**, *100*, 425–430. [CrossRef]
18. Arena, P.; Bernardi, D.; Di Maio, P.A.; Frisoni, M.; Gordeev, S.; Micciché, G.; Nitti, F.S.; Ibarra, A. The design of the DONES lithium target system. *Fusion Eng. Des.* **2019**, *146*, 1135–1139. [CrossRef]
19. Simon, S.; Dézsi, T.; Arbeiter, F.; Tóth, M.; Castellanos, J.; Ibarra, A. Thermal-hydraulic simulation of IFMIF-DONES Test Cell atmosphere. *Fusion Eng. Des.* **2021**, *167*, 112336. [CrossRef]
20. Merrill, B.J. 'Recent Updates to the MELCOR 1.8.2 Code for ITER Applications,' INL/EXT-07-12493, May 2007. Available online: https://inldigitallibrary.inl.gov/sites/sti/sti/3644018.pdf (accessed on 30 September 2021).
21. Merrill, B.; Moore, R.; Polkinghorne, S.; Petti, D. Modifications to the MELCOR code for application in fusion accident analyses. *Fusion Eng. Des.* **2000**, *51–52*, 555–563. [CrossRef]
22. Topilski, L.; Masson, X.; Porfiri, M.; Pinna, T.; Sponton, L.-L.; Andersen, J.; Takase, K.; Kurihara, R.; Sardain, P.; Girard, C. Validation and benchmarking in support of ITER-FEAT safety analysis. *Fusion Eng. Des.* **2001**, *54*, 627–633. [CrossRef]
23. Moore, R.L. *Status Report on an ITER ITA on Comparison of MELCOR 1.8.5 Results to MELCOR 1.8.2 Results for a Selected Set of Accident Analysis Cases Relevant to ITER FEAT*; EDF-5470, Rev. 11; Idaho National Laboratory: Idaho Falls, ID, USA, 2003.
24. Merrill, B.J. *Benchmarking MELCOR 1.8.2 for ITER Against Recent EVITA Results*; Idaho National Laboratory: Idaho Falls, ID, USA, 2007. [CrossRef]
25. Sallus, L.; Van Hove, W. MELCOR Code Validation on HE-FUS3. *Loop* **2008**, 391–404. [CrossRef]
26. Merrill, B.J.; Humrickhouse, P.W.; Moore, R.L. "A Comparison of Modifications to MELCOR Versions 1.8.2 and 1.8.6 for ITER Safety Analysis", INL/EXT-09-16715, June 2010. Available online: https://inldigitallibrary.inl.gov/sites/sti/sti/4536702.pdf (accessed on 30 September 2021).
27. Reyes, S.; Topilski, L.; Taylor, N.; Merrill, B.J.; Sponton, L.-L. Updated Modeling of Postulated Accident Scenarios in ITER. *Fusion Sci. Technol.* **2009**, *56*, 789–793. [CrossRef]
28. Dongiovanni, D.N.; Pinna, T.; Porfiri, M.T. DEMO Divertor preliminary safety assessment. *Fusion Eng. Des.* **2021**, *169*, 112475. [CrossRef]

29. Jin, X.Z. BB LOCA analysis for the reference design of the EU DEMO HCPB blanket concept. *Fusion Eng. Des.* **2018**, *136*, 958–963. [CrossRef]
30. D'Onorio, M.; Giannetti, F.; Caruso, G.; Porfiri, M.T. In-box LOCA accident analysis for the European DEMO water-cooled reactor. *Fusion Eng. Des.* **2019**, *146*, 732–735. [CrossRef]
31. Gonfiotti, B.; Paci, S. Normal and Accidental Scenarios Analyses with MELCOR 1.8.2 and MELCOR 2.1 for the DEMO Helium-Cooled Pebble Bed Blanket Concept. *Sci. Technol. Nucl. Install.* **2015**, *2015*, 1–9. [CrossRef]
32. Moon, S.; Sung, B.; Bang, I.C. Thermal Hydraulic Analysis of K-DEMO Single Blanket Module for Preliminary Accident Analysis using MELCOR. In Proceedings of the KNS 2016 Spring Meeting, Jeju, Korea, 12 May 2016.
33. Murgatroyd, J.T.; Owen, S.; Grief, A.; Panayotov, D.; Saunders, C. Qualification of MELCOR and RELAP5 models for EU HCPB TBS accident analyses. *Fusion Eng. Des.* **2017**, *124*, 1251–1256. [CrossRef]
34. Dobromir Panayotov Brad, J. Merrill, From Fission Nuclear Power Plants to Fusion Power Plant Safety Accident Analyses Challenges. In Proceedings of the 1st International Workshop on Environmental, Safety and Economic Aspects of Fusion Power, Jeju, Korea, 13 September 2015. [CrossRef]
35. Pena, A.; Esteban, G.; Sancho, J.; Kolesnik, V.; Abánades, A. Hydraulics and heat transfer in the IFMIF liquid lithium target: CFD calculations. *Fusion Eng. Des.* **2009**, *84*, 1479–1483. [CrossRef]
36. Gordeev, S.; Gröschel, F.; Heinzel, V.; Hering, W.; Stieglitz, R. Numerical study of the flow conditioner for the IFMIF liquid lithium target. *Fusion Eng. Des.* **2014**, *89*, 1751–1757. [CrossRef]
37. Gordeev, S.; Gröschel, F.; Heinzel, V.; Hering, W.; Stieglitz, R. Numerical Analysis of Unsteady Flow Behavior in Flow Conditioner of IFMIF Liquid-Lithium Target. *Fusion Sci. Technol.* **2015**, *68*, 618–624. [CrossRef]
38. Gordeev, S.; Arena, P.; Bernardi, D.; Di Maio, P.A.; Nitti, F.S. Analytical and Numerical Assessment of Thermally Induced Pressure Waves in the IFMIF-DONES Liquid-Lithium Target. *IEEE Trans. Plasma Sci.* **2020**, *48*, 1485–1488. [CrossRef]
39. Tolli, J.E. Overview of Property Formulations for Helium, Nitrogen, Lithium, and Lithium-Lead in ATHENA/MOD1 with Comparison of Calculated Properties to Measured Properties, EGG-FSP-10245, Idaho National Engineering Laboratory, April 1992. Available online: https://ui.adsabs.harvard.edu/abs/1992opfh.rept.....T/abstract (accessed on 30 September 2021).
40. Merrill, B.J. Modifications Made to the MELCOR Code for Analyzing Lithium Fires in Fusion Reactors, INEEL/EXT-2000-00489, April 2000. Available online: https://www.osti.gov/biblio/764178-modifications-made-melcor-code-analyzing-lithium-fires-fusion-reactors (accessed on 30 September 2021).
41. Jeppson, D.W.; Scoping, D.W. Studies: Behavior and control of lithium and lithium aerosols. *HEDLTME* **1982**, 79–80. [CrossRef]
42. Jeppson, D.W. *Results and Code Prediction Comparisons of Lithium-Air Reaction and Aerosol Behavior Tests*; HEDL-TME 85–25; Hanford Engineering Development Laboratory: Richland, WA, USA, 1986; Available online: https://inis.iaea.org/search/searchsinglerecord.aspx?recordsFor=SingleRecord&RN=18000591 (accessed on 30 September 2021).
43. Merrill, B.; Humrickhouse, P.; Yoon, S.-J. Modifications to the MELCOR-TMAP code to simultaneously treat multiple fusion coolants. *Fusion Eng. Des.* **2018**, *146*, 289–292. [CrossRef]
44. Qiu, Y.; Arbeiter, F.; Fischer, U.; Tian, K. Neutronics analyses for the bio-shield and liners of the IFMIF-DONES test cell. *Fusion Eng. Des.* **2019**, *146*, 723–727. [CrossRef]
45. Frisoni, M.; Bernardi, D.; Nitti, F. Nuclear assessment of the IFMIF-DONES lithium target system. *Fusion Eng. Des.* **2020**, *157*, 111658. [CrossRef]
46. Marsh, D.J. *A Thermal Shock Fatigue Study of Type 304 and 316 Stainless Steels*; UKAEA Risley Nuclear Power Development Establishment (ND-R–606(S)): London, UK, 1982.
47. Bernard, L.J.; Lamain, G.L.; Verzeletti, G. *Crack Initiation and Growth in Stainless Steel Tubes under Thermal Shocks*; Radon, J.C., Ed.; Fracture and Fatigue: Pergamon, Turkey, 1980; pp. 391–400. ISBN 9780080261614. [CrossRef]
48. D'Onorio, M.; Giannetti, F.; Porfiri, M.T.; Caruso, G. Preliminary sensitivity analysis for an ex-vessel LOCA without plasma shutdown for the EU DEMO WCLL blanket concept. *Fusion Eng. Des.* **2020**, *158*, 111745. [CrossRef]

Article

CFD Optimization of the Resistivity Meter for the IFMIF-DONES Facility

Ranieri Marinari *, Paolo Favuzza, Davide Bernardi, Francesco Saverio Nitti and Ivan Di Piazza

ENEA, ENEA Brasimone Research Centre, Department of Fusion and Nuclear Safety Technology, 40032 Camugnano, BO, Italy; paolo.favuzza@enea.it (P.F.); davide.bernardi@enea.it (D.B.); francesco.nitti@enea.it (F.S.N.); ivan.dipiazza@enea.it (I.D.P.)
* Correspondence: ranieri.marinari@enea.it; Tel.: +39-0534-801-114

Abstract: A detailed study of lithium-related topics in the IFMIF-DONES facility is currently being promoted and supported within the EUROfusion action, paying attention to different pivotal aspects including lithium flow stability and the monitoring and extraction of impurities. The resistivity meter is a device able to monitor online non-metallic impurities (mainly nitrogen) in flowing lithium. It relies on the variation of the electric resistivity produced by dissolved anions: the higher the concentration of impurities in lithium, the higher the resistivity measured. The current configuration of the resistivity meter has shown different measuring issues during its operation. All these issues reduce the accuracy of the measurements performed with this instrument and introduce relevant noise affecting the resistance value. This paper proposes different upgrades, supported by CFD simulations, to optimize lithium flow conditions and to reduce measurement problems. Owing to these upgrades, a new design of the resistivity meter has been achieved, which is simpler and easier to manufacture.

Keywords: DEMO-EU fusion reactor; IFMIF-DONES facility; lithium technology; CFD; thermo-fluid dynamics

Citation: Marinari, R.; Favuzza, P.; Bernardi, D.; Nitti, F.S.; Di Piazza, I. CFD Optimization of the Resistivity Meter for the IFMIF-DONES Facility. *Energies* **2021**, *14*, 2543. https://doi.org/10.3390/en14092543

Academic Editor: Hiroshi Sekimoto

Received: 23 March 2021
Accepted: 26 April 2021
Published: 28 April 2021

Publisher's Note: MDPI stays neutral with regard to jurisdictional claims in published maps and institutional affiliations.

1. Introduction

Knowledge regarding the purity of liquid lithium flowing inside metal pipes, such as those typically constituting the loops used in fusion applications, is of primary importance, since it is well known [1–6] that the presence of impurities, particularly non-metallic elements, greatly enhances the corrosion action exerted by lithium itself on wet pipe surfaces. Therefore, it is necessary to keep as low as possible the concentration of non-metallic elements in lithium such as carbon, oxygen, hydrogen and, particularly, nitrogen, which is the most corrosion-affecting element among them. For this purpose, a device or a technique able to quantify their concentrations is necessary.

In the IFMIF (International Fusion Materials Irradiation Facility) project, the maximum allowed concentration value for each of the aforementioned elements had been set to 10 wppm [7]. The same value was also chosen for the IFMIF-DONES (Demo-Oriented NEutron Source) plant [8], even if the possibility to allow a higher nitrogen limit is under evaluation, for instance 30 wppm or more, depending on the results of the corrosion validation activities [9], which are still ongoing.

Unfortunately, the quantification of the non-metals concentration in lithium is not possible for almost all of them. Only for nitrogen has a specific procedure been implemented, but it is in any case an offline procedure, the sample entailing a small amount (a few grams) of lithium from the loop and the performance of a batch chemical analysis on it. This requires many hours and the availability of a chemistry-trained operator [10].

An online (real-time) individual analysis of nitrogen as well as of the other non-metallic elements is not available. Nonetheless, an overall evaluation of the amount of total impurities can be given by the Resistivity Meter (RM). The RM measures the electrical resistance of lithium and relates it to the purity of the alkaline metal, since the higher the concentration of total impurities, the higher the electrical resistance.

The first part of this paper describes the measurement principle of the RM and the issues presently encountered with the instrument. The second part deals with the design and the constructive details of a new, improved RM, possibly giving a more stable, precise and sensitive measurement output; this new RM, once constructed, will be then tested, installed and employed in the IFMIF-DONES Impurity Control System (ICS) loop.

2. The Current Resistivity Meter

The Resistivity Meter is a device able to online monitor the electric resistance of flowing lithium. It was developed at the University of Nottingham in collaboration with ENEA Brasimone and relies on the variation of the metal resistivity produced by the dissolved anions: the higher the concentration of impurities in lithium, the higher the resistivity (α) [11–15]. Clearly, each kind of impurity affects the resistivity value to a different extent. However, taking into account their relative effects and their solubility in lithium, it is possible to almost entirely ascribe the variation of α to hydrogen and nitrogen, the latter being the most corrosion-affecting non-metal impurity.

The current RM, shown in Figure 1, has been installed and operated inside the Lifus 6 plant at the ENEA Brasimone Centre since 2014 [16,17]; its particular location (the purification loop upstream of the Cold Trap), permits one to analyse the less purified lithium flowing throughout the plant so as to avoid underestimation of the total impurities' concentration value.

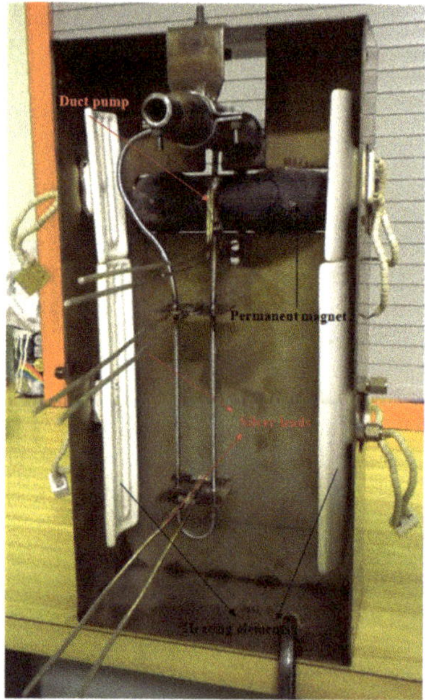

Figure 1. Picture of the current resistivity meter.

Lifus 6 RM is made of AISI 316L and is an all TIG welded assembly. The main components are:

- The source lithium pipe, from which a small fraction is picked for monitoring;
- The electromagnetic pump (EMP), which drives lithium through the capillary tube;
- The capillary tube (φ_o= 4.8 mm, φ_i= 3.0 mm), where the measurement current is applied and has the shape of a loop continuously fed with flowing liquid lithium.

The electrical resistance is measured between two metal plates (40 mm × 40 mm, 2 mm thickness, 150 mm one from the other), which intersect the measurement capillary and where silver electrical leads are connected both to bring the electrical current and to acquire the voltage drop (BINT 20024 Precision Nano-ohmmeter is employed for this purpose). From Figure 1, it is also possible to see that all the apparatus is housed inside a steel parallelepiped box, and that along the lateral inner walls of this box are mounted four heating elements (two on each side), which permit one to control the desired temperature inside. The box, open in the figure, is closed during normal operation in order to minimize the heat losses to the external environment.

During plant operation, the current configuration of the resistivity meter has evidenced two relevant issues [18]:

- The heaters installed inside the box to avoid thermal losses from the capillary are not able to guarantee isothermal conditions around the capillary. The four heating elements heat the air inside the box and not directly the capillary tube. Additionally, due to the vertical orientation of the box, warm air inside the box moves upward originating a convective heat flux and a stratification in the Li flow direction. Since resistivity is highly sensitive to temperature, this thermal instability introduces an important and troublesome contribution to the acquired signal;

- The electromagnetic pump devoted to lithium circulation in the capillary gives rise to parasitic currents, which when added to the measurement current produce a significant noise on the acquired electrical signal. Despite its negative effect, the EMP must in any case be always kept on, otherwise the lithium would be static inside the RM capillary, and the RM output would not be a real-time measurement of the impurity concentration in the flowing lithium.

All of these issues reduce the accuracy of the measurements performed with this instrument and introduce a relevant noise affecting the resistance value.

3. Upgrade Strategy

The component upgrades have followed three parallel lines:

1. The feasibility study of a Venturi suction instead of the electromagnetic pump looking for the elimination of background noise during the measurement;
2. The numerical study of the thermal losses in the air box looking for an optimization of the heating element's location and power rate;
3. The reduction in air thermal stratification in the box.

For the replacement of the electromagnetic pump, an appropriate analytical design of the Venturi suction was performed by keeping the 1 s travel time of lithium inside the capillary (i.e., mass flow in the capillary) and minimizing the irreversible pressure drop through the Venturi. Owing to this effort, the general dimensions of the pipes were fixed (Figure 2). The Venturi geometry foresees an inlet pipe of ¾" sch 40 followed by a sudden restriction to a ¼" sch 40 and a 10° angle divergent to the ¾" sch 40 outlet pipe.

The inlet sudden flow area restriction was chosen instead of a smooth one for reducing both the Venturi length and lithium pressure drop in the component. The adoption of a slow convergent one would have required the adoption of a smaller pipe diameter in the suction region to grant the nominal flow rate in the capillary; on the other hand, the smaller pipe would have increased the pressure drop in the divergent downstream region.

The other two phases (assessment of the thermal losses and reduction of the thermal stratification in the box) require a CFD numerical approach; then, a CFD model of the upgraded component was developed.

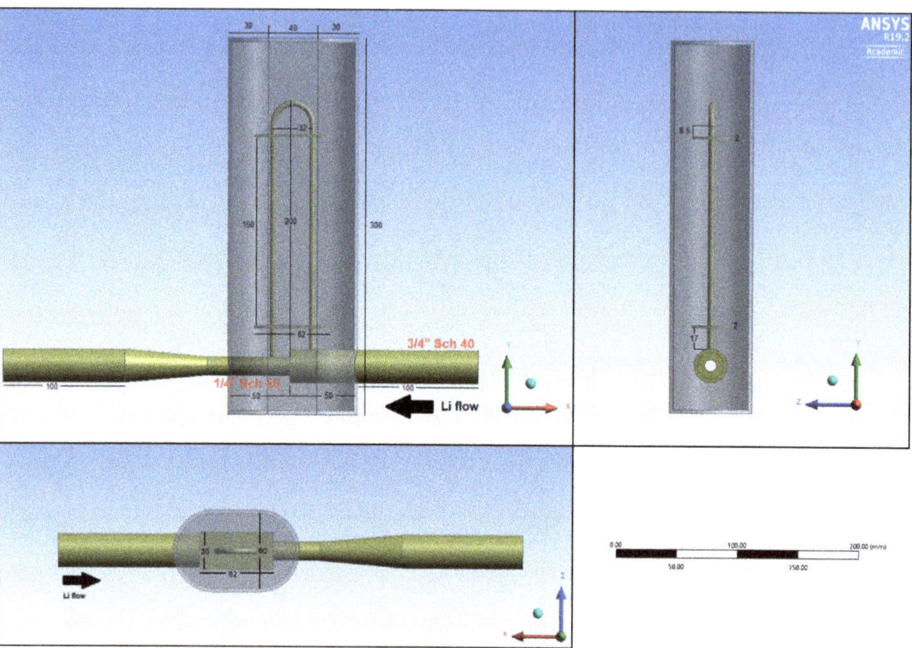

Figure 2. Sketch of the Venturi suction designed for the resistivity meter; pipe capillary with its main dimensions and air box are illustrated.

4. CFD Analysis: Models and Methods

A CFD computational model of the upgraded resistivity meter was built to perform CFD simulations. The general-purpose code ANSYS CFX (V.19.2, Canonsburg, PA, USA) was used for all the numerical simulations presented in this paper [19].

The code employs a coupled technique, which simultaneously solves all the transport equations in the whole domain through a false time-step algorithm. The linearized system of equations is preconditioned in order to reduce all the eigenvalues to the same order of magnitude. The multi-grid approach reduces the low frequency error, converting it to a high frequency error at the finest grid level; this results in a great acceleration of convergence. Although with this method a single iteration is slower than a single iteration in the classical decoupled (segregated) SIMPLE approach, the number of iterations necessary for a full convergence to a steady state is generally of the order of 10^2, against typical values of 10^3 for decoupled algorithms.

The discretization scheme of the convective terms used for energy and momentum equation variables was the high-resolution scheme (second order-like scheme) [20].

The SST (Shear Stress Transport) k-ω model by Menter [21] is extensively used in this paper. It is formulated to solve the viscous sub-layer explicitly and requires several computational grid points inside this latter. The model applies the k-ω model close to the wall, and the k-ε model (in a k-ω formulation) in the core region, with a blending function in between. It was originally designed to provide accurate predictions of flow separation under adverse pressure gradients, but it was applied to a large variety of turbulent flows and is now the default and most-used model in CFX-19 and other CFD codes.

The CFD model developed (Figure 3) includes the Venturi suction described in the previous paragraph, the U tube capillary (the pipe inner/outer diameter of 3/4.8 mm and the curvature radius of 32 mm were kept from the previous configuration) with the two steel measuring plates (again at a distance of 150 mm from each other) and the air in the box surrounding both the capillary and the main pipe. The RM box is a parallelepiped flat oriented 30 cm long and 10 cm × 6 cm rectangular rounded base. The whole model includes three main domains: the lithium flowing in the RM, the piping steel inside the box and the air inside the RM box. The conjugate heat transfer between lithium, steel and air is set owing to the General Grid Interface (GGI) algorithm that links nodes on different domains (solid and fluid) facing each other. No heat source was included in the RM air box looking for an estimation of the thermal losses.

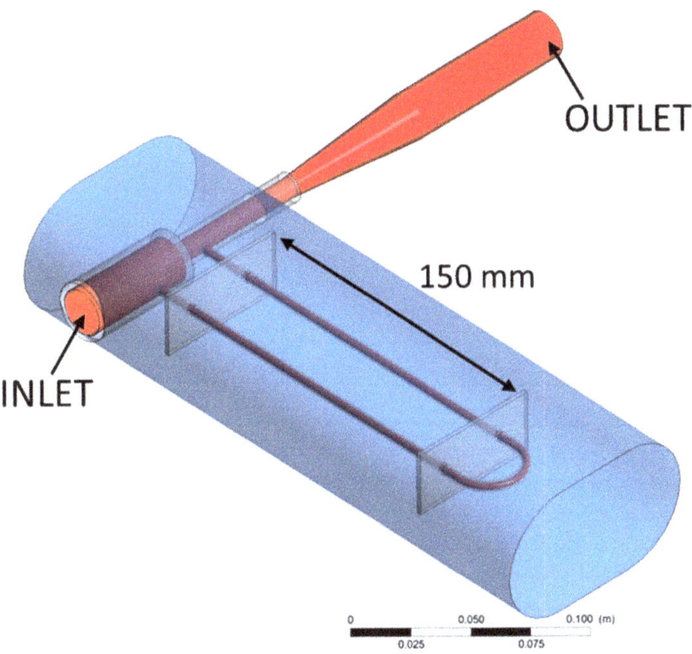

Figure 3. Sketch of the CFD model developed: lithium domain in red, steel domain in grey, and air domain in blue. The measuring length between the two slabs (kept from the former configuration) is highlighted.

Liquid lithium properties were implemented as temperature dependent in ANSYS CFX code according to [22] while air and piping steel were set as constant property materials in the code.

The RM box has a horizontal orientation for limiting the thermal stratification of the air inside it.

On the external surfaces of the air domain, a constant heat transfer coefficient boundary condition of 1.5 $W/m^2/K$ was set (considering 30 mm thermal insulator and stagnant outside air) and an outside temperature of 15 °C. Buoyancy forces were set in the air domain (Boussinesq approximation) for heat exchange between the hot pipes and the cold external walls.

Being interested in average thermal-hydraulic results (pressure drop, heat transfer rate, temperature gradients) and not in point values, full convergence of the simulation is declared when all the Root Mean Square (RMS) residuals reach 1E-5 (according to the CFX User Manual [19] for simulations focused on reliable average quantities but not on detailed point values) and the monitoring points on: lithium pressure drop, thermal loss from air and local velocity values attain a stable value.

Regarding the boundary conditions adopted in the model, it has to be highlighted that the new RM is supposed to be installed inside the future IFMIF-DONES lithium loop. Therefore, its operative parameter will be different from those adopted in Lifus 6 experiments. Unfortunately, a stable definition of the new loop parameters has not been performed yet; thus, two limit cases have been considered here.

The first case assumes the original values set for the IFMIF-DONES plant. These imply a lithium inlet temperature equal to 273 °C (i.e., the same temperature of the plant hot leg, the cold leg being set at 250 °C) and a flow rate of 0.065 L/s (i.e., 10% of the total lithium flow rate trough the ICS, with the residual 90% going through the traps system).

The second case reflects instead the change in the IFMIF-DONES main lithium loop operating parameters, which was proposed to minimize the possible [7]Be accumulation inside the loop. The new values for the temperature and the flow rate of lithium entering the RM circuit are, respectively, 330 °C and 0.108 L/s (1/9 of the lithium flow rate through the trap system).

Prior to any CFD case simulation, an appropriate mesh sensitivity analysis was performed. Three different meshes were generated adopting the same mesh strategy but different element sizes. Mesh A has a coarse grid resolution, both in the lithium and in the air domain; it has about 398,000 elements and 294,000 nodes. Mesh B is an average mesh with 1.4 million elements and 866598 nodes. Mesh C has a fine mesh, both in the air and lithium domain; it has 2.65 million elements and 1.87 million nodes.

The first case boundary conditions are imposed on the three meshes for the mesh sensitivity analysis (0.182 m/s inlet lithium velocity, 273 °C inlet lithium temperature, 0 bar relative pressure on the lithium outlet surface).

Different data were compared: the pressure drop across the whole model (Δp), the heat losses through the RM box (Q), the average lithium velocity inside the capillary (u), the average lithium outlet temperature from the capillary (T_{out}). The general comparison between the meshes is reported in Table 1. As illustrated in the table comparison, the outlet temperature from the capillary is not affected by the mesh. Heat losses and lithium velocity in the capillary are marginally affected with about 1% difference between coarse mesh A and average mesh B. The most interesting comparison seems to be the total pressure drop across the CFD model that tends to converge between average mesh B and fine mesh C with a percentage difference of 1.6. s

Table 1. Comparison of the main mesh data and results between the three grids (A, B and C) developed and run; results are also reported as percentage difference.

Mesh	Number of Elements	Number of Nodes	Δp [Pa]	Δp [%]	Q [W]	Q [%]	u [m/s]	u [%]	T_{out} [°C]	T_{out} [%]
A	398 k	294 k	274.3	-	17.8	-	0.36	-	271.2	
B	1.4 M	866.6 k	308	12	18	1.1	0.356	1.2	271.1	0.03
C	2.65 M	1.87 M	313.3	1.6	18.05	0.3	0.354	0.8	271.09	0.003

Due to the very good convergence of the heat losses, average capillary velocity, outlet temperature and the negligible pressure drop difference (barely measurable or not measurable with the instruments) between mesh B and mesh C, mesh B was selected for all the simulations presented in this paper.

As an insight into the selected mesh B (Figure 4), the grid is fully structured in the lithium domain (red) and in the capillary pipe, while it is unstructured in the solid domain (grey) that includes the pipes and the measuring plates on the capillary.

For the air domain in the box, the mesh is unstructured with node inflation on the surface boundaries. Other interesting mesh details are reported in Table 2. Mesh parameters such as skewness, orthogonal quality and aspect ratio are declared excellent for the structured domain lithium and good for the tetrahedral/unstructured domains air and steel [19].

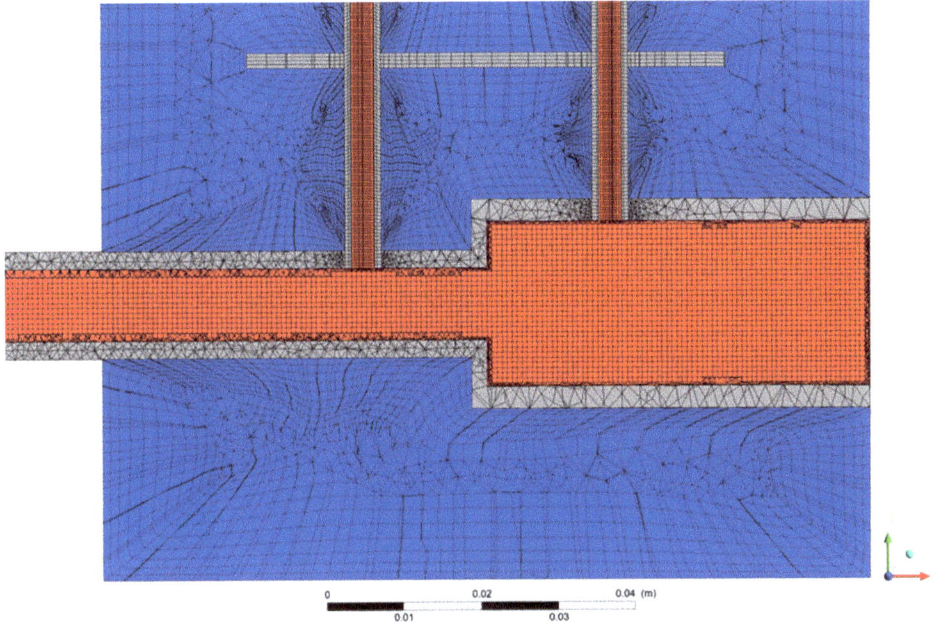

Figure 4. Sketch of the mesh developed for the resistivity meter simulation.

Table 2. Orthogonal quality, skewness and aspect ratio of the mesh developed in the three domains.

Domain	Number of Elements	Number of Nodes	Average Orthogonal Quality	Average Skewness	Average Aspect Ratio
Lithium	694 k	503 k	0.9	0.18	2.48
Steel	98.6 M	92 k	0.85	0.25	1.78
Air	642 k	271 k	0.5	0.45	2.1

5. Results and Discussion

5.1. First Case: Boundary Conditions and Results

The boundary conditions imposed are:

- Inlet lithium temperature: 273 °C.
- Inlet lithium flow rate: 0.065 L/s, corresponding to an average inlet velocity of 0.182 m/s.
- Relative pressure at the outlet: 0 bar.

The first parameter to assess in the post-processing phase is the pressure drop of the upgraded component. As we can see in Figure 5, the overall pressure drop is about 308 Pa (3 mbar), but it is not possible to calculate from the contour plot the different contributions due to the variation of the kinetic/hydrostatic energy component in the pipes. Considering both the pressure and the kinetic energy variation in the lithium between two generic sections i and j, it is possible to obtain the pressure drop Δp with the following expression (Bernoulli equation):

$$\Delta p = p_i + \rho \frac{v_i^2}{2} - p_j - \rho \frac{v_j^2}{2} \tag{1}$$

where p_i (p_j) is the pressure on the generic section i (j), ρ is the lithium density and v_i (v_j) is the lithium velocity. It was found that:

- 111.2 Pa (36% of the total pressure drop) is lost in the sudden flow area reduction;
- 54.6 Pa (18% of the total pressure drop) is lost in the ¼" pipe;

- 142.1 Pa (46% of the total pressure drop) is lost in the 10° enlargement and in the outlet section.

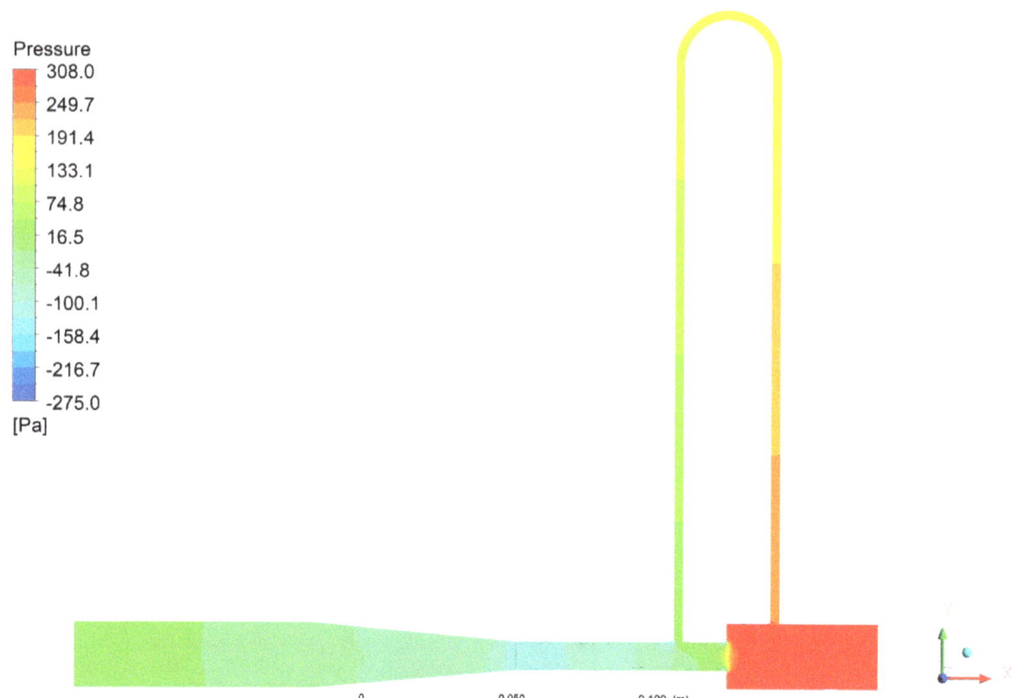

Figure 5. Pressure contour in the lithium domain.

Concerning the velocity field inside the lithium fluid, it must be highlighted that:

- The maximum lithium velocity is 1.3 m/s, and it is reached in the ¼" pipe (Figure 6).
- The average lithium velocity inside the capillary pipe is around 0.35 m/s, meaning that lithium takes about 1 s to fully travel it (in this way, the 1 Hz sampling frequency of the measurement device is kept).
- There is a long vortex near the outlet section due to the entrain of lithium from the capillary.

It must be highlight that the pressure drop inside the model is mainly concentrated in the 10° enlargement section due to the lithium entrainment from the capillary and the recirculation vortex produced (Figure 6). The only possible way to reduce this effect would be to increase the ¼" pipe length downstream of the capillary to reach a fully developed flow (at least tripling the current length), but this solution would increase both the pressure drop inside the ¼" pipe and the resistivity meter length, obtaining a negligible effect on the total pressure drop.

Regarding lithium velocity, the maximum lithium velocity inside the model (1.3 m/s) is not a concern for the IFMIF-DONES loop, where the average lithium velocity in the main loop is 5–6 m/s.

Regarding the temperature field (Figure 7), the lithium flowing inside the main pipe (Venturi suction) is kept at the inlet temperature, while the lithium flowing in the capillary loses about 2 °C because of the thermal losses in the box.

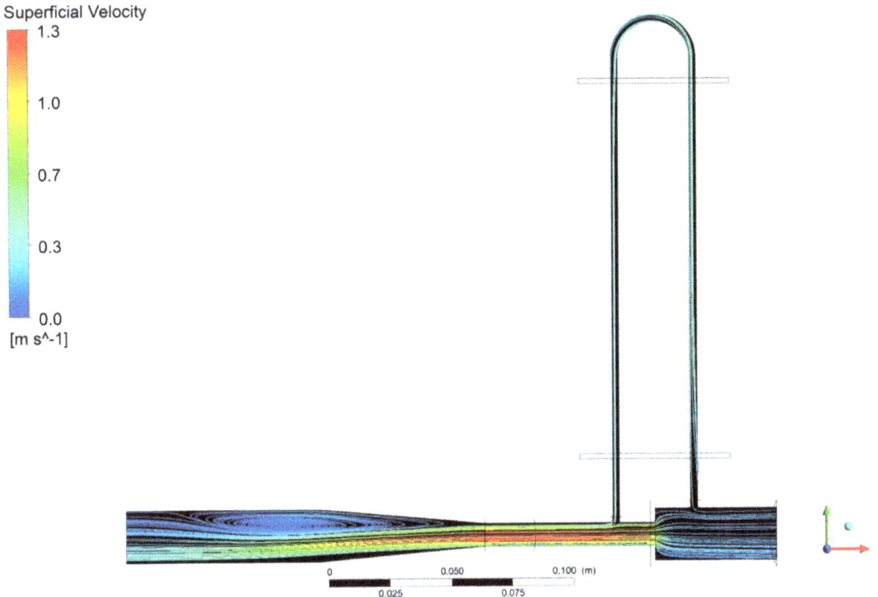

Figure 6. Velocity streamline of lithium flow inside the resistivity meter.

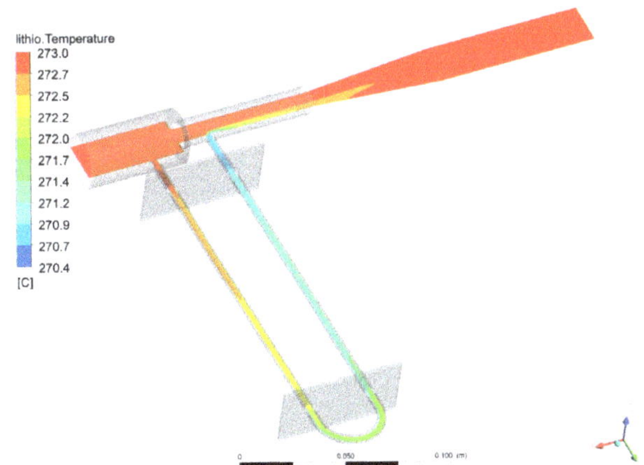

Figure 7. Temperature contour of lithium flow inside the resistivity meter.

The higher lithium cooling inside the capillary has two root causes: one is the reduced dimensions of the capillary pipe that give rise to a remarkable volume to surface ratio for the lithium flowing inside, the other is due to the two measuring plates along the capillary that act as thermal bridge enhancing the heat exchange like a finning surface (Figure 8).

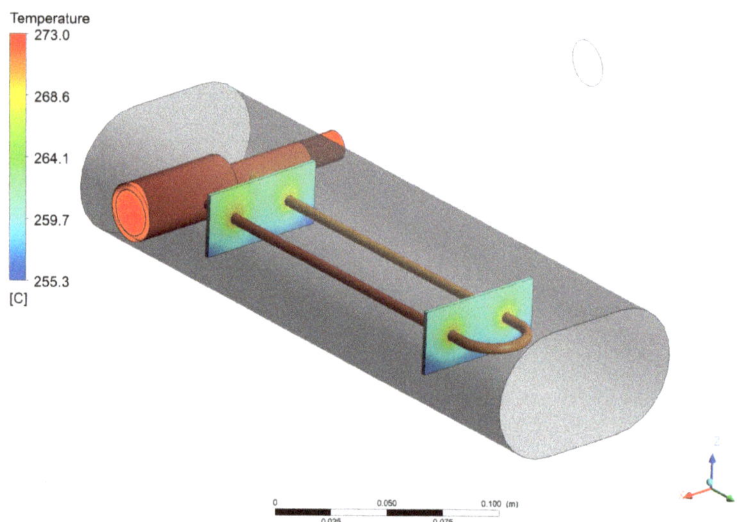

Figure 8. Temperature contour on the steel surface of the capillary.

Regarding the temperature field in the air domain, the temperature contour of Figure 9 highlights a high temperature gradient from 82 °C on the lower surface to about 170 °C on the upper one; this remarkable temperature gradient is a perfect example of how natural circulation works. In the upper air region, the buoyancy forces produce air convective vortices over the capillary (granting a temperature mixing), while the lower air volume under the capillary is stagnant because the cold box surface is under the hot capillary and natural circulation cannot take place.

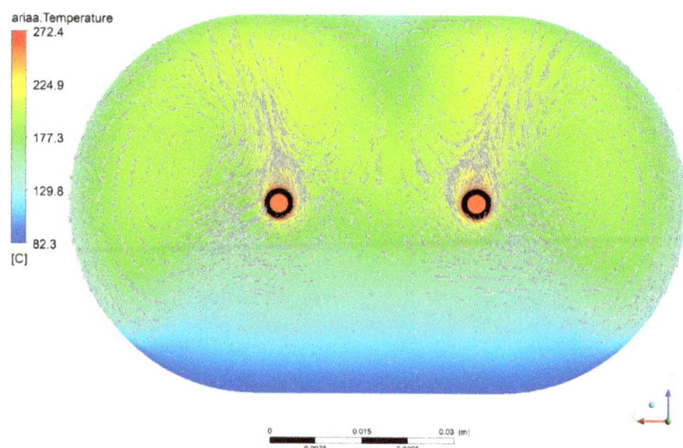

Figure 9. Temperature contour in the air domain (resistivity meter box).

Owing to its horizontal orientation, the capillary pipe is surrounded by a constant temperature of about 175 °C that avoids an uneven heat exchange along the capillary (one of the issues of the previous configuration characterized by a vertical capillary).

The heat losses, mainly concentrated in the upper surface of the box, are estimated as 18 W.

The small heat losses do not require any heating element inside the box; however, a heating cable or infrared device should be installed in the box for the pre-heating (start-up) phase of the facility to avoid lithium freezing inside the capillary.

5.2. Second Case: Boundary Conditions and Results

The boundary conditions imposed in this case are the following:

- Inlet lithium temperature: 330 °C.
- Inlet lithium flow rate: 0.108 L/s, corresponding to an average inlet velocity of 0.303 m/s.
- Relative pressure at the outlet surface: 0 bar.

The total pressure drop of the resistivity meter is about 777 Pa. Applying Equation (1) as in the previous case, it was found that (Figure 10):

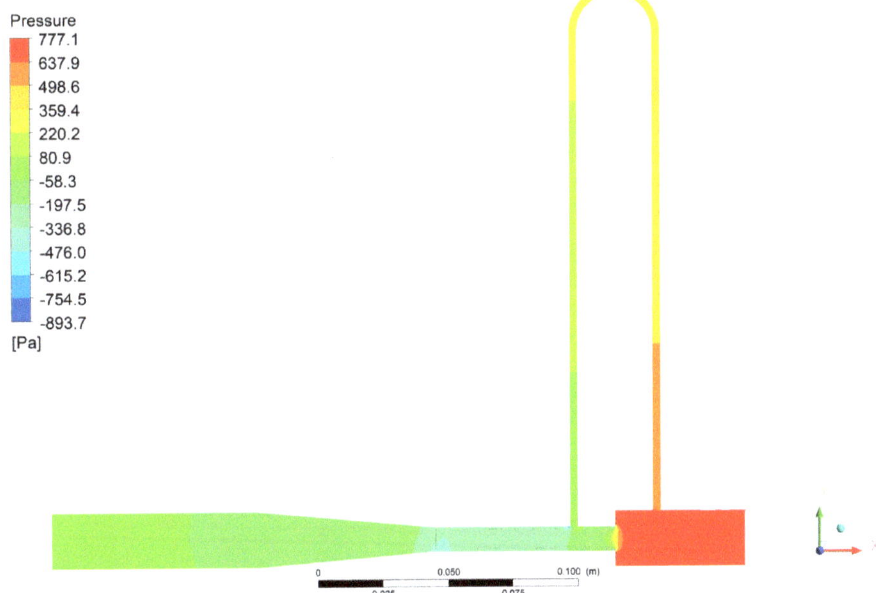

Figure 10. Pressure contour in the lithium domain (new IFMIF-DONES settings).

301.3 Pa (39% of the total pressure drop) is lost in the sudden flow area reduction;

- 139.4 Pa (18% of the total pressure drop) is lost in the ¼" pipe;
- 334.8 Pa (43% of the total pressure drop) is lost in the 10° enlargement and in the outlet section.

If we compare first case and second case pressure drop results, it is possible to preliminarily assess the hydraulic behavior of the component. A velocity/mass flow increase (from 0.182 m/s to 0.303 m/s) through the resistivity meter leads to a redistribution of the percentage pressure drop contributions inside the component. The sudden flow area reduction percentage increases (+3%) against the 10° enlargement one, perhaps because it is more sensitive to mass flow variation, while the distributed pressure loss in the ¼" pipe keeps the same percentage contribution.

Moving to the velocity field inside the lithium fluid (Figure 11), the folowing must be highlighted:

- The maximum lithium velocity inside the model is 2.1 m/s, and it is reached in the ¼" pipe;

- The average lithium velocity inside the capillary pipe is around 0.64 m/s (twice the value of the previous settings), determining half the residence time inside the capillary (0.5 s) if compared to the first case velocity;
- As in the previous case, there is a long vortex near the outlet section due to the entrain of lithium from the capillary.

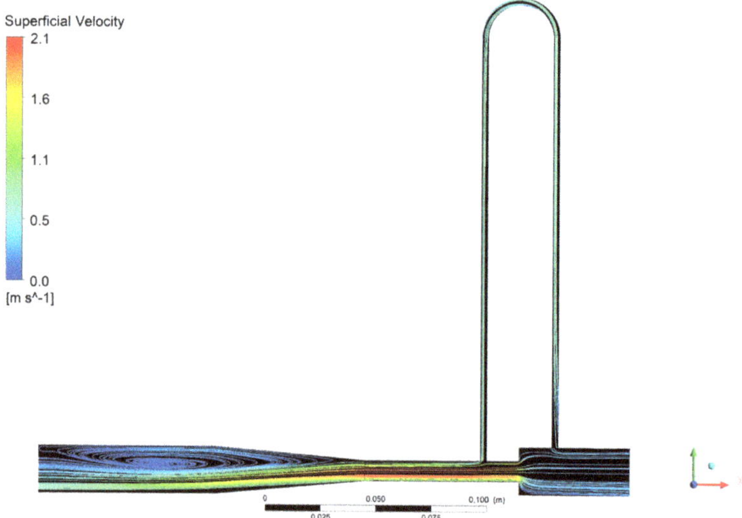

Figure 11. Streamline velocity of lithium flow (new IFMIF-DONES settings).

As previously mentioned, the 2.1 m/s peak velocity inside the ¼″ pipe is not a concern for erosion and corrosion issues for the IFMIF-DONES due to the low lithium velocity inside the capillary (0.64 m/s). This second case set-up shows a favorable effect on the measuring device granting a better lithium sampling; in fact, by keeping the 1 Hz sampling frequency of the current RM configuration, it is possible to always measure a different lithium sample in a normal operation phase.

Regarding the temperature field (Figure 12), lithium flowing inside the main pipe (Venturi suction) keeps its inlet temperature, while lithium flowing in the capillary experiences about 1.5 °C decrease because of the thermal losses in the box, 25% less if compared with previous results but with a higher velocity (mass flow) through the capillary. As for the previous case, in Figure 12 one can clearly observe, from the contour bands across them, the thermal effect of the steel plates on the flowing lithium enhancing thermal losses to the surrounding air.

As for the temperature field in the air domain, the temperature contour of Figure 13 highlights a high temperature gradient (104 °C, +18% increase), ranging from 96 °C on the lower surface to about 200 °C on the upper one. This higher gradient is mainly due to the higher lithium circulation temperature (imposed as a boundary condition) inside the instrument that generates a better natural circulation of heat transfer with the upper "cold" surface of the box; this phenomenon increases the upper air temperature, while it has only minor effect on the lower air temperature because no circulation takes place here. In conclusion, the thermal gradient increase inside the box is not symmetric but unbalanced towards the hot region.

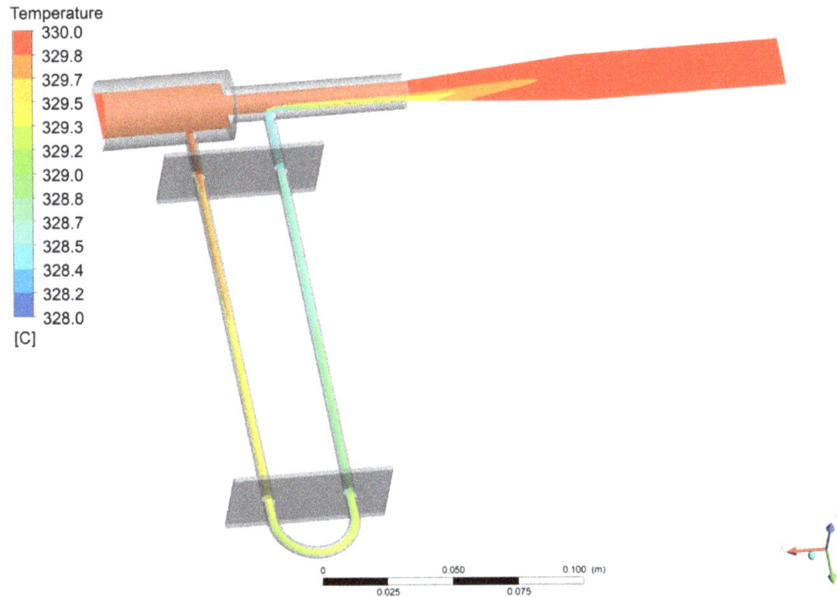

Figure 12. Temperature contour of lithium flow inside the RM pipe (new IFMIF-DONES settings).

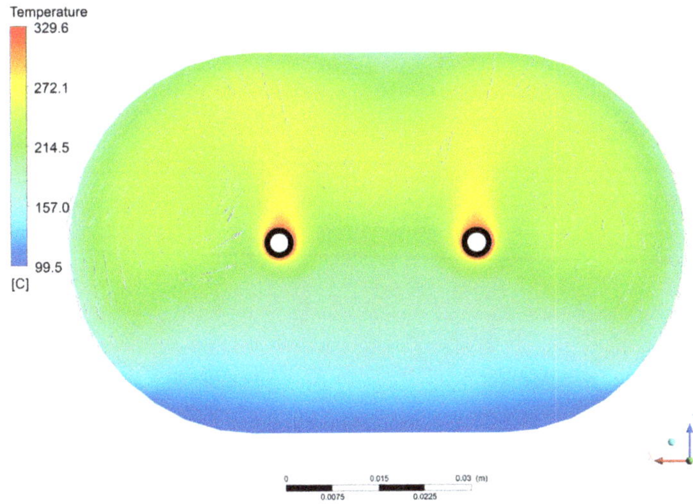

Figure 13. Temperature contour in the air domain inside the RM box (new IFMIF-DONES settings).

The heat losses mainly concentrated in the upper surface of the box (due to the natural circulation of air in the upper air region) are estimated as 19.4 W (7% more than the previous test case); this is due to the combination of three opposed effects (compared to the first case): the higher velocity inside the capillary, the lower density due to the higher inlet/circulation temperature, the lower temperature difference across the capillary.

6. Conclusions

The purity of liquid lithium flowing inside a loop pipe is of primary importance, especially in fusion applications such as the IFMIF-DONES plant, due to corrosion issues on contact surfaces. An appropriate device called the "Resistivity Meter" was developed by the University of Nottingham and procured by ENEA to monitor lithium impurities in Lifus 6 loop. Unfortunately, the device has shown some measuring issues during its operation that reduce its accuracy and introduce relevant background noise.

An upgrade strategy is proposed in this paper to overcome the aforementioned problems, and a CFD assessment of the fluid-dynamic behavior of the new layout was developed. Without fixed parameters for the boundary condition of the device, two limit cases have been considered: a low mass flow and low temperature case (first case) and a high mass flow and high temperature case (second case).

The CFD simulations of both the situations have demonstrated that it is possible to move lithium through the capillary measurement without using a dedicated electromagnetic pump upstream of the capillary. This will permit one to suppress the noxious noise introduced by the pump on the resistivity meter signal, which was largely evidenced with the old configuration. The resulting lithium speed inside the capillary is, in both cases, largely enough to completely replace all the lithium inside the capillary during the time span between two successive signal acquisitions (set as 1 s).

The comparison of the two simulation results was also adopted as sensitivity analysis to better understand the component behavior with different lithium mass flow and inlet temperature. Regarding mass flow, the comparison highlighted that the sudden flow restriction placed at the inlet is more sensitive to the mass flow variation if compared to the $10°$ enlargement placed at the outlet. As for temperature sensitivity, the comparison points out that an inlet temperature increase leads to a higher non-symmetric thermal gradient (+18%) inside the air box.

The simulations have additionally shown that during the operations there is no need to employ an external heating element to keep the proper lithium temperature inside the capillary.

Since the residence time of lithium will be very short (about 1 s or 0.5 s, depending on its flow rate), its thermal losses will be limited to about 2 or 1.5 °C and will be constant over time, provided a good thermal shielding of the RM box may be assured.

Nonetheless, an external heater will be necessary and employed in the following situations:

- During the resistance measurement of the steel capillary alone, i.e., when the capillary has not been filled by lithium yet. In this situation, in fact, due to the absence of lithium, it is necessary to employ an additional heating in order to measure the resistance variation of the steel at the relevant temperature;
- During the first filling with lithium to avoid its freezing inside the pipe;
- Similarly, after stopping lithium circulation inside the capillary, the lithium will probably remain trapped inside the capillary (due to its narrow section, a complete draining cannot be assured), and then it cools down at room temperature to a solid state. When the RM restarts, the external heaters will be necessary in order to melt this trapped lithium and avoid the incoming liquid fluid finding a solid clog blocking its circulation.

Author Contributions: Conceptualization, P.F.; software, R.M. and I.D.P.; data curation, R.M. and P.F.; writing—original draft preparation, R.M. and P.F.; writing—review and editing, R.M.; visualization, P.F.; supervision, D.B. and F.S.N., project administration, D.B.; funding acquisition, D.B. and F.S.N. All authors have read and agreed to the published version of the manuscript.

Funding: This work has been carried out within the framework of the EUROfusion Consortium and has received funding from the Euratom research and training programme 2014–2018 and 2019–2020 under grant agreement No 633053. The views and opinions expressed herein do not necessarily reflect those of the European Commission.

Institutional Review Board Statement: Not applicable.

Informed Consent Statement: Not applicable.

Data Availability Statement: Not applicable.

Acknowledgments: The authors wish to thank all the ENEA's technicians involved in the implementation and operation of the Lifus 6 experimental facility.

Conflicts of Interest: The authors declare no conflict of interest. The funders had no role in the design of the study; in the collection, analyses, or interpretation of data; in the writing of the manuscript, or in the decision to publish the results.

References

1. Natesan, K. Influence of non-metallic elements on the compatibility of structural materials with liquid alkali metals. *J. Nucl. Mater.* **1983**, *115*, 251–262. [CrossRef]
2. Chopra, O.K.; Smith, D.L. Influence of Temperature and lithium purity on corrosion of ferrous alloys in a flowing lithium environment. *J. Nucl. Mater.* **1986**, *141–143*, 584–591. [CrossRef]
3. Klueh, R.L. Oxygen Effects on the Corrosion of Niobium and Tantalum by Liquid Lithium. *Metall. Trans.* **1974**, *5*, 875–879. [CrossRef]
4. Xu, Q.; Kondo, M.; Nagasaka, T.; Muroga, T.; Yeliseyeva, O. Effect of chemical potential of carbon of JLF-1 steel in a static lithium. *J. Nucl. Mater.* **2009**, *394*, 20–25. [CrossRef]
5. Tsisar, V.; Kondo, M.; Xu, Q.; Muroga, T.; Nagasaka, T.; Yeliseyeva, O. Effect of nitrogen on the corrosion behavior of RAFM JLF-1 steel in lithium. *J. Nucl. Mater.* **2011**, *417*, 1205–1209. [CrossRef]
6. Knaster, J.; Favuzza, P. Assessment of corrosion phenomena in liquid lithium at T < 873 K. A Li(d,n) neutron source as case study. *Fusion Eng. Des.* **2017**, *118*, 135–141. [CrossRef]
7. Knaster, J.; Ibarra, A.; Abal, J.; Abou-Sena, A.; Arbeiter, F.; Arranz, F.; Arroyo, J.M.; Bargallo, E.; Beauvais, P.-Y.; Bernardi, D.; et al. The accomplishment of the Engineering Design Activities of IFMIF/EVEDA: The European–Japanese project towards a Li(d,xn) fusion relevant neutron source. *Nucl. Fusion* **2015**, *55*, 086003. [CrossRef]
8. Ibarra, A.; Arbeiter, F.; Bernardi, D.; Cappelli, M.; García, A.; Heidinger, R.; Krolas, W.; Fischer, U.; Martin-Fuertes, F.; Micciché, G. The IFMIF-DONES project: Preliminary engineering design. *Nucl. Fusion* **2018**, *58*, 105002. [CrossRef]
9. Arbeiter, F.; Baluc, N.; Favuzza, P.; Groschel, F.; Heidinger, R.; Ibarra, A.; Knaster, J.; Kanemura, T.; Kondo, H.; Massaut, V.; et al. The accomplishments of lithium target and test facility validation activities in the IFMIF/EVEDA phase. *Nucl. Fusion* **2018**, *58*, 015001. [CrossRef]
10. Favuzza, P.; Antonelli, A.; Furukawa, T.; Groeschel, F.; Hedinger, R.; Higashi, T.; Hirakawa, Y.; Iijima, M.; Ito, Y.; Kanemura, T.; et al. Round Robin test for the determination of nitrogen concentration in solid lithium. *Fusion Eng. Des.* **2016**, *107*, 13–24. [CrossRef]
11. Baley, A.S.; Gregory, D.H.; Hubberstey, P. *Development of a Monitoring System*; Technical Note No. 4; School of Chemistry, University of Nottingham: Notthingham, UK, 2004.
12. Creffrey, G.K.; Down, M.G.; Pulham, R.J. Electrical Resistivity of Liquid and Solid Lithium. *J. Chem. Soc. Dalton Trans.* **1974**, *21*, 2325–2329.
13. Hubberstey, P. Dissolved nitrogen in liquid-lithium a problem in fusion reactor chemistry. In Proceedings of the International Conference on Liquid Metal Engineering and Technology, Oxford, UK, 9–13 April 1984; pp. 85–91.
14. Barker, M.G.; Hubberstey, P.; Dadd, A.T.; Frankham, S.A. The interaction of chromium with nitrogen dissolved in liquid lithium. *J. Nucl. Mater.* **1983**, *114*, 143–149. [CrossRef]
15. Hubberstey, P., Roberts, P.G. Corrosion chemistry of vanadium in liquid lithium containing dissolved nitrogen. *J. Nucl. Mater.* **1988**, *1555157*, 694–697. [CrossRef]
16. Favuzza, P.; Aiello, A.; Tincani, A.; Muzzarelli, M. *Engineering Design Report of Lifus 6 Purification System*; EC H2020 EUROfusion Project; IFMIF/EVEDA, WPENS Deliverable LF 4.4.1; European Commission: Brussels, Belgium, 2014.
17. Favuzza, P.; Mannori, S. *Acceptance Test Report of the Lifus 6 Purification System*; EC H2020 EUROfusion Project; IFMIF/EVEDA, Deliverable LF 4.4.3; European Commission: Brussels, Belgium, 2015.
18. Favuzza, P. *Final Validation Report of the Lifus 6 Purification System*; EC H2020 EUROfusion Project; IFMIF/EVEDA Deliverable LF 4.5.2; European Commission: Brussels, Belgium, 2016.
19. Ansys Inc. *ANSYS User Manual Version. 19.2*; Ansys Inc.: Canonsburg, PA, USA, 2019.
20. Ansys Inc. *Ansys CFX Reference Guide*; Release 19.2; USA, Ansys Inc.: Canonsburg, PA, USA, 2019.
21. Menter, F.R. Two-equation eddy-viscosity turbulence models for engineering applications. *AIAA J.* **1994**, *32*, 1598–1605. [CrossRef]
22. Ohse, R.W. *Handbook of Thermodynamic and Transport Properties of Alkali Metals*; International Union of Pure and Applied Chemistry Chemical Data, Series No. 30; Blackwell Scientific Publishing Ltd.: Oxford, UK, 1985; p. 987.

Article

Wake Shape and Height Profile Measurements in a Concave Open Channel Flow regarding the Target in DONES

Björn Brenneis [1,*], Sergej Gordeev [1], Sebastian Ruck [1], Leonid Stoppel [2] and Wolfgang Hering [1]

[1] Institute for Neutron Physics and Reactor Technology, Karlsruhe Institute of Technology, 76344 Eggenstein-Leopoldshafen, Germany; sergej.gordeev@kit.edu (S.G.); sebastian.ruck@kit.edu (S.R.); wolfgang.hering@kit.edu (W.H.)

[2] Institute for Thermal Energy Technology and Safety, Karlsruhe Institute of Technology, 76344 Eggenstein-Leopoldshafen, Germany; leonid.stoppel@kit.edu

* Correspondence: bjoern.brenneis@kit.edu

Abstract: Wakes appearing downstream of disturbances on the surface of a water flow in a concave open channel were examined experimentally. The investigated channel geometry was similar to the liquid lithium target in DONES (Demonstration fusion power plant Oriented NEutron Source). The objective of the measurements was to analyze the effect of a disturbance on the downstream layer thickness. For measuring the height profiles in the channel, an optical measurement system based on laser triangulation was developed. It was shown that the wake of the undisturbed flow emerged from the nozzle corner, which was in accordance with analytical solutions. For sufficiently large disturbances at the nozzle edge, the height profiles located downstream showed symmetrical minima and maxima on both sides of the disturbance. The wake depth strongly depended on the diameter and penetration depth of the disturbance, as well as the circumferential position in the channel, which yields to a critical wake depth of one millimeter for the lithium target in DONES.

Keywords: wakes; open channel flow; experimental methods; DONES

1. Introduction

For fusion power plant technology, specific materials withstanding the harsh environment of high-energy neutron fluxes are necessary [1]. The neutron irradiation facility for fusion materials, DONES, is intended to be built for this purpose. In DONES, a deuteron beam (125 mA, 40 MeV) hitting the lithium target produces a high-energy neutron flux. The target is a liquid lithium free surface flow on a concave backplate, as depicted in Figure 1. In order to contain the heat released by the beam within the liquid lithium (Bragg peak at 19 mm [2]) and to avoid an introduction of the heat in the backplate, a stable configuration of the free surface flow with a setpoint layer thickness of 25 ± 1 mm is crucial. Stable wave structures, so-called wakes, which occur from accumulated impurities at the nozzle edge [3], can cause a local decrease in the layer thickness of more than one millimeter.

The free surface of the lithium flow without wakes is sufficiently stable [4]. Wakes occuring next to the side walls do not reach the beam footprint (Figure 1) and are therefore not critical [5]. Wake structures caused by stationary lithium droplets sticking at the nozzle were observed by Kondo et al. [3]. Their experimental setup was a horizontal 70 mm wide open channel flow at 15 m/s. The wake structures were temporally and spatially stable and could be described by analytical Kelvin wake equations presented in Section 2, neglecting the gravitational force. The observed area was up to 50 mm downstream of the nozzle.

In a subsequent work by Kondo et al. [6], they used pattern projection to measure the height profile of the liquid lithium flow. The measurements were limited to an area of 5×3 mm. They observed a relatively smooth surface up to a velocity of 6 m/s. At higher flow velocities, the amplitude of the wake increased.

Citation: Brenneis, B.; Gordeev, S.; Ruck, S.; Stoppel, L.; Hering, W. Wake Shape and Height Profile Measurements in a Concave Open Channel Flow regarding the Target in DONES. *Energies* **2021**, *14*, 6506. https://doi.org/10.3390/en14206506

Academic Editors: Alessandro Del Nevo and Marica Eboli

Received: 3 August 2021
Accepted: 6 October 2021
Published: 11 October 2021

Publisher's Note: MDPI stays neutral with regard to jurisdictional claims in published maps and institutional affiliations.

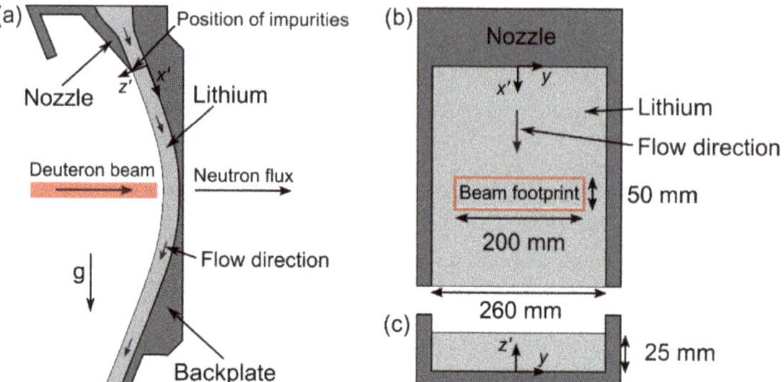

Figure 1. Sketch of the liquid lithium target in DONES. (**a**) Side view on the cross-section through the channel center. (**b**) Front view. (**c**) Cross-section perpendicular to the flow direction at the nozzle outlet. The coordinate axes are y for the channel width, x' parallel to the channel floor in streamwise direction and z' perpendicular to the channel floor. The origin is at the nozzle outlet.

Originally, wakes were examined in the context of ship waves, due to their impact on the wave drag [7]. Kelvin wake equations [8] describe the shape of wakes analytically. The wake shape is defined by the angle of the wave front with maximum amplitude. Measurements showed that the wake angle depends on the Froude number [7,9–11]. Moisy and Rabaud [10] systematically examined wakes behind partially immersed cylinders moving through water at different velocities. The wake angle depended on the Froude number, Bond number and minimal phase velocity. Pethiyagoda et al. [11] showed a dependency of the wake angle on small Froude numbers. Therefore, the originally constant Kelvin angle is limited to Froude numbers around one. Another explanation for these phenomena of smaller wake angles is given by Noblesse et al. [12] considering an interference of multiple wakes.

The deformation of the water surface directly at the cylinder due to run-up upstream and depression downstream was analyzed by Ageorges et al. [13] and Chaplin and Teigen [14]. Koo et al. [15] analyzed the flow past a surface piercing cylinder numerically. They also evaluated the height profile downstream near the disturbance.

An integrated approach was applied to examine the development of wakes downstream of a disturbance in a concave open channel flow. Wake shape and profile measurements were conducted in the test facility FIDES ("Facility for experimental Investigation of a 3D FreE Surface flow") presented in Section 3. The operating point was selected using the analytical equations of Kelvin wakes (Section 2). Therefore, the measured wake shapes (Section 4.3) can be transferred quantitatively to the lithium flow and compared to other published wake shape results. The observed wake profiles in water (Section 4.2) provide a qualitative insight in the development of the wake amplitude along the channel in lithium.

2. Analytical Wake Equations

Waves behind a moving disturbance in calm water or behind a stationary disturbance in a steady flow can be described by the following equations [8]:

$$x' = p \cdot \cos(\theta) - \frac{dp}{d\theta} \cdot \sin(\theta)$$
$$y = p \cdot \sin(\theta) + \frac{dp}{d\theta} \cdot \cos(\theta). \tag{1}$$

The disturbance moves along the negative x'-axis or water streams around the stationary disturbance in the positive x' direction with the velocity, u. A line of length p connects the disturbance with points on the constant phase line with the angle θ in between p and x'.

The points with a constant phase are at the position (x',y). The phase velocity $c = u \cdot \cos(\theta)$ is defined by the dispersion relation [16]:

$$c = \sqrt{\left(\frac{\lambda g}{2\pi} + \frac{2\pi\sigma}{\lambda\rho}\right) \cdot \tanh(2\pi h/\lambda)}. \tag{2}$$

The wavelength is defined as $\lambda = p/n$. The variable n can be chosen as a multiple of 0.5. Wave troughs are given by a positive integer value of $n = 1, 2, 3\ldots$; otherwise, wave crests are described. In addition, the density ρ and surface tension σ of the fluid have an influence.

With regard to the flow in DONES (FIDES), it is assumed that the wavelength is much smaller than the layer thickness of $h = 25$ mm (10 mm). Therefore $\tanh(h/\lambda) \approx 1$, which causes an error smaller than 3% for $\lambda < 3.6\,h$ [17]. Additionally, gravity, g, is substituted by the centrifugal acceleration, u^2/R [3]. This is justified, because gravity is mostly in the streamwise direction and, therefore, the centrifugal acceleration is more than one order of magnitude bigger considering the perpendicular acceleration on the free surface. Without any assumptions about the dominant forces (surface tension, σ, and centrifugal forces) the equations have two solutions, which depend on the curvature radius, R, the Weber number, $We = u\sqrt{\rho L/\sigma}$, and the Froude number, $Fr = \sqrt{R/L}$. The characteristic length, L, is either the nozzle height, h, or for the flow with the cylindrical disturbance, the diameter, D, of the cylinder. For different flow facilities with the same curvature radius, the equations have the same wake curves as long as the product $We \cdot Fr$ is constant, which is equivalent to $2 \cdot U^2$, where $U = u/c_{min}$ is the velocity ratio and $c_{min} = \left(4u^2\sigma/(R\rho)\right)^{1/4}$ is the minimal phase velocity. The flow velocity for the reference lithium flow was $u = 15\text{--}20$ m/s (velocity range in DONES) with a constant curvature radius of $R = 250$ mm for the vertical concave channel, which results in the velocity range $u = 4.7\text{--}6.2$ m/s in water (20 °C).

The solution of Equations (1) and (2) for $n = 1$, $u = 15$ m/s in lithium without the assumption of a dominant force is shown in Figure 2. The blue dashed line in the diagram represents the capillary wake line dominated by the surface tension and the red lines are the centrifugal wake dominated by centrifugal forces. Considering the analytical solutions, it was concluded that wakes from the nozzle corners do not reach the beam footprint [3] and, thus, the accumulation of impurities at the nozzle edge are the main source for a film thickness decrease. For the prediction of the wake shape, the solution for $n = 0.5$ was used, which describes the position of a wave crest and was compared to the experimental results.

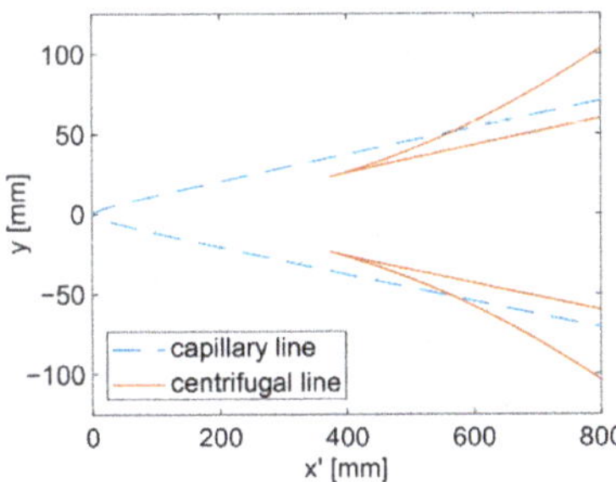

Figure 2. Solution of the Kelvin wake Equations (1) and (2) for $n = 1$ and $u = 15$ m/s in lithium. The blue dashed line (capillary line) represents the wake line dominated by surface tension. The red solid line (centrifugal line) is the wake line dominated by the centrifugal acceleration.

3. Experimental Setup

The experiments were conducted at the water flow test facility FIDES at KIT (Karlsruhe Institute of Technology). The general layout of the facility is shown in Figure 3. A pump provided a constant water mass flow to the top end of the test section, which was measured using a magnetic inductive flow meter. The test section made of plexiglass was geometrically similar to the layout of the liquid lithium target in DONES, as depicted in Figure 4. The water in the test section flowed through a flow straightener into a two-stage nozzle with an outlet cross-section of 100×10 mm. A maximum flow velocity at the nozzle outlet of $u = 20$ m/s could be reached. Downstream of the nozzle, water flowed through the concave open channel with a curvature radius of $R = 250$ mm before it returned to the separator.

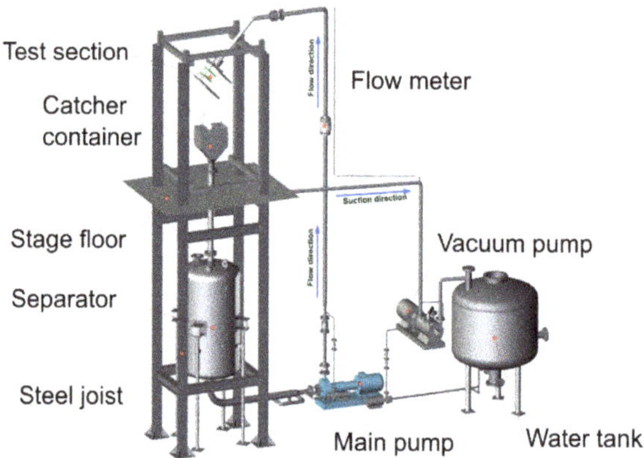

Figure 3. Scheme of the FIDES test facility.

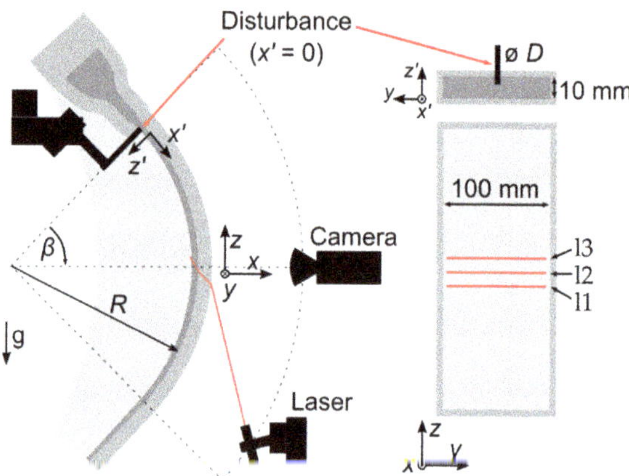

Figure 4. Sketch of the FIDES test section showing the position of the measurement system (camera and laser) and the position of the disturbance with the diameter, D. The test section is a concave open channel with a curvature radius of $R = 250$ mm. The measuring position is described by the angle β. On the right side, the view on the back wall of the test section with the three reflected lines l1, l2 and l3 is shown.

Stable wave structures were generated by an artificial disturbance at the nozzle outlet. This disturbance was a metal cylinder with different diameters D (1.2, 2, 3 mm). The penetration depth of the cylinder into the water could be varied by a linear stage. A camera ("Basler ac2040-55uc" camera [18] and "KOWA LM16JC3M2" objective [19]), placed as depicted in Figure 4 with a back wall distance of 213 mm, was used to observe the wake structures with an exposure time of 1 s. To measure the layer thickness, a laser line was projected in the camera's field of view. The camera observed three reflected lines from the channel outside wall (l1), the channel floor (l2) and the water surface (l3). Figure 4 shows the arrangement of the disturbance, the camera and the laser, as well as the view on the channel back wall with the three reflected lines. To measure the height profile at different positions, the camera and laser could be moved by keeping a constant distance to the back wall. The relative position of the laser and camera remained constant. The measured circumferential positions were $\beta = 35°, 40°, 45°$ and $50°$ downstream of the nozzle. The vertical channel position corresponded to $\beta = 45°$.

MATLAB [20] was used for the image evaluation. Three horizontal areas were defined where the laser lines were expected to be. Those were further subdivided into vertical segments. In every segment, the laser lines were detected by defining threshold values for the pixel intensity. Therefore, the pixel positions of the three lines in the image were known. By using pictures of measuring tapes at the channel outside wall ($R_0 = 270$ mm), the pixel position could be assigned to a circumferential position at the outside wall (the nozzle outlet was zero). The layer thickness was calculated via triangulation with the known circumferential position, the camera position, the refractive indices and the back wall thickness of 20 mm. Laser line l1 was the reflection from the back wall. With the camera position, l2 had to be corrected via refraction through the 20 mm back wall to obtain the position of the laser line on the channel floor. Using the known position of l1 on the back wall, l2 on the channel floor and the laser line through the back wall, the incidence angle on the channel floor and the ray further through the water could be calculated. With the observed reflection on the water surface (l3), the viewing line of the camera to the water surface was calculated. The intersection point of the viewing line and the laser line resulted in the point on the water surface. The difference between the channel floor curvature radius and the distance of the point on the water surface to the center point of the curvature was the output of the algorithm (the thickness of the water flow at this location).

To validate the measurement system, the layer thickness in the area 0–2 mm downstream of the nozzle was measured. The expected thickness was the nozzle height of 10 mm. The average layer thickness along the channel width was 9.99 mm, with a standard deviation of <0.09 mm. In addition, undisturbed profiles were measured with a chromatic confocal sensor (KEYENCE CL-3000 with CL-LP070 optical head [21]). The mean deviation of the undisturbed absolute height profile in between the side wall wakes measured by the two sensors for $\beta = 45°$ ($u = 5$ m/s) was smaller than 0.1 mm.

4. Experimental Results

The presented analytical solutions of Kelvin wakes in Section 2 were used to set the boundary conditions for the tests in water and to be able to transfer the results to DONES. Under the assumption of a constant product $Fr \cdot We$, the velocity range of $u = 15$–20 m/s in lithium (250 °C) is similar to the velocity range of $u = 4.7$–6.2 m/s in water (20 °C). In this range, the wake shape in water is similar to the shape in a lithium flow with the same curvature radius of $R = 250$ mm.

4.1. Undisturbed Height Profile Measurements

In the first step, the layer thickness of the undisturbed water flow in FIDES was analyzed. Figure 5 shows the averaged measured height profile at $u = 5$ m/s, $\beta = 45°$ downstream of the nozzle. In the images of the reflected laser lines, 1 px in the height was equivalent to a height change of 0.043 mm. The standard deviation of the measured undisturbed profile was <0.11 mm. In addition, the calculated systematic error of the absolute height measurements was 0.13 mm. The considered errors were the uncertainty of the camera distance, the measuring tape position and the uncertainty of the threshold value in the laser line detection algorithm. In y-direction, the point to point distance was 0.729 mm, across which the measured heights were averaged.

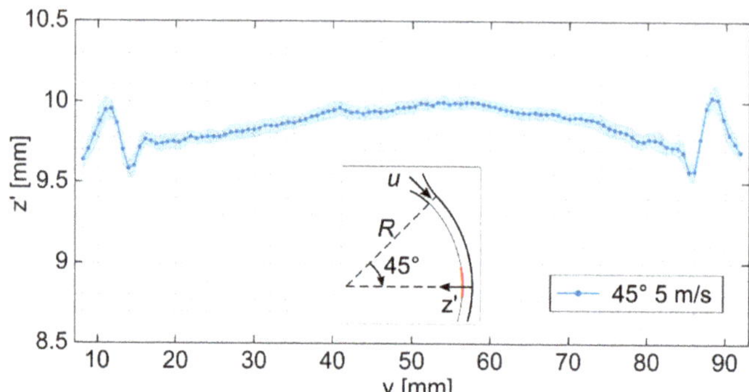

Figure 5. Height profile of the undisturbed flow at $u = 5$ m/s, $\beta = 45°$ downstream of the nozzle. The standard deviation is represented as the area behind the line.

The profile of the undisturbed flow in Figure 5 showed a slight bulge in the center of the channel. Near both side walls, the wakes that occurred from the nozzle corners were visible as a maximum and minimum. The measured maximum had a side wall distance of 11.82 mm, which corresponds to the first maximum ($n = 0.5$) of the centrifugal Kelvin wake line (Section 2). The first 8 mm from both side walls could not be measured. Laser light scattered by the side wall prevented the measurements. Directly at the wall, the film thickness reached another maximum due to wetting. Between the wake maximum and the wall, another minimum was indicated. The average layer thickness of the measured profile shown in Figure 5 was 9.88 mm.

In case of the undisturbed flow, measurements were performed at different flow velocities and circumferential positions. The averaged layer thickness at 45° showed a positive correlation to the flow velocity (Figure 6a). The position of the wake maximum stayed constant. Due to the increasing layer thickness at higher flow velocities, the measured minimum moved slightly in the direction of the nearest side wall. For an increasing nozzle distance and constant flow velocity, the mean layer thickness decreased slightly (Figure 6b). The distance of the side wall wakes to the nearest wall increased with increasing nozzle distances. The variation of the wakes' side wall distance, as well as the constant position of the maximum at different flow velocities, is described by the analytic Kelvin wake equations in Section 2.

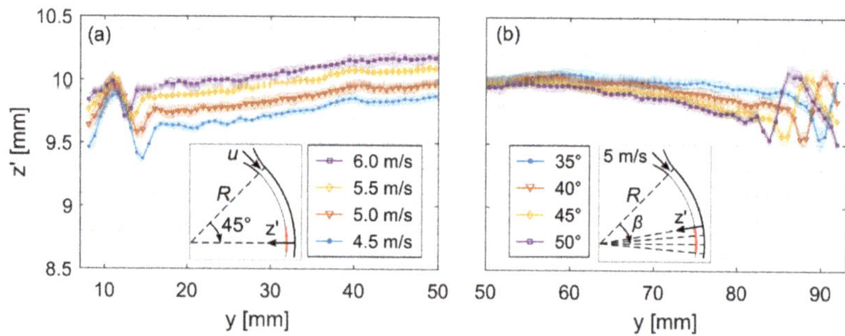

Figure 6. (a) Height profiles at the vertical channel position $\beta = 45°$ at different flow velocities u. (b) Height profiles at $u = 5$ m/s at different circumferential positions (β).

4.2. Disturbance Induced Wake Profile Measurements

Side wall wakes do not extend into the area of the beam footprint in DONES; therefore, the wake structures behind impurities accumulating at the nozzle edge were of particular interest. To study the effect of a disturbance at the nozzle edge, a defined disturbance was placed at the nozzle outlet. The disturbances were realized using metal cylinders of different diameters D (1.2, 2, and 3 mm), which could be moved into the flow by a linear stage. The investigated penetration depths were 0.75, 1.25, 1.5, and 2 mm. Penetration depths were measured from the point when the cylinder first touched the water surface. Besides the parameter variation of the disturbance, the flow velocity and circumferential position of the measurements were varied, as in the undisturbed measurements. The height profiles behind the disturbance were evaluated using their height change in comparison to their corresponding undisturbed profile, which was defined as the relative height profile ($\Delta z'$).

At first, the penetration depth for the diameter $D = 1.2$ mm, which caused a wake minimum larger than 1 mm, $\beta = 45°$ downstream of the nozzle at $u = 5$ m/s, was analyzed. Up to a penetration depth of 0.75 mm, no significant height change was observed. For increasing penetration depths, a clearly visible wave crest and trough were measured on both sides behind the disturbance. The critical wake depth of 1 mm was exceeded with a 2 mm penetration depth. The other way to increase the disturbance size was to use a larger diameter. By increasing the diameter, the penetration depth which caused a critical wake depth decreased. For the disturbance diameter of $D = 2$ mm, the critical penetration depth was reached at 1.25 mm. Further increasing the diameter to $D = 3$ mm had a much smaller impact on the wake amplitude than the change from $D = 1.2$ to 2 mm.

Varying the circumferential position of the measurements showed a decreasing wake depth in the streamwise direction (Figure 7). The last parameter tested for its effect on the wake depth was the flow velocity. In the considered velocity range, no significant height change variation due to the flow velocity could be detected.

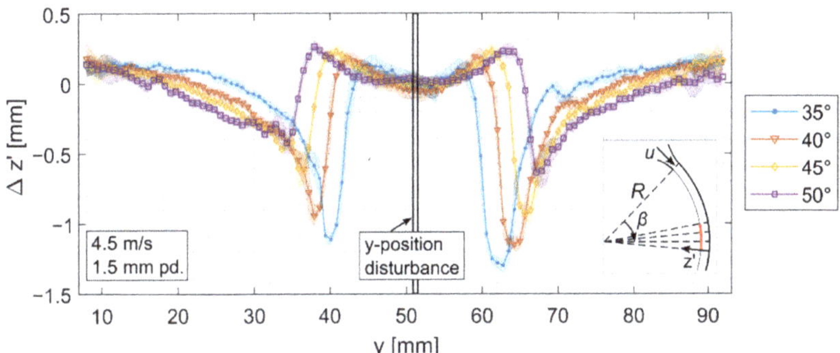

Figure 7. Relative height profile at $u = 4.5$ m/s for four different circumferential positions (β). The disturbance had a diameter of $D = 1.2$ mm and a penetration depth of 1.5 mm.

4.3. Wake Angle Evaluation

In addition to the wake depth, the wake shape could be evaluated using the profile measurements. The wake shape is defined by the position of the maximum wave amplitude. In the profile measurements, the maximum wake amplitude was the dominant minimum downstream and on each side of the disturbance. The wake shape is of particular interest to compare the measured wakes with the analytical results in Section 2 and other published wake angles. For the comparisons, the measurement position, β, was converted to a circumferential position at the water surface with an assumed curvature radius of 240 mm.

Only the circumferential position had an impact on the position of the wake minimum (Figure 7). This was consistent with the analytical results. The position of the dominant wake minimum was between the innermost capillary and outermost centrifugal line. The solution for $n = 0.5$ represents the position of the first maximum. Figure 8 shows the position of the analytical results $\beta = 35°$ downstream of the nozzle in comparison to the measured profile. From the definition of the analytical solution, there are no further solutions defined for $n < 0.5$. Nevertheless, the experimentally measured position of the minimum was described by the centrifugal solution with $n = 0.25$ (vertical dashed line in Figure 8).

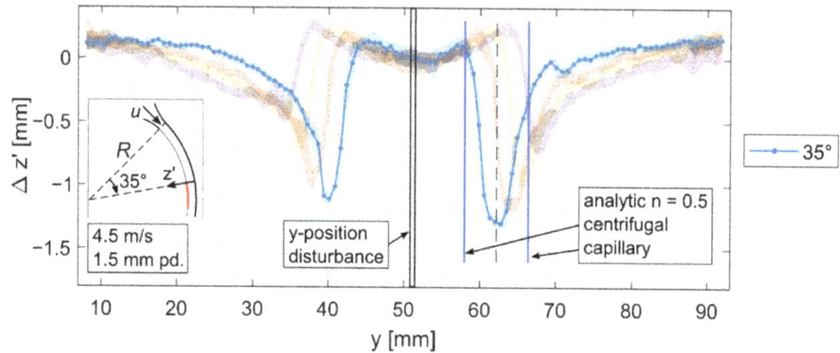

Figure 8. Comparison of the measured wake position with the predicted position of the analytical solutions $\beta = 35°$ downstream of the nozzle. The profiles at the other positions of β from Figure 7 are still indicated in the background as reference (red, yellow and purple).

Following Moisy and Rabaud [10], the Froude number, the Bond number, $Bo = L/\lambda_c$ and the velocity ratio, U, were used to describe the wake angle. The capillary length was $\lambda_c = 2\pi\left(\sigma R/\left(\rho u^2\right)\right)^{1/2}$. Similar to the Froude number, the gravity was substituted by the centrifugal acceleration.

Figure 9 shows the wake angle plotted against the velocity ratio. The horizontal dashed line is the Kelvin angle, which is present at small velocity ratios. A velocity ratio larger than one is needed to create a stable wave pattern. The lower diagonal line is the angle of the inner cusp of the analytical curves (Figure 2). Compared to the experimental results of Moisy and Rabaud [10], this line corresponds to small Bond numbers ($Bo \leq 0.1$). The upper diagonal line fits their measured values for $Bo \approx 0.7$. In the case of the measurements in FIDES, the velocity ratio was $U = 11.5$ to 13.27 and only depended on the flow velocity. The Bond number ranged from $Bo = 0.2$ to 0.66 and depended both on the flow velocity and on the diameter of the disturbance. The measured wake angles in the FIDES facility tended to be in the expected area.

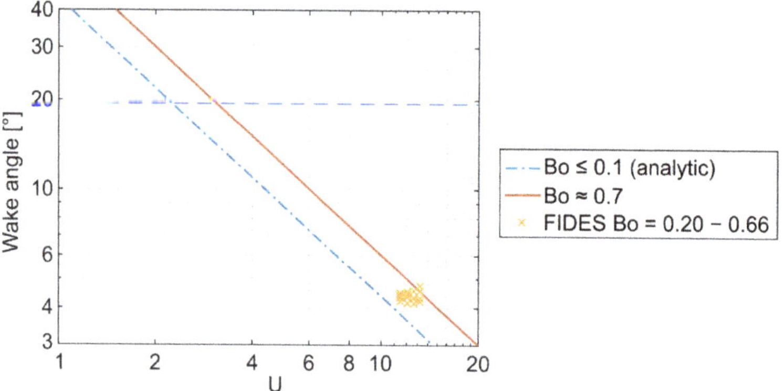

Figure 9. Wake angle plotted against velocity ratio U following Moisy and Rabaud [10]. The horizontal dashed line is the Kelvin angle $19.47°$. The lower diagonal line is the wake angle curve corresponding to small Bond numbers ($Bo \leq 0.1$). The upper diagonal is a fit for $Bo \approx 0.7$. Measured wake angles in FIDES are represented by the yellow crosses.

In the experiments, the flow velocity and the diameter of the disturbance were varied. Increasing the flow velocity caused a larger Bond number and velocity ratio. Due to the diagram, larger Bond numbers at a constant velocity ratio cause a bigger wake angle. An increasing velocity ratio caused smaller angles with constant *Bo*. The experiments did not show any dependency of the flow velocity on the wake angle, which could be caused by the given channel geometry. The increasing flow velocity caused an increase in the centrifugal acceleration.

5. Summary and Conclusions

Measurements on a water flow similar to DONES were performed in order to study the effect of disturbances at the nozzle outlet on the downstream layer thickness of the free surface flow. The height profile in the concave channel was examined by a self-developed optical measurement system based on laser triangulation. The undisturbed flow showed that the wake emerged from the nozzle corners, which was in accordance with analytical solutions.

Symmetrical minima and maxima of the height profile on both sides downstream of a sufficiently large disturbance at the nozzle edge were found. The diameter and penetration depth of the disturbance defined the disturbance size. Both correlated positively with the wake depth and, therefore, the critical wake depth depends strongly on those two parameters. Up to a specific size of the disturbance, changes in the height profile were within the measurement uncertainty. In addition to the disturbance size, the circumferential position affected the wake depth. For an increasing nozzle distance, the wake depth decreased. The critical wake depth at the vertical position was reached for a disturbance with a diameter of $D = 1.2$ mm and a penetration depth of 1.5–2 mm. Concerning the prediction of the wake angle or position of the stable minimum, the constant phase lines of the first wave crests for $n = 0.5$ of the analytical Kelvin wake solutions were the best reference. The curves framed the measured minimum and considered the increasing angle in flow direction and the measured independence of the disturbance size and flow velocity. Nevertheless, there was no direct representation of the wake minimum in the Kelvin wake equations.

The performed height profile measurements of wakes in water provide a first insight into wake structures dominated by centrifugal acceleration at channel positions relevant for DONES. In contrast to gravity-dominated wakes, no impact of the varying flow velocity on the wake angle was measured. The wake shape can be transferred quantitatively to lithium due to $We \cdot Fr$ similarity. The actual height profiles provide a qualitative insight into wake structures on lithium flowing in a concave open channel. It is assumed that impurities at the nozzle in DONES accumulate over time. Therefore, a critical wake is indicated through an increasing wake amplitude. The dependency of the height profile on the circumferential position needs to be considered if the film thickness in the channel is predicted from one measured profile.

Author Contributions: Investigation, B.B. and S.G.; methodology, L.S.; supervision, S.R. and W.H.; writing—original draft, B.B.; writing—review and editing, S.R. and W.H. All authors have read and agreed to the published version of the manuscript.

Funding: This work has been carried out within the framework of the EUROfusion Consortium and has received funding from the Euratom research and training program 2014–2018 and 2019–2020 under grant agreement No 633053. The views and opinions expressed herein do not necessarily reflect those of the European Commission, Fusion for Energy, or of the authors' home institutions or research funders.

Acknowledgments: We acknowledge support by the KIT-Publication Fund of the Karlsruhe Institute of Technology.

Conflicts of Interest: The authors declare no conflict of interest.

References

1. Knaster, J.; Moeslang, A.; Muroga, T. Materials research for fusion. *Nat. Phys.* **2016**, *12*, 424–434. [CrossRef]
2. Ida, M.; Nakamura, H.; Nakamura, H.; Nakamura, H.; Ezato, K.; Takeuchi, H. Thermal-hydraulic characteristics of IFMIF liquid lithium target. *Fusion Eng. Des.* **2002**, *63–64*, 333–342. [CrossRef]
3. Kondo, H.; Fujisato, A.; Yamaoka, N.; Inoue, S.; Miyamoto, S.; Sato, F.; Iida, T.; Horiike, H.; Matushita, I.; Ida, M.; et al. High speed lithium flow experiments for IFMIF target. *J. Nucl. Mater. A* **2004**, *329–333*, 208–212. [CrossRef]
4. Kondo, H.; Kanemura, T.; Furukawa, T.; Hirakawa, Y.; Wakai, E.; Knaster, J. Validation of liquid lithium target stability for an intense neutron source. *Nucl. Fusion* **2017**, *57*, 066008. [CrossRef]
5. Kanemura, T.; Kondo, H.; Furukawa, T.; Hirakawa, Y.; Hoashi, E.; Yoshihashi, S.; Horiike, H.; Wakai, E. Measurement of lithium target surface velocity in the IFMIF/EVEDA lithium test loop. *Fusion Eng. Des.* **2016**, *109–111*, 1682–1686. [CrossRef]
6. Kondo, H.; Kanemura, T.; Yamaoka, N.; Miyamoto, S.; Ida, M.; Nakamura, H.; Matushita, I.; Muroga, T.; Horiike, H. Measurement of free surface of liquid metal lithium jet for IFMIF target. *Fusion Eng. Des.* **2007**, *82*, 2483–2489. [CrossRef]
7. Rabaud, M.; Moisy, F. Ship Wakes: Kelvin or Mach Angle? *Phys. Rev. Lett.* **2013**, *110*, 214503. [CrossRef] [PubMed]
8. Lamb, H. *Hydrodynamics*; Cambridge University Press: Cambridge, UK, 1916.
9. Darmon, A.; Benzaquen, M.; Raphaël, E. Kelvin wake pattern at large Froude numbers. *J. Fluid Mech.* **2013**, *738*, R3. [CrossRef]
10. Moisy, F.; Rabaud, M. Mach-like capillary-gravity wakes. *Phys. Rev.* **2014**, *E90*, 023009. [CrossRef] [PubMed]
11. Pethiyagoda, R.; Moroney, T.; Lustri, C.; McCue, S. Kelvin wake pattern at small Froude numbers. *J. Fluid Mech.* **2021**, *915*, A126. [CrossRef]
12. Noblesse, F.; He, J.; Zhu, Y.; Hong, L.; Zhang, C.; Zhu, R.; Yang, C. Why can ship wakes appear narrower than Kelvin's angle? *Eur. J. Mech. B/Fluids* **2014**, *46*, 164–171. [CrossRef]
13. Ageorges, V.; Peixinho, J.; Perret, G. Flow and air entrainment around partially submerged vertical cylinders. *Phys. Rev. Fluids* **2019**, *4*, 064801. [CrossRef]
14. Chaplin, J.R.; Teigen, P. Steady flow past a vertical surface-piercing circular cylinder. *J. Fluids Struct.* **2003**, *18*, 271–285. [CrossRef]
15. Koo, B.; Yang, J.; Yeon, S.M.; Stern, F. Reynolds and froude number effect on the flow past an interface-piercing circular cylinder. *Int. J. Nav. Archit. Ocean Eng.* **2014**, *6*, 529–561. [CrossRef]
16. Bestehorn, M. *Hydrodynamik und Strukturbildung*; Springer: Berlin/Heidelberg, Germany, 2006.
17. Lighthill, J. *Waves in Fluids*; Cambridge University Press: Cambridge, UK, 1978.
18. acA2040-55uc—Basler Ace. Basler AG, 22926 Ahrensburg, Germany. Available online: https://www.baslerweb.com/de/produkte/kameras/flaechenkameras/ace/aca2040-55uc/ (accessed on 20 September 2021).
19. KOWA LM16JC3M2. Kowa Optimed Deutschland GmbH, 40233 Düsseldorf, Germany. Available online: https://www.kowa-lenses.com/lm16jc3m2-3mp-industrieobjektiv-c-mount (accessed on 20 September 2021).
20. MATLAB R. The MathWorks Inc., Natick, Massachusetts 01760, USA. 2019. Available online: https://www.mathworks.com/products/matlab.html (accessed on 20 September 2021).
21. Confocal Displacement Sensor CL-3000. KEYENCE DEUTSCHLAND GmbH, 63263 Neu-Isenburg, Germany. Available online: https://www.keyence.de/products/measure/laser-1d/cl-3000/models/cl-3000/ (accessed on 22 September 2021).

Article

Hydraulic Characterization of the Full Scale Mock-Up of the DEMO Divertor Outer Vertical Target

Amelia Tincani [1], Francesca Maria Castrovinci [2], Moreno Cuzzani [1], Pietro Alessandro Di Maio [2], Ivan Di Piazza [1], Daniele Martelli [1], Giuseppe Mazzone [3], Andrea Quartararo [2], Eugenio Vallone [2,*] and Jeong-Ha You [4]

[1] ENEA FSN-ING CR Brasimone, Camugnano, 40032 Bologna, Italy; amelia.tincani@enea.it (A.T.); moreno.cuzzani@enea.it (M.C.); ivan.dipiazza@enea.it (I.D.P.); daniele.martelli@enea.it (D.M.)
[2] Department of Engineering, University of Palermo, Viale delle Scienze, Ed. 6, 90128 Palermo, Italy; francescamaria.castrovinci@unipa.it (F.M.C.); pietroalessandro.dimaio@unipa.it (P.A.D.M.); andrea.quartararo@unipa.it (A.Q.)
[3] Department of Fusion and Technology for Nuclear Safety and Security, ENEA C. R. Frascati, Via E. Fermi 45, Frascati, 00044 Roma, Italy; giuseppe.mazzone@enea.it
[4] Max Planck Institute of Plasma Physics (E2M), Boltzmann Str. 2, 85748 Garching, Germany; you@ipp.mpg.de
* Correspondence: eugenio.vallone@unipa.it

Abstract: In the frame of the pre-conceptual design activities of the DEMO work package DIV-1 "Divertor Cassette Design and Integration" of the EUROfusion program, a mock-up of the divertor outer vertical target (OVT) was built, mainly in order to: (i) demonstrate the technical feasibility of manufacturing procedures; (ii) verify the hydraulic design and its capability to ensure a uniform and proper cooling for the plasma facing units (PFUs) with an acceptable pressure drop; and (iii) experimentally validate the computational fluid-dynamic (CFD) model developed by the University of Palermo. In this context, a research campaign was jointly carried out by the University of Palermo and ENEA to experimentally and theoretically assess the hydraulic performances of the OVT mock-up, paying particular attention to the coolant distribution among the PFUs and the total pressure drop across the inlet and outlet sections of the mock-up. The paper presents the results of the steady-state hydraulic experimental test campaign performed at ENEA Brasimone Research Center as well as the relevant numerical analyses performed at the Department of Engineering at the University of Palermo. The test facility, the experimental apparatus, the test matrix and the experimental results, as well as the theoretical model, its assumptions, and the analyses outcomes are herewith reported and critically discussed.

Keywords: DEMO; divertor; plasma facing components; thermal hydraulics

Citation: Tincani, A.; Castrovinci, F.M.; Cuzzani, M.; Di Maio, P.A.; Di Piazza, I.; Martelli, D.; Mazzone, G.; Quartararo, A.; Vallone, E.; You, J.-H. Hydraulic Characterization of the Full Scale Mock-Up of the DEMO Divertor Outer Vertical Target. *Energies* 2021, 14, 8086. https://doi.org/10.3390/en14238086

Academic Editor: Hyungdae Kim

Received: 29 October 2021
Accepted: 29 November 2021
Published: 2 December 2021

Publisher's Note: MDPI stays neutral with regard to jurisdictional claims in published maps and institutional affiliations.

1. Introduction

The pursuit of fusion power is propelled by a call for sustainable and foreseeable low-carbon sources of energy due to the increasing global electricity needs. In this context, Europe has conceived a roadmap [1] that ensures a clear and organized path for proof of commercial electric power production from nuclear fusion facilities, at a reasonable timescale [2]. As an essential element of the European roadmap to the realization of fusion electricity, Europe is leading the design of a DEMO plant, which will be the final pivotal step towards harnessing fusion power after ITER [1].

The DEMO plant is expected to run in the middle of the century, with the main objectives of demonstrating the production of few hundred MWs of net electrical energy, operating with a closed-tritium fuel cycle and maintenance systems capable of achieving acceptable plant availability [3]. The DEMO design and R&D activities in Europe will surely profit from the experiences gained from the design, construction, and operation of ITER. Nevertheless, there are still physics, materials, and engineering challenges that need to be addressed [3].

Among them, the European roadmap to the realization of fusion electricity has defined reliable power exhausting as one of the most decisive missions. The power necessary to maintain plasma at high temperatures has to be exhausted. Heat-exhaust systems have to endure the large heat and particle fluxes of a fusion power plant while ensuring the highest possible performance from the core plasma [1]. The divertor is the fundamental in-vessel component in accomplishing this mission. Being devoted to power exhaust and impurity removal via guided plasma exhaust, the feasibility of fusion power generation depends, to a large extent, on the heat load that can be tolerated by the divertor under normal and off-normal conditions [3].

According to its latest concept, the DEMO divertor is articulated in 48 toroidal assemblies. Each assembly is composed of a cassette body, endowed with a shielding liner, two reflector plates, and two plasma facing components (PFCs), namely an inner vertical target (IVT) and an outer vertical target (OVT) [4]. In the IVT and OVT, there are the plasma facing units (PFUs). These crucial components, adopted to sustain the high superficial thermal loads, are high velocity water cooled tubes, equipped with swirl tapes to increase their heat transfer capabilities. Figure 1 shows the model of the entire divertor assembly.

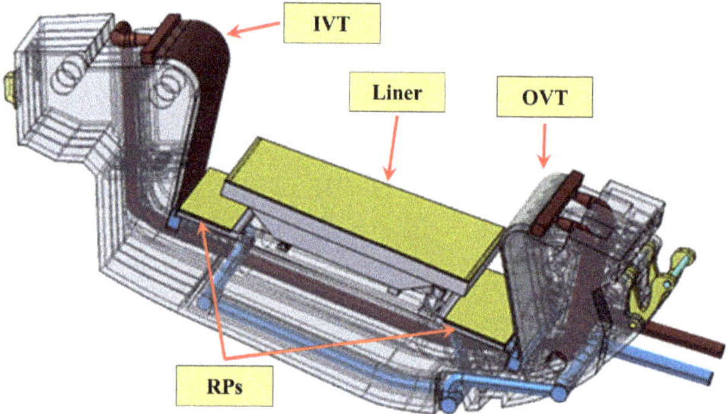

Figure 1. The DEMO divertor assembly.

Given this background, in the frame of the activities of the DEMO work package DIV-1 "Divertor Cassette Design and Integration", of the EUROfusion program aimed at developing a holistic divertor design concept [5], a mock-up of the OVT [6] of the DEMO divertor was built by the ENEA research group, mainly in order to:

- Demonstrate the technical feasibility of manufacturing procedures, focusing the attention on the joining/welding techniques, and optimizing costs of each component applying standard practices;
- Experimentally characterize the hydraulic behavior of the plasma facing components;
- Validate the computational fluid-dynamic (CFD) model developed at the University of Palermo against experimental data.

In this context, a research campaign was jointly carried out by the University of Palermo and ENEA to experimentally and theoretically assess the hydraulic performances of the OVT mock-up, paying particular attention to the coolant distribution among the PFUs and the total pressure drop across the inlet and outlet sections of the mock-up. Both the experimental and the theoretical–numerical campaigns will be extensively described in the following.

In particular, Section 2 is devoted to the description of the experimental test campaign carried out at ENEA Brasimone Labs, providing a brief overview of the CEF1 (Circuito per Esperienze di Fluidodinamica 1) water facility as well as of the OVT full-scale prototype and instrumentation. The attention is then focused on the obtained results, mainly in terms of pressure drops and mass flow rate distribution in the plasma facing cooling tube array.

On the other hand, Section 3 reports the comprehensive validation activity of the 3D CFD calculations carried out at the University of Palermo, based on the data obtained by the ENEA Brasimone Team from the experimental campaign on the OVT mock-up cooling circuit. The computational campaign was performed following a theoretical–numerical approach based on the finite volume method and adopting the well-known ANSYS CFX 2020 R2 commercial CFD code [7].

2. Experimental Test Campaign

In 2020, after the partial upgrade of the experimental hall at ENEA Brasimone laboratories, the complete design and realization of a new test section data and acquisition system and the installation of the OVT mock-up, the experimental test campaign was started by means of the CEF1 water loop, according to the test matrix drafted in [6]. The CEF1 water loop consists of a heat exchanger and two centrifugal pumps, which can be operated in series or in parallel, the test section, and the return line to the pressurized tank [6], as shown in Figure 2.

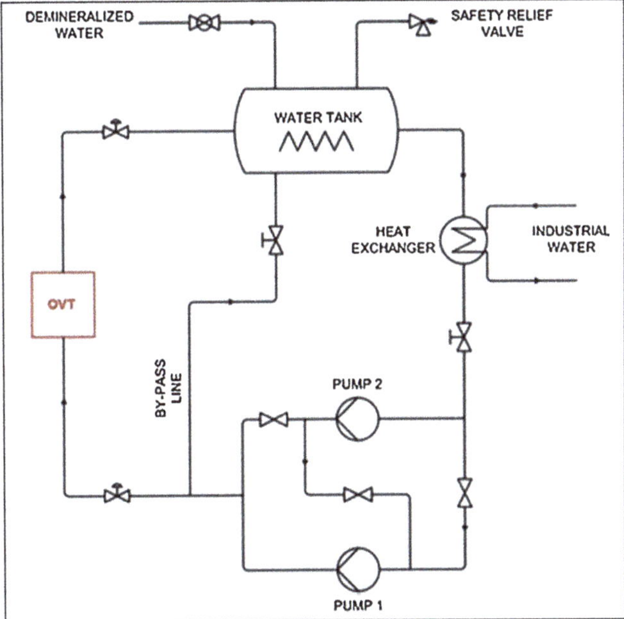

Figure 2. Simplified scheme of the CEF1 facility.

The OVT mock-up main components are the plasma facing channels, consisting of 39 tubes in CuCrZr equipped with internal helical swirl tape made of copper, the inlet/outlet manifolds made of AISI-316L steel, the inlet diffuser made of AISI-316L, the outlet header made of AISI-316L, the transition CuCrZr/AISI-316L tubes, i.e., the connection between the CuCrZr plasma facing tubes and AISI-316L manifolds/diffuser, and the supporting system [6]. Figure 3 shows the CAD model of OVT mock-up and its main components.

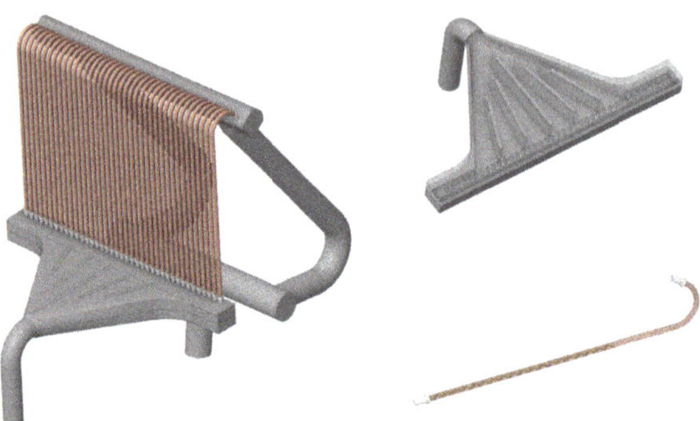

Figure 3. CAD model of the OVT mock-up (**left**), diffuser (**top right**), PFU channel (**bottom right**).

Even if small differences are present in the diffuser and the horizontal headers, mainly due to manufacturing constraints, the OVT cooling circuit exhibits the same structure as the corresponding DEMO divertor OVT [6]. Logically, the tungsten tiles and the copper interlayer were not considered because they were irrelevant for the planned test campaign [6].

Furthermore, certified and calibrated instrumentation was installed in the mock-up and used during the tests. It was mainly composed of two K-type thermocouples, measuring water temperatures at the inlet and outlet manifolds with a precision of $\pm 1\,°C$, a differential pressure transmitter measuring the mock-up total pressure drop with a precision of ± 0.001 bar, and a pressure transmitter at the inlet section with a precision of ± 0.012 bar. The positions of the thermocouples and of the differential pressure transmitter are shown in Figure 4.

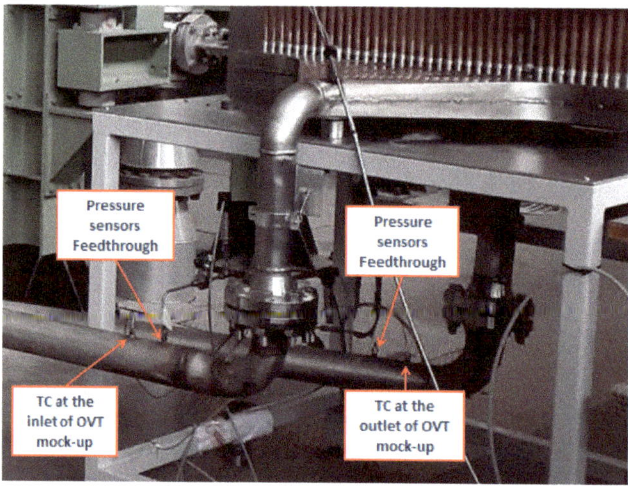

Figure 4. Detail of the thermocouples and of the differential pressure transmitter.

In addition, a new ultrasonic flow meter FLEXIM F601 [8] was acquired to measure coolant mass flow rates inside each of the 39 plasma facing tubes and it was placed downstream of the swirl tape location [6]. Figure 5 shows two pictures of the ultrasonic flow meter sensor installation on the upper part of the tube where the water flow is not disturbed, because it is far enough from the swirl tape, and the measurements are stable and reliable.

Figure 5. Ultrasonic flow meter measurements on the first OVT plasma facing cooling tubes.

The prototype mechanically connected to the water loop, instrumented, and ready for the test campaign is shown in Figure 6.

Figure 6. DEMO divertor OVT prototype installed into the CEF1 building.

Steady-State Tests

As for previous test campaigns performed on the ITER divertor cassette full-scale prototype [9], steady-state tests were carried out in order to determine the hydraulic characteristic function of the OVT mock-up cooling circuit, representing the functional dependence of the total pressure drop along the component, Δp_{tot}, on the corresponding mass flow rate, G, and at a given coolant temperature, T. In particular, tests at four different temperatures (20, 40, 70, and 100 °C) were performed, and for each temperature, five mass flow rate points were acquired from ≈15 kg/s to ≈35 kg/s, while inlet pressure ranged from ≈2 bar to ≈13 bar.

Table 1 summarizes the mock-up total pressure drops measured at different temperatures and mass flow rates. The pressure drop values were averaged over 240 s of data acquisition, in which the mass flow rate, as well as the temperature of the water loop, were kept constant, in order to make sure that quasi steady-state conditions were reached inside the experimental test section. Quasi steady-state conditions were typically reached in a few minutes.

Table 1. Summary of experimental steady-state results.

T [°C]	G [kg/s]	Δp_{Exp} [bar]
20	12.438 ± 0.191	0.510 ± 0.015
	20.223 ± 0.202	1.268 ± 0.013
	25.169 ± 0.188	1.845 ± 0.013
	30.087 ± 0.152	2.567 ± 0.041
	35.041 ± 0.186	3.393 ± 0.058
40	14.513 ± 0.073	0.650 ± 0.006
	22.560 ± 0.113	1.514 ± 0.059
	25.035 ± 0.125	1.821 ± 0.022
	30.135 ± 0.151	2.535 ± 0.040
	34.912 ± 0.175	3.325 ± 0.061
70	15.146 ± 0.076	0.665 ± 0.007
	20.346 ± 0.102	1.162 ± 0.014
	25.521 ± 0.134	1.770 ± 0.026
	30.282 ± 0.152	2.419 ± 0.044
	34.984 ± 0.175	3.212 ± 0.064
100	13.087 ± 0.066	0.498 ± 0.007
	19.705 ± 0.099	1.061 ± 0.023
	25.505 ± 0.128	1.723 ± 0.027
	30.221 ± 0.151	2.384 ± 0.049
	34.974 ± 0.182	3.126 ± 0.062

Afterwards, at each considered temperature, the hydraulic characteristic functions were derived by best fitting these results with the following analytical form:

$$\Delta p_{tot}(G) = \alpha G^{\beta} \tag{1}$$

where α and β are coefficients whose best-fitting values α_{Exp} and β_{Exp}, in SI units, are listed in Table 2.

The overall OVT prototype pressure drop obtained by extrapolating at the reference mass flow rate of 54.95 kg/s, Δp_{Exp}, is reported in Table 2 for each coolant temperature investigated. As it can be observed, they are close to their expected design value of 7.403 bar.

Table 2. OVT extrapolated pressure drops for the different test temperatures at the reference mass flow rate of 54.95 kg/s.

T [°C]	α_{Exp} [bar/(kg/s)$^{\beta}$]	β_{Exp} [-]	Δp_{Exp} [bar]
20	5.20×10^{-3}	1.8253	7.798
40	4.60×10^{-3}	1.8520	7.864
70	4.10×10^{-3}	1.8728	7.437
100	4.00×10^{-3}	1.8720	7.232

In parallel, the flow distributions among the plasma facing channels were experimentally determined by using the ultrasonic flow-meter FLEXIM F601. This detector uses an acoustic signal with ultrasonic frequency, which is sent and received through the liquid and measured by a pair of sensors properly placed on the pipe. The time needed for the acoustic signal to cross the liquid back and forth is directly proportional to the speed of the liquid and, hence, to its flow rate. The coolant mass flow rate was measured within each plasma facing tube only once quasi steady-state conditions were reached.

As shown in Figure 7, the mass flow rate distribution among the plasma facing channels is quite uniform and within the $\pm 10\%$ of its average value, corresponding to maximum axial velocity values in the order of ≈9 m/s at the highest considered mass flow rate.

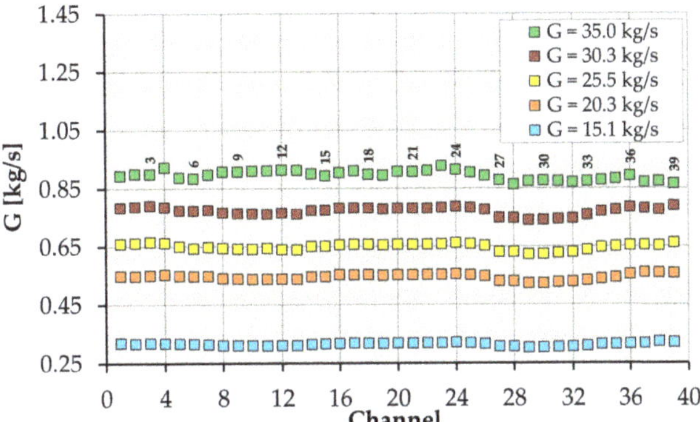

Figure 7. Ultrasonic flow meter measurements on 39 OVT plasma facing tubes at 70 °C (raw data, passed to the numerical simulation team).

3. OVT Mock-Up Cooling Circuit CFD Analysis

One of the main purposes behind the fabrication of the OVT mock-up was the experimental validation of the CFD models developed at the University of Palermo, in order to verify the computationally predicted hydraulic performances. Within this framework, a comprehensive validation activity of the 3D CFD calculations was carried out at the University of Palermo, based on the data obtained at the ENEA Brasimone Labs from the experimental campaign on the OVT mock-up cooling circuit. The computational campaign was performed following a theoretical–numerical approach based on the finite volume method and adopting the well-known ANSYS CFX 2020 R2 commercial CFD code [7].

In particular, in the first phase, a grid independence study was carried out, in order to quantify the discretization errors and to ensure numerical predictions in the asymptotic range of convergence.

Then, computational results were compared to the experimental outcomes, specifically in terms of coolant water velocity distribution in the entire PFU cooling tube array and total pressure drop of the mock-up cooling circuits.

Furthermore, a parametric analysis campaign was performed to evaluate how the uncertainties on the equivalent sand grain wall roughness may affect both total pressure drop and mass flow rate distribution among PFU channels.

It is worth mentioning how the mesh independence study and the uncertainty assessment on the equivalent sand grain wall roughness were performed to provide a measure of the degree of confidence on the outcomes of the numerical campaign.

Results obtained are herewith reported and critically discussed, together with the main assumptions, models, and boundary conditions (BCs).

3.1. Model Setup

To perform a validation campaign on the methodologies adopted for CFD simulation, steady-state isothermal CFD analyses were carried out on the OVT mock-up cooling circuit, whose geometry is reported in Figure 8, considering the assumptions and BCs reported in Tables 3 and 4.

With regard to Table 3, it is worth highlighting that the k-ω based SST model was adopted because it gives an accurate description of the near-wall region [7], especially in the case of flow separation. It is a hybrid model that uses a transformation of the k–ϵ model into a k–ω model in the near-wall region and the standard k–ϵ model in the fully turbulent region far from the wall. Moreover, this model uses the automatic wall treatment that automatically switches from wall-functions to a low-Re near-wall formulation as the mesh is refined. The accuracy and the flexibility of the k-ω based SST model makes it suited for the analysis of very complex domains where a wide range of Reynolds numbers may be encountered and flow separation may occur. For further information, the interested reader is addressed to the code documentation [7].

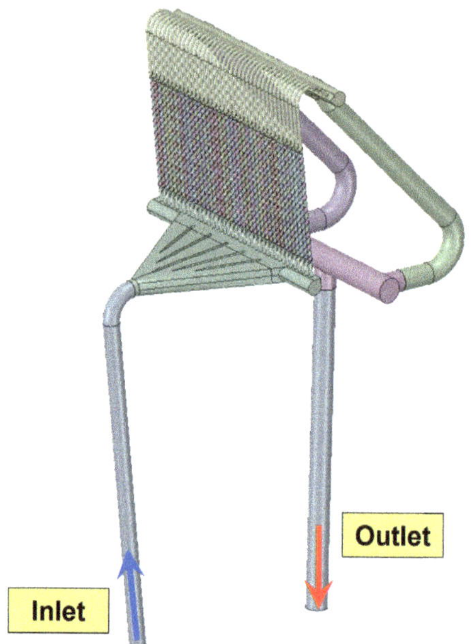

Figure 8. OVT mock-up cooling circuit calculation domain.

Among the wide range of operative conditions tested on the OVT mock-up cooling circuit at ENEA Brasimone Labs, attention was placed on the experimental tests performed at $\approx$20 °C inlet temperature (see Table 4). Under the selected operating conditions, the flow field inside the PFU channels was characterized by values of the Reynolds Re numbers (calculated according to the definition given in [10]), ranging between 3×10^4 and 1×10^5, resulting in a fully turbulent flow. Similar results in terms of Re were obtained for the manifold pipes.

Table 3. Summary of OVT mock-up CFD analysis setup.

	Reference Conditions
Analysis type	Steady-state isothermal
Material library	Water IAPWS IF97 [11]
Turbulence model	k-ω SST [7]
Differencing scheme	High-resolution [7]
Boundary layer modeling	Automatic wall functions [7]
PFU channels absolute wall roughness	2 µm
ST absolute wall roughness	2 µm
Other regions absolute wall roughness	15 µm
Inlet BC (temperature/pressure)	
Outlet BC (mass flow rate)	Variable according to Table 4

Table 4. List of selected experimental tests.

Test # [-]	p_{in} [bar]	G [kg/s]	T_{in} [°C]
1	1.42	12.27	17.85
2	3.92	21.27	20.38
3	5.93	25.29	21.61
4	8.73	30.09	24.06
5	12.09	35.05	25.33

3.2. Mesh Independence Studies

To evaluate and minimize the discretization error related to the CFD simulations, a grid independence study was preliminarily performed and the mesh-related error was assessed by adopting the well-established grid convergence index (GCI), both according to its original derivation, presented in [12], and considering the least-squares (LS) error estimation described in [13].

Four different increasingly refined meshes were considered, whose details are summarized in Table 5. These meshes were created with mixed tetrahedral and prism elements (see Figure 9), maintaining unchanged both the total number of layers and the first layer thickness, and adapting the layer growth rate to have a smooth transition between the bulk mesh and the last prism layer. Despite the limitations imposed by the adoption of tetra elements, the meshes were created with the aim of obtaining computational grids as similar as possible, in compliance with the requirements of ANSYS CFX, in terms of element quality. With regard to Table 5, it is worth highlighting how the average mesh quality metrics are within the acceptable ranges prescribed in [7], and less than 1% of the overall number of cells is characterized by poor metrics. As a consequence, mesh quality is not expected to significantly influence the results.

Figure 9. Finest mesh adopted for OVT mock-up cooling circuit CFD analyses.

Table 5. Adopted mesh parameters and quality metrics.

	Mesh #1	Mesh #2	Mesh #3	Mesh #4
Element size OVT/PFUs [mm]	10.4/1.3	8.0/1.0	6.15/0.77	4.73/0.59
First layer thickness [μm]		20		
Surface with y^+<10 at Test #5 [%]		$\approx$80		
Number of layers [-]		12		
Layer growth rate OVT/PFUs [-]	1.68/1.42	1.65/1.39	1.61/1.36	1.58/1.33
Number of nodes [-]	1.42×10^7	2.09×10^7	3.22×10^7	4.15×10^7
Number of elements [-]	2.05×10^7	2.78×10^7	4.03×10^7	5.14×10^7
Node density [m^{-3}]	3.25×10^8	4.81×10^8	7.40×10^8	9.53×10^8
Orthogonality Factor (average/min)	0.59/0.01	0.61/0.01	0.67/0.01	0.72/0.01
Expansion factor (average/max)	2/617	2/607	2/299	2/258
Aspect ratio (average/max)	38/1830	29/1415	19/1582	17/1077

Additionally, the first layer thickness was selected sufficiently small to have the highest y^+ values in the order of the unity, to capture the entire boundary layer profile. Some details of the resulting finest mesh (Mesh #4) are depicted in Figure 9.

All analyses were run until all residuals reached the iteration convergence control criterion of 10^{-4} and a second-order accurate numerical scheme was adopted, as suggested by [14].

On the other hand, the figure of merit (FoM) monitored and collected to assess the grid convergence is the total pressure drop between inlet and outlet sections of the OVT mock-up cooling circuit. The obtained results are summarized in Table 6.

Table 6. Outlook of the results obtained with the different meshes.

G [kg/s]	Δp_{Num} [bar]			
	Mesh #1	Mesh #2	Mesh #3	Mesh #4
12.27	0.500	0.475	0.462	0.460
21.27	1.394	1.335	1.303	1.296
25.29	1.935	1.851	1.816	1.812
30.09	2.687	2.575	2.509	2.509
35.05	3.597	3.454	3.367	3.358

Starting from the results reported in Table 6, the GCI together with the asymptotic values obtained with the generalized Richardson extrapolation (that will be referred to simply as asymptotic values in the following) were calculated by adopting the procedure described in [12]. In particular, given a general monitored quantity ϕ, it can be related to the average mesh size by the following equation:

$$\phi_i = \phi_0 + \alpha h_i^\rho, \tag{2}$$

where ϕ_i is the measured value obtained with a selected mesh of average linear size h_i, ϕ_0 is the estimate of the asymptotic value, α is a constant and ρ is the observed order of convergence. Given the results obtained with a set of three meshes, it is then possible to obtain, by proper interpolation, the values of ϕ_0, α and ρ.

The GCI can therefore be calculated for the selected i-th grid, as follows

$$GCI_i = F_s \left| \frac{\phi_i - \phi_0}{\phi_i} \right| = F_s \left| \frac{\alpha}{\phi_i} \right| h_i^\rho, \tag{3}$$

where F_s can be interpreted as a safety factor, conservatively chosen here equal to 3.

It is important to remark that the observed order of convergence should be within the acceptable range for a second-order accurate numerical method. A value greater than this formal order of convergence is likely to cause too small error estimates [13], as can be easily deduced from Equation (3), being $h_i \ll 1$.

The outcomes of the analyses showed good monotonically convergent behavior, with the highest GCI calculated according to [12] being $\approx 1.5\%$, suggesting results close to their asymptotic values. Moreover, an observed order of convergence ρ within the range 8–50 was calculated, clearly outside the above-mentioned suitable limits, and possibly related to noisy results. The procedure described in [13] was therefore considered, with the aim to estimate more conservative and realistic GCI values by obtaining a better estimate of ρ. This can be accomplished since this method evaluates the parameters ϕ_0, α and ρ of Equation (2) in an LS sense, performing a best-fitting of the results obtained; thus, limiting the effects of scattered data. In case unduly high or low ρ-values are calculated, this method prescribes that a best-fitting with both linear and quadratic functions is performed. The results obtained by adopting this approach are summarized as follows:

- The observed convergence order ranges between 7.80 and 10.43, lower than the results obtained with the classical GCI formulation, yet outside the prescribed limits;
- GCI values calculated adopting the parameters obtained with the LS approach result between 1 and 2.7% for all of the considered mass flow rates, and are characterized by a lower spread compared to the outcomes of the original GCI formulation;
- The prescribed best-fittings with linear and quadratic functions resulted in unrealistic and underestimated asymptotic values, not compatible with the experimental results and, consequently, in excessively high GCI values.

Therefore, the outcomes of these latter fittings were discarded, and it can be reasonably argued from the results obtained shown that an observed order of convergence of ≈ 8 realistically characterizes the examined problem. To further confirm this assumption, an additional fine mesh of approximately 10^8 elements was tested and, due to the huge computational time required, only one simulation, the one with the lowest mass flow rate, was considered. The results obtained (see Figure 10) prove that the calculated observed order of convergence can be reliably adopted to realistically estimate the GCI, resulting in the additional tested mesh well-predicted by the LS fitting.

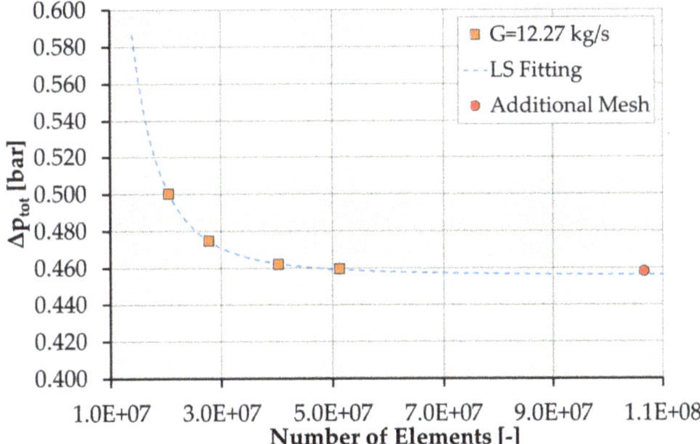

Figure 10. Grid independence results obtained for a mass flow rate of 12.27 kg/s.

The outcomes of the entire mesh independence analysis are summarized in Table 7, while the relative errors with respect to the pertaining asymptotic values are reported in Figure 11. As a closing remark, it can be safely and conservatively stated that the results obtained with Mesh #4 are characterized by a small discretization error, which can be reasonably considered below 3%.

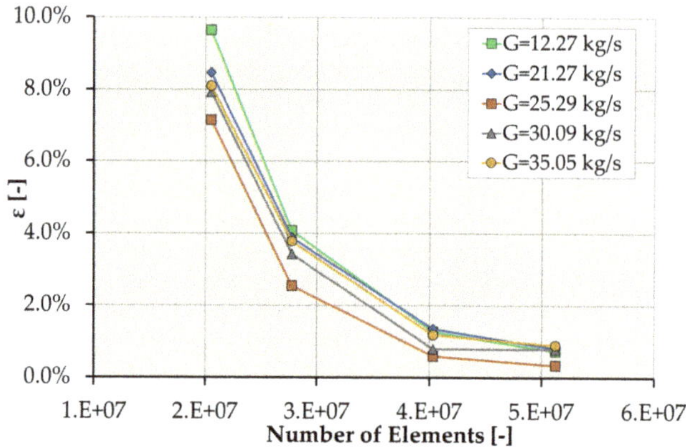

Figure 11. Relative error with respect to the LS asymptotic values.

Table 7. Overview of the grid independence studies.

G [kg/s]	Original GCI Formulation			LS GCI Formulation		
	Asymptotic Values	ρ	GCI [%]	Asymptotic Values	ρ	GCI [%]
12.27	0.458	9.43	1.31	0.456	8.53	2.18
21.27	1.289	8.54	1.52	1.285	7.80	2.40
25.29	1.809	12.31	0.42	1.805	10.43	1.01
30.09	2.509	49.77	0.00	2.489	8.73	2.34
35.05	3.353	13.81	0.38	3.328	7.81	2.67

3.3. Results and Comparison with Experimental Data

The results of the OVT mock-up cooling circuit CFD analysis, considering the previously selected Mesh #4, are herein reported mainly in terms of coolant total pressure drop and mass flow rate distribution among PFU channels, and compared with the experimental data discussed in Section 2.

The coolant total pressure spatial distribution within the OVT mock-up cooling circuit, the details of the coolant flow velocity field inside the manifolds, and the PFU channels assessed for the 35.05 kg/s case are shown in Figure 12.

The comparison between the numerical predictions and the outcomes of the experimental campaign in terms of OVT mock-up characteristic curves, i.e., Equation (1), is reported in Figure 13, with the addition of 10% error bars, so to give an immediate view of the error, and is summarized in Table 8. From the analysis of the results, it may be argued that the maximum error on pressure drop estimation occurs at the lowest investigated flow rate and is equal to −7.22%, showing an overall good agreement between CFD simulations and experimental data.

Table 8. Total pressure drop comparison and errors.

G [kg/s]	Δp_{Num} [bar]	Δp_{Exp} [bar]	ϵ [%]
12.27	0.460	0.495	−7.22%
21.27	1.296	1.354	−4.32%
25.29	1.812	1.859	−2.56%
30.09	2.509	2.567	−2.26%
35.05	3.358	3.384	−0.77%

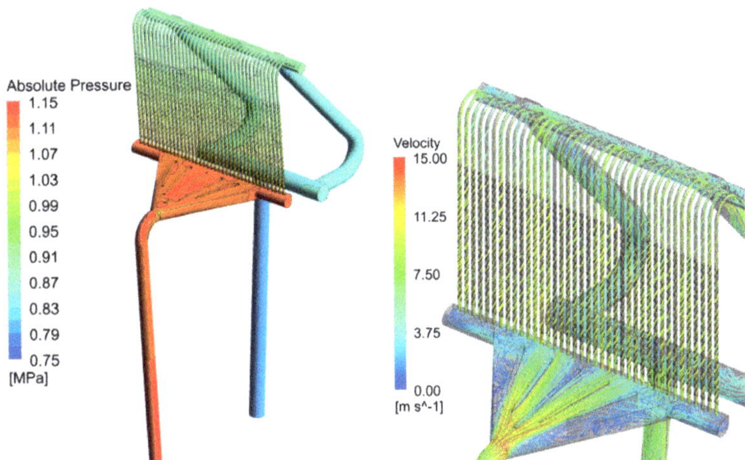

Figure 12. OVT mock-up results at G = 35.05 kg/s. OVT mock-up coolant total pressure field (**left**); OVT mock-up coolant flow velocity field (**right**).

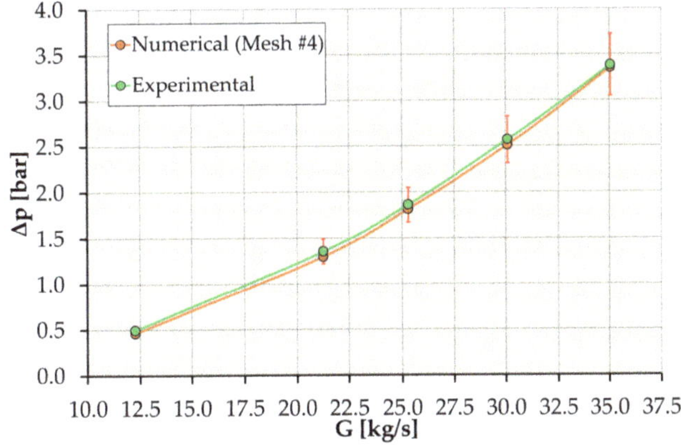

Figure 13. Numerical and experimental characteristic curves of the OVT mock-up cooling circuit.

The comparison between numerical and experimental mass flow rate distributions among the OVT mock-up PFU channels is reported for test #5 of Table 4 (being the one with the worst agreement between experimental and numerical outcomes) in Figure 14, with 5% error bars alongside the key parameters for all of the considered mass flow rates in Table 9 where, together with the maximum and minimum flow rates measured for the PFU channels, the standard deviations are reported σ, and the coefficient of variation, defined as:

$$C_{V,G} = \frac{\sigma}{G_{ave}}.$$

(4)

Additionally, maximum and average relative errors between experimental and numerical distributions among the OVT mock-up PFU channels are reported in Table 10.

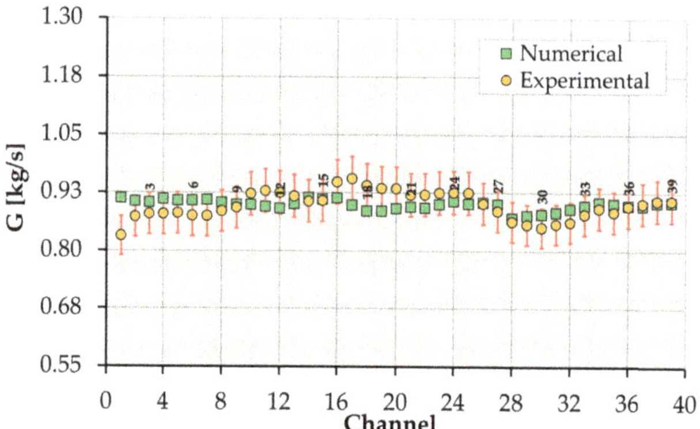

Figure 14. Comparison of OVT coolant mass flow rate distributions at $G = 35.05$ kg/s.

Table 9. Comparison of OVT coolant mass flow rate distribution main parameters.

		G_{Max}/G_{min} [kg/s]	σ [kg/s]	$C_{V,G}$ [%]
$G = 12.27$ kg/s	Numerical	0.323/0.303	0.005	1.59
	Experimental	0.331/0.229	0.008	2.54
$G = 21.27$ kg/s	Numerical	0.560/0.523	0.010	1.83
	Experimental	0.578/0.518	0.015	2.75
$G = 25.29$ kg/s	Numerical	0.666/0.622	0.012	1.85
	Experimental	0.686/0.617	0.014	2.16
$G = 30.09$ kg/s	Numerical	0.791/0.739	0.015	1.94
	Experimental	0.811/0.729	0.021	2.72
$G = 35.05$ kg/s	Numerical	0.921/0.863	0.016	1.78
	Experimental	0.955/0.832	0.029	3.23

Table 10. Maximum and average relative errors on mass flow rate distributions among OVT mock-up PFU channels.

G [kg/s]	ϵ_{Max} [%]	ϵ_{ave} [%]
12.27	5.90	2.29
21.27	8.22	2.23
25.29	4.35	1.40
30.09	7.63	3.11
35.05	9.36	2.20

From the analysis of the results, it might be argued that experimental data show a higher spread of flow distribution among PFU channels and a higher degree of asymmetry with respect to CFD results for all of the cases under investigation. Despite this discrepancy, the agreement between numerical and experimental results is quite good, and the maximum error is predicted for the 35.05 kg/s case simulation and is equal to 9.36% for PFU channel #1, while the average error has a maximum of 3.11% for the 30.09 kg/s case.

3.4. Sensitivity Analysis on the Equivalent Sand-Grain Wall Roughness

One of the major uncertainties in the accurate reproduction of experimental test conditions is the lack of precise knowledge on the equivalent sand grain wall roughness values of PFU channels, swirl tapes, and mock-up pipes and manifolds. Since equivalent sand grain wall roughness may have a strong impact on both total pressure drop and mass flow distribution among the PFU channels, a parametric sensitivity analysis campaign was performed to evaluate how these uncertainties may affect the obtained results, by providing numerical estimates of the local sensitivity measures S:

$$S_{\Delta p_{tot},k_j} = \frac{1}{\Delta p_{tot,0}} \frac{\partial \Delta p_{tot}}{\partial k_j}\bigg|_{k_0}, \quad S_{C_{V,G},k_j} = \frac{\partial C_{V,G}}{\partial k_j}\bigg|_{k_0}, \tag{5}$$

where k_j are referred in turn to PFU channels, swirl tape, and mock-up pipes, and manifolds surface roughness values. The local derivatives of Equation (5) are taken at the baseline configuration, considering the surface roughness values of Table 3, referenced as k_0 in Equation (5).

The baseline analyses were performed considering a surface roughness of 2 μm for both PFU channels and copper swirl tapes. This same value was already adopted for the thermal hydraulic assessment of the DEMO divertor PFC cooling circuit [15] and similar values can be found in literature [16,17]. The response of the OVT mock-up cooling circuit in terms of total pressure drop and mass flow rate distribution among PFU channels was therefore assessed, considering a small variation of this parameter around the baseline value. There were two different equivalent sand grain wall roughness values of 1 and 4 μm (halved and doubled with respect to baseline, respectively), separately for PFU channels and swirl tapes, with the aim to identify the component responsible for higher deviations of the results and to quantify their magnitude.

Regarding mock-up pipes and manifolds—the baseline surface roughness of 15 μm was selected in agreement with the values adopted for the DEMO divertor cassette body (see, for example, [18]). Since no better estimates of the roughness were available, it was deemed necessary to test the response of the CFD mock-up cooling circuit to a broader range of roughness values, and in particular, the two values 1.5 and 150 μm were selected. All calculations were performed considering Mesh #4 discussed in Section 3.2, and the wall treatment presented in [19], necessary when the surface roughness was large compared to the near-wall mesh, was adopted for the 150 μm simulation, thus avoiding grid variations that would have possibly required an additional mesh independence study.

A summary of the cases investigated, together with the baseline equivalent sand grain surface roughness vales, is given in Table 11.

Table 11. Absolute Hydraulic Wall Roughness values investigated.

	k_{pipes} [μm]	k_{ST} [μm]	$k_{manifolds}$ [μm]
Baseline	2	2	15
PFU channels	1–4	2	15
Swirl tapes	2	1–4	15
OVT mock-up pipes and manifolds	2	2	1.5–150

The outcomes of the sensitivity analyses are summarized, for the sake of brevity, as follows.

- Due to the presence of non-stationary features of the fluid flow, such as the strongly detached flow inside the inlet manifold clearly visible in Figure 12 (right), an oscillation of the results is always observed in the simulations. As a consequence, the small variations in terms of total pressure drop and mass flow distribution among PFU channels observed by varying the surface roughness are hardly recognized from the noisy nature of the results. To obtain better estimates of the effects of surface roughness sensitivity, the outcomes reported in the following were calculated by averaging the metrics of Equation (5) obtained for the five different mass flow rate cases.
- Varying the PFU channels surface roughness, a variation of total pressure drop between −0.6% and +1.8% is detected. It results in an average of $S_{\Delta p_{tot},k_{pipes}} \approx 0.8\%/\mu m$.

- Varying the swirl tape surface roughness, a variation of total pressure drop between -0.4% and $+0.5\%$ is obtained. An average of $S_{\Delta P_{tot},k_{ST}} \approx 0.3\%/\mu m$, lower than the sensitivity to PFU channels roughness.
- Varying the OVT mock-up pipes and manifolds surface roughness, an average $\approx -2.5\%$ total pressure drop change is observed with the 1.5 μm equivalent sand grain roughness. The 150 μm case predicted an average 0.4% reduction in total pressure drop with respect to baseline case. This latter result is mainly related to the different wall treatment adopted, whose effect should be further investigated. As a consequence, it is not possible to provide an estimate of the parameter $S_{\Delta P_{tot},k_{manifolds}}$.
- No significant change in flow distribution, in terms of coefficient of variation, is observed modifying the surface roughness of PFU channels, swirl tape, and OVT mock-up pipes and manifolds, always resulting in a lower spread of the mass flow rates if compared to the experimental values. In particular, absolute variations within the range $\pm 0.1\%$ are predicted by the model, resulting in $S_{C_{V,G},k_j} \approx 0.0\%/\mu m$.

In summary, the sensitivity analysis showed a limited dependency of the numerical results on the surface roughness of the various components within the explored range of values, with deviations mostly lower than 3% of the baseline results. Nonetheless, the OVT mock-up cooling circuit showed, in particular, a higher sensitivity to the PFU channel surface roughness that may be related to the higher flow velocities experienced by the fluid inside the channels, and to the centrifugal force caused by the swirl motion responsible for higher local velocities close to the pipe walls.

4. Conclusions

In the frame of the pre-conceptual design activities of the DEMO work package DIV-1 "Divertor Cassette Design and Integration" of the EUROfusion program, a research campaign was jointly carried out by the University of Palermo and ENEA to experimentally and theoretically assess the hydraulic performances of a mock-up of the divertor outer vertical target, paying particular attention to the coolant distribution among the PFUs and the total pressure drop across the inlet and outlet sections of the mock-up.

The paper presents the results of the steady-state hydraulic experimental test campaign performed at the ENEA Brasimone Research Center as well as the relevant numerical analyses performed at the Department of Engineering at the University of Palermo. The test facility, the experimental apparatus, the test matrix and the experimental results are widely described and critically discussed together with the theoretical model, its assumptions, and the analyses outcomes.

In particular, the experimental tests on the OVT prototype showed that the expected overall pressure drop at the reference mass flow rate of 54.95 kg/s is close to the previously calculated design value of 7.403 bar for each coolant temperature investigated. Furthermore, the mass flow rate distribution among the plasma facing channels is sufficiently uniform, being the maximum variation within the $\pm 10\%$ of the average value.

On the other hand, the theoretical–numerical analysis campaign started from a thorough mesh independence study that allowed selecting a mesh characterized by a small discretization error, i.e., below 3%. Afterwards, among the wide range of operative conditions tested on the OVT mock-up cooling circuit at ENEA Brasimone Labs, attention was placed on numerically reproducing the experimental tests performed at $\approx 20\ ^\circ C$ inlet temperature. The results exhibit an overall good agreement with experimental data, being the maximum error on pressure drop estimation equal to -7.22%. Similar conclusion may be drawn from the comparison of the coolant flow distributions among PFU channels since the maximum error is predicted to be 9.36%, while the average error has a maximum of 3.11%. In general, it may be argued that experimental data show a higher spread of flow distribution among PFU channels and a higher degree of asymmetry, with respect to CFD results for all of the cases under investigation. Furthermore, the final sensitivity analyses showed a limited dependency of the numerical results on the surface roughness of the various components within the explored range of values, with deviations in the order of a few percent of the baseline results.

Author Contributions: Conceptualization, methodology, investigation, writing—original draft, A.T., F.M.C., M.C., P.A.D.M., I.D.P., D.M., G.M., A.Q., E.V., J.-H.Y. All authors have read and agreed to the published version of the manuscript.

Funding: This work was carried out within the framework of the EUROfusion Consortium and received funding from the Euratom Research and Training Programme 2014–2018 and 2019–2020, under grant agreement no. 633053. The views and opinions expressed herein do not necessarily reflect those of the European Commission.

Data Availability Statement: The data presented in this study are available on request from the corresponding author.

Conflicts of Interest: The authors declare no conflict of interest.

References

1. Donné, A.J.H.; Morris, W. *European Research Roadmap to the Realisation of Fusion Energy*; EUROfusion: Garching/Munich, Germany, 2018; ISBN 978-3-00-061152-0.
2. Donné, A.J.H. The European roadmap towards fusion electricity. *Philos. Trans. R. Soc.* **2019**, *377*, 20170432. [CrossRef]
3. Federici, G.; Bachmann, C.; Barucca, L.; Baylard, C.; Biel, W.; Boccaccini, L.V.; Bustreo, C.; Ciattaglia, S.; Cismondi, F.; Corato, V.; et al. Overview of the DEMO staged design approach in Europe. *Nucl. Fusion* **2019**, *59*, 066013. [CrossRef]
4. Mazzone, G.; You, J.H.; Bachmann, C.; Bonavolontà, U.; Cerri, V.; Coccorese, D.; Dongiovanni, D.; Flammini, D.; Frosi, P.; Forest, L.; et al. Eurofusion-DEMO Divertor—Cassette Design and Integration. *Fusion Eng. Des.* **2020**, *157*, 111656. [CrossRef]
5. You, J.H.; Mazzone, G.; Bachmann, C.; Coccorese, D.; Cocilovo, V.; De Meis, D.; Di Maio, P.A.; Dongiovanni, D.; Frosi, P.; Di Gironimo, G.; et al. Progress in the initial design activities for the European DEMO divertor: Subproject "Cassette". *Fusion Eng. Des.* **2017**, *124*, 364–370. [CrossRef]
6. Mazzone, G.; You, J.H.; Cerri, V.; Coccorese, D.; Garitta, S.; Di Gironimo, G.; Marzullo, D.; Di Maio, P.A.; Vallone, E.; Tincani, A.; et al. Structural verification and manufacturing procedures of the cooling system, for DEMO divertor target (OVT). *Fusion Eng. Des.* **2019**, *146*, 1610–1614. [CrossRef]
7. ANSYS Inc. *ANSYS CFX-Solver Theory Guide*; Release: 2020 R2; ANSYS Inc.: Canonsburg, PA, USA, 2020.
8. FLEXIM F601. Available online: https://www.flexim.com/en/product/fluxus-f601 (accessed on 1 December 2021).
9. Tincani, A.; Di Maio, P.A.; Dell'Orco, G.; Ricapito, I.; Riccardi, B.; Vella, G. Steady state and transient thermal hydraulic characterization of full-scale ITER divertor plasma facing components. *Fusion Eng. Des.* **2008**, *83*, 1034–1037. [CrossRef]
10. Manglik, R.M.; Bergles, A.E. *Swirl Flow Heat Transfer and Pressure Drop with Twisted-Tape Inserts*; Advances in Heat Transfer; Elsevier: Amsterdam, The Netherlands, 2003; Volume 36, pp. 183–266. [CrossRef]
11. International Association for the Properties of Water and Steam. *Revised Release on the IAPWS Industrial Formulation 1997 for the Thermodynamic Properties of Water and Steam*; IAPWS: Lucerne, Switzerland, 2007.
12. Roache, P.J. Verification of Codes and Calculations. *AIAA J.* **1998**, *36*, 696–702. [CrossRef]
13. Eça, L.; Hoekstra, M. A procedure for the estimation of the numerical uncertainty of CFD calculations based on grid refinement studies. *J. Comput. Phys.* **2014**, *262*, 104–130. [CrossRef]
14. Mahaffy, J.; Chung, B.; Song, C.; Dubois, F.; Graffard, E.; Ducros, F.; Heitsch, M.; Scheuerer, M.; Henriksson, M.; Komen, F.; et al. *Best Practice Guidelines for the Use of CFD in Nuclear Reactor Safety Applications*; NEA/CSNI/R(2014)11; International Atomic Energy Agency (IAEA): Vienna, Austria, 2015.
15. Di Maio, P.A.; Burlon, R.; Mazzone, G.; Quartararo, A.; Vallone, E.; You, J.H. Hydraulic assessment of an upgraded pipework arrangement for the DEMO divertor plasma facing components cooling circuit. *Fusion Eng. Des.* **2021**, *168*, 112368. [CrossRef]
16. Raffray, A.R.; Schlosser, J.; Akiba, M.; Araki, M.; Chiocchio, S.; Driemeyer, D.; Escourbiac, F.; Grigoriev, S.; Merola, M.; Tivey, R.; et al. Critical heat flux analysis and R&D for the design of the ITER divertor. *Fusion Eng. Des.* **1999**, *45*, 377–407. [CrossRef]
17. You, J.H.; Mazzone, G.; Visca, E.; Bachmann, C.; Autissier, E.; Barrett, T.; Cocilovo, V.; Crescenzi, F.; Domalapally, P.K.; Dongiovanni, D.; et al. Conceptual design studies for the European DEMO divertor: Rationale and first results. *Fusion Eng. Des.* **2016**, *109–111*, 1598–1603. [CrossRef]
18. Di Maio, P.A.; Garitta, S.; You, J.H.; Mazzone, G.; Vallone, E. Thermal-hydraulic study of the DEMO divertor cassette body cooling circuit equipped with a liner and two reflector plates. *Fusion Eng. Des.* **2021**, *167*, 112227. 112227. [CrossRef]
19. Aupoix, B. Roughness Corrections for the k–ω Shear Stress Transport Model: Status and Proposals. *J. Fluids Eng.* **2014**, *137*. [CrossRef]

Article

A Validation Roadmap of Multi-Physics Simulators of the Resonator of MW-Class CW Gyrotrons for Fusion Applications

Laura Savoldi [1,*], **Konstantinos A. Avramidis** [2], **Ferran Albajar** [3], **Stefano Alberti** [4], **Alberto Leggieri** [5] and **Francisco Sanchez** [3]

[1] MAHTEP Group, Dipartimento Energia "Galileo Ferraris", Politecnico di Torino, Corso Duca degli Abruzzi 24, 10129 Torino, Italy
[2] IHM, Karlsruhe Institute of Technology, D-76131 Karlsruhe, Germany; konstantinos.avramidis@kit.edu
[3] Fusion for Energy, 08019 Barcelona, Spain; ferran.albajar@f4e.europa.eu (F.A.); francisco.sanchez@f4e.europa.eu (F.S.)
[4] Swiss Plasma Center, EPFL, CH-1015 Lausanne, Switzerland; stefano.alberti@epfl.ch
[5] Microwave & Imaging Solution, THALES, F-78141 Velizy-Villacoublay, France; alberto.leggieri@thalesgroup.com
* Correspondence: laura.savoldi@polito.it; Tel.: +39-011-090-4559

Abstract: For a few years the multi-physics modelling of the resonance cavity (resonator) of MW-class continuous-wave gyrotrons, to be employed for electron cyclotron heating and current drive in magnetic confinement fusion machines, has gained increasing interest. The rising target power of the gyrotrons, which drives progressively higher Ohmic losses to be removed from the resonator, together with the need for limiting the resonator deformation as much as possible, has put more emphasis on the thermal-hydraulic and thermo-mechanic modeling of the cavity. To cope with that, a multi-physics simulator has been developed in recent years in a shared effort between several European institutions (the Karlsruher Institut für Technologie and Politecnico di Torino, supported by Fusion for Energy). In this paper the current status of the tool calibration and validation is addressed, aiming at highlighting where any direct or indirect comparisons with experimental data are missing and suggesting a possible roadmap to fill that gap, taking advantage of forthcoming tests in Europe.

Keywords: gyrotron resonator; multi-physic simulation; thermal-hydraulics; cooling; mini-channels; Raschig rings; validation

Citation: Savoldi, L.; Avramidis, K.A.; Albajar, F.; Alberti, S.; Leggieri, A.; Sanchez, F. A Validation Roadmap of Multi-Physics Simulators of the Resonator of MW-Class CW Gyrotrons for Fusion Applications. *Energies* **2021**, *14*, 8027. https://doi.org/10.3390/en14238027

Academic Editor: Dan Gabriel Cacuci

Received: 28 October 2021
Accepted: 29 November 2021
Published: 1 December 2021

Publisher's Note: MDPI stays neutral with regard to jurisdictional claims in published maps and institutional affiliations.

1. Introduction

In the magnetic confinement machines for nuclear fusion, gyrotrons [1] are employed to heat the plasma [2–5] by injecting a microwave beam at a frequency equal to the electron rotation frequency in the magnetic field which confines the plasma or its harmonics [6]. The microwaves are generated in the so-called resonator or cavity of the gyrotron, see Figure 1, where an electron beam produced by a magnetron-injection gun interacts with an electromagnetic wave, excited therein at its resonance frequency [7]. In the partial conversion of the electrons kinetic energy in microwaves, part of the energy is lost by the system and deposited by induced parasitic currents (Ohmic losses) in the resonator inner wall. Besides the capability of exhausting the power deposited in the cavity, an accurate control of its thermal deformation [8] is needed in order not to lose the optimum resonance regime, avoiding at the same time the plastic (permanent) deformation of the resonator. Those aspects are strictly related. Indeed, the heat released in the resonator, driving the thermal deformation, depends on the interaction between the electrons and the electromagnetic wave, affected in turn by the deformation in a complex thermal-hydraulic, thermo-mechanic and electro-dynamic multi-physics problem, similar to some extent to those encountered in other fields related to nuclear fusion, for instance, that of superconducting magnets [9,10].

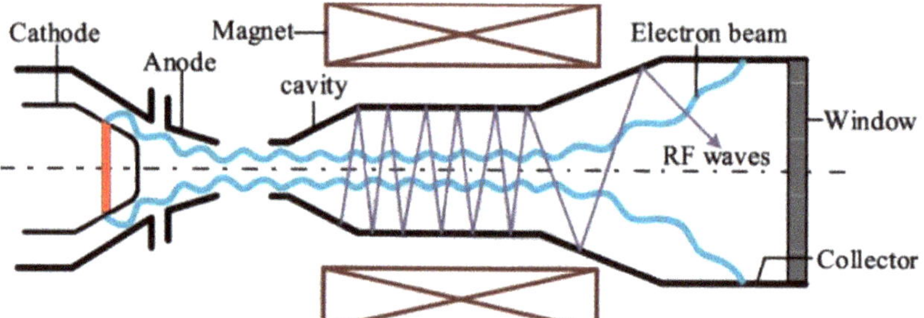

Figure 1. Schematic view of the interaction between the electron beam (blue solid line) and the electromagnetic wave in the resonator, producing the microwaves (in purple).

Since the Ohmic heating of the internal surface of the gyrotron resonator can reach peaks up to 25 MW/m^2 [6], the component needs to be actively cooled. The coolant, which is typically subcooled water at room temperature and weakly pressurized to a few tenths of MPa, flows in an annular space around the cavity, limiting its thermal expansion and the related thermal stresses. Different techniques are currently adopted by different manufacturers to enhance the heat removal, and namely:

- A porous metal matrix made of mm-size copper Raschig rings (RR) [11], brazed together using an Au alloy. This solution has been adopted so far for the tubes manufactured in Europe. Cavities cooled using RR are inserted in the 140 GHz, 1 MW gyrotrons currently installed at W7-X [12], in the dual frequency devices commissioned in the TCV at the EPFL (Lausanne Switzerland) [11,13], and in the EU 1 MW 170 GHz first industrial prototype gyrotron [14] targeting ITER specifications [15,16]. RR-cooled cavities are also under design for the upgraded 1.5 MW gyrotron for W7-X [17] and for the next version of the EU 1 MW 170 GHz gyrotron. A sketch of this cooling configuration is shown in Figure 2a.
- Longitudinal mini-channels (i.e., channels with a characteristic dimension of the cross section of the order of the millimeter, see Figure 2b), probably adopted in the 170 GHz, 1 MW Japanese gyrotrons [18] for ITER [19]. This technique is under consideration in Europe as an alternative for the upgrade of the W7-X gyrotrons to 1.5 MW.
- Azimuthal micro-channels (i.e., channels with a characteristic dimension of the cross section below the millimeter), most likely adopted in the Russian design of the MW-class gyrotrons, on which very rare info in published literature is present. A sketch of this configuration is reported in Figure 2c.

The numerical modelling of the resonator dynamic response to the heat load in operation is one of the key issues for the efficient design and optimization of a new high-performance device [20]. It has already been proven that a proper model for the resonator should be a multi-physics model [21], as it should include at least an electro-dynamic (ED) model to compute the heat load on the cavity surface from the radio-frequency (RF) wave interacting with the electron beam, a thermal-hydraulic (TH) model to compute the temperature of the cavity and a thermo-mechanical (TM) model to compute the cavity deformation. Note that the three ED, TH and TM models are intrinsically connected: the heat load resulting from the ED model is the driver of the TH model, the cavity temperature is the driver of the TM model and the deformation of the cavity affects the RF wave frequency and spatial structure (hence the beam-wave interaction) and becomes the driver of the ED model. Attempts on multi-physics modelling, schematically reported in Figure 3, started some time ago [22]. However, the modeling reached a fully self-consistent status only when the multi-physics tool for the integrated simulation of the cavity (MUCCA) was consolidated [21,23].

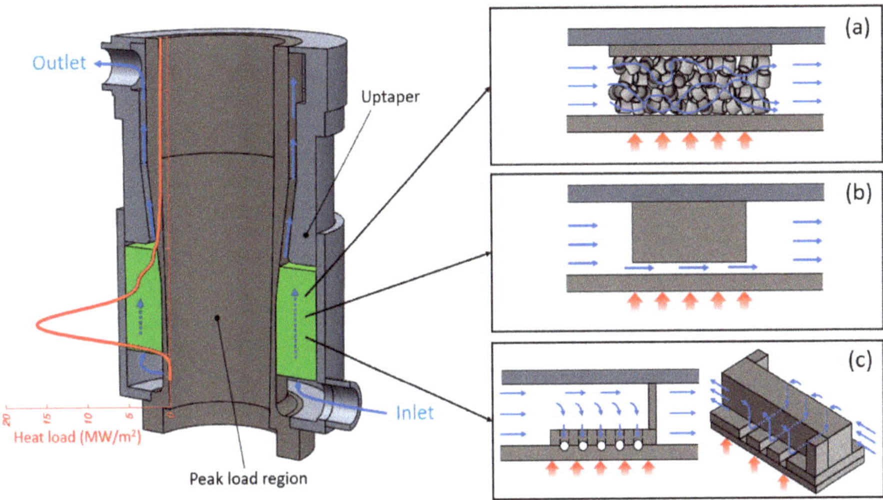

Figure 2. Schematic view of the gyrotron resonator, with different cooling strategies: (**a**) Raschig rings, (**b**) longitudinal mini-channels, (**c**) azimuthal micro-channels. The fluid path is highlighted with blue arrows. An example of a typical heat load distribution along the cavity axis is also reported in red in the left sketch.

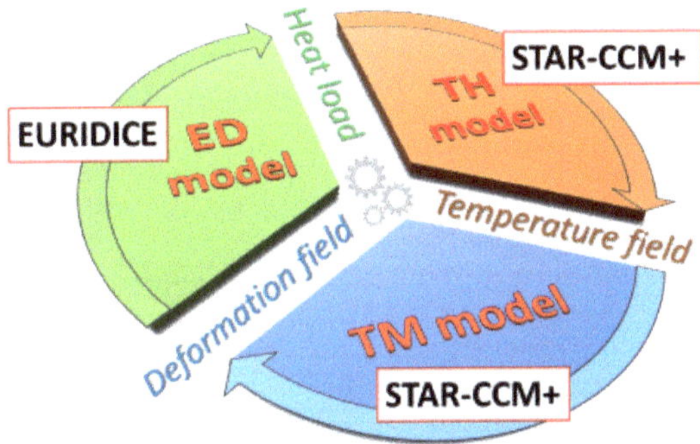

Figure 3. Structure of the MUCCA tool, with the TH, TM and ED modules highlighted, together with their interactions. The software used for the different modules is also reported.

After some calibration of the key parameters was performed in the early stages of development, taking advantage of the few experimental data which were available back then, the tool has been already widely adopted to suggest possible improvements of the design of the cavity of the European gyrotron for ITER [24,25], as well as that of the Swiss TCV. However, systematic validation of the MUCCA tool, both as far as its different modules and the whole coupled models are concerned, has not been fully addressed so far, also because of the lack of a wide and controlled experimental database.

In view of the large number of experimental campaigns addressing different gyrotrons and cavity mock-ups, currently ongoing or foreseen in the near future, a more systematic validation of the tool can now be planned. While several studies were carried out in the past few years on specific aspects of the modelling of the resonator, the aim of this paper is first to perform for the first time a comprehensive evaluation of the calibration and partial validation exercises done so far for the multi-physics modelling of the gyrotron cavity, before moving to the description of present and future tests in regards to their possible contribution to the validation of the model.

2. Status of the Simulator Calibration

Three types of the physical quantities involved in the multi-physics gyrotron model can be encountered, namely: (i) directly measurable, albeit with some uncertainty (e.g., gyrotron output power, inlet-outlet temperature difference), (ii) not directly measurable but calculated (e.g., magnetic field values, electron beam parameters) and (iii) neither directly measured nor strictly calculated (e.g., thermal conductivity of the RR block), whose value can be tuned to fit the measurement/calculation. The calibration procedure described in this section refers to this last type of physical quantities and will be analyzed for each of the models involved in the cavity simulator.

2.1. Calibration of the ED Model

The beam-wave interaction in the gyrotron cavity is simulated using the code package EURIDICE [26], which is based on the slow-variables, self-consistent gyrotron interaction model [27]. The surface density, ρ, of Ohmic losses on the cavity wall (heat flux) is related to the RF wave generated in the cavity through the electrical conductivity, σ, of the wall [28], as shown in Equation (1):

$$\rho = \frac{1}{2\sigma\delta}|\mathbf{H}_t|^2 \tag{1}$$

Here, $\mathbf{H}_t$ is the RF magnetic field component tangential to the cavity wall, $\delta = [2/(\mu\sigma\omega)]^{1/2}$ is the skin depth, $\omega = 2\pi f$ is the angular frequency of the wave, and μ is the permeability of the wall ($\mu = \mu_0 = 4\pi \times 10^{-7}$ H/m for gyrotron cavities). The dependence of the conductivity on the temperature profile along the cavity (provided by the TH simulation) is taken into account by using an appropriate material database, e.g., [29] in the case the cavity wall is made of Glidcop. In addition, a correction factor for the Ohmic loading ρ, due to a typical 0.1 µm rms surface roughness of the cavity wall, is calculated using the Hammerstad/Bekkadal formula [30] and is inserted in Equation (1) as a multiplication factor.

The beam-wave interaction model itself contains physical quantities related to the electrons (energy, velocity etc.) and to the RF wave (amplitude, frequency etc.) as well as universal constants, therefore, a calibration is not actually needed. On the other hand, the electrical conductivity, σ, and its dependence on temperature and surface roughness are nominally known from experimental/empirical data. Consequently, there could be, in principle, room for calibration of the conductivity value. However, this has not been pursued because the used data are considered to be well-established. In addition, the calibration of σ would be challenging because the difference in the results of the ED model incurred using a different conductivity value can be masked by the uncertainty with respect to the internal losses between the gyrotron cavity and the load, where the gyrotron output power is measured.

2.2. Calibration of the TH Model

As far as the TH model is concerned, the conjugate heat transfer in the cavity is solved. Typically, Reynolds-averaged Navier–Stokes (RANS) equations are used for the fluid domain as a fair engineering compromise between accuracy and computational effort and also in view of the fact that the flow thermal-hydraulics is, for the problem at hand, enslaved to the evaluation of the temperature field in the cavity wall. A delicate aspect is always the selection of a suitable turbulence model for the coolant flow within the cooling structure adopted for the refrigeration of the cavity. In the case of RR, mini-channels and micro-channels, the SST-Menter κ–ω model has been widely used [31–33] for the simulation of all the different devices (see more on that in the validation section below). This model is able to work as a standard κ–ω model in the near wall region and as the κ–ε model in the fully turbulent region, through a blending function adding a cross diffusion term in the main stream. No specific calibration has been done on the default parameters of the turbulence model, such as the adjustable constants σ_κ, σ_ω, $\sigma_{\omega 2}$ or β in the transport equations for the turbulent kinetic energy κ and for the specific rate of dissipation of the turbulent kinetic energy, ω; see [34] for more details.

For the simulations of the cavities equipped with RRs, an additional uncertainty concerns the effective thermal conductivity, k_{eff}, of the metal matrix, which is actually unknown. The RR matrix was generated in the simulations by means of the discrete element modeling (DEM) approach, and the rings have point contacts with each other. The soldering material, actually contributing to enlarging the thermal contact between the rings, is typically not modelled (see [31] for more details). A lower bound for the value of k_{eff} to be used in the simulations has been identified via the analysis of the experimental results from a first planar cavity mock-up equipped with RRs, tested in 2015 at the Areva premises in Le Creusot, France [35]. The planar mock-up, shown in Figure 4a,b, had a planar circular target, exposed in a vacuum chamber to the heating provided by an electron beam gun on a 33 mm × 33 mm surface; see Figure 4c. On the back of the target, a RR block was placed. Subcooled water, flowing inside the mock-up, provided the active cooling of the target, passing through the RR matrix. The value of k_{eff} was a key parameter to allow a proper evaluation of the heat removed from the target. The effectiveness of the cooling was assessed via measurement of the temperature of the heated surface using an infrared camera (no accurate calorimetry was possible, unfortunately, due to the lack of temperature measurement at the outlet of the mock-up). The IR camera was calibrated to accurately measure a surface temperature larger than $T_{IR} = 350\,°C$. However, that temperature was never reached in the test campaign, despite the variation of the water flow rate from 90 L/min down to 50 L/min (the minimum value which was tested), even for very large heat loads (up to 25 MW/m^2).

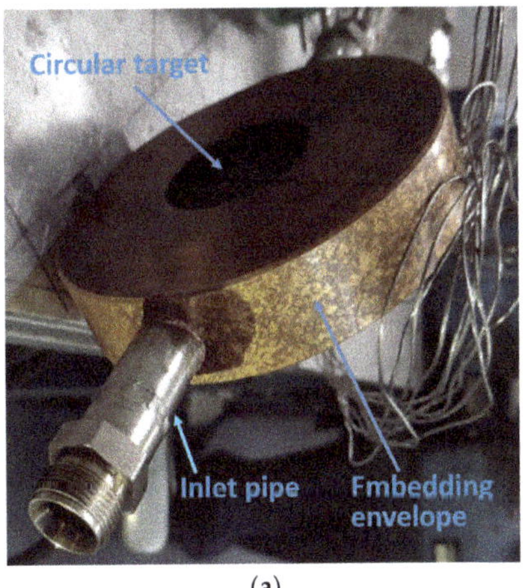

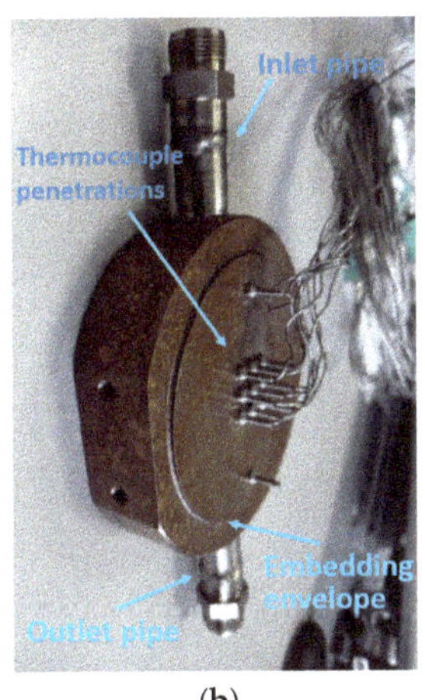

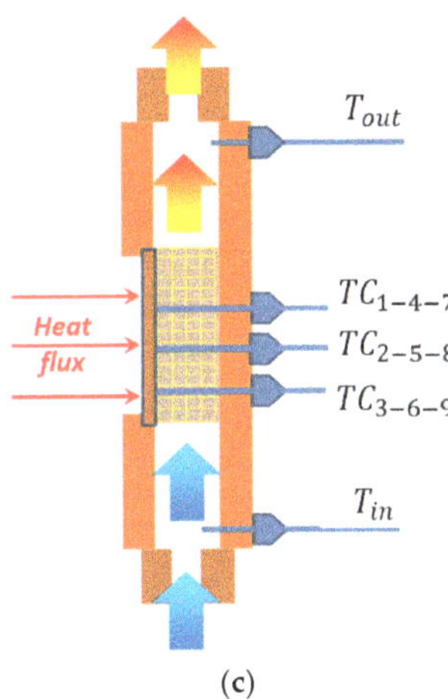

(a)   (b)   (c)

Figure 4. Cavity planar mock-up equipped with RR: (**a**) front/side view with the circular heated target in evidence, (**b**) back view with the instrumentation (thermocouples), (**c**) sample cross section.

The fact that a surface value of T_{IR} was never reached in the tests allowed the lower bound for k_{eff} to be calibrated to 2000 W/m·K in the simulations, which kept the computed surface temperature below T_{IR} in the worst case (lowest flow rate). Such a (unphysically) high value is justified by the fact that, as mentioned already, k_{eff} in the simulations mimics the product of the value of the physical conductivity and of the actual finite area of the thermal contact among the rings, which was not accurately modelled. The simulated results, with the calibrated $k_{eff} = 2000$ W/m·K for the case with the lowest flow rate tested, i.e., 50 L/min, are collected in Figure 5a, showing a good agreement with the experimental points measured with the IR camera, of which, however, the reliability is unknown at such low temperatures. The planar mock-up was also equipped with 9 thermocouples (TC) in the RR matrix, specifically in contact with the back side of the target, but the calibration of the model could not rely on them because of the uncertainty about the exact position of the TC heads (in proper thermal contact with the target, or within a ring, or surrounded by water). Note also that, in the simulations reported in Figure 5a, no radiative losses from the mock-up were considered. Some boiling of the water was also detected in the simulations, due to the high temperature reached also in the RR region, where the saturation temperature was estimated at 169 °C. The boiling regime was handled using the Rohsenow boiling model [36] in a VOF multiphase flow simulation [37], but no specific calibration was possible for the parameters it includes.

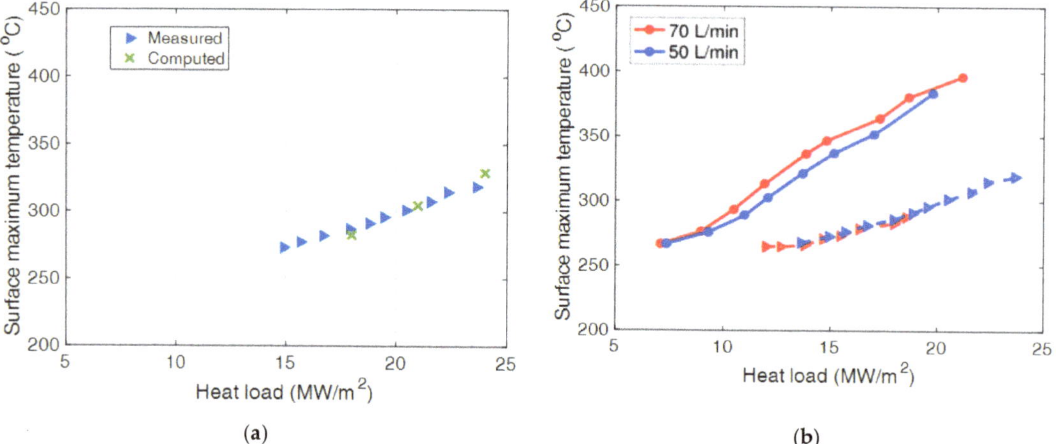

(a) (b)

Figure 5. Cavity planar mock-ups equipped with RR: surface temperature measured as a function of the heating power. (a) 2015 test campaign with a water flow rate of 50 L/min: experimental data measured via the IR camera (blue symbols) and computed data with the calibrated model (green symbols). (b) Comparison between the data from the 2015 test campaign (triangles), measured with the IR camera, and the data from the 2016 test campaign (circles), measured with the pyrometer, with a water flow rate of 50 L/min (blue lines and symbols) and 70 L/min (red lines and symbols).

In view of the issues with the sample instrumentation discussed above [38], a test campaign on a second planar mock-up equipped with RR was performed in 2016, again at the Areva premises, in the same test rig as the first one, to confirm the results obtained in the first test campaign. The instrumentation also included a pyrometer, calibrated against a thermocouple for temperatures above 300 °C. The heat flux was on a square surface slightly smaller than that in the 2015 campaign (28 mm × 28 mm). At a comparable flow rate and heat flux, the values measured in 2016 were expected to be below those of 2015 because of the total deposited power scales with the extension of the heated surface. Unfortunately, the 2016 results turned out to be significantly higher than those of the 2015 test campaign; see Figure 5b. Note that in 2016 the radiative losses from the heated surface were measured calorimetrically (and turned out to be 10–15% of the incident power), and the heat load represented in Figure 5b is the net heat flux entering the sample. A 10–15% increase in the power, however, cannot explain the difference in the two set of measurements. Although nominally identical, the two samples could suffer from a different random distribution of the RR, which could partly explain the difference between the 2015/2016 datasets. The 2016 data look, however, odd in another respect, as the temperature values measured by the pyrometer in 2016 at 70 L/min are higher than those at 50 L/min, and this is very hard to justify. The overall picture confirmed then some issues in the pyrometer measurements in the 2016 test campaign, so that the calibration of k_{eff} was never repeated on the 2016 measured data.

For the simulations of the cavities equipped with longitudinal MCs, the model calibration could benefit from the tests of a planar mock-up, identical to that equipped with RRs but equipped with semi-circular MCs on the back of the target, also tested in 2016 at Areva [38]. In more detail, with reference to the manufacturing process adopted there, which simply put the heated target in contact with a block where the MCs were machined, the analysis of the test results allowed the calibration of the contact thermal resistance between the target and the MC block, as extensively documented in [39]. Note, however, that if a different manufacturing procedure is adopted for the actual cylindrical cavities equipped with MCs, the calibration performed on the planar mock-up is not useful.

As far as the azimuthal micro channels are concerned, no dedicated calibration of the TH model has been performed so far, also in view of the lack of available experimental data on that specific cooling strategy.

2.3. Calibration of the TM Model

As far as the TM model is concerned, the simulations are restricted to the pure solid region of the TH domain, made by Glidcop, for which the appropriate set of material properties (thermal expansion, Young modulus) are adopted. A 3D, steady state, finite-element solid-stress model is adopted, with linear, isotropic and elastic material properties. The constraints applied to the structure vary from case to case. The thermal stress is computed considering the inlet temperature of the fluid as a zero-stress temperature for the assembly.

No specific calibration has been done so far on the TM model. In fact, since the computational domain is typically restricted to the resonator wall and does not include, for instance, the RR matrix, all the parameters of the computational model are nominally known.

2.4. Calibration of the Whole Simulator

Since doubts still remained after 2015 on a proper value of k_{eff} in the simulations of cavities equipped with RRs, its calibration was attempted indirectly by comparing the results computed using the whole multi-physics simulator of the resonator to experimental data coming from different test campaigns on the EU 1 MW 170 GHz ITER gyrotron prototype, the resonator of which was cooled using an RR matrix. A first test campaign was performed at the Karlsruhe Institute of Technology Premises [14,40,41]. The tests were carried out for pulse durations up to 180 s. The results computed by the MUCCA tool for different values of k_{eff}, namely, 2000 W/m·K, 2600 W/m·K and 3600 W/m·K, were compared in terms of the frequency shift, due to thermal expansion of the cavity, to the experimental results, in different operating conditions, resulting in the picture reported in Figure 6a. The k_{eff} values in the range 2000–2600 W/m·K returned a frequency shift slope comparable to the experimental one, when the interpolation error is considered (Figure 6b). The estimation of a more precise value would require more simulation points to reduce the interpolation error. Note that this second calibration of k_{eff} is in line with the first one, in the sense that the lower bound for k_{eff} found in the first calibration is retained. Also in this case, the k_{eff} values remain unphysical though.

Experimental data from the same gyrotron tested for longer pulse durations in a range up to 215 s in the European Gyrotron Test Stand at EPFL, Switzerland [41], could have helped in reducing the range of calibration for k_{eff} [42]. However, the data turned out not to be very useful for additional calibration of the model since strong deviations in the ED model were encountered with respect to the experimental results, most probably due to uncertainties in the actual magnetic field profile and electron beam parameters. (See also next section on the validation of the ED model.) Moreover, there was some uncertainty in the flow rate to the cavity, which increased the inaccuracy of the TH simulations.

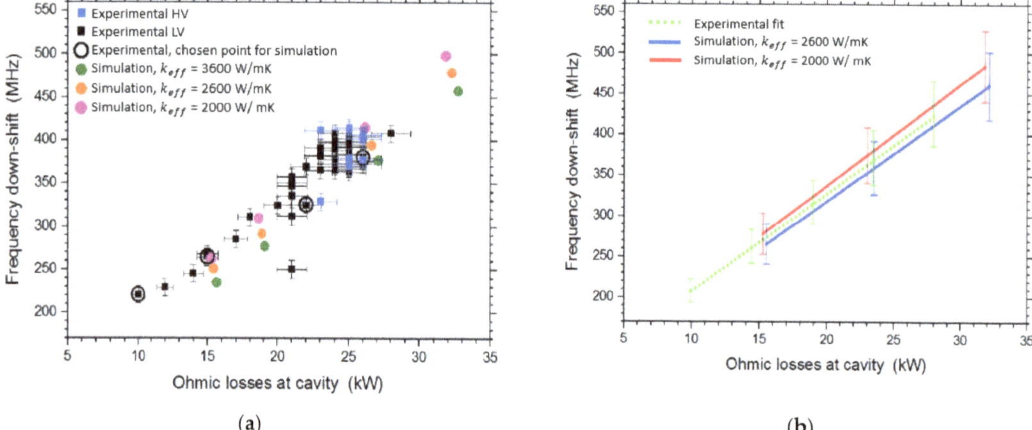

Figure 6. Comparison between the measured and computed frequency shift for several low-voltage (LV) and high-voltage (HV) operating points, as a function of the Ohmic power dissipated at the cavity. (**a**) Detailed map of tested and computed points; (**b**) linear interpolation.

3. Status of the Simulator Validation

Following the discussion on the calibration, the status of validation of the different models of the resonator simulator against experimental results is detailed below.

3.1. Validation of the ED Model

The ED model has already been successfully validated against experimental results for a number of operating points, involving various high-power gyrotrons in short-pulse operation (<10 ms), e.g., the European 170 GHz 1 MW gyrotron for ITER [43], the 170 GHz 2 MW coaxial gyrotron at KIT [44] or the 140 GHz 1.5 MW gyrotron for W7-X [45]. Note that, in short-pulse operation, the cavity thermal expansion is negligible; hence, the ED model can be validated against experimental results without the need of TH and TM simulations. In addition, at the ms time scale, the neutralization of the space charge of the electron beam due to ionization of the residual gas in the gyrotron is still not taking place [46], and consequently, the electron kinetic energy can be calculated without resorting to estimations about the neutralization level.

The critical point with respect to the ED simulations is not the validation of the model itself but the careful assessment and consideration of the inherent experimental uncertainty with respect to the physical quantities used in the model. Apart from the known uncertainty about the operating parameters (e.g., acceleration voltage, electron current, currents of the magnet coils, etc.), there can be factors influencing the electron beam properties that cannot be accurately quantified. These include internal geometrical imperfections, misalignments/tilts with respect to the external magnetic field axis, deviations of the applied magnetic field profile from the nominal one, varying quality of electron emission (as related to the condition of the emitter and the vacuum), etc. It should be also noted here that the electron beam properties (i.e., the velocity and position of electrons), which are used as input to the ED model, cannot be directly measured in the sealed gyrotron and they can only be calculated via beam optics codes. In addition to the above, there can be other effects occurring outside the modelling range of the ED model, which is confined only to the beam-wave interaction in the cavity, that affect the electron beam properties. For instance, there can be parasitic RF oscillations in the gyrotron beam tunnel preceding the cavity (see [47] and references therein). Finally, in longer-pulse operation (>100 ms), the space charge neutralization level in the gyrotron (and hence the electron kinetic energy) can only be estimated. Because of all these aspects, discrepancies between the ED simulation and the experiment are not uncommon, especially in longer-pulse operation of the gyrotron.

3.2. Validation of the TH Model

The validation of the different TH models developed for the resonator equipped with RRs or MCs had different steps. First, the validation exercises of the pure hydraulic model (no heating) are described, before considering the validation in the case of heating.

The validation of the hydraulic model of the cavity equipped with RRs relied on the tests of the planar mock-up with two different fluids (subcooled water tests at Areva [38], air tests in a solar furnace at the Plataforma Solar de Almeria PSA, Spain [32]) and on the test of a dedicated cylindrical mock-up [31]. The interpretation of the entire set of hydraulic tests performed on the planar mock-up is described in [32], in terms of pseudo-dimensionless variables in order to be able to show on the same plot data points coming from tests performed using different fluids. Figure 7a collects the measured and computed results, showing a good capability of the numerical model to reproduce the experimental data, notwithstanding the large error bar affecting them. Moving to the hydraulic characteristic measured at THALES on a cylindrical mock-up, the results of the numerical model with κ–ω STT turbulence closure showed again a very good agreement with the measurements; see Figure 7b.

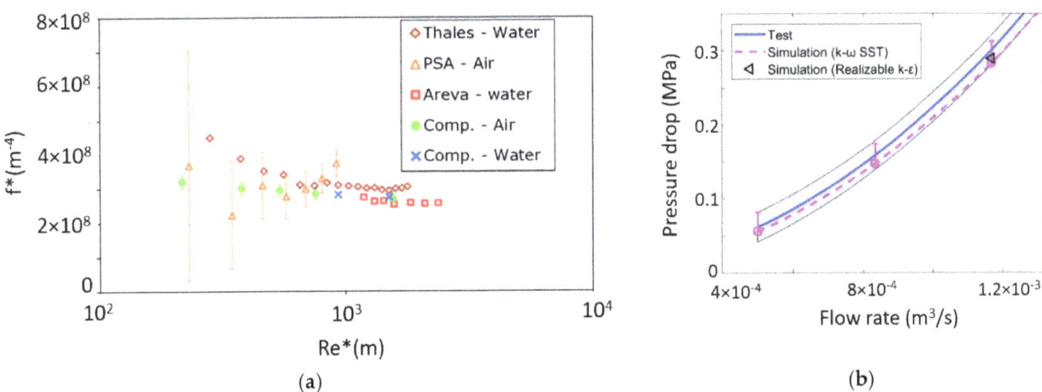

Figure 7. (a) Pseudo-dimensionless hydraulic characteristic (Re* as a function of f*) of the planar mock-up equipped with RR: experimental (open symbols) and computed (solid symbols) (modified from [32]). (b) Comparison between measured and simulated hydraulic characteristic for the cylindrical mock-up equipped with RRs (modified from [31]). The shaded area represents the error band on the experimental results.

A second cylindrical mock-up was tested at the THALES premises in 2019, equipped with circular MCs. A detailed validation of simulations performed with different RANS turbulence closures on the hydraulic characteristic, see Figure 8, with a rigorous assessment of the errors associated to the measured and simulated data through a multivariate metric procedure [48] allowed the identification of the lag EB κ–ε model as the most suitable for the numerical modelling of the case at hand.

Validation of the conjugate heat transfer model for cavities equipped with RRs has been performed so far relying on the TC measurements in the tests of the planar mock-up at the PSA, using air as the coolant. The calibrated value of k_{eff} = 2000 W/m·K was kept frozen. A very good consistency was found between the computed and measured values, in the entire flow rate and heat flux ranges, with the computed values always lying in the error bar of the measured data [32].

For cavities equipped with MCs, some experimental indications of the reliability of the adopted model came recently from the tests of a cavity mock-up at KIT [49]. However, a very limited temperature range was tested in the 2020 test campaign, which cannot be representative of the actual working conditions in the gyrotron cavity.

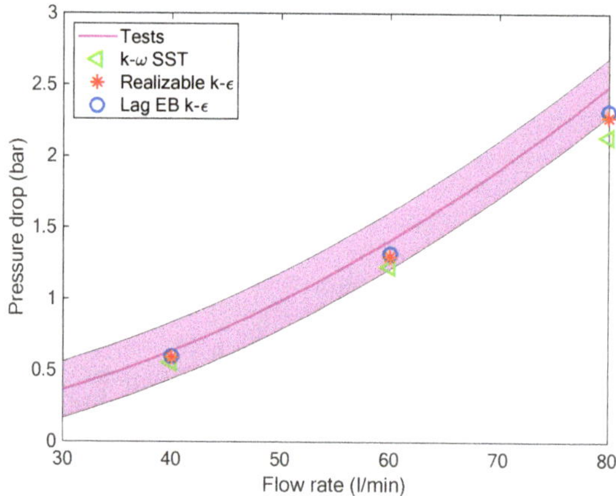

Figure 8. Comparison between the measured hydraulic characteristic of the cylindrical mock-up equipped with MCs (solid line with shaded area) and that computed using different turbulence models (symbols).

3.3. Validation of the TM Model

While no dedicated and independent validation of the TM model has been performed so far, some indirect evidence of the quality of the TM model arrived from post-mortem examination of the resonator of the EU 170 GHz 1 MW CW gyrotron prototype for ITER, which was dismantled and refurbished after the end of the first test campaigns at KIT and EPFL. The simulation performed with MUCCA on the operating points of the gyrotron showed, for a high power level, a peak stress beyond the plasticity level. The post-mortem analysis revealed the presence of some permanent deformation in the cavity, which supports the high stress level computed by the TM model. In more detail, the permanent deformation was found to be anisotropic, and that was attributed to inhomogeneous cooling in the azimuthal direction, leading to inhomogeneous deformation, which was in fact also computed in the simulations.

3.4. Validation of the Whole Simulator

The validation of the multi-physics simulator, with particular reference to the iterative coupling procedure among the different modules of the simulator and its capability to capture the main features of the cavity behavior, has been performed against pulses of about 2 s of the dual frequency gyrotron at TCV, EPFL [3], for operation at a high frequency (i.e., 126 GHz). The gyrotron cavity in that case relies on a slightly different geometry with respect to that of the EU 170 GHz 1 MW CW gyrotron prototype for ITER used for the calibration. In more detail, the cooling section equipped with RR was slightly longer and thinner, with a different uptaper shape. The validation exercise has been performed keeping the calibration parameters, obtained previously with different devices, unchanged. A water mass flow rate of 45 L/min and a reference pressure of 5 bar was adopted in the simulations. The ED settings for the computation of the heat load on the cavity considered an electron beam kinetic energy of 78 keV, a beam current of 40 A, an electron velocity ratio of $\alpha = v_\perp / v_{||} = 1.3$ and an electron beam radius of $R_b = 10.6$ mm. The magnetic field was set to 5 T. At the computed operating point, the peak temperature reached values of ~260 °C for a peak load of ~17 MW/m^2; see Figure 9a. It is evident there that only the first layer of RR actually contributes to the heat removal, calling for a more detailed modelling of the thermal coupling within the porous matrix. The validation exercise returned good agreement between the computed and the experimental results at the highest frequency (see Table 1), with the computed cavity wall power within 10% of that estimated from the experimental calorimetry (Note, however, that the plausible presence of a bypass flow connecting the cavity cooling path to that of the launcher was neglected in the experimental colorimetry.).

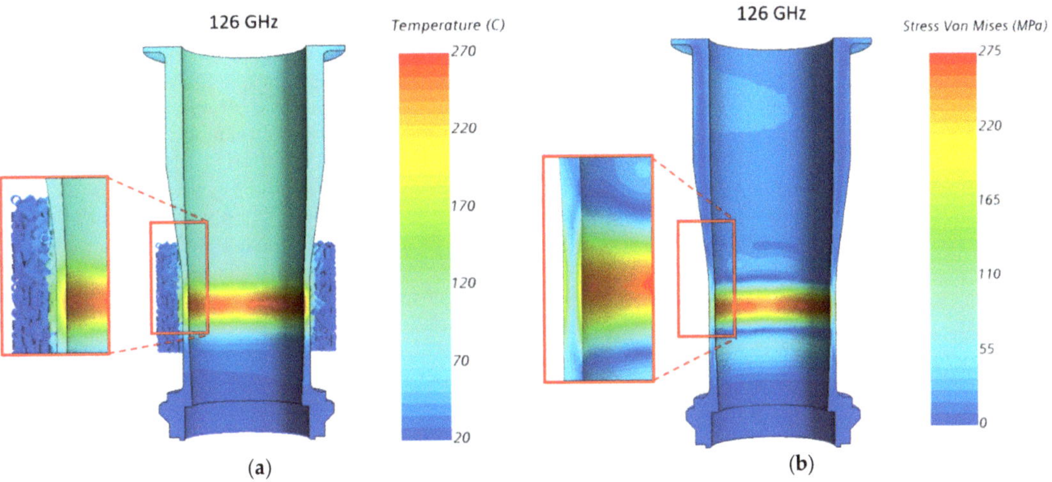

Figure 9. (a) Computed temperature map and (b) computed Von Mises stress map on the solid domain of the resonator of the TCV gyrotron, for the operating point at 126 GHz considered for the simulator validation.

In addition, the computed frequency shift agreed with the measured one within 8%. The computed RF power at the cavity exit lies almost in the experimental error bar; see again Table 1. The analysis of the computed stresses at the operating point, see Figure 9b, shows a peak value which is ~20% higher than the predicted value for the yield strength, according to the ITER database, but below the ultimate strength. The experimental results were quite reproducible at the beginning of the experimental campaign, implying no plastic deformation of the cavity. The comparison above gives then a flavor of the accuracy of the thermo-mechanical simulations performed within this study, which remains on the conservative side. As a caveat, however, please consider that it is still not clear how values of the peak stress, which are very localized axially and radially inside the cavity wall, should be compared to the limiting values.

Table 1. Summary of the main results of the validation of the cavity simulator against the experimental data of the high frequency operating point of the dual frequency gyrotron at TCV, EPFL.

Test Case	Parameter	Experimental Value	Computed Value
126 GHz	RF @ cavity exit [MW]	$1.20 \pm 5\%$	1.07
	Cavity wall power [kW]	$42 \pm 5\%$	38
	Actual frequency [GHz]	125.76	125.89
	Frequency shift [MHz]	360	330 [1]

[1] computed with an estimated 62% neutralization of the beam space charge during the ~2 s pulse.

4. Opportunities for Further Validation

4.1. Available Experimental Data

A first dataset of hydraulic measurement on azimuthal micro-channels is now available from the tests of prototype cavities realized via additive manufacturing at IPP, Greiswald [50]. At the same time, the hydraulic characteristic of a cylindrical mock-up of a 140 GHz cavity equipped with MCs [35], under consideration for the next version of the upgraded 1.5 MW gyrotron for W7-X, has been measured at Thales.

The tests performed at Areva on the planar mock-ups equipped with RRs have only been partially used, and the thermocouple traces could return useful information for a further model validation, complementing the validation performed using air at the PSA, since water was used in the 2015–2016 tests as coolant in a wide range of temperatures, hitting boiling conditions.

An experimental campaign on the planar mock-ups equipped with MCs is currently ongoing in a solar furnace at the PSA, using air as coolant, in a temperature range much wider than what has been tested so far. The planar mock-up with MCs, already tested in water as reported in [39], is, however, not representative of future cavities equipped with longitudinal MCs, in view of the different manufacturing technique.

Data from the EU 1 MW 170 GHz refurbished gyrotron prototype for ITER, which has a different geometry for cavity cooling with respect to the gyrotron cavity used for the model calibration, are available from the recent tests at KIT, however with little information on the frequency shift. Measurements including the frequency shift, as well as the power deposited in the cavity estimated via calorimetry, are available from the operation of the second dual frequency gyrotron at EPFL, for both high-frequency (126 GHz) and low-frequency (84 GHz) operation.

4.2. Planned Experimental Campaigns

An experimental campaign for the measurement of the effective thermal conductivity of the RR matrix, to be performed at Thales, is currently under design at Politecnico di Torino, based on the setup discussed in [51].

More relevant measured data will be available from the tests of the cylindrical mock-up of a 140 GHz cavity equipped with MCs [35]. The tests will be performed at the KIT test bench [49] by the end of 2021, using inductive heating and both water and air as coolant.

As far as data from the gyrotron operation are concerned, additional data from long-pulse operation of CW gyrotrons will also be available. In particular, the EU 1 MW 170 GHz refurbished gyrotron prototype for ITER is currently being tested at SPC, EPFL. The aim is to extend the pulse duration up to 1000 s at nominal power and the tests are expected to be completed by the end of 2021. The first version of the upgraded 1.5 MW 140 GHz gyrotron for W7-X is expected to be delivered at KIT by the end of 2021 and tested during the first quarter of 2022 with pulses up to 180 s.

There are three relevant quantities that can be measured during long-pulse operation of the gyrotron: (i) the temperature difference between the inlet and outlet of the cavity cooling circuit, for a given flow rate; (ii) the drop of the operating frequency (frequency shift) during the first seconds of the pulse, which is caused by the cavity thermal expansion and the neutralization of the beam space charge and (iii) the gyrotron output power measured calorimetrically on a load. The multi-physics simulator is claimed to be validated when its results agree with the three measured quantities. In particular, using the measured temperature difference in the cavity cooling circuit, the Ohmic losses in the cavity can be calculated calorimetrically. From the electron beam parameters, calculated using beam optic codes, and from the cavity inner contour, provided by the TH and TM simulation, the ED simulation calculates the electromagnetic field in the cavity, which should be consistent with the measured Ohmic losses as well as with the measured output power, under the condition that the neutralization level and the additional losses between the cavity and the load are correctly estimated. In addition, the ED simulation provides the operating frequency, which should be consistent with the measured frequency shift.

The final goal for the validation of the multi-physics simulator is the comparison against experimental results coming from various tested gyrotrons, which differ (in terms of design and operating conditions) from those used for the calibration of the simulator. A successful validation against the forthcoming experiments will grant the multi-physics simulator the status of a reliable predictive tool, as far as the behavior of the gyrotron cavity is concerned.

Author Contributions: Conceptualization, L.S. and K.A.A.; methodology, L.S. and K.A.A.; validation, L.S., K.A.A. and S.A.; formal analysis, L.S. and K.A.A.; investigation, L.S. and K.A.A.; resources, L.S., K.A.A., S.A. and A.L.; data curation, L.S. and K.A.A.; writing—original draft preparation, L.S. and K.A.A.; writing—review and editing, F.A., S.A., A.L. and F.S.; visualization, L.S.; supervision, F.A. and F.S.; project administration, F.S.; funding acquisition, F.A., A.L. and F.S. All authors have read and agreed to the published version of the manuscript.

Funding: The work of LS was partially funded by Fusion for Energy, contract number F4E-2021-EXP-376. The views expressed in this publication are the sole responsibility of authors and do not necessarily reflect the views of F4E, the European Commission or ITER.

Institutional Review Board Statement: Not applicable.

Informed Consent Statement: Not applicable.

Data Availability Statement: Not applicable.

Acknowledgments: The authors are grateful to Andrea Allio and Rosa Difonzo for the support in the production of some of the figures.

Conflicts of Interest: The authors declare no conflict of interest.

References

1. Thumm, M. MW gyrotron development for fusion plasma applications. *Plasma Phys. Control. Fusion* **2003**, *45*, A143–A161. [CrossRef]
2. Wolf, R.C.; Bozhenkov, S.; Dinklage, A.; Fuchert, G.; Kazakov, Y.O.; Laqua, H.P.; Marsen, S.; Marushchenko, N.B.; Stange, T.; Zanini, M.; et al. Electron-cyclotron-resonance heating in Wendelstein 7-X: A versatile heating and current-drive method and a tool for in-depth physics studies. *Plasma Phys. Control. Fusion* **2018**, *61*, 014037. [CrossRef]
3. Zohm, H.; Stober, J.; Reisner, M.; Angioni, C.; Navarro, A.B.; Bobkov, V.; Bock, A.; Denisov, G.; Fable, E.; Fischer, R.; et al. Exploring fusion-reactor physics with high-power electron cyclotron resonance heating on ASDEX Upgrade. *Plasma Phys. Control. Fusion* **2020**, *62*, 024012. [CrossRef]
4. Moro, A.; Bruschi, A.; Darcourt, O.; Fanale, F.; Farina, D.; Figini, L.; Gandini, F.; Henderson, M.; Hunt, R.; Lechte, C.; et al. Design of Electron Cyclotron Resonance Heating protection components for first plasma operations in ITER. *Fusion Eng. Des.* **2020**, *154*, 111547. [CrossRef]
5. Ruess, S.; Avramidis, K.A.; Fuchs, M.; Gantenbein, G.; Ioannidis, Z.; Illy, S.; Jin, J.; Kalaria, P.C.; Kobarg, T.; Pagonakis, I.G.; et al. KIT coaxial gyrotron development: From ITER toward DEMO. *Int. J. Microw. Wirel. Technol.* **2018**, *10*, 547–555. [CrossRef]
6. Thumm, M.K.A.; Denisov, G.G.; Sakamoto, K.; Tran, M.Q. High-power gyrotrons for electron cyclotron heating and current drive. *Nucl. Fusion* **2019**, *59*, 073001. [CrossRef]
7. Kuftin, A.N.; Lygin, V.K.; Manuilov, V.N.; Raisky, B.V.; Solujanova, E.A.; Tsimring, S.E. Theory of helical electron beams in gyrotrons. *Int. J. Infrared Millim. Waves* **1993**, *14*, 783–816. [CrossRef]
8. Avramidis, K.A.; Gantenbein, G.; Ioannidis, Z.C.; Pagonakis, I.G.; Rzesnicki, T.; Thumm, M.; Jelonnek, J.; Albajar, F.; Cau, F.; Cismondi, F.; et al. Numerical STUDIES on the influence of Cavity Thermal Expansion on the Performance of a High-Power Gyrotron. In Proceedings of the IVEC 2017–18th International Vacuum Electronics Conference, London, UK, 24–26 April 2017.
9. Zani, L.; Lacroix, B.; Torre, A.; Vallet, J.-C.; Berthinier, C.; Bourcier, C.; Misiara, N.; Nunio, F.; Vallcorba, R.; Van Wambeke, C.; et al. OLYMPE, a multi-physic platform for fusion magnet design: Development status and first applications. *Cryogenics* **2020**, *108*, 103086. [CrossRef]
10. Bottura, L.; Rosso, C.; Breschi, M. A general model for thermal, hydraulic and electric analysis of superconducting cables. *Cryogenics* **2000**, *40*, 617–626. [CrossRef]
11. Marchesin, R.; Albert, S.; Avramidis, K.A.; Bertinetti, A.; Dubrav, J.; Fascl, D.; Gantenbein, G.; Genoud, J.; Hogge, J.P.; Jelonnek, J.; et al. Manufacturing and Test of the 1 MW Long-Pulse 84/126 GHz Dual-Frequency Gyrotron for TCV. In Proceedings of the 2019 International Vacuum Electronics Conference (IVEC), Busan, Korea, 28 April–1 May 2019. [CrossRef]
12. Thumm, M.; Alberti, S.; Arnold, A.; Brand, P.; Braune, H.; Dammertz, G.; Erckmann, V.; Gantenbein, G.; Giguet, E.; Heidinger, R.; et al. EU megawatt-class 140-GHz CW gyrotron. *IEEE Trans. Plasma Sci.* **2007**, *35*, 143–153. [CrossRef]
13. Alberti, S.; Genoud, J.; Goodman, T.; Hogge, J.P.; Silva, M.; Tran, T.M.; Tran, M.Q.; Avramidis, K.; Pagonakis, I.G.; Jin, J.; et al. Progress on the upgrade of the TCV EC-system with two 1MW dual-frequency gyrotrons. In Proceedings of the 2016 41st International Conference on Infrared, Millimeter, and Terahertz waves (IRMMW-THz), Copenhagen, Denmark, 25–30 September 2016. [CrossRef]
14. Ioannidis, Z.C.; Rzesnicki, T.; Avramidis, K.; Gantenbein, G.; Illy, S.; Jin, J.; Kobarg, T.; Pagonakis, I.G.; Schmid, M.; Thumm, M.; et al. First CW experiments with the EU ITER 1 MW, 170 GHz industrial prototype gyrotron. In Proceedings of the 2017 Eighteenth International Vacuum Electronics Conference (IVEC), London, UK, 24–26 April 2017. [CrossRef]

15. Jelonnek, J.; Aiello, G.; Albaiar, F.; Alberti, S.; Avramidis, K.A.; Bertinetti, A.; Brucker, P.T.; Bruschi, A.; Chclis, I.; Dubray, J.; et al. From W7-X Towards ITER and Beyond: 2019 Status on EU Fusion Gyrotron Developments. In Proceedings of the 2019 International Vacuum Electronics Conference, IVEC 2019, Busan, Korea, 28 April–1 May 2019.

16. Darbos, C.; Albajar, F.; Bonicelli, T.; Carannante, G.; Cavinato, M.; Cismondi, F.; Denisov, G.; Farina, D.; Gagliardi, M.; Gandini, F.; et al. Status of the ITER electron cyclotron heating and current drive system. *J. Infrared Millim. Terahertz Waves* **2016**, *37*, 4–20. [CrossRef]

17. Avramidis, K.A.; Ioannidis, Z.C.; Aiello, G.; Bénin, P.; Chelis, I.; Dinklage, A.; Gantenbein, G.; Illy, S.; Jelonnek, J.; Jin, J.; et al. Towards a 1.5 MW, 140 GHz gyrotron for the upgraded ECRH system at W7-X. *Fusion Eng. Des.* **2021**, *164*, 112173. [CrossRef]

18. Sakamoto, K.; Ikeda, R.; Oda, Y.; Kobayashi, T.; Kajiwara, K.; Shidara, H.; Takahashi, K.; Moriyama, S. Status of high power gyrotron development in JAEA. In Proceedings of the 2015 IEEE International Vacuum Electronics Conference (IVEC), Beijing, China, 27–29 April 2015. [CrossRef]

19. Sakamoto, K.; Kajiwara, K.; Takahashi, K.; Oda, Y.; Kasugai, A.; Kobayashi, T.; Kobayashi, N.; Henderson, M.; Darbos, C. Development of high power gyrotron for ITER application. In Proceedings of the 35th International Conference on Infrared, Millimeter, and Terahertz Waves, Rome, Italy, 5–10 September 2010. [CrossRef]

20. Savoldi, L.; Albajar, F.; Alberti, S.; Avramidis, K.A.; Bertinetti, A.; Cau, F.; Cismondi, F.; Gantenbein, G.; Hogge, J.P.; Ioannidis, Z.C.; et al. Assessment and Optimization of the Cavity Thermal Performance for the European Continuous Wave Gyrotrons. In Proceedings of the 27th IAEA Fusion Energy Conference (FEC 2018), Gāndhināgar, Indien, 22–27 October 2018.

21. Bertinetti, A.; Avramidis, K.A.; Albajar, F.; Cau, F.; Cismondi, F.; Rozier, Y.; Savoldi, L.; Zanino, R. Multi-physics analysis of a 1 MW gyrotron cavity cooled by mini-channels. *Fusion Eng. Des.* **2017**, *123*, 313–316. [CrossRef]

22. Cordova, M. Thermo-Mechanical Study of the Cavity of the 1 MW–140 GHz Gyrotron for W7-X. Master's Thesis, Karlsruhe Institute Technology, Karlsruhe, Germany, 2013.

23. Avramidis, K.A.; Bertinetti, A.; Albajar, F.; Cau, F.; Cismondi, F.; Gantenbein, G.; Illy, S.; Ioannidis, Z.C.; Jelonnek, J.; Legrand, F.; et al. Numerical Studies on the Influence of Cavity Thermal Expansion on the Performance of a High-Power Gyrotron. *IEEE Trans. Electron Devices* **2018**, *65*, 2308–2315. [CrossRef]

24. Leggieri, A.; Albajar, F.; Alberti, S.; Allio, A.; Avramidis, K.A.; Bariou, D.; Bin, W.; Bruschi, A.; Chelis, I.; Difonzo, R.; et al. Upgrade of The European ITER 170 GHz 1 MW CW Industrial Gyrotron (TH1509). In Proceedings of the 7th IVEW 2020 and 13th IVeSC 2020, Monterey, CA, USA, 26–29 May 2020.

25. Leggieri, A.; Albajar, F.; Albert, S.; Allio, A.; Avramidis, K.A.; Bariou, D.; Bin, W.; Bruschi, A.; Chelis, J.; Difonzo, R.; et al. TH1509U European 170 GHz 1 MW CW Industrial Gyrotron Upgrade. In Proceedings of the IEEE IVEC 2021, Virtual Event, 27–30 April 2021.

26. Avramides, K.A.; Pagonakis, I.G.; Iatrou, C.T.; Vomvoridis, J.L. EURIDICE: A code-package for gyrotron interaction simulations and cavity design. *EPJ Web Conf.* **2012**, *32*, 04016. [CrossRef]

27. Ginzburg, N.S.; Nusinovich, G.S.; Zavolsky, N.A. Theory of non-stationary processes in gyrotrons with low Q resonators. *Int. J. Electron.* **1986**, *61*, 881–894. [CrossRef]

28. Jackson, J.D. Wave guides and resonant cavities. In *Classical Electrodynamics*; Wiley: Hoboken, NJ, USA, 1975.

29. ITER. *Material Properties Handbook for Glidcop*, available upon request.

30. Tsang, L.; Braunisch, H.; Ding, R.; Gu, X. Random rough surface effects on wave propagation in interconnects. *IEEE Trans. Adv. Packag.* **2010**, *33*, 839–856. [CrossRef]

31. Allio, A.; Difonzo, R.; Leggieri, A.; Legrand, F.; Marchesin, R.; Savoldi, L. Test and Modeling of the Hydraulic Performance of High-Efficiency Cooling Configurations for Gyrotron Resonance Cavities. *Energies* **2020**, *13*, 1163. [CrossRef]

32. Savoldi, L.; Allio, A.; Bonvento, A.; Cantone, M.; Fernandez Reche, J. Experimental and numerical investigation of a porous receiver equipped with Raschig Rings for CSP applications. *Sol. Energy* **2020**, *212*, 309–325. [CrossRef]

33. Savoldi, L.; Bertinetti, A.; Nallo, G.F.; Zappatore, A.; Zanino, R.; Cau, F.; Cismondi, F.; Rozier, Y. CFD Analysis of Different Cooling Options for a Gyrotron Cavity. *IEEE Trans. Plasma Sci.* **2016**, *44*, 3432–3438. [CrossRef]

34. Cantone, M.; Cagnoli, M.; Fernandez Reche, J.; Savoldi, L. One-side heating test and modeling of tubular receivers equipped with turbulence promoters for solar tower applications. *Appl. Energy* **2020**, *277*, 115519. [CrossRef]

35. Savoldi, L. *Support in the Design and Analysis of the Gyrotron Full-Size Cavity and Mock-Ups*, F4E_D_24Q2PG. 2015; available upon request.

36. Rohsenow, W.M. Boiling. *Annu. Rev. Fluid Mech.* **2003**, *3*, 211–236. [CrossRef]

37. Hirt, C.W.; Nichols, B.D. Volume of fluid (VOF) method for the dynamics of free boundaries. *J. Comput. Phys.* **1981**, *39*, 201–225. [CrossRef]

38. Savoldi, L. *Experimental and Simulation Results on the Mock-Ups Equipped with Raschig Rings and Mini-Channels*, F4E_D_28KSJG. 2017; available upon request.

39. Bertinetti, A.; Albajar, F.; Cau, F.; Leggieri, A.; Legrand, F.; Perial, E.; Ritz, G.; Savoldi, L.; Zanino, R.; Zappatore, A. Design, Test and Analysis of a Gyrotron Cavity Mock-Up Cooled Using Mini Channels. *IEEE Trans. Plasma Sci.* **2018**, *46*, 2207–2215. [CrossRef]

40. Ioannidis, Z.C.; Rzesnicki, T.; Albajar, F.; Alberti, S.; Avramidis, K.A.; Bin, W.; Bonicelli, T.; Bruschi, A.; Chelis, I.; Frigot, P.E.; et al. CW Experiments with the EU 1-MW, 170-GHz Industrial Prototype Gyrotron for ITER at KIT. *IEEE Trans. Electron Devices* **2017**, *64*, 3885–3892. [CrossRef]

41. Ioannidis, Z.C.; Albajar, F.; Alberti, S.; Avramidis, K.A.; Bin, W.; Bonicelli, T.; Bruschi, A.; Chelis, J.; Fanale, F.; Gantenbein, G.; et al. Recent experiments with the European 1MW, 170GHz industrial CW and short-pulse gyrotrons for ITER. *Fusion Eng. Des.* **2019**, *146*, 349–352. [CrossRef]

42. Menachilis, P. *Simplified Approach for the Calibration of the RR Conductivity*, F4E_D_2HBAC8. 2019; available upon request.

43. Avramidis, K.A.; Pagonakis, I.G.; Chelis, I.G.; Gantenbein, G.; Ioannidis, Z.C.; Peponis, D.V.; Rzesnicki, T.; Jelonnek, J. Simulations of the experimental operation of the EU 170 GHz, 1 MW short-pulse prototype gyrotron for ITER. In Proceedings of the 2016 41st International Conference on Infrared, Millimeter, and Terahertz Waves (IRMMW-THz), Copenhagen, Denmark, 25–30 September 2016. [CrossRef]

44. Avramidis, K.A.; Aiello, G.; Alberti, S.; Brücker, P.T.; Bruschi, A.; Chelis, I.; Franke, T.; Gantenbein, G.; Garavaglia, S.; Genoud, J.; et al. Overview of recent gyrotron R&D towards DEMO within EUROfusion Work Package Heating and Current Drive. *Nucl. Fusion* **2019**, *59*, 066014. [CrossRef]

45. Avramidis, K.A.; Ioannidis, Z.C.; Illy, S.; Jin, J.; Ruess, T.; Aiello, G.; Thumm, M.; Jelonnek, J. Multifaceted Simulations Reproducing Experimental Results from the 1.5-MW 140-GHz Preprototype Gyrotron for W7-X. *IEEE Trans. Electron Devices* **2021**, *68*, 3063–3069. [CrossRef]

46. Schlaich, A.; Wu, C.; Pagonakis, I.; Avramidis, K.; Illy, S.; Gantenbein, G.; Jelonnek, J.; Thumm, M.; Schlaich, A.; Wu, C.; et al. Frequency-Based Investigation of Charge Neutralization Processes and Thermal Cavity Expansion in Gyrotrons. *JIMTW* **2015**, *36*, 797–818. [CrossRef]

47. Chelis, I.G.; Avramidis, K.A.; Ioannidis, Z.C.; Tigelis, I.G. Improved Suppression of Parasitic Oscillations in Gyrotron Beam Tunnels by Proper Selection of the Lossy Ceramic Material. *IEEE Trans. Electron Devices* **2018**, *65*, 2301–2307. [CrossRef]

48. Difonzo, R.; Allio, A.; Savoldi, L. Multivariate metric assessment of the suitability of different RANS turbulence models for the simulation of mini-channels cooling systems for the fusion gyrotron resonator. In Proceedings of the AMSE V&V Symposium 2021, Virtual Event, 10–13 May 2021.

49. Stanculovic, S.; Difonzo, R.; Allio, A.; Avramidis, K.A.; Brücker, P.; Gantenbein, G.; Illy, S.; Jelonnek, J.; Kalaria, P.C.; Misko, M.; et al. Calibration of the KIT test setup for the cooling tests of a gyrotron cavity full-size mock-up equipped with mini-channels. *Fusion Eng. Des.* **2021**, *172*, 112744. [CrossRef]

50. Laqua, H.P.; Max Planck Institute for Plasma Physics, Garching bei München, Germany. Personal communication, 2021.

51. Abuserwal, A.F.; Elizondo Luna, E.M.; Goodall, R.; Woolley, R. The effective thermal conductivity of open cell replicated aluminium metal sponges. *Int. J. Heat Mass Transf.* **2017**, *108*, 1439–1448. [CrossRef]

Article

Effect of Pebble Size Distribution and Wall Effect on Inner Packing Structure and Contact Force Distribution in Tritium Breeder Pebble Bed

Baoping Gong *, Hao Cheng, Yongjin Feng, Xiaofang Luo, Long Wang and Xiaoyu Wang

Center for Fusion Science, Southwestern Institute of Physics, P.O. Box 432, Chengdu 610041, China; chengh@swip.ac.cn (H.C.); fengyj@swip.ac.cn (Y.F.); luoxiaofang@swip.ac.cn (X.L.); wanglong@swip.ac.cn (L.W.); wangxy@swip.ac.cn (X.W.)
* Correspondence: gongbp@swip.ac.cn; Tel.: +86-28-8285-0366

Abstract: In the tritium breeding blanket of nuclear fusion reactors, the heat transfer behavior and thermal-mechanical response of the tritium breeder pebble bed are affected by the inner packing structure, which is crucial for the design and optimization of a reliable pebble bed in tritium breeding blanket. Thus, the effect of pebble size distribution and fixed wall effect on packing structure and contact force in the poly-disperse pebble bed were investigated by numerical simulation. The results show that pebble size distribution has a significant influence on the inner packing structure of pebble bed. With the increase of the dispersion of pebble size, the average porosity and the average coordination number of the poly-disperse pebble bed gradually decrease. Due to the influence of the fixed wall, the porosity distribution of the pebble bed shows an obvious wall effect. For poly-disperse pebble bed, the influenced region of the wall effect gradually decreases with the increase of the dispersion of pebble size. In addition, the gravity effect and the pebble size distribution have an obvious influence on the contact force distribution inside the poly-disperse pebble bed. The majority of the contact force are weak contact force that is less than the average contact force. Only a few of pebbles have strong contact force that is greater than average contact force. This investigation can help in analyzing the pebble crushing characteristics and the thermal hydraulic analysis in the poly-disperse tritium breeder pebble bed.

Keywords: packing structure; contact force; porosity distribution; tritium breeder pebble bed; breeding blanket; discrete element method

Citation: Gong, B.; Cheng, H.; Feng, Y.; Luo, X.; Wang, L.; Wang, X. Effect of Pebble Size Distribution and Wall Effect on Inner Packing Structure and Contact Force Distribution in Tritium Breeder Pebble Bed. *Energies* **2021**, *14*, 449. https://doi.org/10.3390/en14020449

Received: 5 December 2020
Accepted: 12 January 2021
Published: 15 January 2021

Publisher's Note: MDPI stays neutral with regard to jurisdictional claims in published maps and institutional affiliations.

1. Introduction

Granular matter widely exists in nature and industrial systems, especially in fixed beds or fluidized beds, which are widely used in chemical engineering systems [1,2] or the nuclear energy industry [3–12]. For instance, adsorption beds, chemical catalytic reaction beds and bubbling fluidized beds [13,14] are used in the form of granular fixed beds in the field of the chemical engineering. For nuclear reactor energy systems, the reactor core of the high temperature gas cooled nuclear reactor is formed by fuel pebbles [9–12], and the tritium breeder and the neutron multiplier in the tritium breeding blanket of nuclear fusion reactors [3–8] are also used in form of pebble beds.

For a pebble beds applied to energy-related systems, the heat and mass transfer performance of the pebble bed play an important role in the pebble bed application. However, the characteristics of the packing structure of pebble beds, such as packing fraction, porosity and permeability, the contact state (coordination number), and so forth, have a significant influence on the heat transfer behaviors and the flow characteristic of fluids inside pebble beds [15–28]. For example, the average porosity and porosity distribution in the bed have a significant influence on the pressure drop and fluid velocity distribution inside a pebble bed [23–28]. Further, the effective thermal conductivity of a granular assembly, the

387

thermal diffusion coefficient, the heat transfer coefficient between the pebble bed and the wall [15–22] are all affected by the inner structure of the pebble bed. In addition, the contact state (coordination number) and the contact force of pebbles inside the bed is another very useful and important parameter that characterizes the performance related to the heat transfer and mechanical behaviors of a pebble bed. The heat transfer process of a pebble bed is mainly carried out through the contact heat conduction between pebbles especially the fixed bed with high solid-fluid thermal conductivity ratio and slow fluid flow rate [15–28]. Under the condition, the contact force chain network between pebbles in contact with each other forms a net-like virtual heat transfer path inside a pebble bed [29–35], the size of which will affects the effective thermal conductivity and heat transfer behaviors of pebble bed, while the size of the virtual path is closely related to the magnitude of the contact force and the contact area. Thus, the contact force distribution will affect the effective thermal conductivity of the pebble bed. What's more, the magnitude and the distribution of contact force play an important role for the prediction of the lifetime and the breakage of pebbles inside a fixed pebble bed [29–39]. Too strong a contact force may cause cracks and fragmentation near the contact point of a pebble, affecting the mechanical behaviors of the fixed pebble bed. From this perspective, accurately determining and predicting the pebble bed packing structure (porosity distribution and coordination number) and contact force distribution are of great significance for the analyzing the heat and mass transfer characteristics of a fixed bed, which is the key to designing a fixed pebble bed.

In a fixed pebble bed, the packing structure and the contact force distribution have been extensively investigated experimentally and numerically. Especially the radial porosity distribution in the cylindrical mono-sized pebble bed has been studied by a large number of experiments and simulations [40–43]. In a cylindrical mono-sized pebble bed, the radial porosity distribution of the pebble bed exhibits oscillating and damping characteristics near the fixed wall, the width of the oscillation region is always affected by the ratio of the diameter of the cylinder to the pebbles. This phenomenon of porosity oscillation and damping is the so-called "near-wall channeling effect" or "wall effect". In further, the global porosity and the coordination number of the cylindrical pebble bed are also affected by the diameter ratio of cylinder to pebble. With the increase of the diameter ratio of cylinder to pebble, the global porosity decreases and the coordination increase gradually, and eventually both tend to a constant value [7,8,44]. In addition, numerous numerical simulations and experiments have been conducted on the contact force distribution and force chain inside pebble bed [29–39]. The force chain network, including strong contact force chain and weak contact force chain, formed by the contact force between pebbles gradually evolves with each other under the external excitation, such as gravity, external compression load, thermal expansion, vibration, and so forth. In a fixed pebble bed, although the magnitude of the contact force is distributed in a relatively wide range, the majority of the contact force is weak contact force chain that is less than the average contact force. With the increase of the contact force, the probability of contact force decreases rapidly. Only a small amount of contact force is much larger than the average contact force, which may cause that the pebbles with large strong contact force may be broken due to the large stress concentration near the contact point. Therefore, in-depth investigation of the contact force distribution and the force chains is very important to predict the crushing behavior of the pebbles inside the fixed pebble bed.

In addition, in numerous previous investigations, most of the attention was focused on the study of the macro- and meso-scale packing structure and contact force distribution of the mono-sized pebble bed [45–51]. Such as, the average porosity and radial porosity distribution in the mono-sized cylindrical pebble bed various as the increase of the diameter ratio of cylinder to pebble, and the relation between the coordination number and the porosity in bed. For the progress of the radial porosity variation in mono-sized cylindrical pebble bed can refer to the review literature [40]. For binary-sized pebble beds or poly-disperse pebble beds, the average packing density (or average porosity) and the macroscopic compression response were investigated widely. For the progress of the packing density of the binary-sized pebble bed and the poly-disperse pebble bed one can refer to the literature [52–55]. However, for poly-disperse fixed pebble beds, the inner packing structure features such as the porosity distribution and the coordination number distribution, and so forth, is still insufficiently investigated. Especially there are few reports in the literature about the local porosity distribution close to the fixed wall in poly-disperse pebble bed. In addition, the investigation on the contact force distribution in the poly-disperse pebble bed is insufficient, especially the effect of the pebble size distribution on the contact force distribution in the poly-disperse pebble bed is also few reported in literature.

Therefore, this study is focused on the investigation of the effect of pebble size distribution (discrete normal distribution and discrete uniform distribution) and fixed wall effect on the inner packing structure and the contact force distribution in the poly-disperse fixed pebble bed under gravity packing, based on the discrete element method (DEM). This study can help to analyze the pebble crushing characteristics and the heat transfer behaviors in the poly-disperse tritium breeder pebble bed of a solid tritium breeder blanket in a fusion reactor. In this work, the methodology is mainly concentrated in the Section 2. The results and related discussions are shown in Section 3. Finally, some conclusions obtained in this work are summarized in Section 4.

2. Methodology

2.1. Discrete Element Method

The discrete element method (DEM) was applied to modelling the packing behavior of the poly-disperse pebbles packing, which is an effective numerical modelling method to investigate the dynamic response of granular materials, such as the pebble packing in this work. The motion of pebbles in the DEM simulation follows the Newton's second law of motion, which is driven by the force interactions, such as the contact force between two pebbles and between pebble and wall, the gravity, the cohesive force, the Van der Waals force, and so forth. In this study, due to the investigation was focused on the packing of the millimeter dry pebbles, the gravity effect and the contact force were considered only. Thus, the resultant acceleration of each pebbles is calculated from the gravity and the contact force. The pebble motion can be expressed as the following:

$$m_i \frac{dV_i}{dt} = \sum_{j=1}^{N} \left(F_{n_{ji}} + F_{t_{ji}} \right) + F_g \tag{1}$$

$$I_i \frac{d\omega_i}{dt} = \sum_{j=1}^{N} r_{ij} \times \left(F_{n_{ji}} + F_{t_{ji}} \right) \tag{2}$$

where, m_i and I_i are the pebble mass and the motion of inertia. V_i and ω_i are the velocity of the translational and rotational movement of pebble i, respectively. N is the number of the surrounding pebbles touched the pebble i. F_n and F_t are normal and tangential contact force between two pebbles, respectively. F_g is the gravity force. r_{ij} is the vector pointing from the pebble i to pebble j. Under the influence of the friction interaction between two pebbles, the normal contact force and the tangential contact force satisfy the $|F_t|_{max} \leq \mu \, |F_t|$, where μ is friction coefficient.

The contact force is calculated based on the Hertz-Mindlin [56] contact theory, F_n and F_t can be determined as follows:

$$F_n = k_n \delta_{n_{ij}} - \eta_n v_{n_{ij}} \ \text{and} \ F_t = k_t \delta_{t_{ij}} - \eta_t v_{t_{ij}} \tag{3}$$

For normal contact force, k_n is the elastic constant of normal contact (also known as normal stiffness), η_n is the normal viscoelastic damping coefficient. $\delta_{n_{ij}}$ is the overlap of two normal contact pebbles. v_{nij} is the normal relative velocity of two pebbles. k_n and η_n can be expressed as:

$$k_n = \frac{4}{3}Y^* \sqrt{R^*\delta_{n_{ij}}} \text{ and } \eta_n = -2\sqrt{\frac{6}{5}}\beta\sqrt{S_n m^*} \tag{4}$$

where $S_n = 2Y^* \sqrt{R^*\delta_{n_{ij}}}$, $\eta_n \geq 0$.

For tangential contact force, k_t is the elastic constant of tangential contact (also known as tangential stiffness), η_t is the tangential viscoelastic damping coefficient. δ_{nij} is the tangential relative displacement vector of two contact pebbles. v_{tij} is the tangential relative velocity of two pebbles. k_t and η_t can be shown as:

$$k_t = 8G^* \sqrt{R^*\delta_{t_{ij}}} \text{ and } \eta_t = -2\sqrt{\frac{6}{5}}\beta\sqrt{S_t m^*} \tag{5}$$

where, $S_t = 8G^* \sqrt{R^*\delta_{n_{ij}}}$, $\eta_t \geq 0$. β is a damping constant determined by the restitution coefficient e as follows:

$$\beta = \frac{\ln(e)}{\sqrt{\ln^2(e) + \pi^2}} \tag{6}$$

For the above formulas, where, Y^*, G^*, m^* and R^* are the effective elastic modulus, the effective shear modulus, the equivalent mass and the equivalent radius of two contact pebbles, they are given as follow:

$$\frac{1}{Y^*} = \frac{\left(1 - v_i^2\right)}{Y_i} + \frac{\left(1 - v_j^2\right)}{Y_j} \tag{7}$$

$$\frac{1}{G^*} = \frac{2(2 - v_i)(1 + v_i)}{Y_i} + \frac{2(2 - v_j)(1 + v_j)}{Y_j} \tag{8}$$

$$\frac{1}{R^*} = \frac{1}{R_i} + \frac{1}{R_j} \text{ and } \frac{1}{m^*} = \frac{1}{m_i} + \frac{1}{m_j} \tag{9}$$

where, Y and G are the Young's modulus and the Shear modulus of pebble. v is the Poisson ratio. R and m are the radius and the mass of pebble. In this work, the DEM simulation was carried out by using the open-source software LIGGGHTS [57]. More detailed theory of the DEM can be obtained in [7,56,57].

2.2. Modelling Packing Process and Parameter Setup

In the simulation of the present work, all pebbles were assumed to be completely spherical hard spheres. The packing process in this work can be divided into two stages: the gravity settlement stage and the gradually equilibrium stage. Firstly, a specialized number of pebbles was randomly generated in one moment without overlap between pebbles in the computational region according to the predefined pebble size distribution. There are 30,000 pebbles in all simulation cases in this work. The pebble size distribution in number distribution is shown in Table 1. Then, gravity is applied to all pebble body. Under the gravity effect, the pebbles begin to fall freely and pack in the container bottom gradually. The packing process will turn to the gradually balance stage when all pebbles were packed at the bottom of the container. The kinetic energy of pebbles gradually dissipates in the process of the collision effect and the friction interaction between pebbles. Finally, all pebbles will be packed in the container with a balanced and stabled packing state.

Table 1. Pebble size distribution used in the DEM simulation.

Pebble Size Distribution	Parameter	Values
Mono-sized	$d = d_{avg}$	1 mm
Normal distribution	d_{avg} σ	1 mm 0–0.3 mm
Uniform distribution	d_{avg} Δd	1 mm 0–0.5 mm
Number of pebbles	N	30,000
Average diameter	d_{avg}	1 mm
Time-step	dt	1×10^{-8} s

During the simulation, the interaction between pebbles is simulated by iterative calculation of the contact force between pebbles, displacement, velocity, and so forth. Firstly, according to the physical model of pebbles and the contact state, the contact force is calculated between pebbles. Secondly, the resultant force of the contact force between the pebble and the surrounding pebbles and the gravity force is computed. Thirdly, solve the dynamic equation of the pebble motion in a specific time-step and update the pebble position, translational velocity, angular velocity and other information. According to the updated pebble information, the contact force and the resultant force can be calculated again. Through multiple iteration calculations, a stable packing state of pebbles is achieved finally. The final convergence criterion is determined by monitoring the total kinetic energy of the entire pebble bed. When the total kinetic energy is lower than a specified value, the iterative calculation can be terminated. In this work, the simulation is stopped when the total kinetic energy of the whole bed is less than 10^{-10} J. At this time, the translational and rotational motion of pebbles in the packed bed is very weak, and the position of pebbles almost does not change. It can be considered that the packed bed has reached a stable equilibrium state.

In this work, the material parameters shown in Table 2 are used in the simulation. The average diameters of all simulation cases are equal to 1 mm. There are three kinds of pebble size distribution in number, as follows:

(1) Mono-sized, diameter of all pebbles is equal to 1 mm.
(2) Discrete normal distribution, N (E, σ^2). E is average diameter d_{avg} and also equal to 1 mm. σ is the standard deviation of pebble size varying from 0 to 0.3.
(3) Discrete uniform distribution, U (d_{min}, d_{max}). d_{min} and d_{max} are the minimum and maximum values of pebble diameter. The pebble diameter is discretized with a step of 0.05 mm from d_{min} to d_{max}. the average diameter d_{avg} is equal to the mean of d_{min} and d_{max}. the diameter difference, Δd, defined as the absolute of the difference between the maximum (minimum) diameter d_{max} (d_{min}) and the average diameter d_{avg} was used to represent the dispersion degree of discrete uniform pebble size distribution, which varied from 0 to 0.5 mm.

Table 2. Material parameters used in the DEM simulation.

Parameters	Li_4SiO_4 Pebbles [7,8]	Wall (CLF-1 Steel [58])
Young's modulus	90 GPa	225 GPa
Poisson ratio	0.24	0.33
Density	2323 Kg/m^3	7847 Kg/m^3
Restitution coefficient	0.9	0.9
Static friction coefficient	0.1	0.1
Rolling friction coefficient	0.001	0.001

In addition, for each pebble size distribution, the pebble packing under gravity was carried out with fixed side wall and the periodic boundary, respectively. For the pebble bed with fixed wall boundary, the fixed wall was applied in x-axis and y-axis. For the pebble bed with periodic boundary, the container box is periodic in x-axis and y-axis. But in z-axis, the fixed wall was also adopted to carry the gravity of the pebbles in pebble bed with periodic boundary. The bed dimension is 20 mm $\times$ 20 mm $\times$ 140 mm in all simulation cases. The effect of pebble size distribution and the wall effect on the packing stricture and the contact force distribution was analyzed in detail.

3. Results and Discussions

3.1. Mono-Sized Pebble Bed and Validation

3.1.1. Porosity Distribution

The pebble packing structure and the contact force distribution of mono-sized pebble bed was simulated and analyzed to validate the DEM simulation by comparing with the experimental and numerical results reported in literature. Figure 1a shows the 3D view of the mono-sized pebble bed, in which the average porosity of pebble bed with periodic boundary is 0.3819 $\pm$ 0.0008. For pebble bed with fixed side wall, the average porosity is 0.3981 $\pm$ 0.0007. A relatively lower porosity can be obtained in bed with periodic boundary compared to the pebble bed with fixed wall, which is mainly caused by the influence of fixed wall. There is a relatively larger porosity near the container wall, which can also be revealed from the porosity distribution, as shown in Figure 1b.

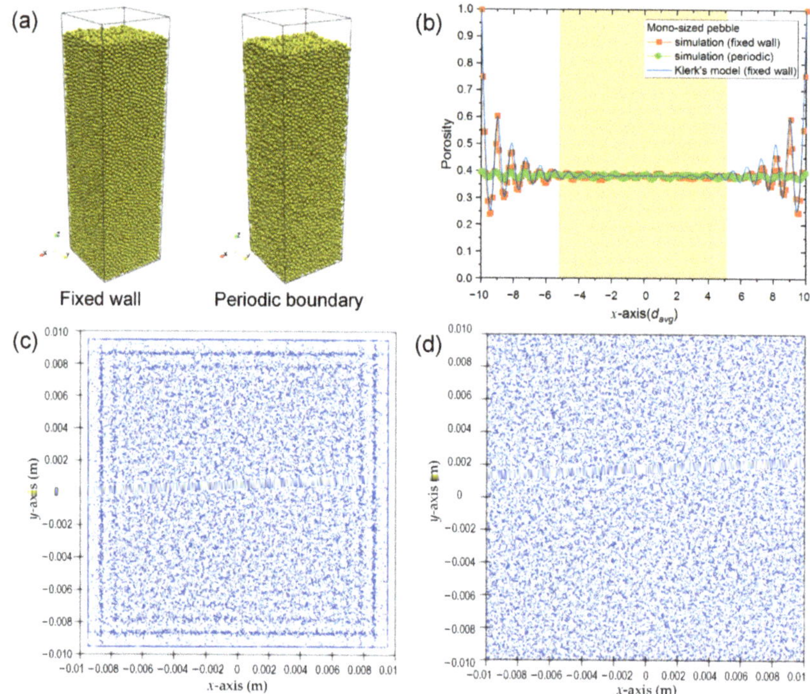

Figure 1. Structures of mono-sized packed pebble bed: (**a**) 3D view, (**b**) local porosity distribution, (**c**,**d**) pebble center distribution bed with fixed wall and periodic boundary.

Figure 1b shows the porosity distribution along the x-axis direction in the pebble bed. A stable porosity can be observed in the whole pebble bed with periodic boundary. The axial porosity is slightly irregularly disturbed around the average porosity due to the effect of random packing structure. However, for pebble bed with fixed wall, it can be revealed from the Figure 1b that the axial porosity in the whole pebble bed shows a damping and oscillating behaviors as the distance to the fixed wall increase. A wall effect of porosity distribution can be observed obviously. In the region adjacent to the fixed wall, the maximum porosity of about 1 is achieved since the pebbles is almost in point contact with the fixed wall. With the increase of the distance to the fixed wall, the porosity decreases rapidly. A minimum porosity of about 0.25 is obtained when the distance to the wall is about 0.5 d. With further increase of the distance to the wall, the porosity of pebble bed shows a characteristic of damping and oscillation. When the distance to the fixed wall is greater than 5 d, the porosity tends to a constant value, namely, the average porosity in the inner region of pebble bed. In this work, the average porosity in the inner region of mono-sized random packed pebble bed with fixed wall is 0.3826 ± 0.0006, which is similar to the global average porosity of the randomly packed mono-sized pebble bed with periodic boundary. In further, a layered distribution of pebble center close to the fixed wall can be clearly observed by projecting the particle sphere center to the bottom wall (x-y plane), as shown in Figure 1c,d which is also corresponding to the axial porosity variation close to the fixed wall. In addition, the axial porosity variation of the mono-sized pebble bed with fixed wall is in line with the results from the Klerk's empirical model [43].

3.1.2. Coordination Number Distribution

The coordination number mainly indicates the number of other surrounding pebbles in contact with the specified pebble in pebble bed. In this work, the average coordination number of the mono-sized pebble bed with fixed wall is 6.3216, which is slightly lower than that of 6.6176 in the bed with periodic boundary. It is mainly because the pebbles directly in contact with the fixed wall and close to the fixed wall have smaller coordination number due to the influence of the fixed wall effect, as shown in Figure 2. While for the pebble bed with periodic boundary, there is no fixed wall effect. Thus, a higher coordination number can be obtained.

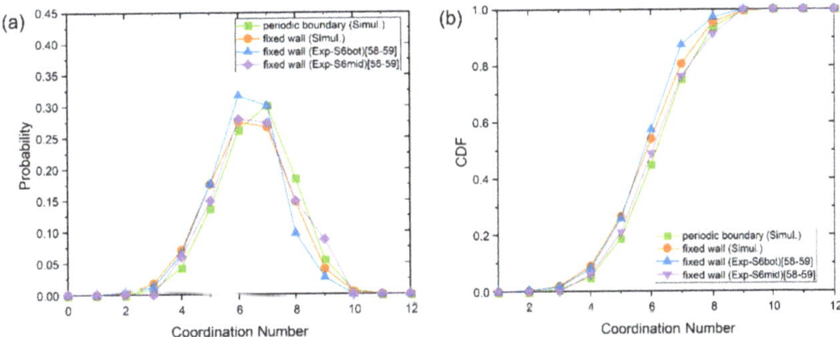

Figure 2. Coordination number distribution of mono-sized pebble bed: (**a**) probability distribution, (**b**) cumulative distribution function.

Figure 2 shows the probability distribution function (PDF) and the cumulative distribution function (CDF) of coordination numbers in a mono-sized pebble bed. It can be seen from the figure that the coordination numbers with larger probabilities in the mono-sized pebble beds with both fixed wall and periodic boundary are 6 and 7, respectively. For the fixed wall pebble bed, the coordination number with the most probability is 6, and for periodic boundary pebble bed, it is 7. The coordination number of the periodic boundary pebble bed with the highest probability is also slightly higher than that of the fixed wall pebble bed, which is also due to the influence of the fixed wall effect. In addition, the coordination number distribution also reaches an agreement with the experiment results from the literature [59,60], which indicates that the reliable results of pebble packing can be obtained by the DEM simulation.

3.1.3. Contact Force Distribution

The contact force distribution in a mono-sized pebble bed with a periodic boundary and fixed wall is shown in Figure 3. If a cylinder is used to connect the center of two contacted pebbles, the diameter and color are used to indicate the magnitude of contact force, the contact force in bed can form a force chain network, as shown in Figure 3a. It is clearly show that due to the influence of the gravity effect, the strong force chain is mainly concentrated close to the bottom in the mono-sized pebble bed. With the increase of the local height, the magnitude of the contact force gradually reduces, the smallest contact force can be observed in the top region of the pebble bed. In addition, the variation of the contact force along the local height of the pebble bed can be obtained by calculating the average contact force in a micro volume with a step of 0.5 d along the height (z-axial) of the pebble bed. The results, as shown in Figure 3b, reveal that the average contact force inside the pebble bed also decreases gradually with the increase of the local height due to the influence of gravity. The normalized contact force is between 2 and 2.5 near the bottom and tends to 0 at the top of the bed. This is consistent with the contact force chain distribution in Figure 3a.

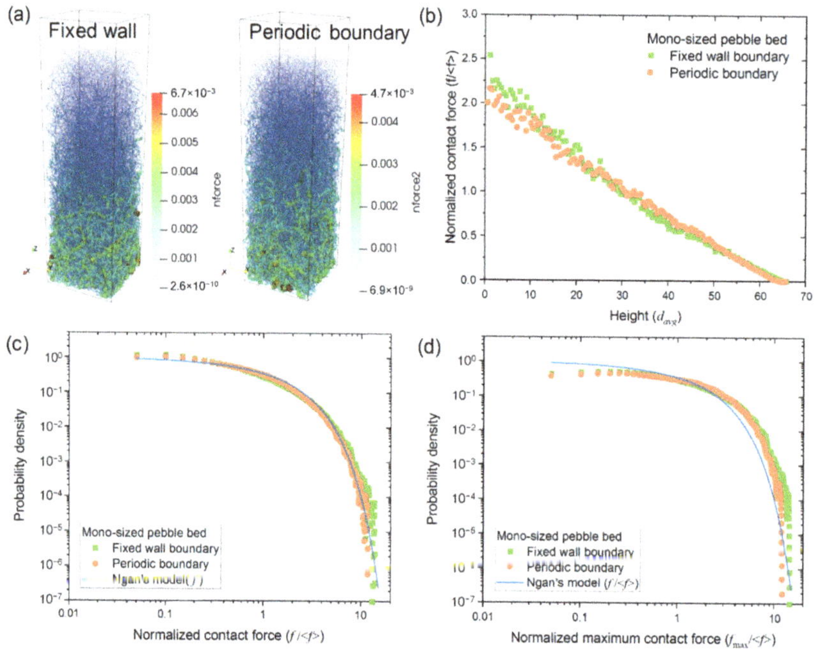

Figure 3. Contact force distribution of mono-sized pebble bed: (**a**) 3D view of contact force chains, (**b**) contact force distribution along local height, (**c**,**d**) probability density distribution of all contact force and the maximum contact force of each pebble.

The probability density distributions of all normalized contact force and the maximum normalized contact force of each pebble are shown in Figure 3c,d. When the normalized contact force is equal to 1, it represents the average contact force inside the pebble bed. It can be found that the weak contact force below the average contact force has the largest probability density. The probability density decreases rapidly with the increase of the contact force. It is demonstrated that the majority of the contact force inside the pebble bed is weak force which maintain the stability of the pebble packing structure, only a few strong contact forces are large than the average contact force, which mainly carry the gravity force and the external load of the pebble bed. In further, the probability density distribution of the uncompressed mono-sized pebble bed is compared with the Ngan's empirical model [34]. It can be seen that the results of this work are in good agreement with the empirical model.

In further, for each individual pebble inside a fixed pebble bed, there are several contact forces due to the several contact between the pebble and several surrounding pebbles. Among these contact forces, some may be greater than the average contact force, others may be less than the average contact force. A feature of particular interest is the maximum contact force of each individual pebble, which is related to the mechanical stability of the tritium breeder pebble. Once the maximum contact force exceeds the crush load of pebbles, the crack and fragmentation of pebbles may occur. For the Li_4SiO_4 pebbles fabricated by melt spraying method, the average crush load of the pebbles with diameter of 1 mm is about 7.0 N, the maximum and the minimum crush load are 16 N and 5.2 N, respectively [61,62]. Therefore, in this study, the probability density distribution of the maximum contact force of each pebble was analyzed and compared with the probability density distribution of all contact force calculated by the Ngan's empirical model [34]. The results show that the maximum contact force of most pebbles in bed is still less than the average contact force of the whole pebble bed. With the increase of the contact force, the probability density distribution of the maximum contact force also decreases rapidly. However, compared with the probability density distribution of all contact force, the probability density of the maximum contact force less than the average contact force is reduced, and the probability density of the maximum contact force greater than the average contact force is increased. It is indicated that all contact forces of most of pebbles is less than the average contact force, which mean that it is difficult to break. Only a small number of pebbles suffer the maximum contact force greater than the average contact force, which means that these pebbles have a higher probability of breaking.

In addition, it can be seen from Figure 3 that the contact force distribution of the pebble bed with fixed wall and periodic boundary conditions is almost same, and the boundary conditions have little influence on the inter-pebble contact force distribution inside the pebble bed in here. Therefore, the contact force in the poly-disperse pebble bed is statistically analyzed only for one kind of boundary condition in the following section.

3.2. Pebble Size Normal Distribution Pebble Bed

3.2.1. Porosity Distribution

The average porosity of the poly-disperse pebble bed is shown in Figure 4 when the pebble size is normally distributed in number. The results reveal that the average porosity gradually decreases as the increase of the standard deviation of the pebble size distribution in poly-disperse pebble bed with both the fixed wall and the periodic boundary. For the poly-disperse pebble bed with periodic boundary, the average porosity gradually decreases from 0.3819 ± 0.0008 when the standard deviation is 0 (namely a mono-sized pebble bed) to 0.3651 ± 0.0007 when the standard deviation is 0.3. The decrease is about 4.4%. For the pebble bed with fixed wall, the overall average porosity also decreases from 0.3981 ± 0.0007 to 0.3865 ± 0.0007 when the standard deviation increases from 0 to 0.3. The decrease is about 2.9%. The decrease of average porosity is mainly because with the increase of pebble dispersion, many small pebbles can be filled in the gap formed between large pebbles resulting in forming a denser packing structure.

Furthermore, it is obviously that the fixed wall has a significant effect on the average porosity of pebble bed. Thus, for a poly-disperse pebble bed with fixed wall, the average porosity both in the overall of the bed and in the inner region of the bed were calculated. The calculation of the overall porosity includes the whole pebble bed, while the calculation of the inner region porosity only considers the inner region of the pebble bed by excluding the region near the vessel wall with drastic change of porosity (such as the region near the fixed wall as shown in Figure 5). It is clearly shown in Figure 3 that the overall average porosity of the bed with fixed wall is higher than that with the periodic boundary. However, after excluding the region close to the fixed side wall, the average porosity in inner region of the fixed wall pebble bed is consistent with that of the periodic boundary pebble bed, which further indicates that the fixed wall effect is on limited to the region near the wall of the pebble bed where with a higher porosity.

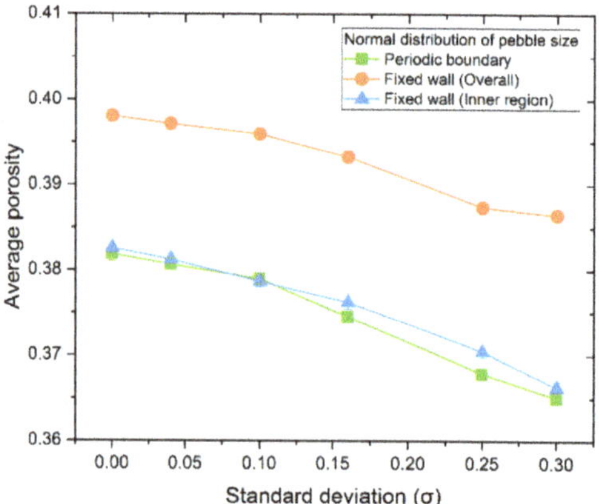

Figure 4. Average porosity of the poly-disperse pebble bed with normally distributed pebble size.

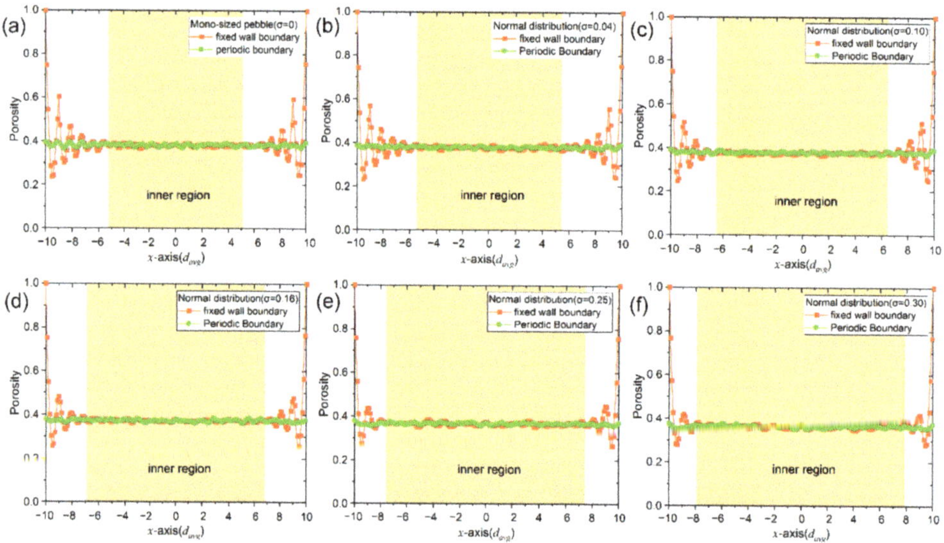

Figure 5. Local porosity distribution in the poly-disperse pebble bed with normally distributed pebble size: (**a**) $\sigma = 0$, mono-sized pebble bed, (**b**) $\sigma = 0.04$, (**c**) $\sigma = 0.1$, (**d**) $\sigma = 0.16$, (**e**) $\sigma = 0.25$, (**f**) $\sigma = 0.3$.

In order to reveal the characteristics of the wall effect in the poly-disperse pebble bed with fixed wall. The local porosity variation along the x-axis was calculated, as shown in Figure 5. Since the curves of the local porosity along the x-axis and the y-axis are basically consistent, the curves along the x-axis are shown only in Figure 5. The local porosity of the poly-disperse pebble bed with periodic boundary was also calculated for comparison. Figure 5 shows that the local porosity of the poly-disperse pebble bed oscillates significantly and dampens in the region close to the fixed wall. However, unlike the mono-sized pebble bed, in which the oscillation and damping is basically limited to the region of about 5 d close to the fixed wall, the influence region of the wall effect in the poly-disperse pebble bed with normally distributed pebble size is smaller than that in mono-sized pebble bed. With the increase of the standard deviation, σ, the influence region of the wall effect gradually decreases. For instance, when the standard deviation of pebble size is 0 or 0.04, the wall effect region is limited within the range of 4.5 d_{avg}–5 d_{avg} close to the wall. When the standard deviation increases to 0.3, the wall effect region is reduced to the region of about 2 d_{avg} close to the fixed wall. It is mainly due to the fact that many smaller pebbles filled into the void space formed inter-pebbles close to the wall and gaps between pebbles and wall, which improves the packing density and decreases the porosity.

In the inner region of the poly-disperse pebble bed with normally distributed pebble size, the local porosity is relatively stable, which is approximately equal to the average porosity of the inner region of the bed. For comparison, the local porosity is always distributed stably in the whole poly-disperse pebble bed with periodic boundary, which is approximately equal to the average porosity of the inner region of the poly-disperse bed with fixed wall. The results further reveal that the effect of the fixed wall on packing structure is always limited close to the wall, the packing of pebbles with normally distributed pebble size can reduce the wall effect by further filling of small pebbles into the void formed between pebbles and wall and formed between pebbles.

3.2.2. Coordination Number Distribution

Figure 6 shows the variation of the average coordination number and the standard deviation of coordination number in the poly-disperse pebble bed with normally distributed pebble size. It can be seen from Figure 6a that the average coordination number of periodic boundary pebble bed is greater than that of fixed wall pebble bed This is because that the pebbles close to the fixed wall or in direct contact with the fixed wall have a smaller coordination number due to the influence of the fixed wall effect. In further, with the increase of the standard deviation of pebble size, the average coordination number of the pebble bed decrease and the standard deviation of coordination increase gradually, which can be attributed to the increase of the probability of the pebbles with smaller coordination number and the decrease of the probability of the pebbles with coordination number of 6 and 7 (seen Figure 7) when the pebble size becomes more and more dispersed. In addition, the maximum coordination number in poly-disperse pebble bed increase gradually with the increase of the are standard deviation of pebble size, which is due to the pebble have larger surface area to contact with the surrounding small pebbles with the increase of pebble size.

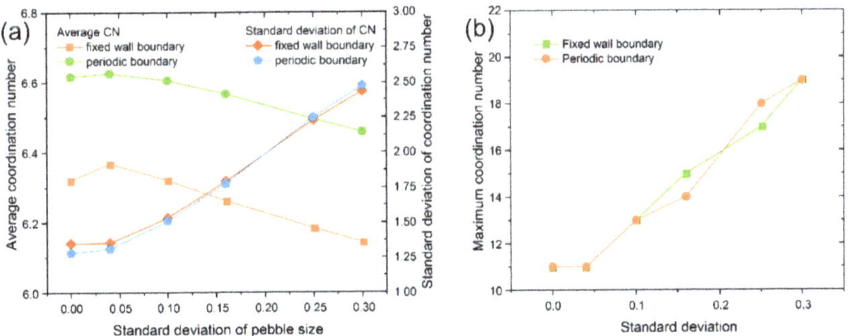

Figure 6. Coordination number of the poly-disperse pebble bed with normally distributed pebble size: (**a**) average coordination number, (**b**) maximum coordination number.

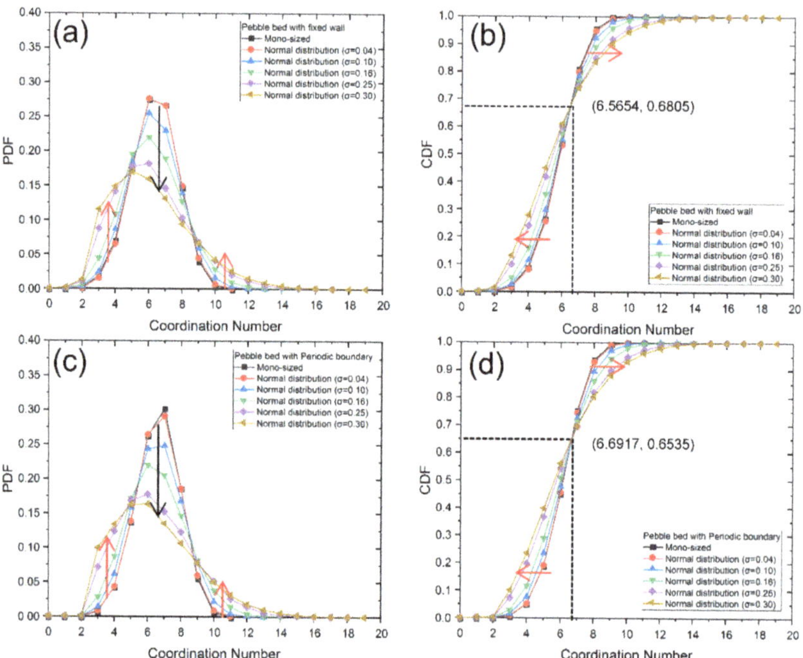

Figure 7. Coordination number distribution of the poly-disperse pebble bed with normally distributed pebble size: (**a,c**) PDF in bed with fixed wall and periodic boundary, (**b,d**) CDF in bed with fixed wall and periodic boundary respectively.

The PDF and the CDF of coordination number in the pebble size normal distributed pebble bed are shown in Figure 7. As can be seen from the figure, the coordination number with the maximum probability decrease with the increase of the standard deviation of pebble size. When the standard deviation is 0 and 0.04, the coordination number with the highest probability is 6, but when the standard deviation increases to 0.3, The coordination number with the largest probability is reduced to 5. In addition, as the standard deviation of pebble size increases, the probability gradually increases when the coordination number is less than 5 and greater than 9. While, when the coordination number is between 5–9, the probability gradually decreases. This is mainly caused by the contact state between large pebbles and small pebbles. In the poly-disperse pebble bed, the large pebble has a higher coordination number, while the small pebbles have a lower coordination number due to the similar "convex wall effect". It can also be found from the CDF of the coordination number. As the standard deviation increases, the cumulative probability profile of low coordination number shifts to left, while the cumulative probability of high coordination number shifts to right. A turning point has occurred at the coordination number of 6.5654 with cumulative probability of about 0.6805 for the fixed-wall pebble bed and of 6.6917 with cumulative probability of about 0.6535 for the periodic boundary pebble bed.

3.2.3. Contact Force Distribution

For pebble beds with normal pebble size distributions, the weak force chain runs through the entire bed. The strong force chains are mainly distributed in the middle and the bottom due to the gravity effect, as shown in Figure 8. The strong force chains are linked to each other to form an arch so as to carry the gravity of the pebble above and the external load. In further, the normalized average contact force distribution along the local position (z-axis) and the horizontal direction (x-axis) of the packed bed with normally distributed pebble size are shown in Figure 9. The results show that with the increase of the local height (*z*-axis), the averaged contact force decreases gradually, and tends to 0 at the top of the bed. However, the average contact force is evenly distributed along the horizontal direction (*x*-axis or *y*-axis) in the bed. This is mostly because the pebble randomly packed in the horizontal direction, while the upper surface of the pebble bed in the vertical direction is a free surface. No additional compression load was applied to the upper surface of pebble bed. Therefore, under the influence of the gravity the contact force decreases with the increase in height until it is 0 at the top of the bed and is evenly distributed in the horizontal direction. In addition, it can be found that the changes of the pebble bed height and the pebble size distribution have little effect on the average contact force near the bottom wall of the pebble bed with the natural packing under gravity and no additional load. The normalized average contact force is between 1.75 and 2.75 close to the bottom wall. This is mainly attributed to the friction interaction between the pebbles and between pebbles and side walls supports part of the gravity force of the pebbles.

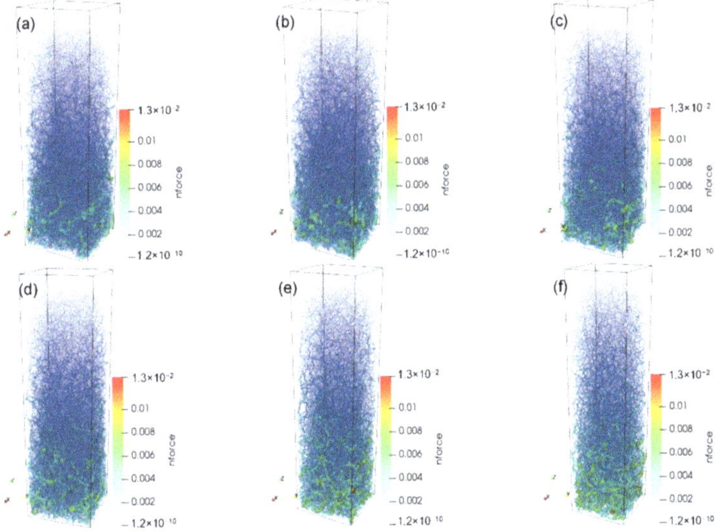

Figure 8. Force chains of contact force in the poly-disperse pebble bed with normally distributed pebble size: (**a**) $\sigma = 0$, (**b**) $\sigma = 0.04$, (**c**) $\sigma = 0.1$, (**d**) $\sigma = 0.16$, (**e**) $\sigma = 0.25$, (**f**) $\sigma = 0.3$.

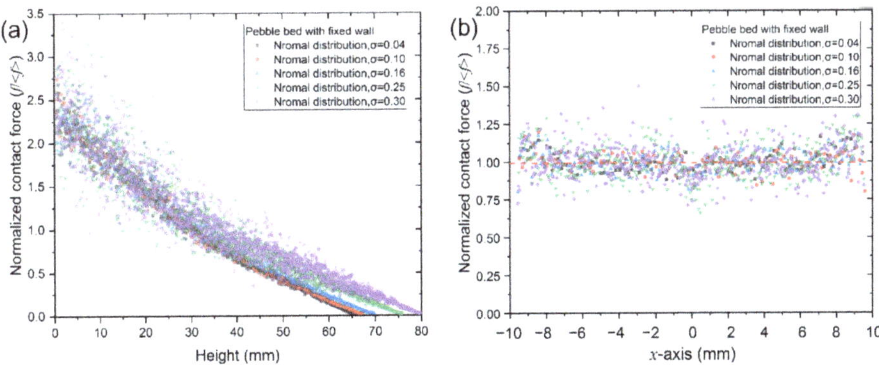

Figure 9. Contact force distribution along (**a**) the height and (**b**) the horizontal of the bed with normally distributed pebble size.

Figure 10 shows the probability density distribution of the normalized contact force in the poly-disperse pebble bed with a normal particle size distribution. The probability density distribution of all contact force is shown in Figure 10a,b. It can be seen from the figure that the probability density of the contact force decreases rapidly as the contact force increases. Most of the contact forces in the pebble bed are weak contact force, which is less than the average contact force. When the contact force less than average contact force, the probability density of the contact force shows a trend of first increasing and then decreasing, as shown in Figure 10b, which also be influenced by the pebble size distribution.

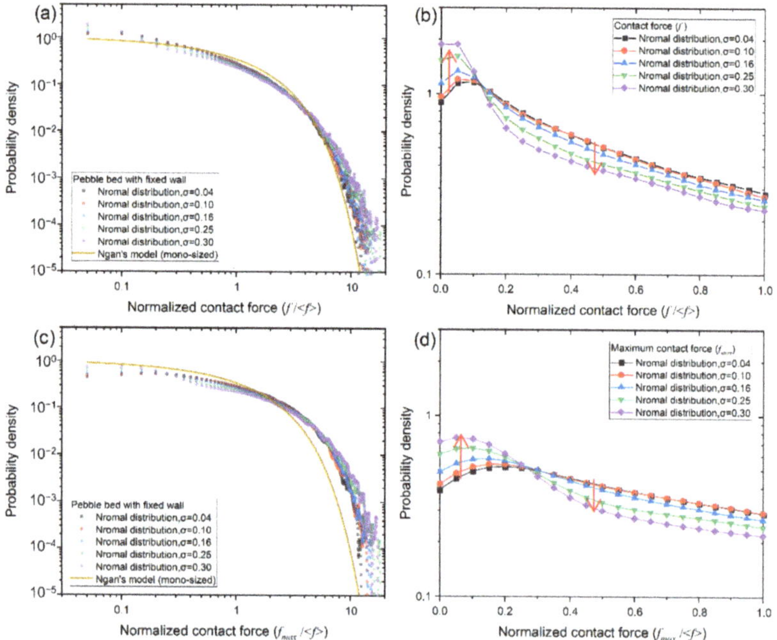

Figure 10. Probability density distribution of contact force in the poly-disperse pebble bed with normally distributed pebble size: (**a,b**) all contact force, (**c,d**) maximum contact force of each pebble.

When the normalized contact force is less than 0.12, the probability density of the contact force gradually increases with the increase of the standard deviation. When the normalized contact force is greater than about 0.12 and less than 1, the probability density of the contact force gradually decreases as the standard deviation increases. With the further increase of normalized contact force, the probability density of strong contact force decreases rapidly. The standard deviation of the normal distribution of particle size has less influence on the probability density distribution of strong contact force. However, compared with the mono-sized pebble bed (solid line in Figure 10), the probability density of weak contact force is reduced, and the probability density of strong contact force is increased. This is mainly because as the standard deviation of pebble size increases, the contact between large pebbles and small pebbles tends to form a strong contact force due to the influence of gravity, especially the contact directly under the large pebbles. Figure 10c,d show the probability density distribution of the maximum contact force of each pebble. It can be seen from the figure that the maximum contact force of each pebbles obtains a probability density distribution which is similar to that of all contact forces. But compared with the probability density distribution of all contact forces, the probability density of the maximum contact forces less than 1 is reduced, and the probability density of strong contact forces greater than 3 is increased.

3.3. Pebble Size Uniform Distribution Pebble Bed

3.3.1. Porosity Distribution

The poly-disperse pebble packing with uniformly distributed pebble size was also simulated and analyzed by DEM modelling. The average porosity variation of the pebble size uniform distribution pebble bed is shown in Figure 11. With the increase of the diameter difference, Δd, the average porosity of the pebble bed decreases gradually. For the pebble bed with periodic boundary, when the Δd is about 0 and 0.05, the average porosity of the pebble bed is about 0.3819 ± 0.0008 and 0.3814 ± 0.0006, respectively. With the increase of the Δd, the average porosity of the poly-disperse pebble bed gradually decreases to 0.3661 ± 0.0008 when the Δd is increased to 0.5. This is also because the small pebbles occupy the voids between large pebbles, which increases the packing density of the poly-disperse pebble bed. For the poly-disperse pebble bed with fixed wall, due to the influence of the fixed wall, the average porosity is greater than that of the periodic boundary. The average porosity of the fixed-wall poly-disperse pebble bed decreases from 0.3977 ± 0.0007 when the Δd is 0.05 to 0.3849 ± 0.0006 when the Δd is 0.5. However, after excluding the influence of the fixed-wall effect, the average porosity in the inner region of the fixed wall pebble bed is consistent with that of the periodic boundary pebble bed. In other words, the influence of the fixed-wall effect is only limited to the area close to the wall.

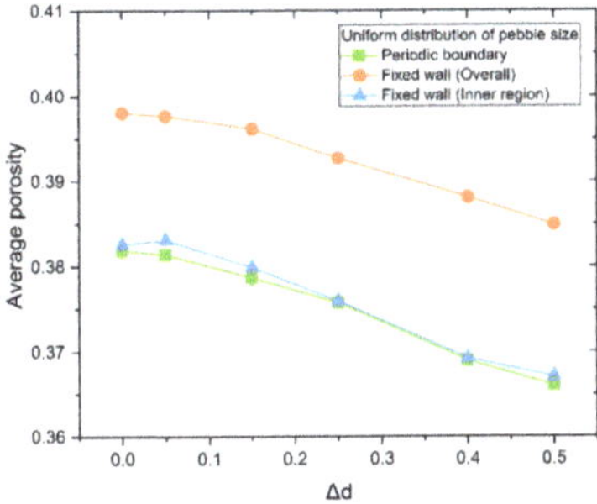

Figure 11. Average porosity of the poly-disperse pebble bed with uniformly distributed pebble size.

To reveal the effect of fixed wall on the packing structures of the pebble size uniform distribution pebble bed, we obtained the local porosity distribution along the x-axis (perpendicular to the wall) inside the pebble bed, as shown in Figure 12. The results indicate that the axial porosity is basically uniformly distributed in the periodic boundary poly-disperse pebble bed. The variation of the pebble side distribution only affects the average porosity of the bed. However, in fixed wall poly-disperse pebble bed with uniformly distributed pebble size, the local porosity distribution along the direction perpendicular to the wall is significantly affected by the fixed wall. As the distance from the fixed wall increase, the axial porosity exhibits the characteristics of damping and oscillation. However, the proportion of the wall affected regions in the uniform particle size distribution pebble bed gradually decreases with the increase of the Δd, compared to the mono-sized pebble bed (see Figure 12a). The fixed wall affects the pebble packing structures in the region of ~5 d close to the wall in mono-sized pebble bed. With the increase of the Δd, the region affected by the wall decreases rapidly. For instance, when the Δd increases to 0.5, the wall effect region is reduced to the range of only ~2 d close to the wall. It is mainly because as the increase of the Δd, smaller pebbles can fill the large pores formed inter-pebbles or between pebble and wall. In the inner region away from the wall, the porosity distribution in the fixed wall bed is similar to that in the periodic boundary poly-disperse pebble bed with the uniformly distributed pebble size, which reveals that without considering or by excluding the wall effect, the pebble size distribution only affects the average porosity of the natural packed pebble bed under the gravity only.

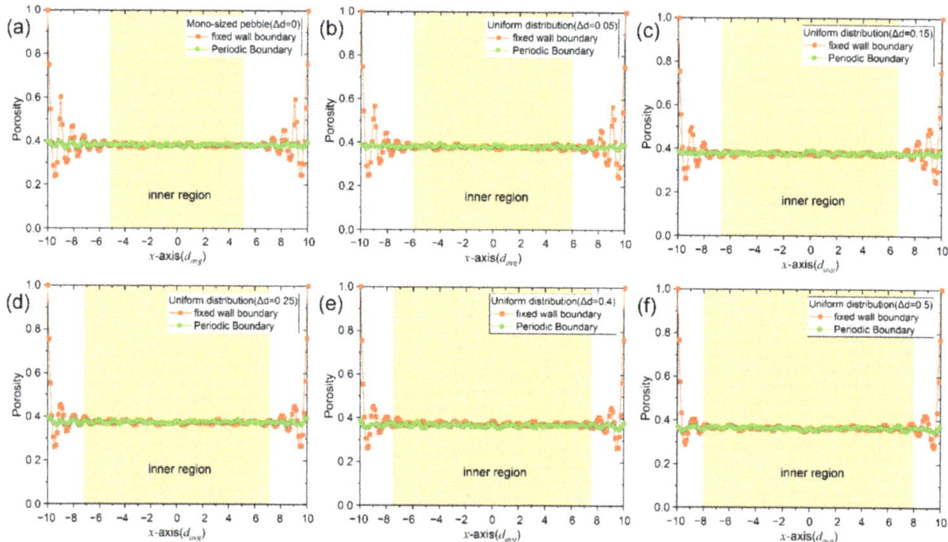

Figure 12. Local porosity distribution in the poly-disperse pebble bed with uniform distributed pebble size: (**a**) $\Delta d = 0$, (**b**) $\Delta d = 0.05$, (**c**) $\Delta d = 0.15$, (**d**) $\Delta d = 0.25$, (**e**) $\Delta d = 0.4$, (**f**) $\Delta d = 0.5$.

3.3.2. Coordination Number Distribution

In the pebble bed with uniform pebble size distribution, the average coordination number of the pebble bed changes with the pebble size difference, Δd, as shown in Figure 13. It can be seen from the figure that as the Δd increases, the average coordination number of the pebble bed gradually decreases, while the standard deviation of the coordination number gradually increases. This is due to the fact that the pebble size becomes more and more dispersed with the increase of the Δd. A small number of large pebbles are in contact with many surrounding small pebbles. Owing to the influence of a similar "convex wall effects", the surrounding small pebbles has a lower coordination number, which results in a decrease in the average coordination number and an increase of the standard deviation of the coordination number as the Δd increases.

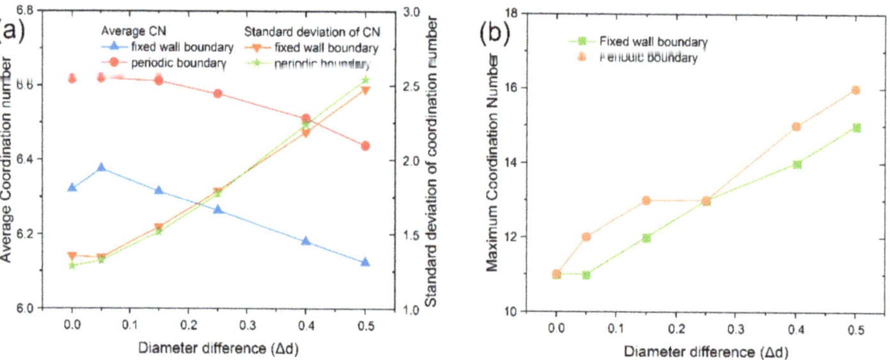

Figure 13. Coordination number of the pebble bed with uniformly distributed pebble size: (**a**) average coordination number, (**b**) maximum coordination number.

In further, it can be seen from Figure 13b that with the increase of the Δd, the maximum coordination number in the pebble bed gradually increases. When the Δd is 0, the maximum coordination number is 11, and when the Δd increases to 0.5, the maximum coordination number of the pebble bed has increased to 15 and 16. It is due to the rapidly increase of the number of small pebbles in contact with the central large pebbles with the increase of pebble size. In addition, it can be seen from the Figure 13 that the average coordination number of the pebble bed with periodic boundary is about 0.3 higher than that of the fixed wall pebble bed, which is also due to the influence of the wall effect. The pebbles in contact with the wall have a lower coordination number.

The PDF and the CDF of the coordination number in the pebble size uniform distribution pebble bed is shown in Figure 14. It can be seen from the distribution that as the particle size difference increases, the coordination number with the highest probability gradually decreases in the pebble size uniform distributed pebble bed. When the Δd is very small (such as 0.05), the maximum probability coordination number for periodic pebble beds is 7 and for fixed-wall pebble beds is 6. The probability distribution of the coordination number is consistent with that of the mono-sized pebble bed due to the pebble size is distributed in a very narrow range. When the Δd increases to 0.15 and 0.25, the coordination number with the maximum probability is 6. As the Δd increases, the highest probability coordination number further decreases to 5 when the Δd is 0.4. There is a similar trend of the probability distribution of coordination number in both the fixed wall pebble bed and the periodic boundary pebble bed. This is mainly attributed to that the influence of the wall effect in the pebble bed gradually decreases with the increase of the Δd, Small, consistent with the change of porosity inside the pebble bed, which corresponds to the variation of the porosity in the pebble size uniform distributed pebble bed.

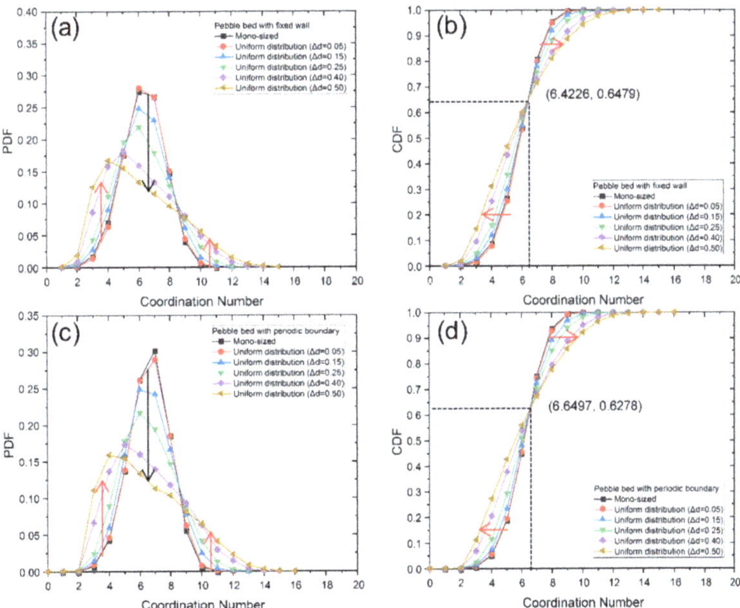

Figure 14. Coordination number distribution in pebble bed with uniformly distributed pebble size: (**a,c**) PDF in bed with fixed wall and periodic boundary, (**b,d**) CDF in bed with fixed wall and periodic boundary respectively.

In addition, the CDF of coordination number shows a similar trend in pebble bed, compared to that of the pebble size normal distributed pebble bed. A turning point exist in the cumulative distribution curve of the coordination number. With the Δd increases, the cumulative probability curves shift to the left when the coordination number is less than 6.4226 for the fixed wall pebble bed and 6.6497 for the periodic boundary bed, while when the coordination number is greater than turning point, the cumulative distribution profile is shifting to the right, which reveal that with the increase of the Δd, the proportion of the pebbles with lower coordination number gradually increase in the poly-disperse pebble bed.

3.3.3. Contact Force Distribution

Figure 15 shows the force chain network in the poly-disperse pebble bed with uniformly distributed pebble size. To clearly show the strong contact force inside the pebble bed, the most of the weak contact forces have been made transparent, as shown in Figure 15. It can be seen from the figure that due to the influence of gravity effect, in the poly-disperse pebble bed with different particle size distribution, the magnitude of the force chain gradually decreases with the increase in local height. The strong contact force chain is mainly distributed in the lower part of the pebble bed, which is consistent with the variation of the local average contact force along the height, as shown in Figure 16a. While the weak contact force is distributed throughout the poly-disperse pebble bed. In further, the normalized average contact force in the bed with different particle size distribution and different bed height is always between 2 and 2.75 in the region near the bottom wall of the pebble bed, which is also owing to the friction interactions between pebbles and between pebble and side walls carry part of the pebble gravity in the poly-disperse pebble bed with uniformly distributed pebble size. In addition, the local average contact force is evenly distributed as shown in Figure 16b owing to the isotropic packing structures along the horizontal direction.

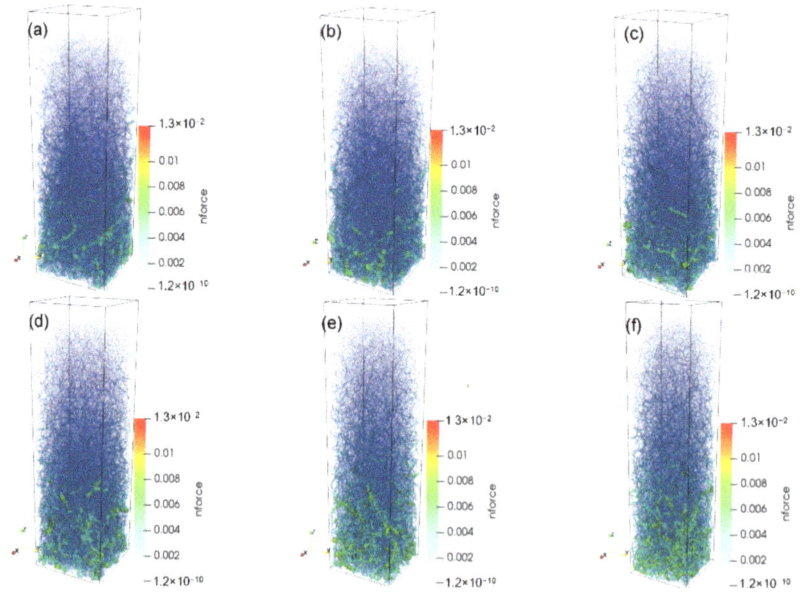

Figure 15. Force chain in the poly-disperse pebble bed with uniformly distributed pebble size: (**a**) $\Delta d = 0$, mono-sized pebble bed, (**b**) $\Delta d = 0.05$, (**c**) $\Delta d = 0.15$, (**d**) $\Delta d = 0.25$, (**e**) $\Delta d = 0.4$, (**f**) $\Delta d = 0.5$.

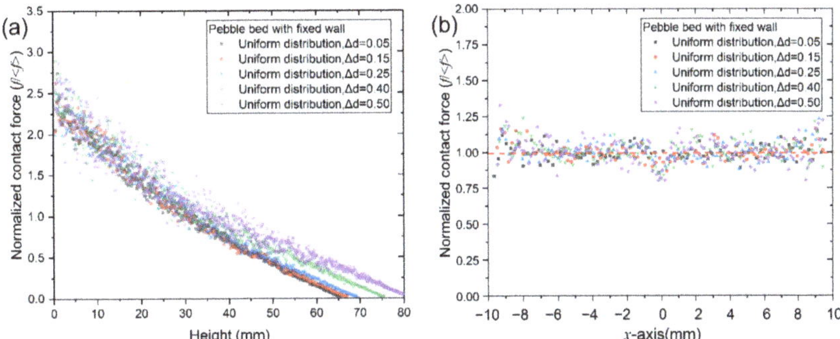

Figure 16. Contact force distribution along (**a**) the height and (**b**) the horizontal of the bed with uniformly distributed pebble size.

The probability density distribution of contact force in the pebble bed with uniform particle size distribution is shown in Figure 17. The results show that the probability density of weak contact force below the average contact force is the largest, and weak force account for the majority of the pebble bed. With the increase of the contact force, the probability density of the contact force drops rapidly. If we enlarge the distribution profile when the normalized contact force is less than 1, the probability density distribution of weak contact force can be obtained as shown in Figure 17b. It can be found that the probability density of weak contact force increases first and then decreases. In further, the pebble size distribution has an influence on the probability density distribution of weak contact force. When the normalized contact force is less than 0.1, the probability density is gradually rising with the Δd increases. While, the probability density reduces gradually with the increase of the Δd when the normalized contact force is greater than 0.1. A turning point occurs when the normalized contact force is equal to 0.1.

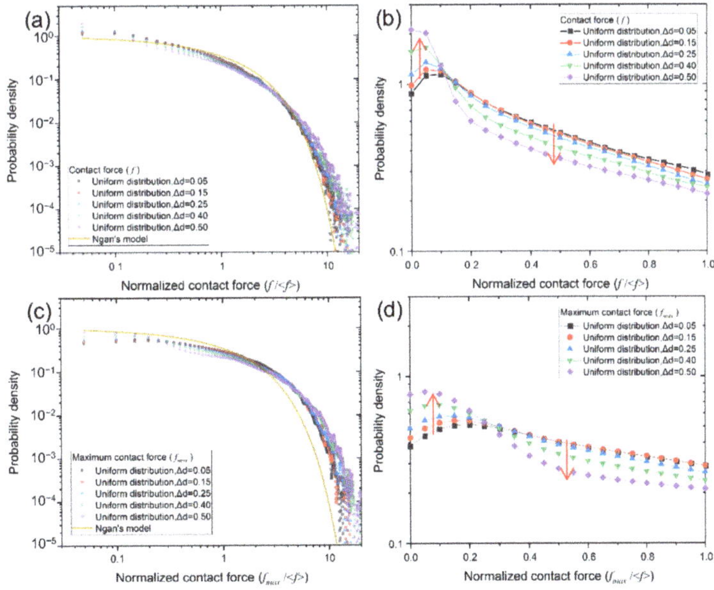

Figure 17. Probability density distribution of the poly-disperse pebble bed with uniformly distributed pebble size: (**a**,**b**) all contact force, (**c**,**d**) maximum contact force of each pebble.

Figure 17c,d shows the probability density distributions of the maximum contact force of each pebble in the poly-disperse pebble bed with uniformly distributed pebble size, which is similar the probability density distribution of all contact force in the pebble size uniform distribution pebble bed. It can be seen that the maximum contact force of the majority of the pebbles is still less than the average contact force. In other word, all contact force of the majority of the pebbles is lower than the average contact force in the poly-disperse pebble bed. Only a few pebbles have a contact force greater than the average contact force, which might break when suffering external load.

4. Conclusions

In terms of the packing structure and the contact force distribution of the poly-disperse pebble bed with normally and uniformly distributed pebble size, respectively, a numerical simulation was conducted based on the DEM method to reveal the influence of the pebble size distribution on the packing structures and the contact force distribution. The results obtained in this work show that the pebble size distribution has a significant effect on the packing structure of the poly-disperse pebble bed. Compared with mono-sized pebble bed, smaller porosity and higher packing fraction can be obtained by the poly-disperse pebbles packing. The average porosity of the poly-disperse pebble bed gradually decreases with the increase of the pebble size dispersion in the poly-disperse pebble bed. In further, compared with the periodic boundary pebble bed, the fixed wall effect can be observed in the local porosity distribution. Close to a fixed wall, the local porosity distribution shows an obvious characteristic of damping and oscillating. The volume fraction of the fixed wall influenced region gradually decrease with the increase of the degree of pebble size dispersion in the poly-disperse pebble bed. The gravity and the pebble size distribution have a significant influence on the contact force distribution of the poly-disperse pebble bed, while the boundary conditions have a relatively small effect on the contact force distribution. In addition, the probability density of contact force decreases rapidly with the increase of the contact force in the poly-disperse pebble beds. the probability density of strong contact force gradually increases with the increase of the pebble size dispersion.

Author Contributions: B.G.: Methodology, Validation, Investigation, Data curation, Visualization, Writing-original draft, review & editing. H.C.: Validation, Visualization, Methodology, Data curation. Y.F.: Conceptualization, Methodology, Supervision, Writing—review & editing. X.L.: Conceptualization, Methodology. L.W.: Conceptualization, Methodology. X.W.: Conceptualization, Resources, Funding acquisition, Writing—review & editing. All authors have read and agreed to the published version of the manuscript.

Funding: This work was funded by the National Key Research and Development Program of China under Grant No. 2017YFE0300602, by the National Natural Science Foundation of China under Grant No. 11905047, and by Sichuan Province Science and Technology Program from the Department of Science & Technology of Sichuan Province of China under Grant No. 2018JZ0014.

Institutional Review Board Statement: Not applicable.

Informed Consent Statement: Not applicable.

Data Availability Statement: The data presented in this study are available on request from the corresponding author.

Acknowledgments: The simulation was supported by the HPC Platform, Southwestern Institute of Physics.

Conflicts of Interest: The authors declare no conflict of interest.

References

1. Lu, J.; Chen, Y.; Ding, J.; Wei, X. High temperature energy storage performances of methane reforming with carbon dioxide in a tubular packed reactor. *Appl. Energy* **2016**, *162*, 1473–1482. [CrossRef]
2. Yang, J.; Bu, S.; Dong, Q.; Wu, J.; Wang, Q. Experimental study of flow transitions in random packed beds with low tube to particle diam-eter ratios. *Exp. Therm. Fluid Sci.* **2015**, *66*, 117–126. [CrossRef]

3. Wang, X.Y.; Feng, K.M.; Chen, Y.J.; Zhang, L.; Feng, Y.J.; Wu, X.H.; Liao, H.B.; Ye, X.F.; Zhao, F.C.; Cao, Q.X.; et al. Current design and R&D progress of the Chinese helium cooled ceramic breeder test blanket system. *Nucl. Fusion* **2019**, *59*, 076019.

4. Chen, H.; Li, M.; Lv, Z.; Zhou, G.; Liu, Q.; Wang, S.; Wang, X.; Zheng, J.; Ye, M. Conceptual design and analysis of the helium cooled solid breeder blanket for CFETR. *Fusion Eng. Des.* **2015**, 89–94. [CrossRef]

5. Liu, S.; Pu, Y.; Cheng, X.; Li, J.; Peng, C.; Ma, X.; Chen, L. Conceptual design of a water cooled breeder blanket for CFETR. *Fusion Eng. Des.* **2014**, *89*, 1380–1385. [CrossRef]

6. Chen, L.; Chen, Y.; Huang, K.; Liu, S. Investigation of effective thermal conductivity for pebble beds by one-way coupled CFD-DEM method for CFETR WCCB. *Fusion Eng. Des.* **2016**, *106*, 1–8. [CrossRef]

7. Gong, B.; Feng, Y.; Liao, H.; Liu, Y.; Wang, X.; Feng, K. Discrete element modeling of pebble bed packing structures for HCCB TBM. *Fusion Eng. Des.* **2017**, *121*, 256–264. [CrossRef]

8. Gong, B.; Feng, Y.; Liao, H.; Wu, X.; Wang, S.; Wang, X.; Feng, K. Numerical investigation of the pebble bed structures for HCCB TBM. *Fusion Eng. Des.* **2018**, *136*, 1444–1451. [CrossRef]

9. Wu, M.; Gui, N.; Wu, H.; Yang, X.; Tu, J.; Jiang, S. Effects of density difference and loading ratio on pebble flow in a three-dimensional two-region-designed pebble bed. *Ann. Nucl. Energy* **2019**, *133*, 924–936. [CrossRef]

10. Gui, N.; Huang, X.; Yang, X.; Tu, J.; Jiang, S. HTR-PM-based 3D pebble flow simulation on the effects of base angle, recirculation mode and coefficient of friction. *Ann. Nucl. Energy* **2020**, *143*, 107442. [CrossRef]

11. Jiang, S.; Tu, J.; Yang, X.; Gui, N. A review of pebble flow study for pebble bed high temperature gas-cooled reactor. *Exp. Comput. Multiph. Flow* **2019**, *1*, 159–176. [CrossRef]

12. Khane, V.; Taha, M.M.; Mueller, G.E.; Al-Dahhan, M. Discrete Element Method–Based Investigations of Granular Flow in a Pebble Bed Reactor. *Nucl. Technol.* **2017**, *199*, 47–66. [CrossRef]

13. Lin, J.; Luo, K.; Wang, S.; Sun, L.; Fan, J. Particle-Scale Simulation of Solid Mixing Characteristics of Binary Particles in a Bubbling Fluidized Bed. *Energies* **2020**, *13*, 4442. [CrossRef]

14. Hofer, G.; Märzinger, T.; Eder, C.; Pröll, F.; Pröll, T. Particle mixing in bubbling fluidized bed reactors with continuous particle exchange. *Chem. Eng. Sci.* **2019**, *195*, 585–597. [CrossRef]

15. Mandal, D.; Sathiyamoorthy, D.; Vinjamur, M. Void fraction and effective thermal conductivity of binary particulate bed. *Fusion Eng. Des.* **2013**, *88*, 216–225. [CrossRef]

16. Mandal, D.; Dabhade, P.; Kulkarni, N. Estimation of effective thermal conductivity of packed bed with internal heat generation. *Fusion Eng. Des.* **2020**, *152*, 111458. [CrossRef]

17. Wongkham, J.; Wen, T.; Lu, B.; Cui, L.; Xu, J.; Liu, X. Particle-resolved simulation of randomly packed pebble beds with a novel fluid-solid coupling method. *Fusion Eng. Des.* **2020**, *161*, 111953. [CrossRef]

18. Zhao, Z.; Feng, K.; Feng, Y. Theoretical calculation and analysis modeling for the effective thermal conductivity of Li4SiO4 pebble bed. *Fusion Eng. Des.* **2010**, *85*, 1975–1980. [CrossRef]

19. Guo, Z.; Sun, Z.; Zhang, N.; Ding, M.; Zhou, Y. Influence of flow guiding conduit on pressure drop and convective heat transfer in packed beds. *Int. J. Heat Mass Transf.* **2019**, *134*, 489–502. [CrossRef]

20. Tian, X.; Yang, J.; Guo, Z.; Wang, Q.; Sunden, B. Numerical Study of Heat Transfer in Gravity-Driven Particle Flow around Tubes with Different Shapes. *Energies* **2020**, *13*, 1961. [CrossRef]

21. Sohn, D.; Lee, Y.; Ahn, M.-Y.; Park, Y.-H.; Cho, S. Numerical prediction of packing behavior and thermal conductivity of pebble beds according to pebble size distributions and friction coefficients. *Fusion Eng. Des.* **2018**, *137*, 182–190. [CrossRef]

22. Panchal, M.; Chaudhuri, P.; Van Lew, J.; Ying, A. Numerical modelling for the effective thermal conductivity of lithium meta titanate pebble bed with different packing structures. *Fusion Eng. Des.* **2016**, *112*, 303–310. [CrossRef]

23. Panchal, M.; Saraswat, A.; Chaudhuri, P. Experimental measurements of gas pressure drop of packed pebble beds. *Fusion Eng. Des.* **2020**, *160*, 111836. [CrossRef]

24. Lee, Y.; Choi, D.; Hwang, S.-P.; Ahn, M.-Y.; Park, Y.-H.; Cho, S.; Sohn, D. Numerical investigation of purge gas flow through binary-sized pebble beds using discrete element method and computational fluid dynamics. *Fusion Eng. Des.* **2020**, *158*, 111704. [CrossRef]

25. Choi, D.; Park, Y.-H.; Han, J.; Ahn, M.-Y.; Lee, Y.; Park, Y.-H.; Cho, S.; Sohn, D. A DEM-CFD study of the effects of size distributions and packing fractions of pebbles on purge gas flow through pebble beds. *Fusion Eng. Des.* **2019**, *143*, 24–34. [CrossRef]

26. Abou-Sena, A.; Arbeiter, F.; Boccaccini, L.V.; Schlindwein, G. Measurements of the purge helium pressure drop across pebble beds packed with lithium orthosilicate and glass pebbles. *Fusion Eng. Des.* **2014**, *89*, 1459–1463. [CrossRef]

27. Bale, S.; Tiwari, S.S.; Sathe, M.; Berrouk, A.S.; Nandakumar, K.; Joshi, J. Direct numerical simulation study of end effects and D/d ratio on mass transfer in packed beds. *Int. J. Heat Mass Transf.* **2018**, *127*, 234–244. [CrossRef]

28. Abdulmohsin, R.S.; Al-Dahhan, M.H. Pressure Drop and Fluid Flow Characteristics in a Packed Pebble Bed Reactor. *Nucl. Technol.* **2017**, *198*, 17–25. [CrossRef]

29. Iikawa, N.; Bandi, M.M.; Katsuragi, H. Force-chain evolution in a two-dimensional granular packing compacted by vertical tappings. *Phys. Rev. E* **2018**, *97*, 032901. [CrossRef]

30. Akella, V.S.; Bandi, M.M.; Hentschel, H.G.E.; Procaccia, I.; Roy, S. Force distributions in frictional granular media. *Phys. Rev. E* **2018**, *98*, 012905. [CrossRef]

31. Kruyt, N.P. On weak and strong contact force networks in granular materials. *Int. J. Solids Struct.* **2016**, 135–140. [CrossRef]

32. Desu, R.K.; Annabattula, R.K. Particle size effects on the contact force distribution in compacted polydisperse granular assemblies. *Granul. Matter* **2019**, *21*, 29. [CrossRef]
33. Minh, N.H.; Cheng, Y.P.; Thornton, C. Strong force networks in granular mixtures. *Granul. Matter* **2014**, *16*, 69–78. [CrossRef]
34. Ngan, A. On distribution of contact forces in random granular packings. *Phys. A Stat. Mech. Appl.* **2004**, *339*, 207–227. [CrossRef]
35. Yang, Y.; Cheng, Y. A fractal model of contact force distribution and the unified coordination distribution for crushable granular materials under confined compression. *Powder Technol.* **2015**, *279*, 1–9. [CrossRef]
36. Annabattula, R.K.; Gan, Y.; Kamlah, M. Mechanics of binary and polydisperse spherical pebble assembly. *Fusion Eng. Des.* **2012**, *87*, 853–858. [CrossRef]
37. Desu, R.K.; Gan, Y.; Kamlah, M.; Annabattula, R.K. Mechanics of binary crushable granular assembly through discrete element method. *Nucl. Mater. Energy* **2016**, *9*, 237–241. [CrossRef]
38. Wang, S.; Wang, S.; Xu, Q.; Chen, H. Crushed model and uniaxial compression analysis of random packed ceramic pebble bed by DEM. *Fusion Eng. Des.* **2018**, *128*, 53–57. [CrossRef]
39. Gan, Y.; Kamlah, M.; Riesch-Oppermann, H.; Rolli, R.; Liu, P. Crush probability analysis of ceramic breeder pebble beds under mechanical stresses. *J. Nucl. Mater.* **2011**, *417*, 706–709. [CrossRef]
40. Van Antwerpen, W.; du Toit, C.G.; Rousseau, P.G. A review of correlations to model the packing structure and effective ther-mal conductivity in packed beds of mono-sized spherical particles. *Nucl. Eng. Des.* **2010**, *240*, 1803–1818. [CrossRef]
41. Mueller, G.E. A modified packed bed radial porosity correlation. *Powder Technol.* **2019**, *342*, 607–612. [CrossRef]
42. Du Toit, C.G. Radial variation in porosity in annular packed beds. *Nucl. Eng. Des.* **2008**, *238*, 3073–3079. [CrossRef]
43. De Klerk, A. Voidage Variation in Packed Beds at Small Column to Particle Diameter Ratio. *AIChE J.* **2003**, *49*, 2022–2029. [CrossRef]
44. Wang, S.; Wang, S.; Chen, H. Numerical influence analysis of the packing structure on ceramic breeder pebble beds. *Fusion Eng. Des.* **2019**, *140*, 41–47. [CrossRef]
45. Wang, X. Computational study of the elastic modulus of single size pebble beds for fusion applications. *Fusion Eng. Des.* **2016**, *112*, 486–491. [CrossRef]
46. Calderoni, P.; Ying, A.; Sketchley, T.; Abdou, M.A. Experimental study of the interaction of ceramic breeder pebble beds with structural materials under thermo-mechanical loads. *Fusion Eng. Des.* **2006**, *81*, 607–612. [CrossRef]
47. Zaccari, N.; Aquaro, D. Mechanical characterization of Li_2TiO_3 and Li_4SiO_4 pebble beds: Experimental determination of the material properties and of the pebble bed effective values. *Fusion Eng. Des.* **2007**, *82*, 2375–2382. [CrossRef]
48. Pupeschi, S.; Knitter, R.; Kamlah, M.; Gan, Y. Numerical and experimental characterization of ceramic pebble beds under cycling mechanical loading. *Fusion Eng. Des.* **2016**, *112*, 162–168. [CrossRef]
49. Pupeschi, S.; Moscardini, M.; Gan, Y.; Knitter, R.; Kamlah, M. Cyclic behavior of ceramic pebble beds under mechanical loading. *Fusion Eng. Des.* **2018**, *134*, 11–21. [CrossRef]
50. Reimann, J.; Fretz, B.; Pupeschi, S. Thermo-mechanical screening tests to qualify beryllium pebble beds with non-spherical pebbles. *Fusion Eng. Des.* **2015**, 1851–1854. [CrossRef]
51. Zhang, C.; Ying, A.; Abdou, M.A.; Park, Y.-H. Ceramic breeder pebble bed packing stability under cyclic loads. *Fusion Eng. Des.* **2016**, *109*, 267–271. [CrossRef]
52. Yu, A.; Standish, N. Porosity calculations of multi-component mixtures of spherical particles. *Powder Technol.* **1987**, *52*, 233–241. [CrossRef]
53. Roquier, G. A Theoretical Packing Density Model (TPDM) for ordered and disordered packings. *Powder Technol.* **2019**, *344*, 343–362. [CrossRef]
54. Kwan, A.; Wong, V.; Fung, W. A 3-parameter packing density model for angular rock aggregate particles. *Powder Technol.* **2015**, *274*, 154–162. [CrossRef]
55. Chan, K.; Kwan, A. Evaluation of particle packing models by comparing with published test results. *Particuology* **2014**, *16*, 108–115. [CrossRef]
56. Kloss, C.; Goniva, C.; Hager, A.; Amberger, S.; Pirker, S. Models, algorithms and validation for open-source DEM and CFD-DEM. *Prog. Comput. Fluid Dyn. Int. J.* **2012**, *12*, 140–152. [CrossRef]
57. Liggghts(R)-Public Documentation, Version 3.X. Available online: http://www.liggghts.com (accessed on 1 October 2020).
58. Liao, H.; Wang, X.; Yang, G.; Feng, Y.; Wang, P.; Feng, K. Recent progress of R&D activities on reduced activation ferritic/ martensitic steel (CLF-1). *Fusion Eng. Des.* **2019**, *147*, 111235. [CrossRef]
59. Reimann, J.; Pieritz, R.; Ferrero, C.; Di Michiel, M.; Rolli, R. X-ray tomography investigations on pebble bed structures. *Fusion Eng. Des.* **2008**, *83*, 1326–1330. [CrossRef]
60. Reimann, J.; Pieritz, R.; Di Michiel, M.; Ferrero, C. Inner structures of compressed pebble beds determined by X-ray tomography. *Fusion Eng. Des.* **2005**, 1049–1053. [CrossRef]
61. Feng, Y.; Feng, K.; Cao, Q.; Zhang, J.; Hu, J. Current Status of the Fabrication of Li_4SiO_4 and Beryllium Pebbles for CN HCCB TBM in SWIP. *Plasma Sci. Technol.* **2013**, *15*, 291–294. [CrossRef]
62. Feng, Y.; Feng, K.; Cao, Q.; Hu, J.; Tang, H. Fabrication and characterization of Li_4SiO_4 pebbles by melt spraying method. *Fusion Eng. Des.* **2012**, *87*, 753–756. [CrossRef]

MDPI

St. Alban-Anlage 66

4052 Basel

Switzerland

Tel. +41 61 683 77 34

Fax +41 61 302 89 18

www.mdpi.com

Energies Editorial Office

E-mail: energies@mdpi.com

www.mdpi.com/journal/energies